SUSTAINABLE ENGINEERING, ENERGY, AND THE ENVIRONMENT

Challenges and Opportunities

SUSTAINABLE ENGINEERING, ENERGY, AND THE ENVIRONMENT

Challenges and Opportunities

Edited by
Kailas L. Wasewar, PhD
Sumita Neti Rao, PhD

AAP | APPLE
ACADEMIC
PRESS

First edition published 2022

Apple Academic Press Inc.
1265 Goldenrod Circle, NE,
Palm Bay, FL 32905 USA

4164 Lakeshore Road, Burlington,
ON, L7L 1A4 Canada

CRC Press
6000 Broken Sound Parkway NW,
Suite 300, Boca Raton, FL 33487-2742 USA

2 Park Square, Milton Park,
Abingdon, Oxon, OX14 4RN UK

© 2022 by Apple Academic Press, Inc.

Apple Academic Press exclusively co-publishes with CRC Press, an imprint of Taylor & Francis Group, LLC

Library and Archives Canada Cataloguing in Publication

Title: Sustainable engineering, energy, and the environment : challenges and opportunities / edited by Kailas L. Wasewar, PhD, Sumita Neti Rao, PhD.

Names: Wasewar, Kailas L., editor. | Rao, Sumita Neti, editor.

Description: First edition. | Includes bibliographical references and index.

Identifiers: Canadiana (print) 20220136947 | Canadiana (ebook) 20220136963 | ISBN 9781774910009 (hardcover) | ISBN 9781774910016 (softcover) | ISBN 9781003277484 (ebook)

Subjects: LCSH: Sustainable engineering. | LCSH: Renewable energy sources. | LCSH: Environmental protection.

Classification: LCC TA170 .S87 2023 | DDC 628—dc23

Library of Congress Cataloging-in-Publication Data

....

CIP data on file with US Library of Congress

....

ISBN: 978-1-77491-000-9 (hbk)
ISBN: 978-1-77491-001-6 (pbk)
ISBN: 978-1-00327-748-4 (ebk)

About the Editors

Kailas L. Wasewar, PhD
Associate Professor and former Head,
Chemical Engineering Department,
Visvesvaraya National Institute of Technology
(VNIT), Nagpur, India

Kailas L. Wasewar, PhD, has been in the Chemical Engineering Department of the Visvesvaraya National Institute of Technology (VNIT), Nagpur, India, for the last 12 years and was Head of the department for two tenures (2008–2010 and 2013–2016).

Presently he is Associate Dean, Education Technology and Library at VNIT. He was formerly a visiting professor at Kyung Hee University, South Korea. He has more than 20 years of industrial, research, and teaching experience from various reputed organizations such as the Indian Institute of Technology (IIT), Roorkee, India, and Birla Institute of Technology and Science (BITS), Pilani, India.

Dr. Wasewar has more than 200 journal publications with Google Scholar citations: 6078, h-index: 48, and i-10 index: 124. He has been granted three patents. Recently, Dr. Wasewar has been included in the world's top 2% scientists in chemical engineering, as per a survey done by Stanford University. His world rank is 369 (top 0.66%) and 18th rank in India in chemical engineering. The names of the scientists from across the world were included based on their citations, h-index, and number of published papers classified in chemical engineering and allied fields.

Dr. Wasewar has guided 15 PhD with five ongoing and 15 MTech dissertations and 80 BTech students. His research interest is in biotechnology, reaction engineering, process intensification, separation technology, environmental engineering, ionic liquids, nanotechnology, CFD, modeling and simulation, and reliability engineering. He has the highest publication record in the field of reactive extraction (more than 60). He has delivered more than 55 invited expert lectures at various conferences, workshops, and seminars. He has been a reviewer for various international journals. He is a member of the editorial boards and advisory boards of various national/

international journals. He has been on various committees, including the National Board of Accreditation (India), All India Council for Technical Education (AICTE), and other expert committee members. He is a member of the boards of studies of the departments of chemical engineering of various universities. He is a member of the Indian Institution of Chemical Engineers (IIChE); Institution of Engineers India (IE(I)); International Congress of Chemistry and Environment (FICCE); Indian Water Association; and World Academy of Science and Engineering. He was an Executive Member of the Uttarakhand and Maharashtra State Center of Institution of Engineers, India. He has been keenly occupied in social activities through Vijyana Bharati, Druva (Prayas) club, and Auropath Society. He has been actively involved in academic, administrative, and extracurricular activities.

Dr. Wasewar earned his Bachelor of Technology in Chemical Engineering (1995) from Laxminarayan Institute of Technology (LIT), Nagpur, and Master's (2000) and full-time PhD (2004) from the Institute of Chemical Technology (ICT), Mumbai, India. He has completed projects from the Department of Science and Technology (DST) Young Scientist, Council of Scientific and Industrial Research (CSIR), Defense Research and Development Organization (DRDO), Department of Biotechnology (DBT), and Institution of Engineers India (IEI), and one industrial project. He is presently working on one DBT and one DST.

Sumita Neti Rao, PhD

Associate Professor,
Department of Applied Chemistry,
Dean Academics, Priyadarshini College of
Engineering, Nagpur, India

Sumita Neti Rao, PhD, is an Associate Professor and Dean Academics at Priyadarshini College of Engineering, Nagpur, India, run under the banner of Lokmanya Tilak Jankalyan Shiksha Sanstha. She has 28 years of research experience and 21 years of teaching experience. Her main areas of research are homogeneous/heterogeneous catalysis, synthesis, and characterization of coordination compounds and their zeolite variants for applications in catalysis and green synthesis, wastewater treatment applications of advanced oxidation processes, regeneration of spent activated carbon (SAC), and waste utilization. She has published 44 research papers in international and national research journals of high repute. She also has four patents to her credit. She has participated and presented her research at several international and national conferences and won best paper prizes. She has worked as a principal investigator for industrially funded projects and consultancies. She is a life member of various professional societies, namely, The Indian Society for Technical Education, Society of Environmental Chemistry and Allied Sciences (SECAS), Indian Association for Environmental Management (IAEM), Indian Society of Solid-State Chemists and Allied Sciences, Society for Technologically Advanced materials of India (STAMI), and Association of Chemistry Teachers (ACT). She has visited Germany, Singapore, USA, Malaysia, Portugal, and Tashkent and has fostered research collaboration.

After completing her MSc in Chemistry, she pursued her PhD in Central Salt and Marine Chemicals Research Institute (a CSIR Laboratory), Bhavnagar, Gujarat, India.

Contents

Contributors

Y. S. Bais
Electrical Engineering Department, Govindrao Wanjari College of Engineering and Technology, R.T.M.N.U. Nagpur, Maharashtra, India, E-mail: yogeshbais@yahoo.co.in

P. P. Bajpai
Mechanical Engineering Department, Priyadarshini Institute of Engineering and Technology, Nagpur–4400019, Maharashtra, India, E-mail: poonambajpai.2008@rediffmail.com

Monita A. Bedmohata
Department of Applied Chemistry, G H Raisoni Institute of Engineering and Technology, Nagpur, Maharashtra, India, E-mail: monitabaidmohta@gmail.com

Archana Bele
Priyadarshini Institute of Architecture and Design Studies, Nagpur, Maharashtra, India, E-mail: archana.piads@gmail.com

A. M. Bhake
Applied Physics Department, Priyadarshini Institute of Engineering and Technology, Nagpur, Maharashtra, India, E-mail: aparna.dhond@gmail.com

Vivek P. Bhange
Department of Biotechnology, Priyadarshini College of Engineering, Nagpur–19, Maharashtra, India, E-mail: vivekbhange@gmail.com

Kiran D. Bhuyar
Department of Chemical Engineering, Priyadarshini College of Engineering, Nagpur–19, Maharashtra, India, E-mail: kiranbhuyar@gmail.com

Alefiyah S. Bohra
Department of Biotechnology, Priyadarshini Institute of Engineering and Technology, Nagpur, Maharashtra–440019, India, E-mail: bohraalefiyah@gmail.com

Mahesh Bundele
Poornima College of Engineering, Jaipur, Rajasthan, India, E-mail: maheshbundele@poornima.org

Bharati Burande
Department of Applied Chemistry Priyadarshini College of Engineering, Nagpur, Maharashtra, India, E-mail: bharatiburande@gmail.com

Ujwala Chakradeo
Smt. Manoramabai Mundle College of Architecture, Nagpur, Maharashtra, India, E-mail: uchakradeo@gmail.com

Shikhar Chakravorty
Research Scholar, Shri Ramdeobaba College of Engineering and Management, Nagpur, Maharashtra, India, E-mail: shikharchakravorty@gmail.com

Archana R. Chaudhari
Department of Applied Chemistry, Priyadarshini Bhagwati College of Engineering, Nagpur, Maharashtra, India, E-mail: arcbce@gmail.com

Jayashri Chopade
Assistant Professor, Department of Mechanical Engineering, Pimpri Chinchwad College of Engineering and Research Ravet, Pune, Maharashtra, India, E-mail: chopadejv91@gmail.com

Trupti J. Dabe
Priyadarshini Institute of Architecture and Design Studies, Maharashtra, India,
E-mail: truptidabe78@gmail.com

Rashmi Dande
Priyadarshini Institute of Architecture and Design Studies, Nagpur, Maharashtra, India,
E-mail: rashmi.piads@gmail.com

Sandhya Deole
Department of Home-Economics, Vasantrao Naik Government Institute of Arts and Social Science, Nagpur, Maharashtra, India, E-mail: sandhyadeole@gmail.com

A. Waheed Deshmukh
Department of Chemical Engineering, Priyadarshini Institute of Engineering and Technology, Near CRPF Campus, Hingna Road, Nagpur–440019, Maharashtra, India, E-mail: awasdk@gmail.com

S. B. Deshpande
Electrical Engineering Department, Priyadarshini Institute of Engineering and Technology, R.T.M.N.U. Nagpur, Maharashtra, India, E-mail: sbd119@gmail.com

M. S. Dhande
Mechanical Engineering Department, Priyadarshini Institute of Engineering and Technology, Nagpur–4400019, Maharashtra, India, E-mail: mahendra.shivaji@gmail.com

G. A. Dhomane
Department of Electrical Engineering, Government College of Engineering (GCOE), Amravati, Maharashtra, India, E-mail: gadhomane@gmail.com

Mangesh S. Dhore
Department of Forensic Science, G.H. Raisoni University, Amravati, Maharashtra, India,
E-mail: mangeshdhore2@gmail.com

Ronen P. Dutta
Department of Electronics, Sibsagar College Joysagar, Assam, India,
E-mail: ronenprokashdutta@gmail.com

Vinay Dwivedi
Executive Vice President (Works), JK Paper Mills, Unit: JKPM, Rayagada, Odisha, India,
E-mail: vinay.dwivedi@jkpm.jkmail.com

J. D. Ekhe
Department of Chemistry, Visveswaraya National Institute of Technology, Nagpur, Maharashtra, India,
E-mail: j_ekhe@yahoo.com

S. V. Gaikwad
Retired Assistant Professor, Dr. Babasaheb Ambedkar College of Engineering, Nagpur, Maharashtra, India, E-mail: swatigaikwad71@gmail.com

U. V. Gaikwad
Priyadarshini Bhagwati College of Engineering, Nagpur, Maharashtra, India,
E-mail: umagaikwad651@gmail.com

Kanchan D. Ganvir
Assistant Professor, Department of Mechanical Engineering, Priyadarshini Bhagwati College of Engineering, Nagpur, Maharashtra, India, E-mail: kanchan.ganvir100@gmail.com

Leena Ganvir
Priyadarshini Institute of Architecture and Design Studies, Nagpur, Maharashtra, India,
E-mail: leenaganvir2000@yahoo.co.in

S. R. Ghatole
Mechanical Engineering Department, Priyadarshini Institute of Engineering and Technology,
Nagpur–4400019, Maharashtra, India

Swapna Ghatole
Research Scholar, Poornima University, Jaipur, Rajasthan, India, E-mail: swapnaghatole72@gmail.com

S. R. Gupta
Department of Biotechnology, Priyadarshini Institute of Engineering and Technology, Nagpur,
Maharashtra–440019, India, E-mail: sushgupta1109@gmail.com

Rakesh Himte
Mechanical Engineering Department, Priyadarshini Institute of Engineering and Technology,
Nagpur–4400019, Maharashtra, India, E-mail: rakeshhimte@gmail.com

K. K. Hirekar
Electrical Engineering Department, Govindrao Wanjari College of Engineering and Technology,
R.T.M.N.U. Nagpur, Maharashtra, India, E-mail: k.hirekar@rediffmail.com

Sachin P. Jolhe
Assistant Professor, Department of Electrical Engineering, Government College of Engineering,
Nagpur, Maharashtra, India, E-mail: spjolhe@gmail.com

Gautami Kant
Department of Chemical Engineering, Visvesvaraya National Institute of Technology, Nagpur,
Maharashtra, India, E-mail: kantgautami@gmail.com

M. D. Karalkar
Assistant, Professor, Department of Electrical Engineering, Priyadarshini J.L. College of Engineering,
Nagpur, Maharashtra, India, E-mail: minalkaralkar@gmail.com

Ruta D. Khonde
Department of Chemical Engineering, Priyadarshini Institute of Engineering and Technology,
Near CRPF Campus, Hingna Road, Nagpur–440019, Maharashtra, India, E-mail: rutakhonde@gmail.com

A. P. Kopulwar
Department of Biotechnology, Priyadarshini Institute of Engineering and Technology, Nagpur,
Maharashtra–440019, India, E-mail: ashkopulwar31@gmail.com

Nandini Kulkarni
Symbiosis School of Planning, Architecture, and Design, Nagpur, Maharashtra, India,
E-mail: nandini.kulkarni@sspad.edu.in

Shivrai Kulkarni
Department of Mechanical Engineering, Shri Ramdeobaba College of Engineering and Management,
Nagpur, Maharashtra, India, E-mail: shivrai.nandini.shashank@gmail.com

Niranjan Kumar
Department of Electrical Engineering, National Institute of Technology, Jamshedpur, Jharkhand, India,
E-mail: nkumar.ee@nitjsr.ac.in

Sachin A. Mandavgane
Department of Chemical Engineering, Visvesvaraya National Institute of Technology, Nagpur,
Maharashtra, India, E-mail: sam@che.vint.ac.in

J. P. Modak
JD College of Engineering, Nagpur, Maharashtra, India, E-mail: jpmodak@gmail.com

S. V. Moharil
Retired Professor and Head, Department of Physics, R.T.M. Nagpur University, Nagpur, Maharashtra, India, E-mail: sanjiv.moharil@gmail.com

Swaroop Laxmi Mudliar
Department of Chemistry, Shri Ramdeobaba College of Engineering and Management, Nagpur-13, Maharashtra, India, E-mails: swarooplaxmim71@gmail.com; mudliarsl@rknec.edu

Rupali A. Nandanwar
Department of Applied Chemistry, Priyadarshini Bhagwati College of Engineering, Nagpur, Maharashtra, India, E-mail: rupalianandanwar@gmail.com

V. M. Nanoti
Priyadarshini Institute of Engineering and Technology, Nagpur, Maharashtra, India, E-mail: viveknanoti@gmail.com

Nimisha Neog
Department of Electronics, Sibsagar College Joysagar, Assam, India, E-mail: nimi.neog2004@gmail.com

Rashmi P. Nimbalkar
Department of Chemical Engineering, Priyadarshini Institute of Engineering and Technology, Rashtrasant Tukadoji Maharaj Nagpur University, Nagpur, Maharashtra, India, E-mail: rashminagmote25@gmail.com

Nikhil D. Pachkawade
Assistant Professor, Department of Mechanical Engineering, Priyadarshini Bhagwati College of Engineering, Nagpur, Maharashtra, India, E-mail: ndpachkawade@gmail.com

Maithili Paikane
Humanities and Social Science Department, *Visvesvaraya* National Institute of Technology, Nagpur, Maharashtra, India, E-mail: maithilibarahate@gmail.com

Dharm Pal
Department of Chemical Engineering, National Institute of Technology Raipur, Raipur–492010, Chhattisgarh, India, E-mail: dpsingh.che@nitrr.ac.in

Aditi S. Pandey
Department of Applied Chemistry, Priyadarshini Institute of Engineering and Technology, Nagpur, Maharashtra, India, E-mail: aditipandey2016@gmail.com

Babubhai C. Patel
Nandesari Industries Association-CETP, Vadodara, Gujarat, India, E-mail: nia_cetp@yahoo.co.in

Pravin D. Patil
Department of Basic Science and Humanities, NMIMS Mukesh Patel School of Technology Management and Engineering, Mumbai–400056, Maharashtra, India, E-mail: pravinpbt@gmail.com

Dhruti. S. Pattanayak
Department of Chemical Engineering, National Institute of Technology Raipur, Raipur–492010, Chhattisgarh, India, E-mail: dhrutisundar23@gmail.com

D. S. Pattanayak
Department of Chemical Engineering, National Institute of Technology Raipur, Raipur–492010, Chhattisgarh, India

Pradeep P. Pipalatkar
Pollution and Ecological Control Services, Nagpur, Maharashtra, India, E-mail: pipalatkar@gmail.com

Ashish P. Pradhane
Department of Chemical Engineering, Visvesaraya National Institute of Technology, Nagpur,
Maharashtra, India, E-mail: ashishpradhane281@gmail.com

Naval Rajas
Department of Mechanical Engineering, Shri Ramdeobaba College of Engineering and Management,
Nagpur, Maharashtra, India, E-mail: rajasnd24@gmail.com

N. N. Rao
Former Chief Scientist and HOD, Wastewater Technology Division, CSIR-National Environmental
Engineering Research Institute (NEERI), Nagpur–440020, Maharashtra, India,
E-mail: nnrao.neeri@gmail.com

Sumita N. Rao
Department of Applied Chemistry, Priyadarshini Institute of Engineering and Technology (PIET),
Hingna Road, Nagpur–440019, Maharashtra, India, E-mail: sumitarao2000@rediffmail.com

Prerana Ratnaparkhi
Department of Home-Economics, C.P. and Berar College, Nagpur, Maharashtra, India,
E-mail: preranaratnaparkhi@gmail.com

Pranay Raut
Department of Chemical Engineering, National Institute of Technology Raipur, Raipur–492010,
Chhattisgarh, India; Department of Biotechnology, Priyadarshini College of Engineering,
Nagpur–19, Maharashtra, India, E-mail: pranay.raut@gmail.com

Shuvam Sahay
Department of Electrical Engineering, National Institute of Technology, Jamshedpur, Jharkhand, India,
E-mail: shuvam.sahay90@gmail.com

S. M. Sawde
Applied Physics Department, Priyadarshini Institute of Engineering and Technology, Nagpur,
Maharashtra, India, E-mail: suwarnasawde@gmail.com

Erena Sayankar
Bhavans B.P. Vidya Mandir, Trimurti Nagar Nagpur–440022, Maharashtra, India, E-mail:
erenasayankar@gmail.com

Abdul Rahim Sheikh
Department of Chemical Engineering, Priyadarshini Institute of Engineering and Technology,
Near CRPF Campus, Hingna Road, Nagpur–440019, Maharashtra, India,
E-mail: rahimabdul914@gmail.com

Mayuresh V. Shivramwar
Department of Chemical Engineering, Priyadarshini Institute of Engineering and Technology, Near
CRPF Campus, Hingna Road, Nagpur–440019, Maharashtra, India, E-mail: mayuresh7414@gmail.com

Seema Shrivastava
Department of Applied Chemistry, Priyadarshini College of Engineering, Nagpur–440019, Maharashtra,
India, E-mail: seemashrivastava@rediffmail.com

Vishal Shukla
Associate Professor, Department of Mechanical Engineering, Shri Ramdeobaba College of Engineering
and Management, Nagpur, Maharashtra, India, E-mail: shuklavv2@rknec.edu

Shripal P. Singh
CSIR-Central Institute of Mining and Fuel Research Regional Center, Bilaspur, Chhattisgarh, India,
E-mail: shripal_singh@yahoo.com

V. K. Singh
Department of Chemical Engineering, NIT Raipur, Raipur–492010, Chhattisgarh, India,
E-mail: vksingh.che@nitrr.ac.in

Manju A. Soni
Department of Biotechnology, Priyadarshini College of Engineering, Nagpur–19, Maharashtra, India,
E-mail: manjuasoni2016@gmail.com

Vinayak Sood
Research Scholar, Shri Ramdeobaba College of Engineering and Management, Nagpur, Maharashtra,
India, E-mail: vinayaksood23@gmail.com

P. Suryanarayana
Chief Manager (Technology) Development, JK Paper Mills, Unit: JKPM, Rayagada, Odisha, India,
E-mail: p.suryanarayana@jkpm.jkmail.com

Megha H. Talwekar
Engineer (Technology Development), JK Paper Mills, Unit: JKPM, Rayagada, Odisha, India,
E-mail: talwekarmegha@gmail.com

Aparna Tarar
Priyadarshini Institute of Architecture and Design Studies, Nagpur, Maharashtra, India,
E-mail: tararaparna7@gmail.com

Chandrakant Thakur
Department of Chemical Engineering, National Institute of Technology Raipur, Raipur–492010,
Chhattisgarh, India, E-mail: cthakur.che@nitrr.ac.in

Rupali Thokal
Priyadarshini Institute of Architecture and Design Studies, Nagpur–440019, Maharashtra, India,
E-mail: rupali.piads@gmail.com

K. L. Wasewar
Department of Chemical Engineering, VNIT, Nagpur–440010, Maharashtra, India,
E-mail: k_wasewar@rediffmail.com

Yashpal
Poornima University, Jaipur, Rajasthan, India, E-mail: yashpal.kaushik1986@gmail.com

Shrutee Dhanorkar Yeolekar
Priyadarshini Institute of Architecture and Design Studies, Nagpur, Maharashtra, India,
E-mail: shrutee20@gmail.com

Poonam Yeul
Department of Applied Chemistry, Priyadarshini College of Engineering, Nagpur–440019, Maharashtra,
India, E-mail: poonamyeul@gmail.com

Laxmi A. Zadgaonkar
Department of Chemical Engineering, Visvesvaraya National Institute of Technology, Nagpur,
Maharashtra, India, E-mail: laxmi.zadgaonkar@gmail.com

Abbreviations

ABE	acetone-butanol-ethanol
AC	activated carbons
AgNPs	silver nanoparticle
ANOVA	analysis of variance
ASP	activated sludge process
ASTM	American Society for Testing and Materials
BOD	biochemical oxygen demand
BRICS	Brazil, Russia, India, China, and South Africa
CAS	climb assist system
CASP	cyclic activated sludge process
CBD	chemical bath deposition
CCH	*Cajanus cajan* husk
CNCG	concentrated non-condensable gases
CO_2	carbon dioxide
CSE	center for science and environment
c-Si	c-silicon
CSS	closed space sublimation
DC	direct current
DCO	dehydration of castor oil
DeWATS	decentralized water treatment systems
DG	distributed generation
DGA	dissolved gases analysis
DMARDs	disease-modifying anti-rheumatic drugs
DPS	duckweed pond systems
DSSC	dye-sensitized solar-cell
EBE	electron beam evaporation
EEF	energy ecological footprint
EFI	ecological footprint index
ERC act	Electricity Regulatory Commission Act
ET	evapotranspiration
EU	European Union
FAB	fluidized aerated bed
FAC	fresh activated carbon
FAL	facultative aerated lagoon

FAMEs	fatty acid methyl ester
FFA	free fatty acid
FWS	flow systems
G	glycerol
GC-MS	gas chromatogram mass spectrometer
GHG	greenhouse gases
GOI	Government of India
HDPE	high-density polyethylene
IC	indigo carmine
IGBT	insulated gate bipolar transistors
ISO	International Standards Organization
IWL	industrial waste lignin
LCA	life cycle assessment
LFAC	low frequency alternating current
L-G	line to ground
LLMT	live line maintenance techniques
M	methanol
MB	methylene blue
MBBR	moving bed biological reactor
MBN	methylene blue number
MBR	membrane bioreactor
ME	methyl esters
MG	malachite green
MICs	minimum inhibitory concentration
MR	methyl red
MR	moisture ratio
MSW	municipal solid waste
MVC	mechanical vapor compression
MVR-E	mechanical vapor recompression type evaporator
NAAC	National Assessment and Accreditation Council
NIR	near-infrared
NOx	nitrogen oxides
NPP	net primary productivity
NSAID's	non-steroidal anti-inflammatory drugs
PCB	pollution control board
PL	photoluminescence
PSC	perovskite solar-cell
PSPWM	phase shift PWM
PTC	power trading corporation

PV	photovoltaic
R.C.C.	reinforced concrete cement
RA	rheumatoid arthritis
RAC	regenerated carbon
RDF	refuse-derived fuel
RED	regional energy density
RMSE	root mean square error
ROTFA	rope type fall arrester
ROW	right of way
RTFA	rail type fall arrester
SA	surface area
SAC	spent activated carbon
SAFF	submerged aeration fixed-film
SBR	sequential batch reactor
SERC	state electricity regulatory commission
SFS	subsurface flow systems
SnO_2	tin oxide
SOG	stripped off-gas
SPSS	statistical package for the social science
TDP	tones per day
TDS	total dissolved solids
Te	tellurium
TE	thermal evaporation
TF	trickling filter
TNPCB	Tamil Nadu Pollution Control Board
TSA	total surface area
TSI	tropical summer index
UASB	up-flow anaerobic sludge blanket
UEV	unit energy value
VSC	voltage source converters
WCAR	window to carpet area ratio
WF	water footprint
WSPS	waste stabilization pond systems
WTE	waste to energy
ZLD	zero liquid discharge

Acknowledgments

Academicians, engineers, industrialists, scientists, and students were invited to give their contributions in the form of chapters for this edited book. We, the editors, acknowledge the generous support given by contributors who spared their precious time in scripting the chapters for this edited book and genuinely thank the publisher for their cooperation in making this book a reality. We thank the reviewers for their comments and suggestions, which have been very helpful in improving the manuscripts.

We would like to extend our gratitude to Dr. Satish Chaturvedi, Chairman, Lokmanya Tilak Jankalyan Sikshan Sanstha (LTJSS), Nagpur, Mrs. Abha Chaturvedi, Secretary, LTJSS, Nagpur, Mr. Dushyant Chaturvedi, Director, (GB) LTJSS, Nagpur, and Mr. Abhijeet Deshmukh, Director, LTJSS, Nagpur, the management of Priyadarshini Institute of Engineering and Technology (PIET), Nagpur, for their constant support.

Our special thanks to Dr. V. M. Nanoti, Principal, and Dr. G. M. Asutkar, Vice-Principal, P.I.E.T., Nagpur, for their constant guidance and motivation. Sincere gratitude to the deans and heads of all departments for their support. We are grateful to Dr. Soni Chaturvedi, Mayuri A. Chandak, Dr. A. Waheed Deshmukh, Dr. Aditi S. Pandey, and Dr. Aniruddha C. Kailuke for their devoted assistance and cooperation. We also express our sincere appreciation to Dr. Deepali Marghade for her help with proofreading the contributed papers. It is our pleasant duty to acknowledge the support extended by Dr. Deepali Marghade, Dr. Aditi Pandey, Dr. Suwarna Sawde, and Mrs. Sarika Vithalkar for helping us compile the chapters in the form of a book.

—**Editors**

Preface

Sustainability necessitates bridging social science with civic engineering, environmental science, and technologies for a promising future. It exists as a common domain created by intersections of environment, economic, and social development. The word sustainability is generally related to energy sources, reducing carbon emissions, protecting the environment, and balancing delicate ecosystems of planet Earth. Scientists, engineers, and academicians need to find holistic and effective solutions to protect our vital life-supporting ecosystems and, at the same time, meet the needs of a growing human population. Sustainability policy is also focused on maintaining the sustainability of economic, environmental, and social equity goals for the good life of all on earth.

It is being realized that wind, solar, geothermal, hydropower, and biomass are clean sources of energy with the potential to substitute grid-based electrical energy. It is widely believed that slow introduction and integration of renewable energy technologies into existing infrastructure, together with improved use of non-renewable energy sources and finally phasing out carbon-based energies, may ensure 'sustainability.' The thrust should be on steering lifestyles that depend more on renewable energy technologies. Sustainable building design has the potential to reduce the operation and maintenance costs of a building while protecting the environment. The energy needs of a building can be met by harnessing power from solar, wind/air, and water.

The goal of researchers in different fields is to accelerate the process of attaining sustainability. Researchers are studying and analyzing different approaches and ways by which nanotechnology and its compartments can become useful in this endeavor. Thus, green nanotechnology aims to create eco-friendly designs and use them to minimize environmental and health hazards by creating new methodologies.

This book comprising of the latest developments and advances in renewable and green energy, green nanotechnology, green processing, and solar energy, sustainable energy policies, biofuels, energy, and environment, waste to energy, renewable energy, economics, business, policy, and management, clean energy technology, fuel cells, and advanced engineering technology is intended to serve the cause of sustainability. All the chapters were written by academicians and researchers. Each contributed paper was duly reviewed before acceptance for publication in this book. The chapters were accepted for publication based on their interest, relevance, innovation, and application to the themes of the book. The book values both research and practice proportionately and chapters are written with considerable simplicity for the benefit of relevant professionals, academicians, and researchers.

—Editors

PART I
Sustainable Engineering

CHAPTER 1

Triggering Sustainable Urbanism Through Space Planning and Design

SHRUTEE DHANORKAR YEOLEKAR and APARNA TARAR

Priyadarshini Institute of Architecture and Design Studies, Nagpur, Maharashtra, India, E-mail: shrutee20@gmail.com (S. D. Yeolekar)

ABSTRACT

Urban areas keep on changing from time to time; and every change holds an opportunity for improved design, and anything that is not upgrading with time is in the state of decline. The ignorance towards interconnected, multidimensional, and interrelated perspectives of the design and planning of spaces has destabilized the eventual need of immoderate communal and environmental facilities in initiating urban sustainability intents. Answering to these flaws, this chapter used both qualitative and quantitative approaches and amalgamated various communal and environmental aspects to instruct the growth and progression of design and planning of spaces. The chapter aimed at triggering the improvement of the socio-economic and environmental health of urban areas.

At first, the chapter tries to examine and categorize neighborhood parks depending upon various parameters such as environmental, communal, urban fabric, land-use, land cover, greenness, site amenities and built spaces of the neighborhood park. In a later section of the chapter, the chapter assesses the current neighborhood park system to draw recommendations for planning and design of these spaces by keeping sensitive approach towards its social and built context. The chapter concludes with the design guidelines for triggering sustainable urbanism through space planning and design and encourages to design across the urban and natural gradient amplified multiple ecosystems that are intertwined together.

1.1 INTRODUCTION

Urban areas keep on changing from time to time; and every change holds an opportunity for improved design, and things that are not upgrading with time are in the condition of downswing. Over the last few years, the architects have retrieved the possibilities of urban plazas, neighborhood parks and other outdoor public areas to upgrade urban quality of life by overcoming the social and environmental ills of recent times. Entry to civic areas has been shown to enhance the quality of human lifestyle [1], facilitate social cohesion, harmony, and bonding amongst man and surrounding environment, and enrich mental health, physical health, and spiritual health of human beings [2]. These urban areas take an active part in protecting and promoting ecosystem, multifariousness, environmental functions, and other services in the urban areas [3].

In spite of continuous and frequent researches on neighborhood parks, there is a major disparity between understanding and knowledge. Most researchers do not consider the diversity in communal, ecological, and contextual aspects of these spaces; specially the features of communal urban fabric, vigorous requirements and choices of people living around. Researchers generally consider these spaces either as "parks" or as "green spaces," neglecting diversity in the characteristics of the space and confluence of environmental and communal facilities available in the public spaces. These loopholes in the neighborhood parks indicate the requirement of actual investigation, which increases the concern of these integrated and composite urban components so that they can come up with initiating sustainable urbanism through design and planning of spaces.

1.2 SUSTAINABLE URBANISM

Sustainability is an interdisciplinary and integrative field for research that tends to amplify the social health, economic development, and ecosystems simultaneously. Despite the fact that accurate significance of sustainability is still a controversial affair, the multidimensional definition discusses about common concerns. Notwithstanding the fact that urban sustainability is the target which is to be achieved through a continuous process which may never end but will eventually lead towards betterment, the final goal is a town with a good standard of living that is eugenically unbiased, economically dynamic, biologically varied and environmentally utilitarian.

Enlightened by the description of urban sustainability and motivated by concept and conjecture covering various aspects on sustainability, this investigation suggests an imagination for good neighborhood park structure, which itself is developing in the direction of increased sustainable circumstances, while coming up with the comprehensive sustainability of the urban area. However, not every park will be able to or should be able to keep up all the targets and actions, as suitable to geological factors, public spaces should stimulate communal interconnection, solidarity, and the cohort of communal assets and help for environmental processes. These spaces should have the pleasing environment and should be approachable to the variety of generations and communities, through multiple ways of transportation, which includes vehicular, non-vehicular, public, and private transportation. The planning of neighborhood park has to be contextual in accordance with the surrounding urban fabric as well as the natural resources of the adjacent areas, having various activities and crowded residential areas around tiny parks in the heart of the city, and lesser developed areas around big parks, untouched landscapes in the outskirts of the city and in the villages.

1.3 MAINTAINING THE BALANCE BETWEEN VARIOUS SUSTAINABILITY TARGETS IN NEIGHBORHOOD PARKS

The major area of concern in the neighborhood park system is maintaining the balance between various parameters for achieving sustainability in public spaces in terms of its design and space planning. According to the idea that every park cannot and rather should not facilitate all the advantageous amenities, but finally it should be the most ideal example and should set up a benchmark. Although there are certain proofs that sustainability targets can most of the times be collaborative, emphatically strengthening and holding up one another. The "Sustainable Neighborhood Park" example combines both communal and environmental principles, amalgamating parameters for achieving sustainable development with quality of life for human beings. These areas highlight the conservation of landscape by using local and non-encroaching plants, natural repairing process, stormwater recycling and reusing, natural shelter for animals, use of recycled and sustainable materials for construction and appropriate ways and means of maintaining the park. The proposed notion encourages quality of life for human beings by encouraging integration of nature in the built environment for communal interaction, ecological awareness. These upcoming park standards encourage a more ecological, maintenance-free neighborhood park which is esthetically appealing and radically transforms the impact of nature in the urban spaces as well as the pre-conceived notions about urban nature. In such a manner, sustainable neighborhood parks make us rethink about the responsibility of urban designers and urban planners in the development process of neighborhood park because it needs input from society, context, and region.

Forsyth and Musacchio [11], developed detailed park design guidelines with respect to park size, shape, number, context, location, and trade-offs. Their guidelines emphasized the importance of connectivity, diversity, and access for both human and non-human life. In addition, authors noted, the relationship of park boundaries or edges to the natural areas is critical [5]. Hard edges are those that do not connect to other vegetation, while soft edges provide a transition zone for wildlife to travel from one patch of habitat to the next.

1.4 MULTIDIMENSIONAL ASSESSMENT OF NEIGHBORHOOD PARKS FOR ENHANCED SUSTAINABILITY

Cities across India are retrieving the futuristic design of neighborhood parks and public spaces which includes urban plazas, green spaces, natural green belts, and additional open-air public areas so as to enhance quality of life in the urban context and resist the ill effects caused by communal and environmental aspects [6]. The need of the present situation and the urban components should comprehend to intensify the economic, environmental, and societal sustainability of urban areas. Approach towards public spaces of the city should comprehend to enhance quality of life for human beings, and should promote community bonding, justice, and democracy, and should also improve spiritual, mental, and physical health of human beings [7]. Such spaces also contribute in preserving the ecosystem, its functions, and its processes within the urban areas, also landscapes that are imported from the other places are majorly affected by the human intervention. In spite of the reality that there is enough research on neighborhood parks, but still there is disparity in knowledge and understanding about it. Maximum studies and researches lack in considering the variation in ecological, communal, and contextual aspects of these spaces, especially the features of surrounding urban fabric, vigorous needs and demands of the society, strategic segregation, and environmental features of the place. Urban parks that are socially and ecologically sustainable must fulfill the frequently changing needs and demands of the people living in the city with the ecosystem, at different levels. Because of this, the ratio of people and parkland hardly facilitate the ecologically and socially operational public areas; moreover, such neighborhood park examples have caused an increase in unutilized spaces and neglected city landscapes [8]. These parameters keep on changing with every change in the demographics of the city. The lacunas in neighborhood parks indicate the requirement of thorough research which will improve the understanding of such complicated urban features so as to take part in the development of sustainable urban design and urban planning.

1.5 STUDY FOR THE DESIGN CONSIDERATIONS

Diversification of the public areas having different types of communal and environmental advantages is the core problem of Indian parks, but the system to categorize the urban areas is random and superficial. Furthermore, up till now, there is no major valuation of such spaces; however, parks of major cities in India have undergone inspection. Ignorance about the multiple features of Indian parks indicates the major disparity in the theory and practice, and it therefore restricts the prospects of planning of public spaces which targeted at improving the sustainability of the region. Corresponding to the various needs, this investigation was done analytical approach to formulate a multi-faceted arrangement of Indian urban parks, facilitating additional minute, formulated, and planned comprehension of such complicated urban facilities. The outcome constitutes strong guidelines for evaluation Neighborhood Park and a trigger point for advancement of policies for people targeted at improving urban sustainability by planning and designing urban spaces.

1.6 OBSERVATIONS OF THE STUDY

1.6.1 PHYSICAL

Large neighborhood parks were methodically more presumably to have play areas, sandpits, water as landscape element, public washrooms, recreational areas and all in all bigger amount and variety of total facilities. Bigger neighborhood parks mostly seem to be underdeveloped in the larger perspective, having very few commercial activities around it. The urban fabric adjacent to the bigger neighborhood parks seem to be of high-profile income group and having low population density. Usually, bigger neighborhood parks were situated away from the heart of the city. Neighborhood parks having a good variety of facilities were usually big and underdeveloped.

1.6.2   ECOLOGICAL AND ENVIRONMENTAL

Greener parks were having more recreational amenities. Such neighborhood parks were usually in areas with high population density and having few industrial and commercial areas. Neighborhood parks with a larger amount of landscape cover were away from the heart of the city, having a larger number of pathways and lesser amount of impermeable surface covers more and these were usually situated in neighborhoods having high profile income groups and non-crowded areas.

1.6.3   SOCIAL

Neighborhood parks in the locality of high-profile income groups were undoubtedly bigger, away from the heart of the city and in non-crowded areas. Such neighborhood parks also had larger landscape covers and were underdeveloped in the broader perspective. Neighborhood parks in crowded places were having lower-income groups in their surroundings and had few commercial and retail areas around them.

1.6.4   BUILT ENVIRONMENT

Highly developed neighborhood parks had fewer amenities overall. Such parks were situated near the heart of the city, usually in the areas of lower-income group and having a greater number of retail as well as commercial places.

1.7   DISCUSSION

The following study has helped to improve perception of the environmental, physical, communal, and other features of the built environment of the public spaces in India with the help of quantitative analysis and ways of categorization. The outcome uncovers the various trigger points for developing communal and environmental sustainability of the public spaces and also by extending the city boundaries. The implementation of policies and suggestions for enhancement are discussed in this study.

Various outcomes indicate that the neighborhood park is playing a major role in making the city more and more sustainable and is also contributing in enhancing communal and environmental health. The existence of many neighborhood parks, outside the heart of the city, underdeveloped, and not properly watered indicates that these neighborhood parks are preserving the natural ecosystem of the place along with the provision of relaxation activities for people living in the city [9, 10].

Multiple important outcomes also suggest justice with respect to the distribution of parks and the amount of green cover in the city. The truth about neighboring areas adjacent to the parks were seemed to have people from lower-income groups apart from the city, which suggests that these facilities are not appropriately situated in the rich areas.

1.8   CONCLUSION

The following study has gone to the next level of basic categorization criteria to generate exact, multi-faced comprehension of the environmental, physical, communal, and spatial features of the built environment of neighborhood parks which can help in achieving the goals for achieving sustainability of the area by planning the public spaces with appropriate design. The study also facilitated the area using exact methods of quantitative analysis for neighborhood parks which can be used for other cities on the basis of their economic, communal, environmental, and other conditions depending upon the circumstances. Also, the following study gave a trigger point for advancement, rationalization, and investigation of policies for public and various steps towards making sustainable urban spaces in India.

The outcome of the research indicates that the neighborhood parks in India are highly functional with respect to environment and society. Bigger, inappropriately distributed neighborhood parks are helping in the crucial function of the ecosystem by preserving local biodiversity and further facilitating the basic relaxation activities for the people living around. Regularly maintained landscapes and parks full of recreational facilities are giving major protection from urban heat island effect which otherwise hampers the community integration and physical activities of the people living around. The study outcome also indicates that entry to the parks through lower income group localities is almost unbiased. Although the connections amongst rich neighborhood with bigger neighborhood parks having more recreational activities and esthetic qualities suggest a biased approach. As rich people can maintain their own private gardens as compared to the poor people living in the crowded areas and hence this seems unfair and objectionable.

Many other crucial spaces for developing neighborhood parks were also pointed out by this investigation. The absence of public toilets, walking tracks, drinking water, and interactive spaces in the desired amount should be provided properly because the amenities like these help in improving the condition of neighborhood park, specifically in the areas having hot climatic conditions. And there are many neighborhood parks which are having less crowded areas in the surrounding also, there are cases of densely populated areas which are having limited facilities of play areas for children and other recreational activities for other age groups. Responding to such study outcomes, the access to the neighborhood parks can be increased not only by creating new recreational spaces in the surrounding but also by adding the new facilities in the existing neighborhood parks. This will encourage high density population areas and mixed-use development surrounding a smaller public space. Also, the parks in the core of the city were identified to have a highly clustered composition, which resulted in a lack of access to such important amenities. The positioning of recreation centers to be proposed in the future should take into consideration the geographic conditions and socio-economic state of neighborhood for low-income group localities which are currently neglected.

In concluding the chapter, the study presents a ray of hope for the present and futuristic approach for neighborhood parks. In the coming time, the goals for the development keeping in mind the environmental, communal, contextual, and geographical conditions of the city will help to improve the contribution for sustainability of civic spaces. And will undoubtedly set a bench mark for other civic spaces. To increase the possibility of success, précised policy should be formulated and implemented after thorough study and investigation of each park on the basis of their environmental and communal features, and planners must try to have a participatory design approach at every stage of design decision.

KEYWORDS

- economic development
- geological factors
- neighborhood park
- space planning
- sustainability
- sustainable urbanism

REFERENCES

1. Baker, L., Brazel, A., Selover, N., Martin, C., Nelson, A., & Musacchio, L., (2002). Urbanization and warming of Phoenix (Arizona, USA): Impacts, feedbacks, and mitigation. *Urban Ecosystem, 6*(3), 183.
2. Barradas, V., (1991). Air temperature and humidity and human comfort index of some city parks of Mexico City. *Int. J. Biometeoro., 35*, 24–28.
3. Bolund, P., & Hunhammer, S., (1999). Ecosystem services in urban areas. *Ecological Economics, 29*(2), 293–301.
4. Esbah, H., Cook, E., & Ewan, J., (2009). Effects of increasing urbanization on the ecological integrity of open space preserves. *Environmental Management, 43*, 846–862.
5. Harnik, P., (2000). *Inside City Parks.* Washington, D.C., Urban Land Institute.
6. Harnik, P., & Welle, B., (2009). *Measuring Economic Value City Park System.* Chicago: Graham Foundation for Advanced Studies in the Fine Arts.
7. Mass, J., Verheij, R., Groenewegen, P., Sjerp De, V., & Spreeuwenberg, P., (2006). Green space, urbanity, and health: How strong is the relation? *Journal of Epidemiology and Community Health, 60*(7), 587.
8. Massey, D., (1994). *Space, Place and Gender.* Minneapolis: University of Minnesota Press.
9. Mitchell, D., (1995). The end of public space? People's park, definitions of the public and democracy. *Annals of the Association of American Geographers, 85*, 108–133.
10. Sherer, P., (2003). *The Benefits of Parks: Why America Needs More City Parks and Open Space.* San Francisco: Trust for Public Land.
11. Forsyth, A., & Musacchio, L. (2005). *Designing Small Parks: A Manual for Addressing Social and Ecological Concerns.* (1 edition ed.) Wiley.

CHAPTER 2

Generalized Data Base Model for Enhancement in Failure Productivity of Industry 4.0

M. S. DHANDE, R. L. HIMTE, S. R. GHATOLE, and V. M. NANOTI

Priyadarshini Institute of Engineering and Technology,
Nagpur, Maharashtra, India,
E-mail: mahendra.shivaji@gmail.com (*M. S. Dhande*)

ABSTRACT

Today the automation and modernization of industry, industrial layout, productivity, and quality production are the major issues in the technological field. It has been observed in Murli Agro Solvent Plant, G. S. Oil, Adilabad, Jaiswal NECO Wadoda and collected data from number of industries. Productivity is the major issue in a lot of industries such as, in Madhya Pradesh (Pithampur, Devas, Indor), in Kerala (Selam), North America, Africa, Netherland, South Korea, Australia, China, and Canada. Now building construction, industrial development work is going on very fast. Hence manufacturing industry and process industry require enhancing the productivity. Maximum production is needed from industry. Building construction can be done in a short period by using automated machines for wall construction. To increase the rate of construction fast and to improve the productivity of human being, new advanced development is technique essential. Today the problems of the sick industry are major. Mass and quality production is needed by the manufacturing and process industry. Now the question arose why the industry is slowed down. When is the industry established? In the beginning, the owner is very active, but after a certain period of time, he faces some critical problems. Because of this, he has to bear huge losses. After all, why such problems are generated? Now a day's human life has become luxurious. Everybody depends on industrial product. In this situation no one wants to face the industrial problems. These industrial problems should be solved socially or by any other mechanism. In this chapter, the solution is provided to improve the productivity in industry and also stated that how to apply them. Some mathematical techniques can be applied here. If required SWOT (Strength-Weakness-Opportunities-Threat) analysis can also be used.

2.1 INTRODUCTION

Everybody wants to enhance the industrial layout to improve the production rate, but due to more time loss, tedious work, and labor problems, owners' loss the interest is boredom to run the industry. Industrial layout plays a major role in increasing the production rate. Try to design the layout very easily with minimum effort. If productivity is enhanced in an economical way, it will be better for industry owners so that they will be in good profit. The contractor wants to construct the building, in less finance. Labor is tired in construction site. Same way industry owner desires to erect and run it, to increase is given productivity. Now a day's industry 4.0 in industrial sector is work. Again, issue creates here industry 4.0 need to improve the productivity. The industry practically fails to procure the data available in Chapter 9 and the same scope found in Chapter 10; the industry is going on slow down. Therefore, researcher takes decision to do the work on the failure of industrial productivity.

2.2 LITERATURE SURVEY

The researcher visited many industrial sectors to know the industrial plotting layout and industrial function. Researcher personally visited number of industries and observed that there is very unhealthy situation now a days. To enhance the industrial standard, target to erect number of industries in short period. With taking into consideration the above facts we started our research work. Some research papers collected on this subject, for how to design the industrial layout. Personally, visited solvent plants at Umrer industrial M. I. D. C. area [1996], Panchgaon, Umred Road, Dist. Nagpur [2000], Bhilai-Durg city and Bhilai Steel Plant, Chhattisgarh in 2003, Bhandara district Sunflag Steel Industry, Bhandara (M.S.) [2004], Neco-Jaiswal Industries,

Facor steel, Paramount conductors, Mahananda dairy, VIP Industries, Hingna M. I. D. C., Nagpur. Also visited in technological exhibition personally at Bangalore (Karnataka) [2014] and consulted other countries industrialists and major industries in Metropolitan's city. In industrial areas of number of towns personally visited and observed that in a very critical situation people are leaving. To erect the plant is huge amount is needed. Previously machineries were used. But now, a day's fully automated system is used. The steel plants are using some new techniques to improve the productivity through mathematical model discussed in Ref. [9]. With taking these references, a researcher works over the same title.

2.3 OBJECTIVES

1. Work measurement technique;
2. Use method study;
3. Continuous quality improvement techniques;
4. Material transportation process.

2.4 FORMULATION

For finding the output of industry in paper [8]:

$$Y = K\left[A^a \times B^b \times C^c \times D^d \times E^e \times F^f \times G^g \times H^h \times I^i \times J^j \right] \tag{1}$$

Taking log both sides:

$$\log_{10} Y = \log_{10} K \begin{bmatrix} a\log_{10} A \times b\log_{10} \times c\log_{10} C \\ \times d\log_{10} D \times e\log_{10} E \\ \times f\log_{10} F \times g\log_{10} G \\ \times h\log_{10} H \times i\log_{10} I \times j\log_{10} J \end{bmatrix} \tag{2}$$

where; Y is the output of production; K is the proportionality constant.

A, B, C, D, E, F, G, H, I, J are the values of plant need. Such as processing material, electricity, manpower, finance, water, transportation facility, land, machinery, maintenance system.

Table 2.1 shows the constant variable and movable variables with classification.

TABLE 2.1   Constant and Movable Variables [1]

		Industry Model			
SL. No.	**Constant Variables**	**Movable Variables**		**Classification of Movable Variables**	
Variables: 1	Operator [A]	Education	A1	Technical	Non-technical
		Age	A2	20–30 year	30–58 year
		Place	A3	Native	Transferable
		Wages	A4	High	Medium
		Experience	A5	Higher	Lower
		Mentality	A6	Sound	Normal
		Family background	A7	Industrial oriented	Non-industrial
		Intellectual capability	A8	High	Low
		Add-on program attended	A9	High number	Low number
		Software awareness	A10	Modern software	Normal software
		Vehicles using	A11	Own	Industrial vehicles
		Higher study deserving	A12	Part-time	Study leave
		Categories of operator	A13	Permanent	Contract
		Working capability	A14	Hardcore	Software
		Operator availability	A15	Mass quantity	Limited
Variables: 2	Location [B]	Hill area	B1	High hill	Lower hill
		Distance from city	B2	Nearby	Not possible to operator up-down
		Quality of land	B3	Dry	Stone oriented
		Facility	B4	Gardening	Robust area
		Water facility	B5	Ample	Only working
		Land ownership	B6	lease	Owen purchased
Variables: 3	Transport [C]	Raw material and final goods	C1	Road	Rail
		By trucks	C2	Logistic	Daily/weekly/ monthly

TABLE 2.1   *(Continued)*

		Industry Model			
SL. No.	Constant Variables	Movable Variables		Classification of Movable Variables	
Variables: 4	Working environment [D]	Ergonomics	D1	Light system	Color light system
		Seating arrangement	D2	Suitable m/c operating oriented chair	Abe dent
		Air system	D3	Air conditioning	Normal
		Suitable to body	D4	Healthy	Tedious
Variable: 5	Electrification [E]	Power consumption	E1	Heavy	Normal
		Current flowing capability	E2	A/C	D/C
Variable: 6	Finance [F]	Finance mode	F1	Bank	Own finance
		From market collection	F2	Share	Bond
Variable: 7	Machine [G]	CNC M/c	G1	Automatic	Semi-automatic
		M/C orientation	G2	Traditional	Unconventional
Variable: 8	Innovation [H]	Place of invention	H1	Institute invention	Industrial invention
		Invention	H2	Industry incubation	Business incubation
		Idea	H3	New idea	Research idea
		Enhancement level	H4	Corporation level	Institute level
		Opinion invention	H5	Interactive invention	Experimental invention
		Categories of invention	H6	Number of ideas	Complete invention
Variable: 9	Marketing [I]	Spread in market	I1	Order base	Door to door
			I2	Mass quantity	Retailer
			I3	Digital	Communication
Variable: 10	Feedback [J]	Intentional product	J1	Individual	Group
			J2	Product improved	Product rejected
			J3	Product limited demand	Product heavy demand

Table 2.1 variables values as written in following equations:

$$A = A_1^a, A_2^a, A_3^a, A_4^a, A_5^a, A_6^a, A_7^a, A_8^a, A_9^a, A_{10}^a, A_{11}^a, A_{12}^a, A_{13}^a, A_{14}^a, A_{15}^a \tag{3}$$

$$B = B_1^b, B_2^b, B_3^b, B_4^b, B_5^b, B_6^b \tag{4}$$

$$C = C_1^c, C_2^c \tag{5}$$

$$D = D_1^d, D_2^d, D_3^d, D_4^d \tag{6}$$

$$E = E_1^e, E_2^e \tag{7}$$

$$F = F_1^f, F_2^f \tag{8}$$

$$G = G_1^g, G_2^g \tag{9}$$

$$H = H_1^h, H_2^h, H_3^h, H_4^h, H_5^h, H_6^h \tag{10}$$

$$I = I_1^i, I_2^i, I_3^i \tag{11}$$

$$J = J_1^j, J_2^j, J_3^j \tag{12}$$

2.5   RESULT AND DISCUSSION

According to Eqns. (3) to (12) it is observed that constant and movable variables are effectively worked in the model. The number of variables taken into consideration for finding the failure productivity of industry. There are so many problems on industrial layout are up till now. The problems can be solved by filling the Google form from vendors and analyze the results. Asked the people what are the queries are in the industrial field? Modernization of industries is needed now. Computational fluid dynamics, advanced software are used in industries. Total quality policies are not used earlier in the factories. There should be a contentious process to improve the techniques. Ergonomics system, maintenance technique, Time measurement, Method engineering and system engineering are not using effectively. All these protocols not implemented effectively earlier, therefore industrial productivity failure.

2.6   CONCLUSION

From the above mathematical formulation, analysis, result, and discussion, it is clear that the industrial productivity failed due to the advanced techniques are not used in industry. For the enhancement of industrial productivity, need of highly well qualified technological experienced staff, Soft skill trainer, motivational lectures, automated industrial visits for employees and are also needed, Advanced recent updated software, organizational management database, time record system, staff recreational club needed, residential industry facility, school, college, post office facility is needed in industry campus, Better transportation facility, medical benefits, guest hospitality, add-on program should be provided. Ergonomics factor must take into consideration, computer numerical control system, innovation, and invention laboratories needed, also the recent international market demand survey is essential. Advanced journals, books, newspaper should be in industrial library for new advancement awareness among the staff. The productivity increases by organizing Google form quiz to increase technology pocket of staffs and save the time. Collect the feedback of the product produced by applying new product development design. There are different variables given in Table 2.1. If these variables implemented in industry, then productivity will be improved by 25% as compared to previous productivity. The mathematical equation is generated and verified with practically with the productivity before applying new measures. With considering the above formulation, productivity difference between original industry and after these variables, model is applied in industry, larger gap will be found. Hence through this system not only enhance the production rate, quality assurance but also maintenance problems will also be reduced.

KEYWORDS

- **computational fluid dynamics**
- **industrial layout**
- **industrial problem identification**
- **industrial productivity**
- **technical problems**
- **time study**

REFERENCES

1. Chakravorty, S. S., (2009). Six Sigma programs: An implementation model. *International Journal of Production Economics, 119*, 1–16.

2. Meng, Y., & Thomas, B. G., (2003). Heat transfer and solidification model of continuous slab casting. *Metallurgical and Materials Transactions B Con1D., 34B*, 685–705.
3. Beckermann, C., & Viskanta, R., (1993). Mathematical modeling of Transport phenomena during alloy solidification. *Appl. Mech. Rev., 46*, 1–27.
4. Forestier, R., Costes, F., Jaouen, O., & Bellet, M., (2009). Finite element thermo-mechanical simulation of steel casting. *Proceedings MCWASP XII, 12*th *International Conference on Modeling of Casting Welding are Advanced Solidification Process* (pp. 1–12). Vancouver, Canada.
5. Naveen, B., & Babu, R. (2017). Productivity improvement in manufacturing industry using industrial engineering tools. *IOSR Journal of Mechanical and Civil Engineering (IOSR-JMCE)*, 11–18.
6. Lahutre, N. K., Sharma, A. K., & Kumar, P., (2014). Distortions in hole and tool during microwave drilling of Perspex in a customized applicator. *IEEE MTT-S International Microwave Symposium Digest.*
7. Dhande, M. S., Himte, R. L., Nanoti, V. M., & Modak, J. P., (2018). Survey with design & development of mathematical modeling for ash brick production hydraulic machine. *IJIRSET, 7*, 9604–9610. doi: 10.15680/IJIRSET.2018.0709.
8. Vikhar, A. D., & Modak, J. P., (2013). Formulation of field database model: A case study at PVC pipe manufacturing industries. *International Journal of Mechanical Engineering and Technology, 40*, 94–99.
9. Dhande, M. S., & Khandare, S. S., (2012). Mathematical modeling on effective heat energy radiated in steel plant. *International Journal on Computer Application,* 21–25. ISSN-0975-8887.
10. Dhande, M. S., & Khandare, S. S., (2011). Energy consumption in steel plant by SWOT analysis. *IJMRAE, 3*, 165–177.

CHAPTER 3

Empirical Modeling and Kinetic Study of Microwave Drying Process

RUTA D. KHONDE, A. WAHEED DESHMUKH, ABDUL RAHIM SHEIKH, and MAYURESH V. SHIVRAMWAR

Department of Chemical Engineering, Priyadarshini Institute of Engineering and Technology, Near CRPF Campus, Hingna Road, Nagpur–440019, Maharashtra, India,
E-mail: rutakhonde@gmail.com (R. D. Khonde)

ABSTRACT

Drying is one of the prime preservation techniques for extending the shelf life of food items. The present investigation is focused on the development of a mathematical model which will predict the kinetics of thin-layer drying at different conditions. Drying characteristics of prawns in a laboratory-scale microwave dryer with variable wattages were investigated. The moisture content was successfully reduced up to 12% from the initial moisture content of 84% on a wet basis. A significant influence of microwave power on drying rate and subsequently the drying time had been observed. Most of the drying was observed to be in a falling rate period; however constant rate period was absent at all power levels. The study of transport properties revealed increasing effective diffusivity during drying with increment in the wattages and was found in a range of 10^{-11} to 10^{-12} m²/s. An increase in the microwave power also increased the drying efficiency, but decreased the consumption of specific energy. Modeling and simulation of experimental data had also been performed in MATLAB by using eight different kinetic models for the drying process. The Parabolic and Midilli both models of drying kinetics had shown good fitting with the experimental data, which can be very useful for analyzing the transfer processes during drying.

3.1 INTRODUCTION

Drying is one of the prime preservation techniques for extending the shelf life of many food items and all the agricultural products [1], since drying prevents decay of material by micro-organisms. The drying process is a combination of thermo-physical and biochemical processes consisting of heat and mass transfer that takes place simultaneous between the material surface and surrounding medium. Generally, drying refers to an operation during which water content is removed from a wet solid material by thermal means. The process evolves the water content from a wet solid into an unsaturated gas phase. The main reasons for drying food are to prevent the growth and activity of micro-organisms and hence to preserve the food, and to reduce the weight and bulk density of food for cheaper transport and storage.

Sun-drying and hot air drying are widely used methods for industrial drying, but have disadvantages like unable to handle huge quantity of food materials, unable to maintain consistent quality standards, contamination problems, very long drying times, low energy efficiency and high costs. However, microwave drying is an alternative method as it provides uniform energy and high thermal conductivity to the inner sides of the material being dried, low space requirement, sanitation, energy savings, precise process control, and fast start-up and shutdown conditions. Researchers reported that microwave drying reduces the drying time with increasing power rating that prevents the food material from decomposition [2–4]. When drying is carried out in a correct way, the nutritional quality, color, flavor, and texture of rehydrated foods are only slightly less than fresh food. However, if drying is carried out incorrectly, there is a greater loss of nutritional and eating qualities and, more seriously, a risk of microbial spoilage and possibly even food poisoning [4]. Microwave heating takes place in a non-conductor due to polarization effects at frequencies between 300 MHz to 300 GHz (wavelength range from 1 m to 1 mm), which is dielectric heating. While microwave drying, microwaves carrying energy radiates from a source in all directions. The material absorbs this energy and converts it to heat by polar molecules like water molecules present in food material. Hence, the water molecules start to evaporate and the material becomes dried [4, 5].

Prawns and other seafoods are dried as its preservation is necessary due to its huge production and consumption. Many researchers studied the drying of a variety of seafoods, vegetables, fruits, flowers, and agricultural items. The effective control and uniform distribution of drying air and temperature conditions over the material can be achieved by thin layer convective drying technique, which thereby improves the overall quality of the dried material. The drying kinetics for various types of food and agricultural materials are generally described by thin-layer drying model equations. These models describe the drying phenomena by three types, namely, theoretical, semi-theoretical, and empirical. The semi-theoretical and empirical models have been developed to describe the drying process and to predict the drying kinetics of a wide range of food materials. The process of drying involves simultaneous heat and mass transfer, which is a complex process. Hence some researchers developed semi-theoretical and empirical models to explain the moisture removal process from the solid food items [6]. Kinetic study is necessary for modeling the drying process. Mathematical models have proved to be widely useful for analyzing the transfer processes during drying. The objective of this investigation is to develop a mathematical model which will predict the kinetics of thin-layer drying at different drying conditions. Drying characteristics of prawns in a laboratory-scale microwave dryer with variable wattages were investigated in the present work.

Many researchers studied drying rate, drying time, and its kinetics of a variety of vegetables, leaves, and flowers by conventional and non-conventional process. However, very few literatures reported drying of seafood items such as prawns. Hence, this gap motivated to study drying of prawns, which was reported in this study along with kinetics. Several researchers have proposed a number of mathematical models for the thin layer drying of many food materials, however, very rare studies using microwave drying have been reported as summarized below.

Demirhan and Ozbek [5] studied microwave drying of purslane leaves and reported the influence of microwave on drying parameters such as moisture content, moisture ratio (MR), drying time, and effective diffusivity. The authors determined the kinetic parameters by fitting the drying data to various models based on the initial, final, and equilibrium moisture contents against drying time, and reported the best fit Midilli model for all the drying conditions. Various investigators studied the kinetics of food materials using microwave drying such as garlic [3], apple [7], wheat [8], yellow pea [9], carrot [10], peach [11], parsley [12], black tea [13], mushroom [14], potato [15, 16], cabbage [17], millet [18], tobacco [19], okra [20], spinach [21] and basil [22]. However, microwave drying of prawns had not been reported yet. The objective of this study includes the investigation of drying characteristics and kinetics of prawns in microwave system.

Pandey and co-researchers reviewed mathematical models of thin-layer drying process [23]. As has been reported by authors, the theoretical models may cause considerable error as these include assumption of removal of moisture. Based on Fick's second law, semi-theoretical models are derived and modified into simple forms. The semi-theoretical and empirical models consider the external resistance to the transfer of moisture between the material and drying agent. Theoretical equations give a better understanding of the transport process of drying; however, empirical equations give a better fit to the experimental data. Hence, most of the literature reported empirical equations for prediction of MRs with a variety of operating conditions.

3.2   MATERIALS AND METHODS

The prawns used in this investigation were obtained from a local market in Nagpur city during summer season in India. Prawn samples were cleaned by using tap water many times to make samples free from dust and unwanted foreign materials; and then stored in a refrigerator to preserve its original quality, till further drying experiments. In a convective oven, the initial moisture content of the sample was measured to be 84% on wet basis. Then microwave drying was performed at different wattages on fresh samples of prawn in the laboratory oven for kinetic study and modeling. All the experiments were performed at least twice to ensure repeatability of the drying data.

As shown in Figure 3.1, the microwave oven has variable power adjustment and a digital balance. Drying was carried out at five different microwave wattages (140, 175, 230, 350 and 700 W). The experiment set up was associated with digital weighing balance and sensors to measure the relative humidity, temperature, and moisture content for accurate recording. The microwave oven was operated by microwave power control and microwave emission time control. In order to maintain weighing accuracy, the digital balance was integrated within the oven, having a sensitivity of 0.01 g. While experimentation, prawns samples were placed in a single layer, each piece not touching each other, and its moisture loss was recorded with time intervals of 15 seconds until no further remarkable change in weight was observed.

The moisture content of the sample at any instant (t) can be converted to MR as follows:

$$MR = \frac{M_t - M_e}{M_o - M_e} \tag{1}$$

FIGURE 3.1 Microwave system used for drying experiments.

Here MR is dimensionless, M_t is the moisture content (weight of water per unit weight of dry material) at any instant t, M_e is the equilibrium moisture content (weight of water per unit weight of dry material) and M_o is initial moisture content (weight of water per unit weight of dry material). The subscripts t, e, and o denote at any instant t, at equilibrium and initial values, respectively. Effective moisture diffusivity for unsteady-state condition can be calculated by Fick's second law, given by Eqn. (2).

$$\frac{\delta M}{\delta t} = D_{eff} \times \nabla^2 M \qquad (2)$$

Assuming mass transfer by diffusion and constant diffusion coefficient, the solution of Eqn. (2) in thin layer is given as follows [24]:

$$MR = \frac{8}{\pi^2} \sum_{n=0}^{\infty} \frac{1}{(2n+1)^2} \exp\left(\frac{-(2n+1)^2 \pi^2 D_{eff}}{4H^2} t\right) \qquad (3)$$

Here D_{eff} is the effective diffusivity (m²/s), t is the drying time (s) and H is the thickness of sample layer (m). Eqn. (3) can be simplified to Eqn. (4) when the Fourier number is greater than 0.2 for thin layer samples.

$$t = \left(\frac{H^2}{\pi^2 D_{eff}} \right) \ln \left(\frac{8}{\pi^2} * \frac{M_t}{M_0} \right) \tag{4}$$

From above Eqn. (4), a normalized plot of log MR against the drying time can be plotted to obtain effective moisture diffusivity from its slope, where the slope is given by:

$$K = \frac{\pi^2 D_{eff}}{4H^2} \tag{5}$$

Next, an Arrhenius type of equation can be used to express the effect of the temperature on the effective diffusivity as follows:

$$D_{eff} = D_0 e^{(-E/RT)} \tag{6}$$

In terms of microwave power, above equation can be written as:

$$D_{eff} = D_0 e^{(-E\,m/P)} \tag{7}$$

Here, E is the energy of activation (kJ/mol or W/g), D_0 is the pre-exponential factor (m²/s), R is the universal gas constant (kJ/mol.K), T is the drying temperature (Kelvin), m is sample weight (g), P is microwave power (Watt). Knowing the effective diffusivity from Eqn. (4), the activation energy can be determined from Eqn. (7) for variable wattages of microwave system.

3.3 KINETIC MODELING AND MATLAB SIMULATION

Mathematical modeling is one of the significant processes in most of the engineering applications such as process optimization, design, control, and more recently in the area of risk assessment. It represents the behavior of the system at given operating conditions that can be used to predict the steady-state and dynamic behavior of the system. In the present investigation, kinetic modeling, and MATLAB simulation of drying of prawns at 175 watts only was carried out. The kinetic models are useful for the design of the process and the simulation is for predicting the process parameters for given conditions. However, a wide variety of software is being used for process modeling and simulation purposes like ASPEN, CHEMCAD, CFD, COMSOL, Excel, and other software and coding. Many numerical techniques are available to solve linear/non-linear and differential equations. A number of kinetic models have been reported by researchers [23, 25]. Drying kinetics relates the moisture removal process to the process variables. In this study, eight kinetic models were employed, as given in Table 3.1, for experimental data fitting to predict kinetic parameters from the best suitable model, and the simulation was carried out by MATLAB coding. Model prediction and goodness of fit were identified by comparing experimental and simulated values of MRs and calculating the root mean square error (RMSE) for each data point using Eqn. (8).

TABLE 3.1 Thin Layer Drying Kinetic Models

SL. No.	Model	Equation	References
1.	Newton	$MR = \exp(-k\,t)$	[26, 27]
2.	Henderson and Pabis	$MR = a \times \exp(-k\,t)$	[28, 29]
3.	Page	$MR = \exp(-k\,t^n)$	[30, 31]
4.	Modified Page	$MR = \exp(-k\,t)^n$	[32, 33]
5.	Wang and Singh	$MR = 1 + b\,t + a\,t^2$	[34]
6.	Parabolic	$MR = c + b\,t + a\,t^2$	[35]
7.	Logarithmic	$MR = a \times \exp(-k\,t) + b$	[36]
8.	Midilli	$MR = a \times \exp(-k\,t^n) + (b\,t)$	[37]

3.4   RESULTS AND DISCUSSION

The MRs with increasing drying time and different wattages of microwave oven are shown in Figure 3.2. It was observed that increasing the drying time decreased the MR for all wattages. The drying rate also decreases with drying time at all the different wattages as shown in Figure 3.3. As the microwave power is increased, its operating temperature increases, which causes the drying of material in lesser time. Hence for higher microwave wattages, the drying time reduces than for lower wattages. Similar observations for microwave drying system had been reported by Kutlu and Isci [24].

Table 3.2 summarizes the statistical results from 8 different thin layer models where k is drying constant and a, b, n are equation constants. The best fit model for thin-layer drying characteristics of prawn were described by Logarithmic and Midilli models, which has the highest coefficient of determination (R^2) and lowest RMSE values given by Eqn. (8):

$$RMSE = \sqrt{\frac{\sum \left(MR_{exp} - MR_{sim} \right)^2}{n}} \qquad (8)$$

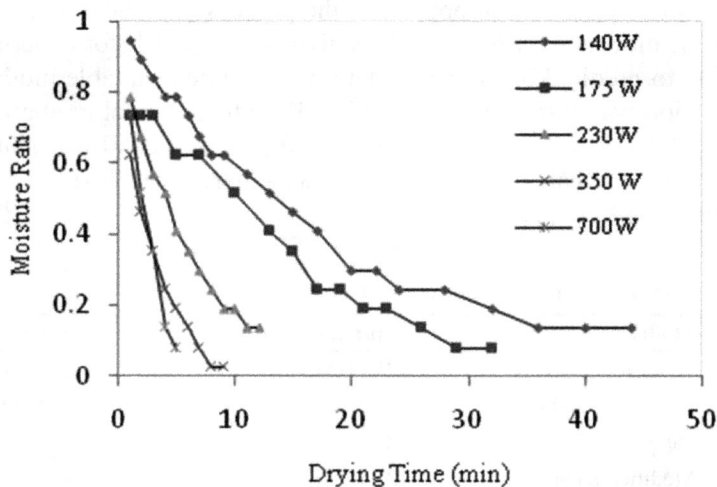

FIGURE 3.2   Moisture ratios at different microwave wattages.

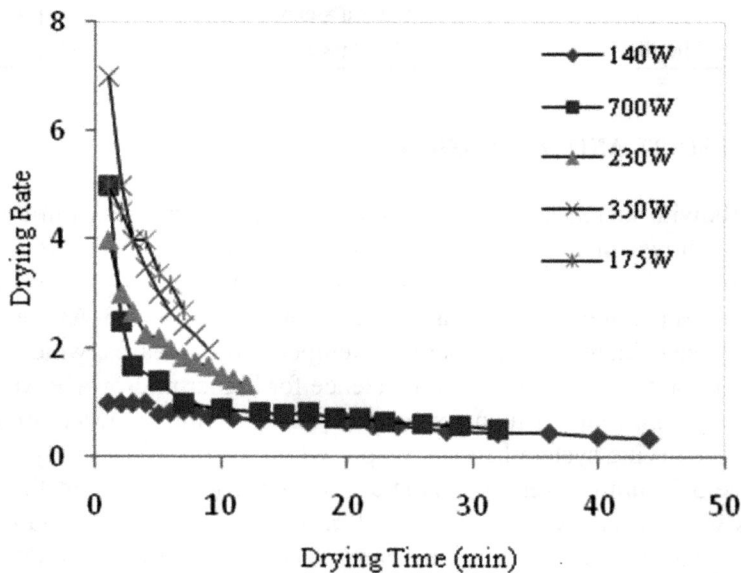

FIGURE 3.3   Drying rates at different microwave wattages.

MR_{exp} is the experimentally determined MR and MR_{sim} is the simulated MR for n number of observations. The comparison of all the eight models on the basis of RMSE and R^2 are given in Table 3.2.

TABLE 3.2   Comparison of RMSE and R^2 Values of 8 Different Models

SL. No.	Model	RMSE	R^2	Constants
1.	Newton	0.220031	0.983	$k = -3.000$
2.	Henderson and Pabis	0.003178	0.837	$k = 0.1219; a = 0.0512$
3.	Page	0.00638	0.795	$k = 2.9033; n = 0.1601$
4.	Modified Page	0.02119	0.198	$k = 2.9699; n = 1.0247$
5.	Wang and Singh	0.37579	0.991	$b = -0.2094; a = 0.0093$
6.	Parabolic	0.00075	0.997	$c = 0.0491; b = -0.0041; a = 7.41 \times 10^{-5}$
7.	Logarithmic	0.000852	0.996	$a = 0.0804; k = 0.0536; b = -0.0309$
8.	Midilli	0.000498	0.998	$a = 0.0453; k = 0.0294; b = -0.0002;$ $n = 1.5014$

The study of transport properties revealed that effective diffusivity increases with increasing the power wattages and was found in a range of 10^{-11} to 10^{-12} m²/s, as shown in Figure 3.4. The effect of changing the microwave power on effective diffusivity had also been studied which reveals that increasing the wattages of microwave, increases the effective diffusivity due to increase in drying temperature as shown in Figure 3.4.

Kinetic study of drying of prawns in microwave shown by Arrhenius plot in Figure 3.5 indicates that increasing the microwave power increases the drying temperature which ultimately decreases the activation energy required for moisture transport process.

3.5   CONCLUSION

In this investigation, a mathematical model for predicting thin layer drying kinetic parameters at different drying conditions in a microwave system is developed. Thin layer drying characteristics of prawns in a laboratory microwave dryer was investigated at different wattages. The moisture content was successfully reduced up to 12% from the initial moisture content of 84% on wet basis. It was investigated that microwave power has a significant influence on drying rate and subsequently the drying time. Hence, increasing the microwave power reduces the drying time. The study of transport properties revealed that effective diffusivity increases with increasing the power wattages and was found in a range of 10^{-11} to 10^{-12} m²/s. Although, increasing the microwave power increased the drying efficiency, but decreased the specific energy consumption. Modeling of experimental data was performed using eight different kinetic models for above drying process and the simulation was done using MATLAB by applying numerical techniques. It revealed that the Parabolic and Midilli both models of drying kinetics shown good fitting with the experimental data. Hence this method can be very useful for analyzing the transfer processes during drying.

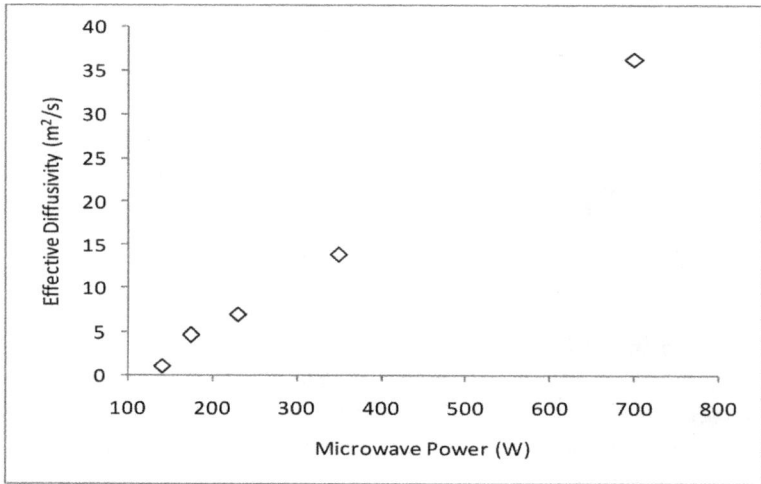

FIGURE 3.4   Influence of microwave powers on effective diffusivity.

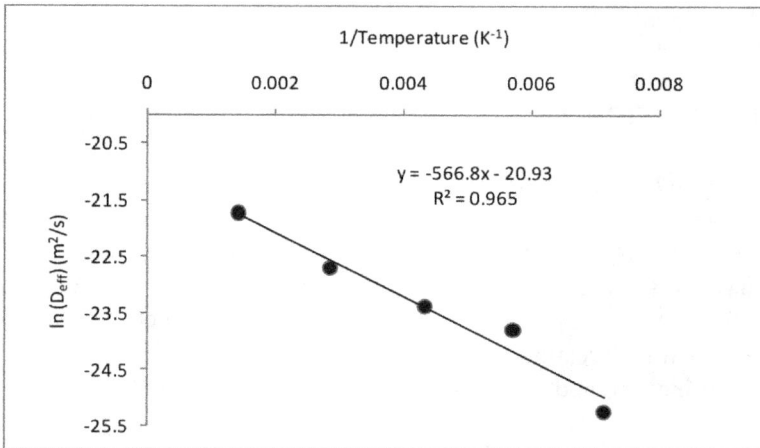

FIGURE 3.5   Arrhenius plot of microwave drying of prawns.

ACKNOWLEDGMENT

The authors highly acknowledge the Institute for providing the necessary experimental facilities to carry out this work.

KEYWORDS

- **energy efficiency**
- **industrial drying**
- **micro-organisms**
- **moisture ratio**
- **relative humidity**
- **root mean square error**

REFERENCES

1. Doymaz, I., (2007). Air-drying characteristics of tomatoes. *J. Food Eng., 78*(4), 1291–1297.
2. Decareau, R. V., (1985). *Microwaves in the Food Processing Industry* (1st edn.). Academic Press: Orlando, FL. ISBN:9780122084300, Elsevier.
3. Sharma, G. P., & Prasad, S., (2006). Optimization of process parameters for microwave drying of garlic cloves. *J. Food Eng., 75*(4), 441–446.
4. Decareau, R. V., (1992). *Wiley Encyclopaedia of Food Science and Technology, 3*. John Wiley & Sons: USA.
5. Demirhan, E., & Ozbek, B., (2010). Drying kinetics and effective moisture diffusivity of purslane undergoing microwave heat treatment. *Korean J. Chem. Eng., 27*, 1377–1383.
6. Erbay, Z., & Icier, F., (2010). A review of thin-layer drying of foods: Theory, modeling, and experimental results. *Critical Reviewers in Food Science and Nutrition, 50*(5), 441–464.
7. Feng, H., Tang, J., & Cavalieri, R. P., (1999). Combined microwave and spouted bed drying of diced apples: Effect of drying conditions on drying kinetics and product temperature. *Drying Technol., 17*(10), 1981–1998.
8. Walde, S. G., Balaswamy, K., Velu, V., & Rao, D. G., (2002). Microwave drying and grinding characteristics of wheat (*Triticum aestivum*). *J. Food Eng., 55*(3), 271–276.
9. Kadlec, P., Rubecova, A., Hinkova, A., Kaasova, J., Bubnik, Z., & Pour, V., (2001). Processing of yellow pea by germination, microwave treatment and drying. *Innov. Food Sci. Emerg. Technol., 2*(2), 133–137.
10. Wang, W., Thorat, B. H., Chen, G., & Mujumdar, A. S., (2002). Simulation of fluidized-bed drying of carrot with microwave heating. *Drying Technol., 20*(9), 1855–1867.
11. Wang, J., & Sheng, K., (2006). Far-infrared and microwave drying of peach. *LWT Food Science and Technology, 39*(3), 247–255.

12. Soysal, Y., (2004). Microwave drying characteristics of parsley. *Biosyst. Eng., 89*(2), 167–173.
13. Temple, S. J., & Van, B. A. J. B., (1999). Modeling of Fluidized-bed drying of black tea. *J. Agric. Eng. Res., 74*(2), 203–212.
14. Funebo, T., & Ohlsson, T., (1998). Microwave-assisted air dehydration of apple and mushroom. *J. Food Eng., 38*(3), 353–367.
15. McMinn, W. A. M., Khraisheh, M. A. M., & Magee, T. R. A., (2003). Modeling the mass transfer during convective, microwave and combined microwave-convective drying of solid slabs and cylinders. *Food Res. Int., 36*, 977–983.
16. Khraisheh, M. A. M., McMinn, W. A. M., & Magee, T. R. A., (2004). Quality and structural changes in starch foods during microwave and convective drying. *Food Res. Int., 37*, 497–503.
17. Xu, Y., Min, Z., Mujumdar, A. S., Le-Qun, Z., & Jin-Cai, S., (2004). Studies on hot air and microwave vacuum drying of wild cabbage. *Drying Technol., 22*(9), 2201–2209.
18. Choi, K. B., Park, S. I., Park, Y. S., Sung, S. W., & Lee, D. H., (2002). Drying characteristics of millet in a continuous multistage fluidized bed. *Korean J. Chem. Eng., 19*(6), 1106–1111.
19. Yi, S. C., & Moon, S. K., (1997). Reconstitute tobacco product drying model. *Korean J. Chem. Eng., 14*(2), 141–145.
20. Dadali, G., Apar, D. K., & Ozbek, B., (2007). Microwave drying kinetics of okra. *Drying Technol., 25*(5), 917–924.
21. Dadali, G., Demirhan, E., & Ozbek, B., (2007). Microwave heat treatment of spinach: Drying kinetics and effective moisture diffusivity. *Drying Technol., 25*(10), 1703–1712.
22. Demirhan, E., & Ozbek, B., (2010). Microwave drying characteristics of basil. *J. Food Process. Pres., 34*, 476–494.
23. Pandey, S. K., Diwan, S., & Soni, R., (2015). Review of mathematical modeling of thin-layer drying process. *International Journal of Current Engineering and Scientific Research, 2*(11), 96–107.
24. Kutlu, N., & Isci, A., (2017). Drying characteristics of zucchini and empirical modeling of its drying process. *Int. J. Food Studies, 6*, 232–244.
25. Maisnam, D., Rasane, P., Dey, A., Kaur, S., & Sarma, C., (2016). Recent advances in conventional drying of foods. *J. Food Technol. Pres., 1*(1), 25–34.
26. Ayensu, A., (1997). Dehydration of food crops using a solar dryer with convective heat flow. *Solar Energy* (Vol. 59, No. 4–6, pp. 121–126). Proceedings of ISES 1995: Solar World Congress, HARARE, ZIMBABWE.
27. Lewis, W. K., (1921). The rate of drying of solid materials. *Journal of Industrial Engineering and Engineering Chemistry, 13*, 427–432.
28. Bengtsson, G., Rahman, M. S., Stanley, R., & Perera, C. O., (1998). Effect of specific pretreatment on the drying behavior of apple rings. In: *Proceedings of the New Zealand Institute of Food Science and Technology and the Nutrition Society of New Zealand Conference*. Nelson, New Zealand.
29. Wang, Z., Sun, J., & Liao, X., (2007). Mathematical modeling on hot air drying of thin layer apple pomace. *Food Res. Int., 40*(1) 39–46.
30. Page, G., (1949). *Factors Influencing the Maximum Rates of Air Drying Shelled Corn in Thin Layers [M.S. Thesis]*. Purdue University, West Lafayette, Ind, USA.
31. Sarsavadia, P. N., Sawhney, R. L., Pangavhane, D. R., & Singh, S. P., (1999). Drying behavior of brined onion slices. *J. Food Eng., 40*, 219–226.
32. Henderson, S. M., & Pabis, S., (1961). Grain drying theory: Temperature effect on drying coefficient. *Journal of Agricultural Engineering Research, 6*, 169–174.

33. Yaldiz, O., & Ertekin, C., (2001). Thin layer solar drying of some vegetables. *Drying Technol., 19*(3, 4) 583–597.
34. Wang, C. Y., & Singh, R. P., (1978). *A Single Layer Drying Equation for Rough Rice*. ASAE Paper No. 78-3001, ASAE, St. Joseph, MI.
35. Sharma, G. P., & Prasad, S., (2004). Effective moisture diffusivity of garlic cloves undergoing microwave convective drying. *J. Food Eng., 65*(4), 609–617.
36. Sacilik, K., (2007). The thin-layer Modeling of tomato drying process. *Agriculture Conspectus Scientificus (ACS), 72*(4), 343–349.
37. Midilli, A., Kucuk, H., & Yapar, Z., (2002). A new model for single-layer drying. *Drying Technol., 20*(7), 1503–1513.

CHAPTER 4

Analysis of the Relation Between Building Design Elements and Appearance of Natural Light and Internal Temperature in Residential Apartments

TRUPTI J. DABE

Priyadarshini Institute of Architecture and Design Studies, Maharashtra, India, E-mail: truptidabe78@gmail.com

ABSTRACT

The present study has analyzed the effect of building design elements on the appearance of natural light and internal temperature in the low-rise residential apartments through statistical analysis to achieve a comfort visual and internal environment for inhabitants. The data for statistical analysis has collected from the simulation experiments are inferred to know the measure of building design variables and their effect on natural light and internal temperature outcome. For this study, the case of residential apartment typology selected from the Nagpur district, of Central India, having climatic conditions as hot and dry. The field measurements of the selected case have taken to regulate the simulation model. In simulation experiments, the assessment of the level of natural light is done with finalized variables like window to carpet area ratio (WCAR) and orientation of building, by using natural light standard of measurement, namely useful natural light illuminance (UDI). For this research, the statistical analysis methods such as multiple variable regression analysis, analysis of variance (ANOVA), and correlation coefficients, are used to investigate the relationship between the building design elements and natural light performance, internal temperature under the statistical computer program, i.e., statistical package for the social science (SPSS). Based on the statistical analysis the chapter concludes that the area of window, number of windows, depth of room, carpet area of room, percentage of WCAR, and orientation of building have the strong positive relationship with the appearance of natural light and internal temperature in enclosed livable spaces in residential apartment.

4.1 INTRODUCTION

Appearance of internal vigor and atmosphere has become leisurely significant in all type of structure design. People need good natural lighting in their living and working surroundings as natural light provides a more pleasant and attractive internal environment [1]. In residential apartments, it has been observed that good natural light possibly increases the performance of inhabitants and pays to a healthier enclosed working space [2]. The internal atmosphere is naturally lighted and ventilated buildings greatly influenced by the microclimate and the way environmental controls are used. The inhabitants have been offered by main controls like openable windows and doors to improve the visual and thermal environment, in their pursuit to comfort [3]. Due to lack of consideration of building design parameters associated with these environmental controls in the design process of residential apartments may negatively cause visual and thermal comfort to the inhabitants [4]. In current trends, it observed that in the residential apartment typology, the architects or designers accommodate 10–12 flats on each floor of six to seven floored building, so ignorance has made towards the availability of natural light and thermal comfort in the internal space of the building. Conversely, during the design process of residential apartment, it is difficult for the architects or designers to evaluate each building design parameters independently by doing experiments before construction. In the published literature, Architects or designers often tried to limit the number of building design parameters in their simulation experiments because it is harder to detect the reason and effect of relationships between these building design variables and appearance of natural light and internal comfort [5].

Therefore, in the residential apartment design process needs a diagnostic tool has the capability to co-operate with the architects or designers at the initial phase of the design development and reduces the computation time. And this need fulfills by developing the statistical model functionality from the viewpoint of the behavior of natural light and processed with the statistical relationship between the building design parameters, should provide the reliability, accuracy, and simplicity in natural light analysis that will suitably help architects and designers in their design process. The substantial studies on statistical analysis have been done by the researchers like to find out the important architectural building design parameters and predict healthy building enactment related with the green building assessment [6, 7], statistical method can be used to evaluate the effect of the size and the type of a window on internal comfort [8], to evaluate the relationship between the level of natural light and performance of students in school building [9, 10], effect of natural light on productivity [10], and, effect of natural light on performance of workers in office building with the multiple regression as research method [11] and, effect of natural light on behavior of patients health [12, 13]. And, very few statistical analysis researches have been done on the relationship between the building design parameters like size of window, placement of window, type of window, type of window glazing, type, and size of shading devices, orientation of building, length, and depth of interior space (room), area of interior enclosed space and the appearance of natural light and internal temperature in the residential apartments. Hence, the study concentrated on the investigation of the bond between these building design parameters, and the level of natural light and internal temperature in the livable enclosed places of the residential apartment. This investigation shall important to make the design decision to architects or designers at the early phase of construction to achieve optimal natural light and internal temperature in the livable enclosed places of the residential apartment.

The aim of this study is to identify the building design parameters which are strongly affected by the appearance of natural light and internal temperature in the livable enclosed spaces of the residential apartment. Through this research architects and designers may derive the adequate values (dimensions) of each building design parameters before construction, to achieve the optimum natural light and internal temperature for the visual and thermal comfort of inhabitants.

4.1.1   THE CASE STUDY

The residential apartment has been selected for this study as an illustrative case from Nagpur region (21.14° N, 79.08° E), Maharashtra, Central India. The selected case has a warm and arid climate that is alike in case of tropical climate experiential at the worldwide level. The selected residential apartment has ground plus two floors and two flats on every floor and carpet area of every flat as 50.07 m^2. Each flat of a residential apartment having a living room, two bedrooms, and kitchen and dining; as livable enclosed spaces along with service area and is facilitated with lobby, a staircase, and a meter room (on the ground floor as shown in Figure 4.1) which form a service core of the apartment. In this study, the flat on the ground floor was selected for investigation.

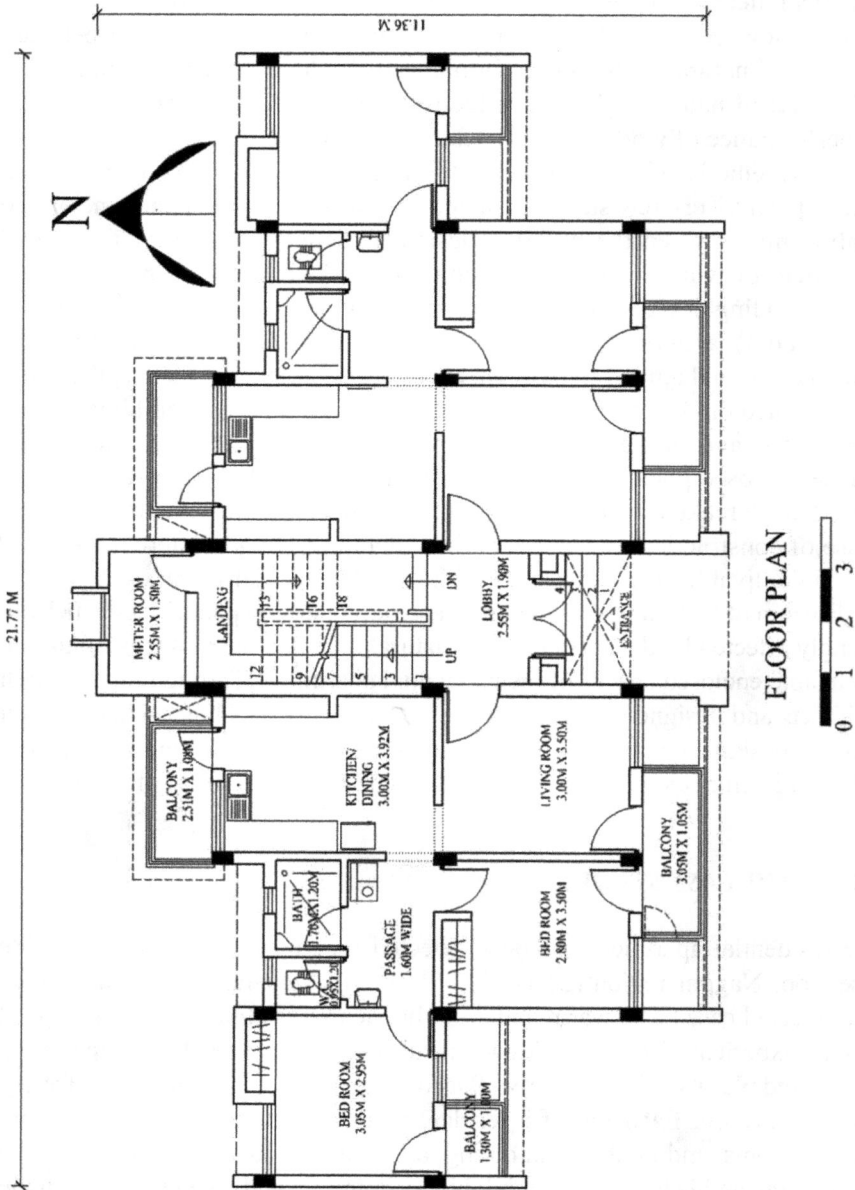

FIGURE 4.1 Typical floor plan of the selected residential apartment.

The overall climate of Nagpur is hot and dry all over the year, excluding rainy season, i.e., between June to September and, a very warm weather conditions during the summer between March to May, which reaches the high point in the month of May, i.e., 48°C. The winter season of Nagpur district has a lowest temperature about 10–12°C and often dips below that level [14].

4.2   RESEARCH METHODOLOGY

In this study, the research methodology has been conducted on the basis of a literature review [15, 16]. The research methodology regarding simulation experiments and statistical analysis is summarized below (refer to Figure 4.2).

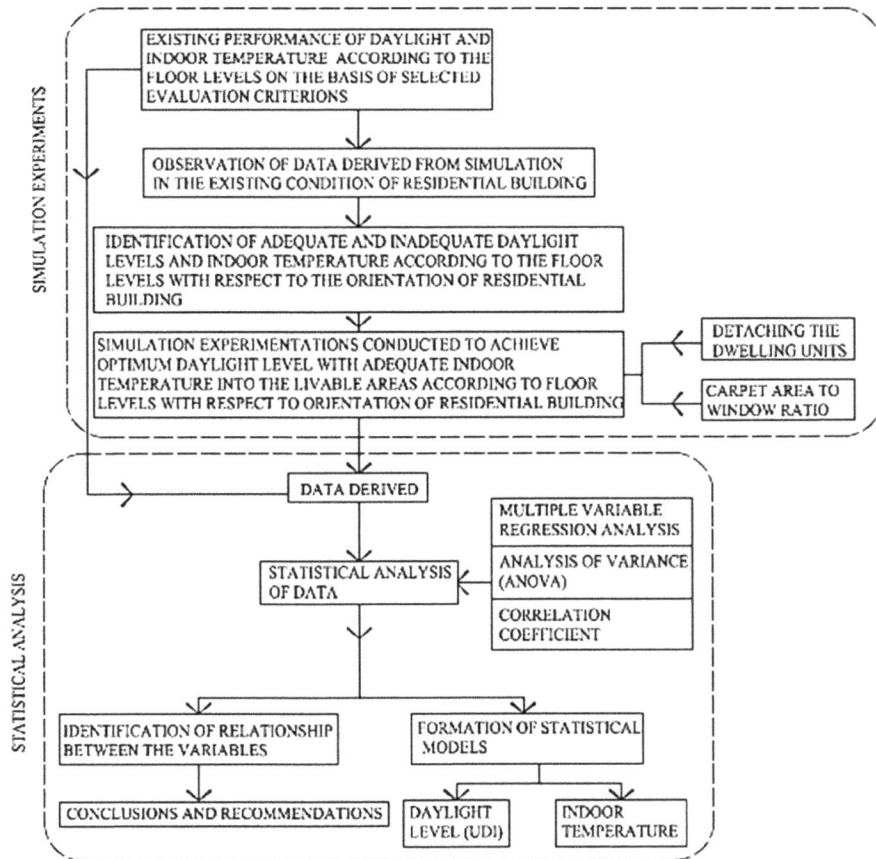

FIGURE 4.2   Workflow diagram.

The existing appearance of natural light and the internal temperature of the livable enclosed spaces of the selected case has evaluated on the basis of a regulated simulation model of the case. The existing field measurements of the natural light and internal temperature of selected case are used for the calibration of simulation model [15]. The observations of the adequate and inadequate appearance of natural light in existing condition have done based on recommended criteria (Table 4.1) of level of natural light (measured in percentages of useful natural light illuminance (UDI) in this study context). According to that, the simulation experiments have been conducted to achieve an excellent range of level of natural light and internal temperature. In this research total 144 simulation experiments (refer Tables 4.2 and 4.3) have conducted on the regulated simulation model of the selected residential apartment by using Daysim (Radiance based natural light analysis tool) plugin Ecotect, developed by the National Research Council of Canada and the Fraunhofer Institute for Solar Energy Systems in Germany [17] with natural light metrics, i.e., useful natural light illuminance (UDI).

The simulation experiments have mainly included the detaching of flats from the main service core of the residential apartment (Table 4.1) and changing in percentages of window to carpet area ratio (WCAR) respecting the orientation of the apartment (Table 4.3). Some experiments aspects have considered for the simulation experiments, to get the results in an appropriate manner as summarized in the following section:

- Simulation experiments have conducted on ground floor respecting the cardinal orientations of a residential apartment.
- Single window unit is measured as 45 centimeters × 120 centimeters, and its recurrence was limited up to three parts due to the overlying of neighboring area/rooms.
- One/two/three window units are used for the addition of new window to the room or for increasing the width of the existing window as per the requirement.
- The experiment also includes detaching of the flat from the common main service core area (Table 4.1) up to maximum of 3.00 meters to enhance natural light and ventilation of livable enclosed spaces of the residential apartment.

In the current study, founded on the studied literature [18], the assessment standards of the range of percentage for UDI (100–3000 lux) is suggested (Table 4.2) for judging the output result obtained by statistical models. It may help the architects and designers to achieve the optimal level of natural lights in the livable enclosed spaces of the building.

TABLE 4.1 Plan of a Residential Apartment in the Current and Investigational Condition

Existing Condition	Experimental Condition (Detaching from Service Core)
FLOOR PLAN SERVICE CORE	FLOOR PLAN SERVICE CORE

TABLE 4.2 Recommended Criteria for Evaluation of the Variety of Proportion (in %) for UDI (100–3000 lux) [19]

Standards	Poor	Average	Good	Excellent
Percentage of UDI (100–3000 lux)	0–50.0%	51.0–70.0%	71.0–85.0%	Above 86.0%

Some variables are frozen during the simulation experiments as follows:

1. **Wall:** External: 0.23 m thick brick wall in cement mortar with sand faced/roughcast cement plaster (reflectivity of color: 0.964). Internal: 0.15 m thick brick wall in cement mortar with cement plaster (reflectivity of color: 0.944).
2. **Floor:** Living room, bedroom, and kitchen: tiles (reflectivity of color: 0.907).
3. **Ceiling:** Flat reinforced concrete cement (R.C.C.) slab (reflectivity of color: 0.901).
4. **Shading Device (Overhang):** About 0.60 m (depth) horizontal structural (R.C.C.) projection usually providing over window on exterior walls to provide safety from sun and rain and for purpose of architectural look.
5. **Window Glazing Type:** 6 mm thick clear glass.

6. **Window Type:** Aluminum frame double shutter sliding window.
7. **Windowsill and Head Level:** 0.9 m sill and 2.1 m head height of window from floor level.
8. **Useful Natural Light Illuminance (UDI):** It is defined as the yearly occurrence of illuminance through the work plane, that is, within a range measured 'Useful' by occupants that are between 100 lux and 3000 lux. The UDI standard measurements have been applied by decisive the incidence of the natural light illuminance at each control point. The UDI between the range of 100 lux and 3000 lux is considered as the minimum and maximum limits as an adequate level for useful natural lighting in residential apartments [20]. The UDI is applied by determining each calculation point the incidence of level of natural lights where: The illuminance is less than 100 lux, i.e., darkness. The illuminance is greater than 100 lux and less than 300 lux, i.e., UDI supplementary. The illuminance is greater than 300 lux and less than 3,000 lux, i.e., UDI autonomous. The illuminance is greater than 3,000 lux, i.e., glare [21]. Many contrasting results can be seen in the applicable literature on illuminance optimal values, [21, 22] but a point is measured to receive good natural lighting as the illuminance level, i.e., between 100 lux and 3000 lux for at least 50% of the time per year. Therefore, for this research, UDI (100–3000 lux) of 50% of the time per year is measured as a verge value.

Permitting to the handbook of the functional requirement of the building (other than industrial buildings) [23], 19°C and 34°C tropical summer index (TSI) are the lower and upper limits of easily tolerable cold, and warm conditions, respectively. In the NBC-2005, the TSI is mentioned as 25°C and 30°C [24]. With the reference to these standards, the adaptive comfort temperature range was set to 18°C–32°C for the purpose of this research. The internal temperature study is done with Ecotect software [25] by using 'Temperature distribution tool.' This tool is used for the study of the internal air temperature, which shows the distribution of temperature in terms of hours of total annual hours, i.e., hours out of 8760 (365 × 24) and can show a separate zone or all zone's air temperature values of the building.

The simulation experiments of the addition of new/extra window have done in the livable enclosed spaces of flat (Table 4.3) with two window units (size of window – 0.90 meters × 1.20 meters) and three window units (size of the window – 1.35 meters × 1.20 meters).

The details of selected variables considered during the simulation experiments as mentioned in Table 4.4.

The statistical analysis is done with numerical data derived from the above 144 numbers of simulation experiments. From the data analysis, it is found that there are two groups of variables one is dependent variable, i.e., level of natural light (UDI) and internal temperature (Internal temperature hours per year) which changes according to the independent variables, i.e., Window Area, number of the window, depth of the room, carpet area of the room, the percentage of WCAR, and orientation of the building. For the calculation of percentages of WCAR, area of window and carpet area of room are used. Therefore, number of the window, depth of the room, the percentage of WCAR, and orientation of the building, these independent variables are finalized for this research.

TABLE 4.3 The Percentages of Window to Carpet Area Ratio in the Current and Investigational Condition of the Residential Apartment (E-Existing Window)

Room	Existing	Experiment-1	Experiment-2
Living Room			
WCAR in %	15.43	25.71	30.86
Bedroom-1			
WCAR in %	16.53	27.55	33.06

TABLE 4.3   *(Continued)*

Room	Existing	Experiment-1	Experiment-2
Bedroom-2			
WCAR in %	30.0	42.01	48.01
Kitchen and Dining			
WCAR in %	13.84	22.96	27.55

On the basis of two groups of variables, i.e., dependent and independent variables, the multivariable regression analysis [26–28]. The study of variance approach is used to quantify the importance of individually selected variables on the appearance of natural light and internal temperature [29, 30] and correlation coefficient [31, 32]; these statistical methods are selected to examine the relationship between dependent and independent variables.

4.3   STATISTICAL ANALYSIS AND RESULTS

For this research, the simulation model study has selected because it represents the simplest, most acceptable, and most reliable technique for simulating

TABLE 4.4 Details of Simulation Experiments (North-N, East-E, South-S, and West-W)

Carpet Area (Sq M)	Room Width (M)	Room Length (M)	Numbers of Windows	Window Width W1 (M)	Window Height W1 (M)	Window Width W2 (M)	Window Height W2 (M)	Total Window Area (Sq M)	WCAR in %	Orientation	Condition
						Living Room					
10.5	3	3.5	1	1.35	1.2	0	0	1.62	15.43	N	Existing
			2	1.35	1.2	0.9	1.2	2.7	25.71	N	Experiment-1
			2	1.35	1.2	1.35	1.2	3.24	30.86	N	Experiment-2
			1	1.35	1.2	0	0	1.62	15.43	E	Existing
			2	1.35	1.2	0.9	1.2	2.7	25.71	E	Experiment-1
			2	1.35	1.2	1.35	1.2	3.24	30.86	E	Experiment-2
			1	1.35	1.2	0	0	1.62	15.43	S	Existing
			2	1.35	1.2	0.9	1.2	2.7	25.71	S	Experiment-1
			2	1.35	1.2	1.35	1.2	3.24	30.86	S	Experiment-2
			1	1.35	1.2	0	0	1.62	15.43	W	Existing
			2	1.35	1.2	0.9	1.2	2.7	25.71	W	Experiment-1
			2	1.35	1.2	1.35	1.2	3.24	30.86	W	Experiment-2
						Bedroom-1					
9.8	2.8	3.5	1	1.35	1.2	0	0	1.62	16.53	N	Existing
			2	1.35	1.2	0.9	1.2	2.7	27.55	N	Experiment-1
			2	1.35	1.2	1.35	1.2	3.24	33.06	N	Experiment-2
			1	1.35	1.2	0	0	1.62	16.53	E	Existing
			2	1.35	1.2	0.9	1.2	2.7	27.55	E	Experiment-1
			2	1.35	1.2	1.35	1.2	3.24	33.06	E	Experiment-2
			1	1.35	1.2	0	0	1.62	16.53	S	Existing
			2	1.35	1.2	0.9	1.2	2.7	27.55	S	Experiment-1
			2	1.35	1.2	1.35	1.2	3.24	33.06	S	Experiment-2
			1	1.35	1.2	0	0	1.62	16.53	W	Existing
			2	1.35	1.2	0.9	1.2	2.7	27.55	W	Experiment-1
			2	1.35	1.2	1.35	1.2	3.24	33.06	W	Experiment-2

TABLE 4.4 *(Continued)*

Carpet Area (Sq M)	Room Width (M)	Room Length (M)	Numbers of Windows	Window Width W1 (M)	Window Height W1 (M)	Window Width W2 (M)	Window Height W2 (M)	Total Window Area (Sq M)	WCAR in %	Orientation	Condition
						Bedroom-2					
9.0	3.05	2.95	1	1.35	1.2	0	0	2.7	30.01	N	Existing
			2	1.35	1.2	0.9	1.2	3.78	42.01	N	Experiment-1
			2	1.35	1.2	1.35	1.2	4.32	48.01	N	Experiment-2
			1	1.35	1.2	0	0	2.7	30.01	E	Existing
			2	1.35	1.2	0.9	1.2	3.78	42.01	E	Experiment-1
			2	1.35	1.2	1.35	1.2	4.32	48.01	E	Experiment-2
			1	1.35	1.2	0	0	2.7	30.01	S	Existing
			2	1.35	1.2	0.9	1.2	3.78	42.01	S	Experiment-1
			2	1.35	1.2	1.35	1.2	4.32	48.01	S	Experiment-2
			1	1.35	1.2	0	0	2.7	30.01	W	Existing
			2	1.35	1.2	0.9	1.2	3.78	42.01	W	Experiment-1
			2	1.35	1.2	1.35	1.2	4.32	48.01	W	Experiment-2
					Kitchen and Dining Room						
11.76	3	3.92	1	1.35	1.2	0	0	1.62	13.84	N	Existing
			2	1.35	1.2	0.9	1.2	2.7	22.96	N	Experiment-1
			2	1.35	1.2	1.35	1.2	3.24	27.55	N	Experiment-2
			1	1.35	1.2	0	0	0	13.84	E	Existing
			2	1.35	1.2	0.9	1.2	2.7	23.08	E	Experiment-1
			2	1.35	1.2	1.35	1.2	3.24	27.69	E	Experiment-2
			1	1.35	1.2	0	0	1.62	13.84	S	Existing
			2	1.35	1.2	0.9	1.2	2.7	23.08	S	Experiment-1
			2	1.35	1.2	1.35	1.2	3.24	27.69	S	Experiment-2
			1	1.35	1.2	0	0	1.62	13.84	W	Existing
			2	1.35	1.2	0.9	1.2	2.7	23.08	W	Experiment-1
			2	1.35	1.2	1.35	1.2	3.24	27.69	W	Experiment-2

natural light and internal temperature performance. In this research statistics analysis includes four major variables such as number of the window, depth of the room, the percentage of WCAR, and building orientation, it is detected that as per changes in the values of these variables during the experiments major variations are found in the percentage of UDI (level of natural light) and internal temperature of the livable enclosed spaces. The percentage of WCAR is the window area divided by the carpet area of the room. The multi-variable regression models were established under the computer program, i.e., statistical package for the social science (SPSS).

From Table 4.5, it observed the values of percentage of UDI (100 lux–3000 lux), i.e., 60.45%–92.93% recorded in the living room with mean value of 84.42% and standard deviation of ± 10.10%. Similarly, values of % of UDI (100 lux–3000 lux), i.e., 54.88%–91.54% recorded in the bedroom-1 with mean value of 80.71% and standard deviation of ± 12.39, values of % of UDI (100 lux–3000 lux), i.e., 73.85%–88.58% recorded in the bedroom-2 with mean value of 79.59% and standard deviation of ± 4.41. In addition, values of % of UDI (100 lux–3000 lux), i.e., 65.78%–93.04% recorded in the kitchen with a mean value of 83.60% and standard deviation of ± 9.09. The living room, bedroom-1, and kitchen and dining data had standard deviation of ± 10.10%, ± 12.39, and ± 9.09 whereas the bedroom-2 data had standard deviation of ±4.41. The living room, bedroom-1, and kitchen and dining data having higher standard deviations means the data of these livable enclosed spaces is more spread out or dispersed than the data of bedroom-2. The bedroom-2 data had a standard deviation of ±4.41 due to it has three window openings from three sides and helps to penetrate optimum natural light into the internal space of bedroom-2.

TABLE 4.5 The Mean Percentage of UDI (100–3000 lux) Recorded in the Livable Enclosed Spaces of Selected Residential Apartment

Room	Minimum of % of UDI (100 lux–3000 lux)	Maximum of % of UDI (100 lux–3000 lux)	± Standard Deviation	Mean of % of UDI (100 lux–3000 lux)
Living room	60.45	92.93	±10.10	84.42
Bedroom-1	54.88	91.64	±12.39	80.71
Bedroom-2	73.85	88.58	±4.41	79.59
Kitchen and dining	65.78	93.04	±9.09	83.60

From Table 4.6, it observed that the values of internal temperature hours, i.e., 3501–6794 recorded in the living room with a mean value of

5967.1 and a standard deviation of $\pm$ 1469.6. Similarly, the values of internal temperature hours, i.e., 3723–6794 recorded in the bedroom-1 with a mean value of 6017.5 and standard deviation of $\pm$ 1367.3, the values of internal temperature hours, i.e., 6776–6794 recorded in the bedroom-2 with mean value of 6788.1 and standard deviation of $\pm$ 6.5. And, the values of internal temperature hours, i.e., 6542–6739 recorded in the kitchen and dining with mean value of 6638.2 and standard deviation of $\pm$ 77.9. Similarly, the living room, bedroom-1, and kitchen and dining data had standard deviation of $\pm$ $\pm$1469.6, $\pm$1367.3, and $\pm$77.9 whereas the bedroom-2 data had standard deviation of $\pm$ 6.5. The living room, bedroom-1, and kitchen and dining data having higher standard deviations means the data of these livable enclosed spaces is more spread out or dispersed than the data of bedroom-2. The bedroom-2 data had a standard deviation of $\pm$ 6.5 due to it has three window openings from three sides and creates cross ventilation and helps in makes internal temperature comfort.

TABLE 4.6   The Internal Temperature Recorded in the Livable Enclosed Spaces of the Selected Residential Apartment

Room	Minimum Internal Temperature Hours Per Year	Maximum Internal Temperature Hours Per Year	$\pm$Standard Deviation	Mean of Internal Temperature Hours Per Year
Living room	3501	6794	$\pm$1469.6	5967.1
Bedroom-1	3723	6794	$\pm$1367.3	6017.5
Bedroom-2	6776	6794	$\pm$6.5	6788.1
Kitchen and dining	6542	6739	$\pm$77.9	6638.2

4.4   RELATIONSHIP BETWEEN DAYLIGHT LEVEL AND VARIABLES

In this research, the multiple regression analysis was used to study the numerous building design variables, which significantly predicted the level of natural light [UDI (100 lux–3000 lux)]. The results of the regression analysis indicated the four independent variables explained 80.1% of the variance [$R^2 = 0.621$, $F (6, 57) = 21.587$, $p<0.01$].

The 'R' column in Table 4.7 represents the value of R, the multiple correlation coefficients; it can be considered to be one measure of the quality of the prediction of the dependent variable [% of UDI (100 lux–3000 lux)]. A value of R = 0.801 shows a good level of prediction. The 'R Square' column represents the R^2 value (the coefficient of determination), which is the quantity of variance in the dependent variable that can be explained by the independent variables (number of windows, depth of room, percentage of WCAR, and orientation of building) (precisely, it is the proportion of variation accounted for by the regression model above and beyond the mean model). Based on the results it was observed to be 0.621 (R^2), i.e., the selected independent variables (number of the window, depth of the room, the percentage of WCAR, and orientation of building) explain 62.1% of the predictability of the dependent variable [UDI (100–3000 lux)]. However, the 'Adjusted R Square' found to be 0.605, which means that the selected independent variables contributed 60.50% UDI (100–3000 lux) of the room. The remaining 39.50% have contributed by the other factors, which are not included in this study.

TABLE 4.7   Summary of Model

Model	R	R Square	Adjusted R Square	Standard Error of the Estimate
1	0.801	0.621	0.605	5.40200

*Independent Variables (Constant):** Number of the window, depth of the room, the percentage of WCAR, and orientation of the building.

Dependent Variable: Percentage of UDI (100 lux–3000 lux) (level of natural light).

The F-ratio in Table 4.8 test whether the overall regression model was a good fit for the data. Table 4.8 shows that the independent variables statistically significantly predict the dependent variable [Percentage of UDI (100–3000 lux)], $F (5, 51) = 20.217$, $p < 0.0001$ (i.e., the regression model was a good fit of the data).

TABLE 4.8   Results of ANOVA

Model	Sum of Squares	DF	Mean Square	F	Sig.
Regression	3779.509	5	625.835	20.217	0.000
Residual	1512.101	51	29.101	–	–
Total	5291.61	56	–	–	–

In Table 4.9, the unstandardized coefficients indicate how much the dependent variable [% of UDI (100 lux–3000 lux)] varies with an independent variable when all other independent variables are held constant. From the p-value of the t-test for each predictor, it found that each of the independent variables contributes to the statistical model, but the depth of room and orientation of the building does not contribute to the statistical model.

TABLE 4.9   Correlation Coefficients

Model	Unstandardized Coefficients		Standardized Coefficients	t	Sig.
	B	Standard Error	Beta		
(Constant)	170.611	28.757	–	5.933	0.000
Depth of room	−16.888	8.988	−0.173	−1.879	0.065
Numbers of windows	18.802	4.311	1.229	4.362	**0.000**
Percentage of WCAR	−4.586	0.953	−4.937	−4.812	**0.000**
Orientation of building	−0.963	0.614	−0.115	−1.568	0.122

In this case, the variable like numbers of Windows (unstandardized coefficient of 18.802, p<0.01) have a significant positive relationship with the % of UDI (100 lux–3000 lux). Thus, a multiple regression predicts the percentage of UDI (100 lux–3000 lux) from the depth of the room, numbers of windows, percentages of WCAR, and the orientation of the building. The results shows that these variables statistically significantly predicted UDI, F $(5, 51) = 20.217$, $p < 0.0001$, $R^2 = 0.621$. Moreover, except depth of room and orientation of the building, all the other independent variables, i.e., numbers of windows, percentages of WCAR showed significant (P<0.05) relationship with the % of UDI (100 lux–3000 lux). In simulation experiments, it is observed that the % of UDI (100 lux–3000 lux) vary according to the directions (North, East, South, and West) towards residential apartment oriented but not much variation in values between two directions.

4.5   RELATIONSHIP BETWEEN THE PERFORMANCE OF INTERNAL TEMPERATURE AND VARIABLES (NUMBER OF WINDOWS, DEPTH OF ROOM, PERCENTAGE OF WCAR, AND ORIENTATION OF BUILDING)

Multiple regression study was carried out to assess if the various variables of the room significantly predicted its internal temperature. The results of the regression analysis show that the four independent variables explained 74.1% of the variance ($R^2 = 0.530$, F $(5, 51) =12.272$, p<0.01.

In Table 4.10, a value of R = 0.741 shows a good level of prediction. The 'R Square' column represents the R^2 value, which is the quantity of variance in the dependent variable that is explained by the independent variables (number of the window, depth of the room, the percentage of WCAR, and orientation of building). Based on the results, it was observed to be 0.530, i.e., the selected independent variables explain 53.0% of the variability of the dependent variable (Internal temperature hours per year). However, the 'Adjusted R Square' was found to be 0.511, which means that the selected independent contributed 51.1% in internal temperature hours per year of the room. The remaining 54.10% had contributed by other factors, which are not included in this study.

TABLE 4.10 Model Summary

Model	R	R Square	Adjusted R Square	Standard Error of the Estimate
1	0.741	0.530	0.511	701.20721

*Independent Variables (Constant):** Number of the window, Depth of the room, the percentage of WCAR, and orientation of the building.

Dependent Variable: Comfort hours of internal temperature per year.

The F-ratio in Table 4.11 tests whether the total regression model was a good fit for the data. Table 4.11 displays that, the independent variables (number of windows, depth of room, percentage of WCAR, and orientation of building) statistically considerably predict the dependent variable (Internal temperature hours per year), $F(5, 51) = 12.272$, $p < 0.0001$ (i.e., the regression model is a good fit of the data).

TABLE 4.11 ANOVA Results

Model	Sum of Squares	DF	Mean Square	F	Sig.
Regression	40370843.479	5	6728473.913	12.272	**0.000**
Residual	28673771.458	51	503048.622	–	–
Total	**69044614.938**	**56**	–	–	–

Unstandardized coefficients in Table 4.12 show how much the dependent variable (Internal temperature hours per year) differs with independent variables when all other independent variables are detained constant. Observing at the p-value of the t-test for each predictor, it was found that numbers of windows (independent variable) contribute to the statistical model, but other variables, i.e., depth of room, percentage of WCAR, and orientation of building does not contribute to the statistical model. In this case, the effect of a number of windows on internal temperature hours per year can be understood from the unstandardized coefficient, B_1, for the same, i.e., numbers of windows are equal to 2133.262 (see Table 4.12), which means that the increase in number of windows will result in the significant ($P<0.05$) increase in comfort hours of internal temperature per year. However, the remaining variables do not have a significant effect on the internal temperature hours per year.

TABLE 4.12   Correlation Coefficients

Model	Unstandardized Coefficients		Standardized Coefficients	t	Sig.
	B	Std. Error	Beta		
(Constant)	4444.763	3717.197	–	1.196	.237
Depth of room	−132.695	1161.799	−0.197	−1.836	0.072
Numbers of windows	2133.262	557.212	1.258	3.828	0.000
Percentage of WCAR	−14.736	123.188	−0.143	−0.120	0.905
Orientation of building	3.354	79.366	0.004	0.042	0.966

Thus, a multiple regression predicts the internal temperature hours per year from a number of the window, depth of the room, the percentage of WCAR, and orientation of the building. The results show that these variables statistically significantly predicted internal temperature hours per year, F $(5.51) = 12.272$, $p < 0.0001$, $R^2 = 0.530$ but only number of windows have a significant ($p <0.05$) positive relationship with the internal temperature hours per year.

4.6   CONCLUSIONS

This study includes statistical analysis by using multiple regression analysis tests, ANOVA, and correlation coefficients for the dependent variables, i.e., level of natural light [UDI (100 lux–3000 lux)] and internal temperature (Internal temperature hours per year) and independent variables, i.e., number of windows, depth of room, percentage of WCAR, and orientation of building. From the statistical analysis, it is concluded that the numbers of window building design variables have a significant ($P<0.05$) positive relationship with the percentage of UDI (100 lux–3000 lux) (natural light) and internal temperature hours per year (internal temperature). The percentages of WCAR showed significant ($P<0.05$) relationship with only the percentage of UDI (100 lux–3000 lux). Therefore, the adequate calculation of these building design variables for the livable enclosed spaces are very important at the initial stage of the residential apartment designs process.

Moreover, the selected independent variables like numbers of window, depth of room, percentage of WCAR, and orientation of building having a strong positive relationship with the dependent variable such as natural light and internal temperature. And helpful for the architects and designer for the calculation of the level of natural lights and internal temperature at the initial conceptual stage of the design process of residential apartments under warm and arid climatic zones.

The future work included the formation of a statistical model with the numbers of cases of residential apartments from different climatic conditions.

ACKNOWLEDGMENTS

The authors would like to acknowledge Visvesvaraya National Institute of Technology, Nagpur, India, and its facilities. This research work is the outcome of a full-time PhD program of the Department of Architecture and Planning, Visvesvaraya National Institute of Technology, Nagpur, India.

KEYWORDS

- **multiple regression analysis**
- **natural light simulation**
- **residential typology**
- **standard deviation**
- **statistical analysis**
- **tropical summer index**

REFERENCES

1. Plympton, P., Conway, S., & Epstein, K., (2000). *Natural Lighting in Schools Improving Student Performance and Health at a Price School Scan Afford* (pp. 487–492). American Solar Energy Society, American Institute of Architects.

2. Li, D. H. W., & Lam, J. C., (2003). An investigation of natural lighting performance and energy saving in a daylit corridor *Energy and Buildings, 35*, 365–373.
3. Indraganti, M., (2010). Adaptive use of natural ventilation for thermal comfort in Indian apartments. *Building and Environment, 45*, 1490–1507.
4. Ishtiaque, A., & Ali, A. R., (2017). a study on window configuration to enhance natural light performances in apartments of Dhaka. *International Journal of Science and Research (IJSR), 6*(3), 2193–2199.
5. Hester, H., (2013). *Natural Light and View Wöhrmann Print Service*. Zuthpen, Netherlands.
6. Chia-Peng, C., (2004). The performance of natural lighting with shading device in architecture design. *Tamkang Journal of Science Adaylightnd Engineering, 7*(4), 205–212.
7. Xi, C., Hongxing, Y., & Ke, S., (2016). Developing a meta-model for sensitivity analyses and prediction of building performance for passively designed high-rise residential buildings. *Applied Energy*.
8. Sicurella, F., Evolaa, G., & Wurtz, E., (2012). A statistical approach for the evaluation of thermal and visual comfort in free-running buildings. *Energy and Buildings, 47*, 402–410.
9. Patricia, P., Susan, C., & Kyra, E., (2000). *Natural lighting in Schools: Improving Student Performance and Health at a Price Schools Can Afford in National Renewable Energy Laboratory*. Madison, Wisconsin.
10. Heschong, M., (2003). *Windows and Classrooms: A Study of Student Performance and the Indoor Environment*. California Energy Commission, California.
11. Siyu, Z. H. O. U., & Neng, Z. H. U., (2013). Multiple regression models for energy consumption of office buildings in different climates in China front. *Energy, 7*(1), 103–110.
12. Aries, M. B. C., Aarts, M. P. J., & Hoof, V. J., (2015). Natural light and health: A review of the evidence and consequences for the built environment. *Lighting Research and Technology, 47*, 6–27.
13. Mustafa, K., & Levent, D., (2005). Natural light exposure and the other predictors of burnout among nurses in a University Hospital. *International Journal of Nursing Studies, 42*, 549–555.
14. Nandankar, P. K., Dewangan, P. L., & Surpam, R. V., (2011). *Climate of Nagpur, Regional Meteorological Centre Airport Regional Meteorological Centre Airport*. Nagpur (Maharashtra State), Nagpur region.
15. Dabe, T. J., & Dongre, A. R., (2016). Validation of methodology for natural light analysis through field measurements and simulation model - a case of residential building from Nagpur region. *Research Journal of Fisheries and Hydrobiology, 14*(4), 19–26.
16. Dabe, T. J., & Dongre, A. R., (2016). Analysis of performance of the natural light into critical liveable area of 'type design' dwelling unit on the basis of natural light metrics for hot and dry climate. *Indoor and Built Environment*.
17. Chien, S., & Tseng, K. J., (2014). Assessment of climate-based natural light performance in tropical office buildings a case study. *International Journal of Low-Carbon Technologies*, 1–9.
18. Berardi, U., & Anaraki, H. K., (2015). Analysis of the impacts of light shelves on the useful natural light illuminance in office buildings. In: *Proceeding 6th International Building Physics Conference*. Energy Procedia.
19. Dabe, T. J., & Adane, V. S., (2019). Assessment of Natural light into the Residential Building According to the Floor Levels for Hot and Dry Climatic Zone *Journal of Architectural Environment & Structural Engineering Research, 2*(2), 18–29.

20. Nabil, A., & Mardaljevic, J., (2005). Useful natural light illuminance: A new paradigm for assessing natural light in buildings. *Lighting Research and Technology, 37*, 41–59.

21. Mardaljevic, J., Andersen, M., Roy, N., & Christoffersen, J., (2011). Natural lighting metrics for residential building. In: *Proceeding of the 27th Session of CIE*. South Africa.

22. Hansen, V. G., Kennedy, R., Sanders, P., & Varendoff, A., (2012). Natural lighting performance of subtropical multi-residential towers; simulations tools for design decisions. In: *Proceeding of 28th Conference of Opportunities, Limits & Needs Towards an Environmentally Responsible Architecture*. Lima, Perú.

23. Bureau of Indian Standards, (1987). *Handbook on Functional Requirements of Buildings (Other than Industrial Buildings)*. Bureau of Indian standard, New Delhi.

24. Bureau of Indian Standards, (2005). *National Building Code of India*. Bureau of Indian Standards, Manak Bhavan, 9 Bahadur Shah Zafar Marg, New Delhi – 110002.

25. Abdullah, A. H., Abu, B. S. K., & Rahman, I. A., (2013). Simulation of office's operative temperature using Ecotect model. *International Journal of Construction Technology and Management, 1*, 2289–4454.

26. Nebia, B., & Aoul, K. T., (2017). Overheating and natural lighting; assessment tool in early design of London's high-rise residential *buildings. Sustainability, 9*, 1–23.

27. Regis, S., Fernando, S. W., & Roberto, L., (2001). Regression analysis of electric energy consumption and architectural variables of conditioned commercial buildings in 14 Brazilian cities. In: *Seventh International IBPSA Conference*, Rio de Janeiro, Brazil.

28. Catalina, T., Virgone, J., & Iordache, V., (2011). Study on the impact of the building form on the energy consumption. In: *Proceedings of Building Simulation 2011: 12th Conference of International Building Performance Simulation Association*. Sydney.

29. Lama, T. C., Hua, G., & Paul, F., (2015). Impact of curtain wall configurations on building energy performance in the perimeter zone for a cold climate. *Energy Procedia, 78*, 352–357.

30. Amirkhani, M., & Garcia-Hansen, V., (2017). An energy efficient lighting design strategy to enhance visual comfort in offices with windows. *Energies, 10*, 1–16.

31. Peng, X., & Mak, C. M., (2014). The effects of natural lighting and human behavior on luminous comfort in residential buildings: A questionnaire survey. *Building and Environment, 81*, 51–59.

32. Davarpanah, S., (2017). The impact of light in interior architecture of Residential building. *International Journal of Scientific & Engineering Research, 8*(6)1059–1065.

CHAPTER 5

An Attempt to Convert Vibrations to Mechanical Work to Generate Power: A Theoretical Approach

NANDINI KULKARNI,[1] SHIVRAI KULKARNI,[2] and NAVAL RAJAS[2]

[1]Symbiosis School of Planning, Architecture, and Design, Nagpur, Maharashtra, India, E-mail: nandini.kulkarni@sspad.edu.in

[2]Department of Mechanical Engineering, Shri Ramdeobaba College of Engineering and Management, Nagpur, Maharashtra, India

ABSTRACT

In the era where meeting the energy requirements of the world seems to be a Herculean task, every possible solution, innovative techniques for generation, and harnessing the energy adds to the growth of resources for power generation. Vibration energy harvesters have emerged as a very promising solution for catering to the needs of small-scale electricity generation [5]. Devices such as sensors and monitoring devices can easily be powered by the efficient use of vibration energy harvesters. In this chapter, one such novel idea is presented, in which vibrations of any system are taken as input by a cantilever beam converted into rotary motion with the help of crank and connecting rod and converted into electricity with the help of a generator.

5.1 INTRODUCTION

The war of the century is to find sustainable sources of energy for posterity [5]. A very important battle of this war is to tap into every source of energy possible. Energy conservation is one of the major concerns of scientists, researchers all around the globe. They are making significant efforts to harness, save, and reproduce energy from whichever source possible and affordable to humankind. Around the mid-1990s, researchers realized that the energy dissipated from any vibrating body can be converted directly into electricity with the help of innovative energy harvesting techniques, such as piezoelectric systems [3, 5]. One such mechanism for converting the vibrations into electricity is proposed, prototyped, and studied in this chapter.

With this proposed model, we tried to convert vibrations which are usually go unattended for useful work. Our objective was the conversion of vibrations to useful mechanical energy and in turn, converted to electrical energy. Basically, we used a cantilever-like arrangement to transfer amplified vibrations from a source to a connecting rod and a crank, which drives a small generator. Two springs are used in the parallel configuration, one over the other to increase the efficiency and for the smooth functioning of the model. According to the model, the vibrations are given directly to the cantilever, but in practice, we expect to extend beyond the stationary end, so that motion can be transferred to the device in a non-invasive way. This was used as a simulation model of a real time set-up, which would be used to be operated for large scale vibrations such as the vibrations caused due to hammer blow of the locomotive engine. Most of the model is made with cast iron angles, this is a very crude prototype. The model is based on calculations using ideal cases, to avoid complexity at primary stages. The viability of the design is proved by the glowing of a 3V LED connected to the generator.

5.2 METHODOLOGY

In this chapter, literature related to various techniques of conversion of mechanical vibrations to a useful form of energy was studied. A detailed prototype was fabricated and experimented on, by giving vibrations manually. The model was first made to be in mechanical constrained motion. The basic and primary objective of this study is to analyze, study, and innovate various methods of harvesting power by using the vibrational energy and converting it to useful mechanical work.

5.3 BASIC STRUCTURE

The proposed project/model consists of a frame of cast iron plates, welded together using arc welding. Crank and connecting rod are made up of acrylic sheets. Selection of acrylic sheet was motivated with a cause of low weight and high strength. Motion from the crank to the generator is transmitted via a belt drive. Redundant vibrations from the source brought by frames can be avoided by using truces and dampers. This can also be done by increasing the weight of the frame. Also, tighter fastening to the base can also help.

Due to random vibrations, swiveling of the cantilever is caused. This can also be avoided by using a proper bearing at the pivot of the cantilever. If swiveling is prevented further losses due to movement of spring is avoided. Also, reducing the height of the model can help as it would bring the center of gravity of the model down. Figure 5.1 gives a rough idea of the proposed model of the proposed project, a line diagram of which is illustrated in Figure 5.2. Table 5.1 gives the details of the parts of the model.

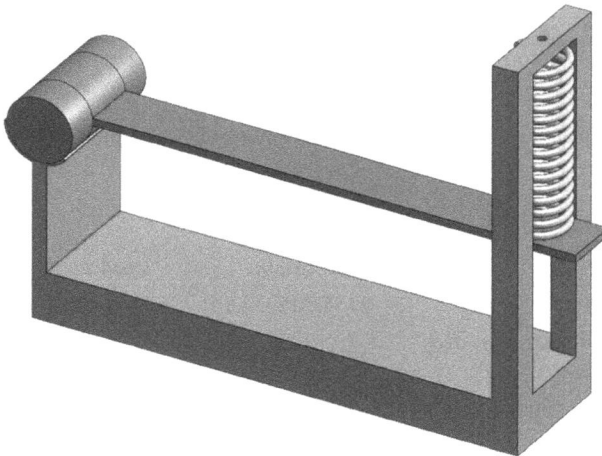

FIGURE 5.1 Schematic (version 1) diagram of proposed model.

5.4 POWER GENERATION

The oscillatory motion achieved from the beams vibrations is then transferred in the to and fro motion of the connecting rod, connected to the horizontal beam. This rotates the crank fitted with vertical support (6). The belt drive transfers the motion from the crank to the axis of the generator. The crank-belt drive is for getting more output RPM than input RPM. A complex, robust, and intricate gearbox was used to required RPM and torque to drive the generator. The differential motor of the type given in Figure 5.3 was used.

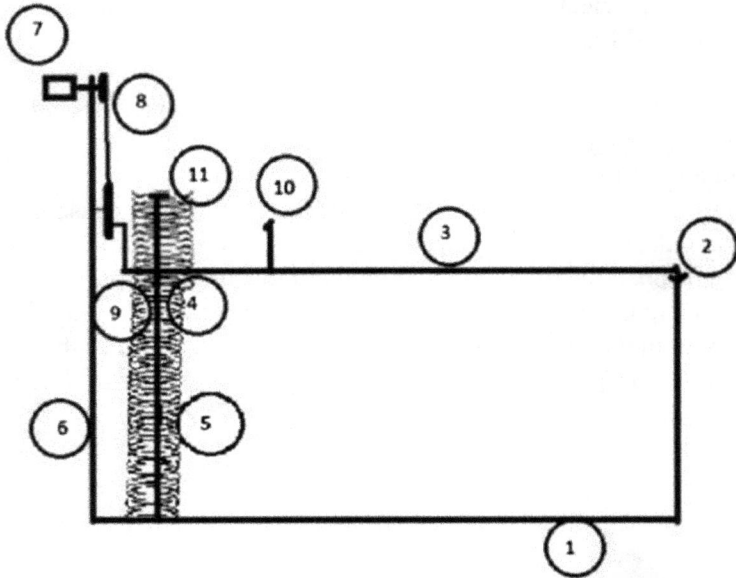

FIGURE 5.2   Line diagram of proposed model (fabricated for actual working).

TABLE 5.1   Description of Model

Part Number	Name	Description	Used for
1	Stand support	24 in. MS angles	To support the model
2	V-shaped grooves/ bearings	–	To bear the vibrations and pass them to the bar
3	Horizontal beam	22 in. MS bar	Carry transverse vibrations
4	Lower spring	K = 400 N/m	Carry longitudinal vibrations
5	Support for spring	18 in MS angle	To support the spring
6	Support for crank	24 in. MS bar	On which crank and generator shaft are mounted
7	DC generator with LED	100 rpm; 12 V	To give required potential difference
8	Belt drive	Rubber drive 18.11 in.	Transform motion from crank to generator shaft
9	Connecting rod with crank	Crank Dia = 3.9 in. Rod Length = 1.98 in.	Transform motion from horizontal beam to crank
10	Vertical bolt	–	Loading weight
11	Upper spring	K = 400 N/m	Carry longitudinal vibrations

FIGURE 5.3   Internal gearing arrangement of the generator.

In electricity generation, a generator is a device that converts motive power (mechanical energy) into electrical power for use in an external circuit [1]. A very generator of sorts is used in the model, as only powerful enough to light a LED was required.

5.5   WORKING

The vibrations from the source are gained by the horizontal beam (3) and transverse vibrations are induced in it. These vibrations are converted to oscillations of the spring (11 and/or 4). The longitudinal vibrations in the spring are responsible for to and fro motion of the connecting rod (9). Thus, the crank rotates with low RPM. This rotation is transferred to the shaft of the generator by a belt drive. As the diameter of the crank is more than the diameter of the shaft of the generator, the RPM of the generator increases. This RPM is responsible for the generation of an electric potential difference of 3 Volts, indicated by the glowing of the 3 volts LED bulb. Hence, the vibrational energy is converted into useful mechanical energy, and hence, power is generated. Figures 5.4 and 5.5 together illustrate the working of the proposed model.

5.6   RESULT

Thus, the way of converting the mechanical vibrations to useful mechanical work and intern generation of power was investigated. No use of piezoelectric

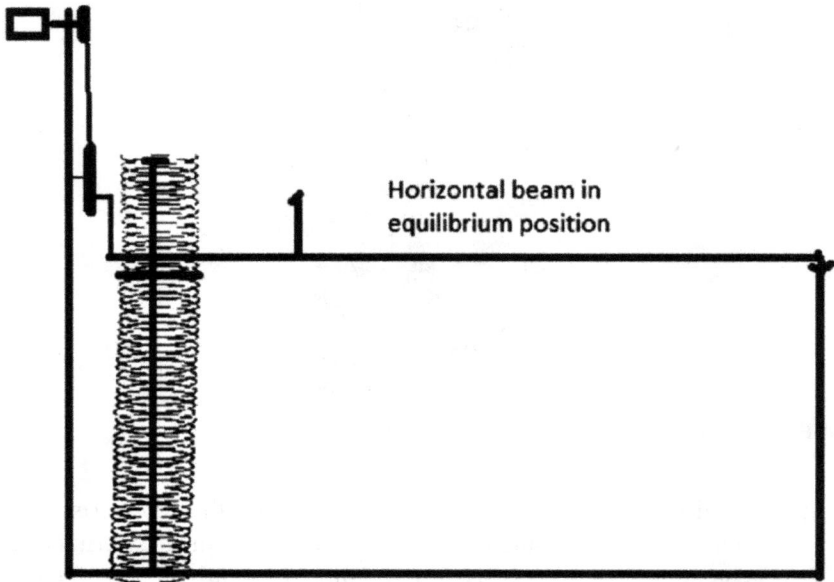

FIGURE 5.4   Beam without vibrations.

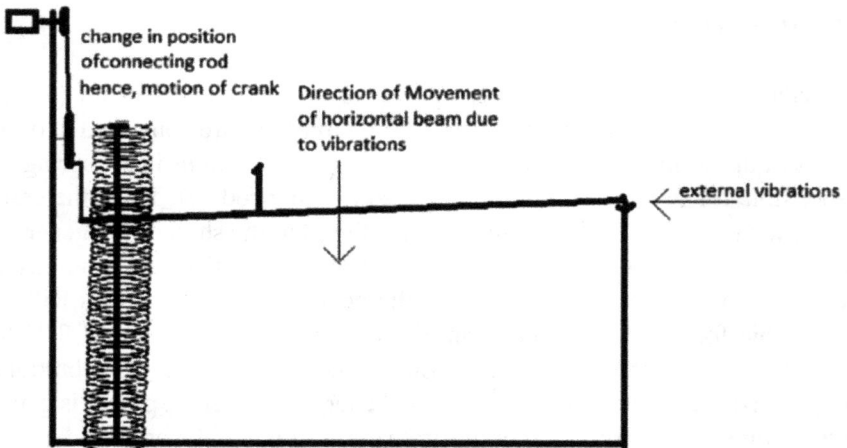

FIGURE 5.5   Deflection of beam post vibrations.

material was done. The proposed model is efficient in generating power. As the vibrations produced in the lab are very low, the power generated was also as low as 3 volts. However, the actual vibrations occurring at the construction sites, railway tracks, heavy industries, mines, etc., can be easily encashed by the proposed model [3, 4]. The novelty of the model is:

- No use of piezoelectric material;
- Most of the vibrations can be encashed;
- Model easy to construct and set up at the required sites such as construction sites, railway;
- tracks, heavy industries, etc.;
- Generation of energy without adding any pollutant to the environment;
- Use of vibrational 'energy' which are going to wastem, however.

5.7   CONCLUSION

Vibrational energy in mechanical systems mostly goes in vain [3, 4]; hence it becomes very crucial to tap those vibrations and convert them into useful electrical energy, given the present scenario where resources are depleting at astonishing rates. A model for the same is proposed in the chapter, and the theoretical aspects related to the manufacturing, design, and construction are presented in a lucid manner.

The proposed model, as the name suggests is made to regain the lost vibrations and draw some useful work from it. The high amount of vibrations goes in vain in the hammer blow produced by locomotives, explosions carried out at mines, excavation processes, etc. This energy is needed to be reused as they have consumed a lot of energy, somewhere desirably, somewhere undesirable. Thus, the model, after overcoming the losses in it, can be used for doing so.

5.8   FUTURE SCOPE

The proposed model is for large scale vibrations ranging from 50 to 200 Hz. These vibrations can be achieved by a train engine or a heavy-duty truck [3]. The fabricated prototype in a laboratory cannot be given these much of vibrations, generation of high amplitude vibrations being a limitation of the setup. Hence, a prototype can be simulated in any simulation software and given vibrations of the scale up to 200 Hz. This data can help in giving the relation of the number of volts of the potential difference generated for the given amount of vibrations.

KEYWORDS

- **electrical energy**
- **mechanical energy**
- **methodology**
- **piezoelectric material**
- **piezoelectric systems**
- **useful mechanical work**
- **vibrations**

REFERENCES

1. Krishnan, R., (2001). *Electric Motor Drives, Modeling, Analysis, and Control.* Prentice-Hall.
2. Say, M. G., (1978). *Alternating Current Machines.* Pitman Publishing.
3. Krieev, A. B., (2017). *Energy-Regenerative Shock Absorber Mathematical Mode.* International conference on industrial engineering.
4. Thomson, W. T. (1993). *Theory of Machines.* CBS Publications.
5. https://www.thegreenage.co.uk/tech/piwzoelectric-materials (accessed on 31 July 2021).

CHAPTER 6

Mathematical Modeling and Analysis of Time of Roof Bolting Operation in an Underground Coal Mine

RAKESH HIMTE, S. R. GHATOLE, M. S. DHANDE, and P. P. BAJPAI

Mechanical Engineering Department, Priyadarshini Institute of Engineering and Technology, Nagpur–4400019, Maharashtra, India, E-mail: rakeshhimte@gmail.com (R. Himte)

ABSTRACT

In the underground coal mine, the roof bolting operation is performed to give support to the roof so that coal can be extracted easily from the mine [1, 2, 4]. In this research work, the effort is made to formulate the arithmetical model for the time of roof bolting operation. The data of the independent and dependent parameters has been gathered during the observation. It is necessary to correlate various independent and dependent pi terms quantitatively involved in this operation. This correlation is nothing, but a mathematical model as a design tool for such work station. The optimum values of the independent pi terms can be decided by optimization of these models for minimum time of drilling operation. 2D graphical analysis of the mathematical models showed that the influence of relative humidity on the time of bolting operation is significant. From the studies on the model of time of bolting operation, it is found that the influence of operator and humidity is predominant over the specification of drill rod, specification of bolt, and illumination. The optimum values of independent pi terms for minimum time of bolting operation are found to be $\pi_1 = 5.4638$, $\pi_2 = 9.20\mathrm{e}{-}11$, $\pi_3 = 0.004788$, $\pi_4 = 2.535$, $\pi_5 = 92$.

6.1   INTRODUCTION

The process of operation of the underground mine comprises of the following operations [1, 2, 4]:

- Deciding the location of blasting;
- Preparing the drilling layout;
- Drilling of the holes;
- Filling of the explosive in the holes;
- Assembly of the components of the ignition system;
- Firing of the explosives using the ignition system;
- Planning of the collection of coal;
- Roof supporting operation (roof bolting);
- Executing the manual coal collection by shoveling;
- Transferring the collected coal in the trolleys of the haulage system;
- Pulling the complete haulage system out of the mine.

Roof bolting is one of the primary tasks and makes up 50–60% of the total time for the three-man crew on the miner. In this chapter roof supporting operation (roof bolting) is selected for the best performance efficiency and it is a man-machine operation. It is a repetitive task that involved awkward postures and can be physically demanding on the neck, shoulder, back, and forearms, and the consumption of human energy is substantial. Here attempt is made to formulate the mathematical model for the time of roof bolting operation. The data of the independent and dependent parameters of the system has been gathered during the observation. The optimum values of the independent pi terms can be decided by optimization of these models for minimum time of drilling operation.

6.2   FORMULATION OF PROBLEM

Roof bolting is one of the primary tasks and consumes of the total time for the three-man crew on the miner. It is repetitive work that involves awkward postures and can be very strenuous, demanding on the neck, shoulder, back, and forearms, and the consumption of human energy is substantial. Therefore, it is required to formulate the mathematical model for roof bolting operation so as to identify the factors which are most influencing on the entire operation [29]. The proposed research work attempts to establish the correlation between independent variables and dependent variables and their influence on the entire roof bolting operation (Tables 6.1–6.4).

TABLE 6.1   Independent and Dependent Variables

SL. No.	Variables Description	Variable Type	Symbol	Dimension
1.	Hole diameter	Independent	Dh	L
2.	Hole length	Independent	Lh	L
3.	Diameter of bolt	Independent	Db	L
4.	Length of bolt	Independent	Lb	L
5.	Length of spanner	Independent	L_s	L
6.	Weight of spanner	Independent	Ws	MLT^{-2}
7.	Diameter of spanner	Independent	Ds	L
8.	Weight of bolt	Independent	Wb	MLT^{-2}
9.	Weight of resin tube	Independent	Wt	$L\, MLT^{-2}$
10.	Length of resin tube	Independent	Lt	L
11.	Diameter of resin tube	Independent	Dt	L
12.	Length of push rod	Independent	Lr	L
13.	Weight of push rod	Independent	Wr	MLT^{-2}
14.	Diameter of push rod	Independent	Dr	L
15.	Air velocity	Independent	Ar	LT^{-1}
16.	Relative humidity	Independent	–	–
17.	Illumination	Independent	I	MT^{-3}
18.	Anthropometric data of roof bolter	Independent	A_d	–
19.	Time of bolting operation	dependent	T_m	T^{-1}

TABLE 6.2   List of Independent and Dependent Pi Terms

SL. No.	Pi Terms	Pi Terms Equations
1.	Pi term relating anthropometric data of roof bolter	$\pi1 = [(a{\times}c{\times}e{\times}g{\times}Wl)/(b{\times}d{\times}f{\times}h{\times}Ww)]$
2.	Pi term relating specification of drill rod and drill bit	$\pi2 = [Dh/Lb, Db/Lb, Ls/Lb, Ds/Lb,\ Lt/Lb, Lr/Lb, Lh/Lb, Dt/Lb, Dr/Lb,]$
3.	Pi term relating specification of bolt	$\pi3 = [Ws/Wb, Wt/Wb,]$
4.	Pi term relating illumination	$\pi4 = [I{\times}Lb2/Ar{\times}Wb]$
5.	Pi term relating relative humidity	$\pi5 = \varphi\ \%$

TABLE 6.3   Dependent Pi Terms

Pi Terms Description	Pi Terms Equations
Pi term relating response variable time of drilling operation	$\pi = Tm \times Ar/Lb$

TABLE 6.4   Observation Table of Bolting Operation

Work-station	Pi 1 (A)	Pi 2 (B)	Pi 3 (C)	Pi 4 (D)	Pi 5 (E)	Pi 6 (Z4)	Pi 7 (Z5)	Pi 8 (Z6)
1	6.520413	9.20E-11	0.0048	1.458	96	6667	1.00E-05	10.66667
2	5.464069	9.20E-11	0.0048	1.32545	94	7211	1.00E-05	10.88889
3	6.366155	9.20E-11	0.0048	1.76727	95	7456	1.00E-05	11.11111
4	7.813795	9.20E-11	0.0048	2.31429	93	7233	1.00E-05	11.33333
5	8.400302	9.20E-11	0.0048	2.53565	94	7411	1.00E-05	11.11111
6	7.430672	9.20E-11	0.0048	1.215	93	8000	1.00E-05	11.55556
7	6.423497	9.20E-11	0.0048	1.5552	92	8056	1.00E-05	11.33333
8	6.380652	9.20E-11	0.0048	2.08286	95	8867	9.00E-06	11.55556
9	6.552107	9.20E-11	0.0048	1.04143	94	9256	9.00E-06	11.33333
10	7.659435	9.20E-11	0.0048	1.296	95	10000	8.00E-06	10.88889
11	6.686785	9.20E-11	0.0048	1.458	93	6722	1.00E-05	11.55556
12	6.277261	9.20E-11	0.0048	1.38857	94	7117	1.00E-05	11.55556
13	7.83045	9.20E-11	0.0048	1.76727	92	7456	1.00E-05	11.33333
14	7.21667	9.20E-11	0.0048	2.11304	95	7411	1.00E-05	11.11111
15	7.445194	9.20E-11	0.0048	2.43	94	7600	1.00E-05	11.11111
16	5.510431	9.20E-11	0.0048	1.32545	93	6906	1.00E-05	10.88889
17	7.527177	9.20E-11	0.0048	1.69043	92	7411	1.00E-05	11.33333
18	6.955619	9.20E-11	0.0048	2.3328	95	8194	1.00E-05	11.55556
19	6.563262	9.20E-11	0.0048	1.26783	96	7667	1.00E-05	10.66667
20	7.220841	9.20E-11	0.0048	1.34069	93	9828	9.00E-06	10.88889
21	6.812328	9.20E-11	0.0048	1.1664	94	8625	1.00E-05	10.88889
22	7.580212	9.20E-11	0.0048	1.12154	92	8956	1.00E-05	10.66667
23	6.261204	9.20E-11	0.0048	1.44	94	8850	9.00E-06	11.33333
24	7.172113	9.20E-11	0.0048	1.73571	93	9567	9.00E-06	11.55556
25	7.104163	9.20E-11	0.0048	2.01103	95	9828	9.00E-06	10.88889
26	7.162886	9.20E-11	0.0048	0.972	92	10250	8.00E-06	10.88889
27	7.167324	9.20E-11	0.0048	1.38857	95	9489	9.00E-06	11.55556
28	6.614744	9.20E-11	0.0048	2.16	93	9000	9.00E-06	11.33333
29	6.693198	9.20E-11	0.0048	1.12154	94	8811	1.00E-05	11.33333
30	6.885589	9.20E-11	0.0048	1.5552	95	8333	1.00E-05	10.88889

6.3   DEVELOPMENT OF MODEL FOR TIME OF DRILLING OPERATION

The matrix method of solving equations using 'MATLAB' is given below [27–30].

Let, A = 6 × 6 matrix of the multipliers of K_4,' a_4, b_4, c_4, d_4 and e_4.
B = 6 × 1 matrix of the terms on L.H.S. and
C = 6 × 1 matrix of solutions or values of K_4,' a_4, b_4, c_4, d_4 and e_4.

Then, Z = inv (A) × B gives the unique values of K_4,' a_4, b_4, c_4, d_4 and e_4.

In our case:

A = [30, log(207.6985), log(2.76E-09), log(1.44E-1), log(48.377), log(2815); log(207.6985), log(1450.492), log(1.91E-08), log(0.9969), log(337.9597), log(19484.7); log(2.76E-09), log(1.91E-08), log(2.54E-19), log(1.32E-11), log(4.45E-09), log(2.59E-7); log(1.44E-1), log(9.97E-1), log(1.32E-11), log(0.000691), log(0.232211), log(13.512); log(48.377), log(337.9597), log(4.45E-09), log(0.232211), log(83.76075), log(4541); log(2815), log(19484.07), log(2.59E-7), log(13.512), log(4541.69), log (264183)] >> B = [log(248175); log(1722103); log(2.28E-5); log(1191.24); log(396619.2); log(23285183)] >> Z = inv(A) × B.

Antilog K_1,' a_1, b_1, c_1, d_1, and e_1 will be the solution for the equation.

The model for π_6 after substituting these values is as under:

$$\pi_6 = 1.000 \times (\pi_1)^{0.4940} \times (\pi_2)^{-0.1445} \times (\pi_3)^{-0.1008} \times (\pi_4)^{-0.1907} \times (\pi_5)^{0.9420}$$

6.4   ANALYSIS FROM THE INDICES OF INDEPENDENT PI TERMS OF THE MODELS FOR BOLTING OPERATION

6.4.1   EFFECT OF Π1 ON TIME OF BOLTING OPERATION

It is found that as the values of π_1 is relating anthropometry of roof bolters increases-response variable time of operation Tm increases significantly [29].

6.4.2   EFFECT OF Π2 ON TIME OF BOLTING OPERATION

It is found that as the value of π_2 is relating specification of drill rod increases-response variable time of operation Tm decreases significantly.

6.4.3   EFFECT OF Π3 ON TIME OF BOLTING OPERATION

It is found that as the value of π_3 is relating specification of drill machine increases-response variable time of operation Tm decreases significantly.

6.4.4   EFFECT OF Π4 ON TIME OF BOLTING OPERATION

It is found that as the value of π_4 is relating illumination increases. Response variable time of operation Tm decreases significantly.

6.4.5   EFFECT OF Π5 ON TIME OF BOLTING OPERATION

It is found that as the value of π_5 is relating the speed of drill machine increases. Response variable time of operation Tm decreases significantly.

6.4.6   EFFECT OF Π6 ON TIME OF BOLTING OPERATION

It is found that as the value of π_3 is relating relative humidity increases response variable time of operation Tm Increases significantly. The indices of the model are the indicator of how the phenomenon is getting affected because of the interaction of various independent pi terms in the models. The influence of indices of the various independent pi terms on each dependent pi term is shown in Figure 6.1. The sequence of influence of independent pi terms on dependent pi terms is given in Table 6.4.

6.5   ANALYSIS OF THE MODEL FOR DEPENDENT PI TERM Π6

The model for the dependent pi term π_6 is as under:

$$\pi_6 = 1.000 \times (\pi_1)^{0.4940} \times (\pi_2)^{-0.1445} \times (\pi_3)^{-0.1008} \times (\pi_4)^{-0.1907} \times (\pi_5)^{0.9420}$$

The deduced equation for this pi term is given by:

$$\pi_6 = Tm \times Ar/Lb$$

It can be seen from the equation that this is a model of a pi term containing time of bolting operation Tm as response variable. The following primary conclusions appear to be justified from the above model [29, 30]:

i.   The absolute index of π_5 is highest viz. 0.9420. Thus π_5, the term related to the relative humidity is the most influencing pi term in this model. The value of this index is positive indicating π_6 is directly varying with respect to π_5.

ii. The absolute index of π_3 is lowest viz. −0.1008. Thus π_3, the term related to the specification of roof bolt is the least influencing pi term in this model. The value of this index is negative indicating π_6 is inversely varying with respect to π_3.

iii. The sequence of influence of the other independent pi terms present in this model is π_1, π_2, π_4, having absolute indices 0.4940, −0.1445, −0.1907, respectively [29]. The index of π_2, π_3, π_4, are also negative indicating that π_6 varies inversely with respect to π_2, π_3, π_4. The value of index π_1, π_5 are negatives indicating that π_7 varies inversely with $_3$. Thus, π_5 is the most sensitive pi term and π_3 is the least sensitive pi term. The sequence of the various pi terms in the descending order of sensitivity is π_5, π_1, π_4, π_2, and π_3.

6.5.1 GRAPHICAL REPRESENTATION OF MODELS (3D) FOR BOLTING OPERATION

The above analysis helps to get the variation of bolting operation with the help of 3D plots as shown in Figures 6.1–6.3.

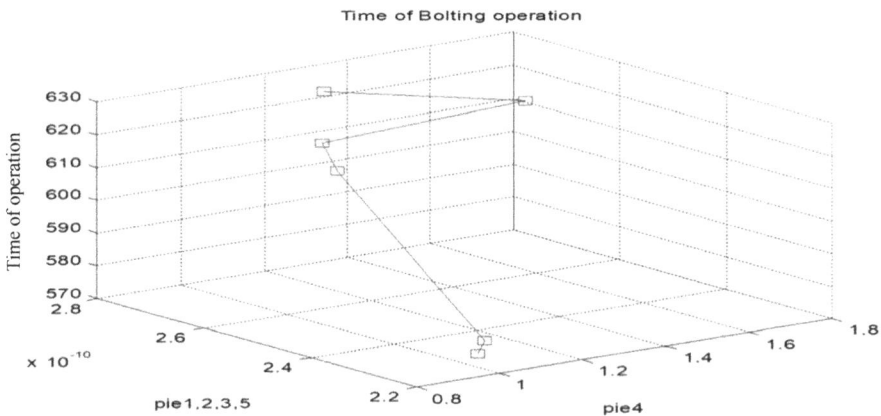

FIGURE 6.1   3D plot showing variation of time of bolting operation with prime pi terms.

Human Energy of Bolting Operation

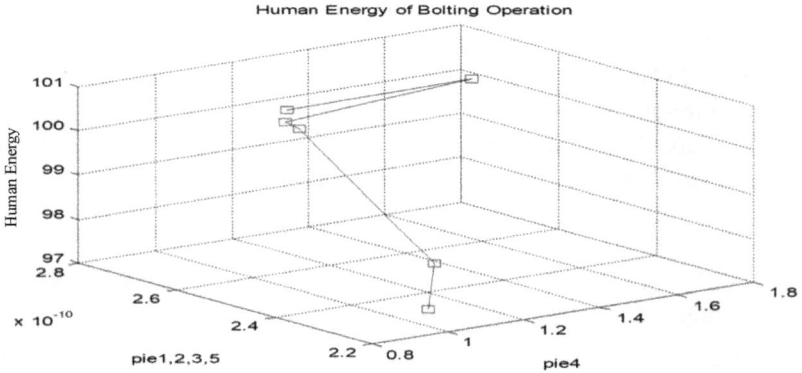

FIGURE 6.2   3D plot showing variation of productivity of bolting with prime pi terms.

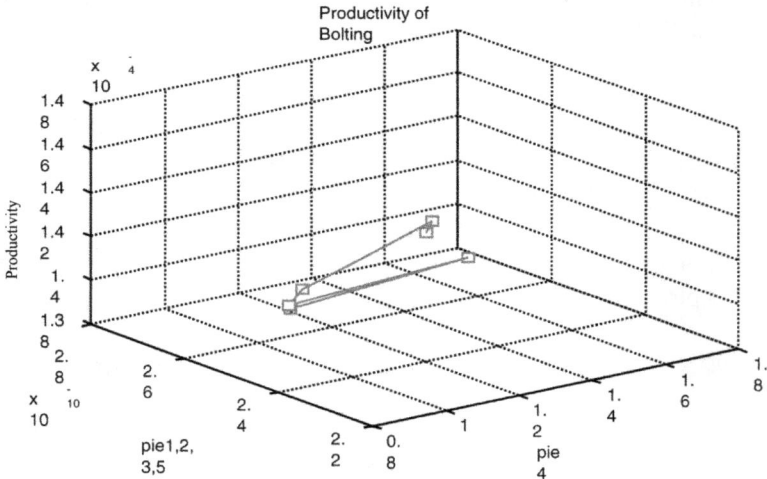

FIGURE 6.3   3D plot showing variation of human energy with prime pi terms.

6.6   CONCLUSIONS

1. Roof bolter experience postural discomfort while performing the task. They are not aware as to, to what extent ergonomic intervention can elevate their drudgery [29]. Secondly, the relationship between various inputs (anthropometry of operator, specification of drill rod, specification of bolt, illumination, and environment) and the outputs (time of bolting operation, productivity, human energy) of the system is not known to them quantitatively.

2. The mathematical models and ANN developed for the phenomenon truly represent the degree of interaction of various independent variables [29, 30]. This is done with the approach adopted in this work. The standard error of estimate of the predicted/computed values of the dependent variables is found to be very low. This gives authenticity to the developed mathematical models and ANN.

3. 2D graphical analysis of the mathematical models showed that the influence of relative humidity on time of bolting operation and human energy is significant, the influence of the operator on productivity is significant.

4. From the studies on the model of time of bolting operation, it is found that the influence of operator and humidity is predominant over the specification of drill rod, specification of bolt, illumination.

5. From the studies on the model of productivity, it is found that the influence of specification of bolt and humidity is predominant over the specification of drill rod, operator, and illumination.

6. From the studies on the model of human energy, it is found that the influence of humidity and specification of bolt is predominant over the specification of drill rod, operator, and illumination.

7. Optimum values of independent pi terms for minimum time of bolting operation are found to be $\pi 1 = 5.4638$, $\pi 2 = 9.20\mathrm{e}{-}11$, $\pi 3 = 0.004788$, $\pi 4 = 2.535$, $\pi 5 = 92$.

8. Optimum values of independent pi terms for maximum productivity are found to be $\pi 1 = 8.4638$, $\pi 2 = 9.200\mathrm{e}{-}11$, $\pi 3 = 0.004788$, $\pi 4 = 2.535$, $\pi 5 = 92$.

9. Optimum values of independent pi terms for minimum human energy are found to be $\pi 1 = 5.4638$, $\pi 2 = 9.20\mathrm{e}{-}11$, $\pi 3 = 0.004788$, $\pi 4 = 0.97200$, $\pi 5 = 92$.

KEYWORDS

- **haulage system**
- **human energy**
- **manual coal**
- **mathematical model**
- **relative humidity**
- **roof bolting operation**
- **time of bolting operation**
- **underground coal mine**

REFERENCES

1. Shuttle, P. C., & Smith, J. R., (2001). Practical ergonomics in mechanized mining, 6th international symposium on mine. Mechanization and automation. *South African Institute of Mining and Metallurgy, 1*, 61–65.
2. Das, B., & Grady, R. M., (1983). Industrial workplace layout design: An application of engineering anthropometry. *Ergonomic, 26*(5), 433–447.
3. Das, B., & Sengupta, A. K. (1996). Industrial workstation design: A systematic ergonomics approach. *Applied Ergonomics, 27*(3), 157–163.
4. Krajewski, J. T., (2007). Implementation of an ergonomics process at US surface coal mine. *International Journal of Industrial Ergonomics, 37*, 157–167.
5. Mason, S., (1992). Improving the ergonomics of British coal's mining machinery. *Applied Ergonomics, 23*(4), 233–242.
6. Robbins, R. J., (2000). Mechanization of underground mining: A quick look backward and forward. *International Journal of Rock Mechanics and Mining Sciences, 37*(1), 413–421.
7. Francis, J., & Winn, Jr., (1996). Exposure probabilities to ergonomics hazards among miners. *International Journal of Industrial Ergonomics, 18*, 417–422.
8. Shikdar, A. A., & Al- Hadhrami, M. A., (2007). Smart workstation design: Ergonomics and methods engineering approach. *International Journal of Industrial and System Engineering, 2*, 67–79.
9. Robin, B. L., & Leon, S., (2007). Implementation of the participative ergonomics for manual task (Perform) program at four Australian underground coal mines. *International Journal of Industrial Ergonomics, 37*, 145–155.
10. Fraser, P., (2001). *Appropriate Mechanization for Underground Mining in Southern Africa 6th International Symposium on Mine Mechanization and Automation.* South African Institute of Mining and Metallurgy.
11. Dunlap, J. W., (1947). Men and machines. *International Journal of Applied Psychology, 31*(6), 565–579.
12. Johnson, A. T., Benjamin, M. B., & Silverman, N., (2002). Oxygen consumption, heat production, and muscular efficiency during uphill and downhill walking. *Applied Ergonomics, 33*(5), 485–491.
13. Forsman, G. A., & Medbo, H. L., (2002). A method for evaluation of manual work using synchronized video recordings and physiological measurements. *Applied Ergonomics, 33*(6), 533–540.
14. Davies, B. T., (1972). Moving loads manually. *Applied Ergonomics, 3*(4), 190–194.
15. Mital, A., & Kumar, G., (1997). Cardiac rehabilitation (CR): Use of manual materials handling. *International Journal of Industrial Ergonomics, 20*(2), 93–99.
16. Edward, M., & John, R. B., (1986). Backward walking practice decreases oxygen uptake, heart rate and ratings of perceived exertion. *Applied Ergonomics, 17*(2), 105–109.
17. Gedliezka, A., (1977). The shape of workspace data transformation on a model basis. *Applied Ergonomics, 8*(1), 31–34.
18. Patrick, J. M., (1985). This month surprise and delight: Statistical methods in ergonomics. *Ergonomics, 28*(3), 529–530.
19. John, S., (1992). Technical change and productivity growth in the British coal industry, 1974–1990. *Technovation, 12*(1), 15–39.
20. Delela, S., (1990). *A Text-Book of Work Study and Ergonomics Standard* (4th edn.). Publishers Distributions, Delhi.

21. Hilbert, S. Jr., (1961). *Theories of Engineering Experimentation*. Mc Graw Hill Book Company, New York.

22. Murrel, K. F. H., (1986). *Ergonomics: Man in His Working Environment*. Chapman and Hall, London.

23. Malhotra, M. S., (1996). Physical work capacity as influenced by age. *Ergonomics, (4)*, 305–316.

24. Modak, J. P., (2004). *Specialized COURSE ON Research Methodology in Engineering*. Priyadarshini College of Engineering and Architecture, Nagpur.

25. Murrell, K. F. H., (1965). *Practical Ergonomics*. Chapman and Hall, New York.

26. Coffelt, R. J., & Wikman, J., (1985). Anthropometer a versatile design. *Applied Ergonomics, 16*(2), 147–149.

27. Awari, G. K., Sakhale, C. N., Modak, J. P., & Kadu, R. S., (2014). Formulation of mathematical model for the investigation of tool wears in boring machining operation on cast iron using carbide and CBN tools. *Procedia Materials Science, 6*, 1710–1724.

28. Waghmare, S. N., Undirwadec, S. K., Sonded, V. M., Singhe, M. P., & Sakhalea, C. N., (2014). Formulation and comparison of experimental based mathematical model with artificial neural network simulation and RSM (Response Surface Methodology) model for optimal performance of sliver cutting operation of bamboo. *Procedia Materials Science, 6*, 877–891.

29. Himte, R. L., & Modak, J. P., (2010). An approach to form field data based model for roof bolting operation in underground mine. *International Journal of Engineering Research and Industrial Applications, 3*(1).

30. Himte, R. L., Modak, J. P., & Agrawal, S. H., (2009). Formulation of mathematical model for roof bolting operation- underground mine. *Proceeding of the International Conference on Emerging Research and Advances in Mechanical Engineering*. Velamal Engineering College, Chennai, India.

31. Himte, R. L., & Modak, J. P. (2019). Implementation of participative ergonomics for improving the productivity of roof bolter in underground coal mine. In: *51ˢᵗ National Convention of Industrial Engineers & National Seminar on Industrial Productivity; Safety and Disaster Management*, RKNEC, Nagpur.

32. Himte, R. L., & Modak, J. P. (2010). Modeling and analysis of roof bolting in underground coal mine. *National Conference on High Tech Materials, Charastictics, Synthesis and Applications, DIMAT*. Raipur.

33. Himte, R. L., & Modak, J. P. (2010). *Formulation of Field Data-Based Model for Roof Bolting Operation*. NIRMITI, PIET, Nagpur.

CHAPTER 7

Relationship Formed for Process Time by Using An Approach of a Mathematical Model for Sewing Machine Operation

SWAPNA GHATOLE,[1] YASHPAL,[2] MAHESH BUNDELE,[3] and J. P. MODAK[4]

[1]Research Scholar, Poornima University, Jaipur, Rajasthan, India, E-mail: swapnaghatole72@gmail.com

[2]Poornima University, Jaipur, Rajasthan, India

[3]Poornima College of Engineering, Jaipur, Rajasthan, India

[4]JD College of Engineering, Nagpur, Maharashtra, India

ABSTRACT

Pedal sewing operation is a man-machine system. This operation takes place at a workstation usually known as single-person-owned premises meant for stitching clothes of the clients of the shop owner. The data collected comprises 30 sewing machine operators working on standalone pedal-driven sewing machines. The field study indicated that there is a considerable Musculoskeletal disorder on the part of the operator. Based on this data, mathematical co-relationships are established between various causes and effects of this activity. The chapter reports on: (i) planning of collection of data; (ii) execution towards collection of data; (iii) presentation of collected data; (iv) processing of the data for formulation of cause-effect relationships. Upon formulation of model, the reliability of model and its optimization is executed. The collection of data was done on 40 causes/input parameters/ independent variables and 1 effects/output parameters/dependent variables. The causes are clubbed and converted into five independent dimensionless terms known as pi terms such as Π_1, Π_2, Π_3, Π_4, Π_5. The effects/output/ dependent parameter is nomenclated as Z1.

7.1 INTRODUCTION

This work is the actual working of sewing operator in his workstation. The operator's work is observed and noted down. This concept is a man-machine system where the operator is working in his environment taking place in the society at various locations. The data of the thirty-man machine working system is compiled. Data comprises of independent parameters (causes) in activity, dependent parameters (effects) of activity and extraneous variables of the activity. The various responses of the system are identified for every workstation. Based on this gathered data, the analytical co-relationships between causes and effects are formulated. These analytical relationships are considered here as field database modeling. This concept of field database modeling is developed by Modak and employed in various studies [1–30] in his past endeavors through several attempts. These analytical relationships are termed in this investigation as Field Databased Modeling for sewing machine operation. The optimization of these models is logically expected to indicate magnitudes of causes/independent variables/inputs to be set during operation of pedal-driven sewing machine so as to realize the targeted performance/output of operation.

7.2 MATERIALS AND METHODS

Independent and dependent parameters are identified and nomenclated. In this work, dependent quantities are grouped into dimensionally homo-geneous groups. As per these groups, independent, and dependent Π terms are defined. Rewriting all the quantities into dimensionless formula, total 5 independent Pi terms, and 1 dependent Π term is defined in Table 7.1.

TABLE 7.1 Pi Terms

SL. No.	Pi Terms
1.	Π_1 related to anthropometric data of operator
2.	Π_2 related to personal data of operator
3.	Π_3 related to environmental conditions of workstation
4.	Π_4 related to specification of workstation
5.	Π_5 related to related to specifications of
6.	$Z1$ related to process time

7.3 MATHEMATICAL MODELLING

The clubbing of all independent Π terms into one form of model is worked out. In this chapter, field data is presented by clubbing independent data. Further presenting clubbed independent quantities on X-axis and presenting corresponding values of dependent quantities (pi terms) on the Y-axis. All five independent pi terms have been identified as Π_1, Π_2, Π_3, Π_4, Π_5 are clubbed together to form Z1. For Z1 = f (Π_1, Π_2, Π_3, Π_4, Π_5), various combinations of independent Pi terms are considered by multiplication and/or division of independent Pi terms. Based on the collected data, the models are established. The various combinations formed are listed in Table 7.2.

TABLE 7.2 Combinations of Independent Pi Terms

SL. No.	Combinations of Independent Pi Terms
1.	$\Pi 1 \times \Pi 2 \times \Pi 3 \times \Pi 4 \times \Pi 5$
2.	$\Pi 1 \times \Pi 2 \times \Pi 3 \times \Pi 4 / \Pi 5$
3.	$((\Pi 1 \times \Pi 2 \times \Pi 3 \times \Pi 5) / \Pi 4)$
4.	$((\Pi 1 \times \Pi 2 \times \Pi 4 \times \Pi 5) / \Pi 3)$
5.	$((\Pi 1 \times \Pi 3 \times \Pi 4 \times \Pi 5) / \Pi 2)$
6.	$((\Pi 2 \times \Pi 3 \times \Pi 4 \times \Pi 5) / \Pi 1)$
7.	$((\Pi 1 \times \Pi 2 \times \Pi 3) / (\Pi 4 \times \Pi 5)$
8.	$((\Pi 2 \times \Pi 3 \times (\Pi 4) / (\Pi 1 \times \Pi 5)$
9.	$(\Pi 3 \times (\Pi 4 \times \Pi 5) / (\Pi 1 \times \Pi 2)$
10.	$((\Pi 4 \times \Pi 5 \times \Pi 1) / (\Pi 3 \times \Pi 2)$
11.	$((\Pi 5 \times \Pi 1 \times \Pi 2) / (\Pi 3 \times \Pi 4)$

7.3.1 FORMULATION OF MATHEMATICAL MODEL FOR Z1

All 11 combinations of independent pi terms are plotted on X-axis and corresponding variation of Z1 on Y-axis, i.e., dependent Pi terms are plotted. The first combination of Z1 against 5 independent pi terms is shown in Table 7.2. Similarly, other combinations against Z1 are shown with their interpretation in Tables 7.3–7.5. The graphical interpretation of all combinations is depicted in Figures 7.1–7.11.

TABLE 7.3 Combination of Values in Eqns. (1)–(4)

SL. No.	1		2		3		4	
	$Z1$	$\mathrm{Log}\,(\Pi_1 \times \Pi_2 \times \Pi_3 \times \Pi_4 \times \Pi_5)$	$Z1$	$\mathrm{Log}\,(\Pi_1 \times \Pi_2 \times \Pi_3 \times \Pi_4)/\Pi_5$	$Z1$	$\mathrm{Log}\,(\Pi_1 \times \Pi_2 \times \Pi_3 \times \Pi_5)/\Pi_4$	$Z1$	$\mathrm{Log}\,(\Pi1 \times \Pi2 \times \Pi4 \times \Pi5)/\Pi3$
1.	0.8751	−0.008	0.8451	−42.3377	0.8751	−0.0447	0.7782	−0.0032
2.	0.7782	−0.0074	0.7782	−8.6201	0.9542	−0.0192	0.8751	−0.0029
3.	0.7782	−0.0031	0.9542	−1.8296	0.7782	−0.0176	0.9542	−0.0011
4.	0.9542	−0.003	0.8751	−1.1221	0.7782	−0.0159	0.7782	−0.0011
5.	0.9542	−0.0023	0.8751	−0.7722	0.9542	−0.0148	0.8451	−0.0009
6.	0.8451	−0.002	0.7782	−0.7118	0.8451	−0.0142	0.9542	−0.0008
7.	0.9542	−0.001	0.7782	−0.4281	0.9542	−0.005	0.9542	−0.0004
8.	1.0212	−0.0008	0.9542	−0.4227	0.7782	−0.004	1.0212	−0.0003
9.	0.7782	−0.0006	1.0212	−0.1145	1.0212	−0.0039	0.7782	−0.0002
10.	0.7782	−0.0005	0.9542	−0.0961	0.9542	−0.0034	0.7782	−0.0002
11.	0.8751	−0.0004	0.7782	−0.075	0.7782	−0.0026	0.8751	−0.0002
12.	0.9542	−0.0004	0.699	−0.0536	0.699	−0.0014	0.9542	−0.0001
13.	0.699	−0.0003	0.9542	−0.0521	0.8751	−0.0012	0.699	−0.0001
14.	0.7782	−0.0002	0.7782	−0.008	0.7782	−0.0008	0.7782	−0.0032
15.	0.8451	0.0001	0.9542	0.0361	0.9542	0.0005	0.8751	−0.0029
16.	0.9542	0.0001	0.8451	0.2036	0.8451	0.0005	0.9542	0.0001
17.	0.8451	0.0002	0.7782	0.2997	0.8451	0.0008	0.8451	0.0001
18.	0.8451	0.0003	0.8451	0.3086	0.8451	0.0009	0.8451	0.0001

TABLE 8.3 (Continued)

SL. No.	1		2		3		4	
	Z1	Log $(\Pi_1 \times \Pi_2 \times \Pi_3 \times \Pi_4 \times \Pi_5)$	Z1	Log $(\Pi_1 \times \Pi_2 \times \Pi_3 \times \Pi_4)/\Pi_5$	Z1	Log $(\Pi_1 \times \Pi_2 \times \Pi_3 \times \Pi_5)/\Pi_4$	Z1	Log $(\Pi1 \times \Pi2 \times \Pi4 \times \Pi5)/\Pi3$
19.	0.8451	0.0003	0.8451	0.5021	0.8451	0.0009	0.8451	0.0001
20.	0.7782	0.0006	0.7782	0.8777	0.7782	0.0023	0.7782	0.0002
21.	0.9542	0.0009	0.9542	1.1504	0.7782	0.0043	0.9542	0.0003
22.	0.8451	0.0012	0.7782	1.1731	0.9542	0.0047	0.8451	0.0004
23.	0.9542	0.0013	0.8451	1.3453	0.8451	0.0063	0.9542	0.0005
24.	0.7782	0.0018	0.9542	1.4859	0.9542	0.0064	0.7782	0.0007
25.	0.9542	0.0036	0.8451	2.3357	0.9542	0.011	0.9542	0.0015
26.	0.9542	0.0046	0.8751	2.6983	0.8751	0.0171	0.8751	0.0018
27.	0.8751	0.0055	0.9542	2.7909	0.9542	0.0262	0.9542	0.0019
28.	0.9542	0.0059	0.7782	3.4029	0.9542	0.0267	0.9542	0.0026
29.	0.7782	0.0081	0.9542	3.6673	0.7782	0.031	0.9542	0.0001
30.	0.7782	0.0084	0.9542	39.2553	0.7782	0.0342	0.8451	0.0001

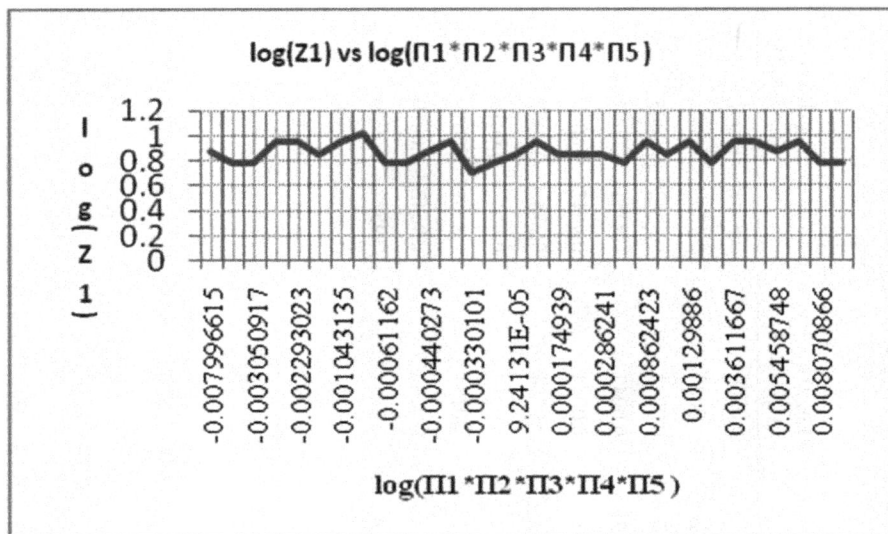

FIGURE 7.1   The variation indicates 12 peaks.

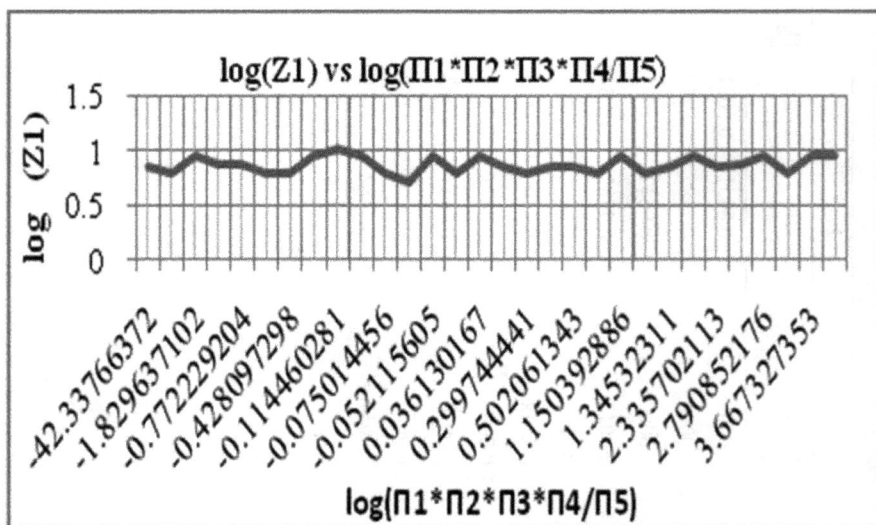

FIGURE 7.2   The variation shown indicates 12 peaks.

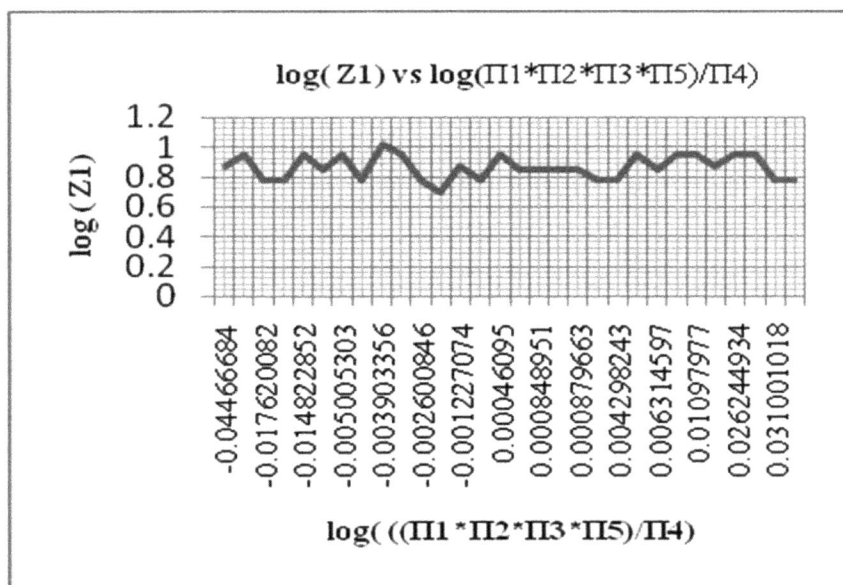

FIGURE 7.3 The variation indicates 11 peaks.

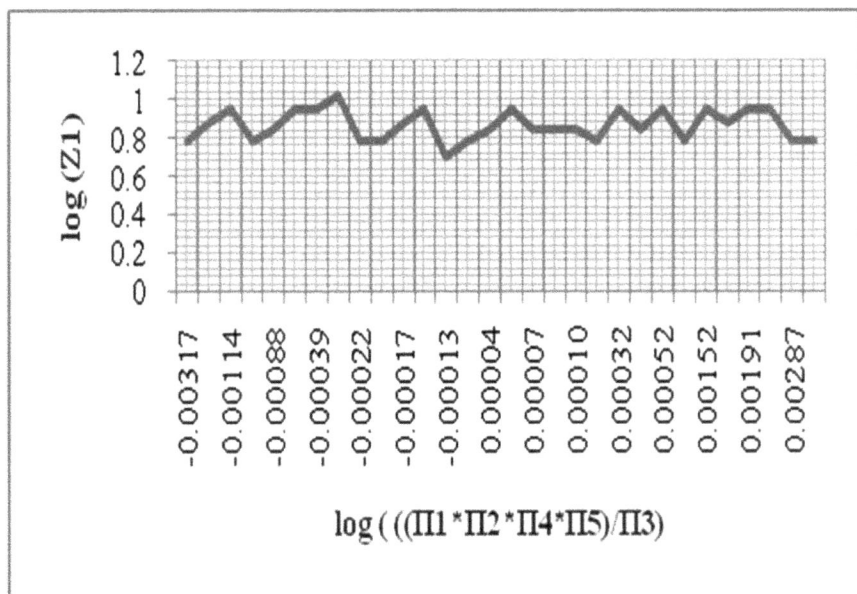

FIGURE 7.4 The variation indicates 10 peaks.

TABLE 8.4 Combination of Values in Eqns. (5)–(8)

SL. No.	Z1	5 $\log\!\left(\dfrac{\Pi_1 \times \Pi_3 \times \Pi_4 \times \Pi_5}{\Pi_2}\right)$	Z1	6 $\log\!\left(\dfrac{\Pi_2 \times \Pi_3 \times \Pi_4 \times \Pi_5}{\Pi_1}\right)$	Z1	7 $\log\!\left(\dfrac{\Pi_1 \times \Pi_2 \times \Pi_3}{\Pi_4 \times \Pi_5}\right)$	Z1	8 $\log\!\left(\dfrac{\Pi_1 \times \Pi_2 \times \Pi_3 \times \Pi_4}{\Pi_1 \times \Pi_5}\right)$
1.	0.7782	−0.0479	0.7782	−4.9903	0.8451	−305.749	0.8751	−1954.04
2.	0.8751	−0.0232	0.7782	−1.1767	0.7782	−18.5486	0.7782	−700.233
3.	0.7782	−0.0214	0.9542	−1.1449	0.8751	−6.2675	0.7782	−489.511
4.	0.9542	−0.0091	0.699	−1.1337	0.9542	−5.9306	0.8451	−247.504
5.	0.9542	−0.0084	0.8751	−1.1141	0.7782	−4.7008	0.9542	−186.496
6.	1.0212	−0.0071	1.0212	−0.5791	0.9542	−2.6936	0.699	−184.05
7.	0.9542	−0.0044	0.7782	−0.4206	0.7782	−2.4724	0.7782	−165.118
8.	0.9542	−0.0039	0.9542	−0.3881	0.8751	−2.1523	1.0212	−81.2582
9.	0.699	−0.0034	0.9542	−0.3528	0.9542	−0.7023	0.9542	−54.4538
10.	0.7782	−0.003	0.7782	−0.2925	0.9542	−0.621	0.7782	−50.7155
11.	0.8451	−0.0013	0.8751	−0.1534	1.0212	−0.5477	0.9542	−49.4975
12.	0.7782	−0.0008	0.9542	−0.1063	0.7782	−0.3649	0.9542	−47.963
13.	0.7782	−0.0008	0.7782	−0.0436	0.699	−0.2196	0.8751	−21.52
14.	0.8751	−0.0005	0.8451	−0.0115	0.7782	−0.0354	0.7782	−12.2521
15.	0.8451	0.0003	0.9542	0.0255	0.9542	0.1953	0.9542	12.5441
16.	0.8451	0.0007	0.7782	−0.0436	0.8451	0.9881	0.7782	14.1139
17.	0.9542	0.0007	0.8451	−0.0115	0.8451	1.0231	0.9542	37.6211
18.	0.8451	0.0008	0.9542	0.0255	0.7782	1.1353	0.7782	39.8086

TABLE 8.4 (Continued)

SL. No.		5		6		7		8
	Z1	$Log (\Pi_1 \times \Pi_3 \times \Pi_4 \times \Pi_5)/\Pi_2$	Z1	$Log (\Pi_2 \times \Pi_3 \times \Pi_4 \times \Pi_5)/\Pi_1$	Z1	$Log (\Pi_1 \times \Pi_2 \times \Pi_3)/(\Pi_4 \times \Pi_5)$	Z1	$Log (\Pi_1 \times \Pi_2 \times \Pi_3 \times \Pi_4)/(\Pi_1 \times \Pi_5)$
19.	0.9542	0.0014	0.7782	0.0335	0.8451	1.5429	0.8751	91.4627
20.	0.9542	0.0014	0.9542	0.0387	0.7782	2.1247	0.7782	146.0685
21.	0.8451	0.002	0.9542	0.0501	0.7782	4.7969	0.7782	160.0262
22.	0.9542	0.0028	0.8451	0.0574	0.9542	6.5756	0.8451	201.9291
23.	0.8451	0.0042	0.7782	0.0805	0.9542	6.7003	0.8451	255.9741
24.	0.7782	0.0045	0.9542	0.1325	0.8451	7.3493	0.9542	277.3734
25.	0.7782	0.0055	0.9542	0.1502	0.8751	8.4519	0.9542	364.1856
26.	0.8751	0.0104	0.8451	0.1735	0.9542	11.6503	0.9542	374.133
27.	0.9542	0.011	0.8751	0.185	0.7782	13.0708	0.8451	679.204
28.	0.7782	0.028	0.8451	0.2199	0.8451	13.6513	0.9542	978.3917
29.	0.9542	0.0408	0.7782	0.2955	0.9542	17.9577	0.8451	1330.494
30.	0.7782	0.0423	0.8451	0.5836	0.9542	119.3395	0.8451	1451.667

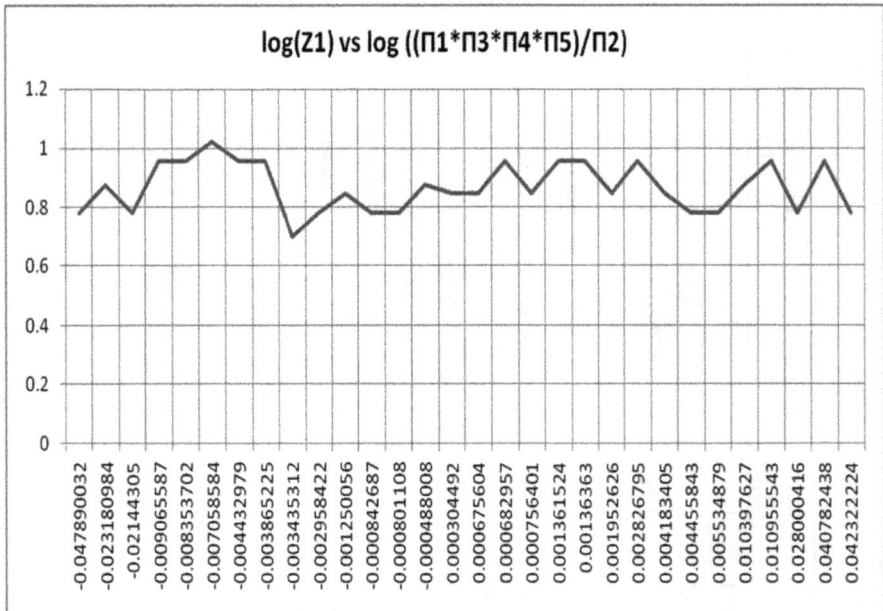

FIGURE 7.5   The variation indicates 12 peaks.

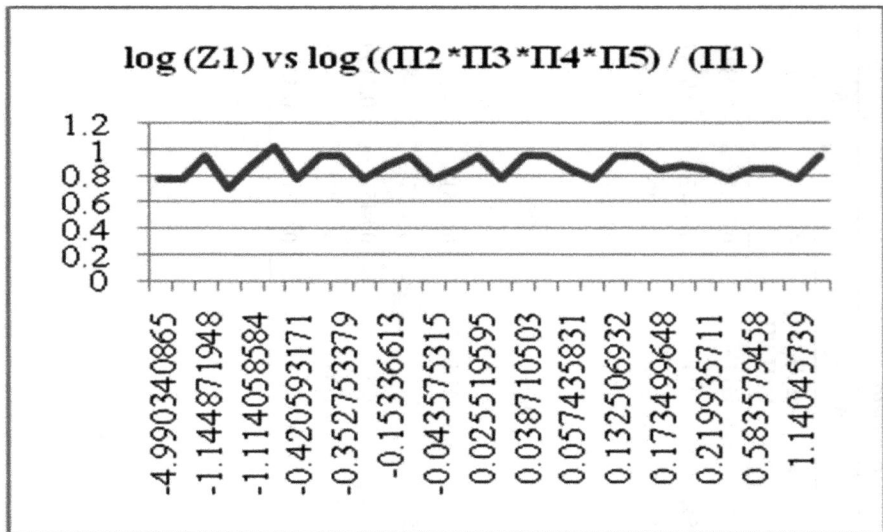

FIGURE 7.6   The variation indicates 13 peaks.

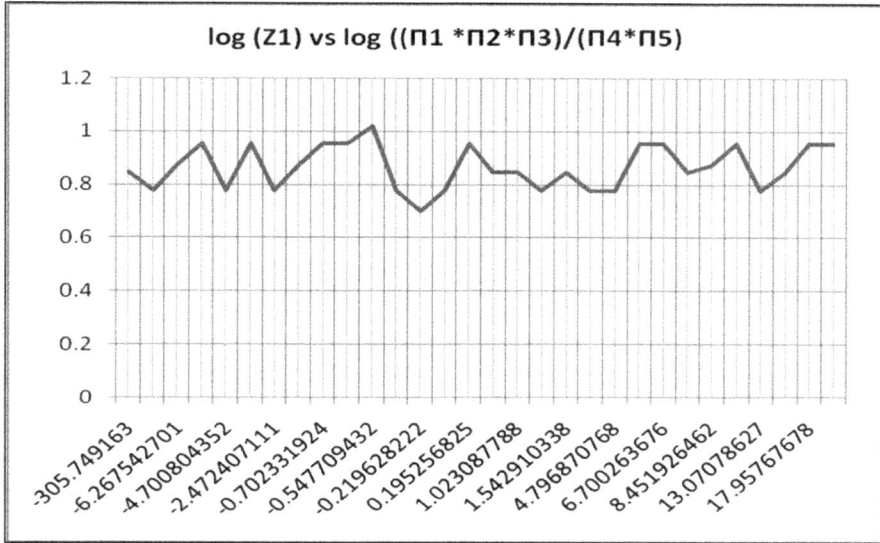

FIGURE 7.7   The variation indicates 12 peaks.

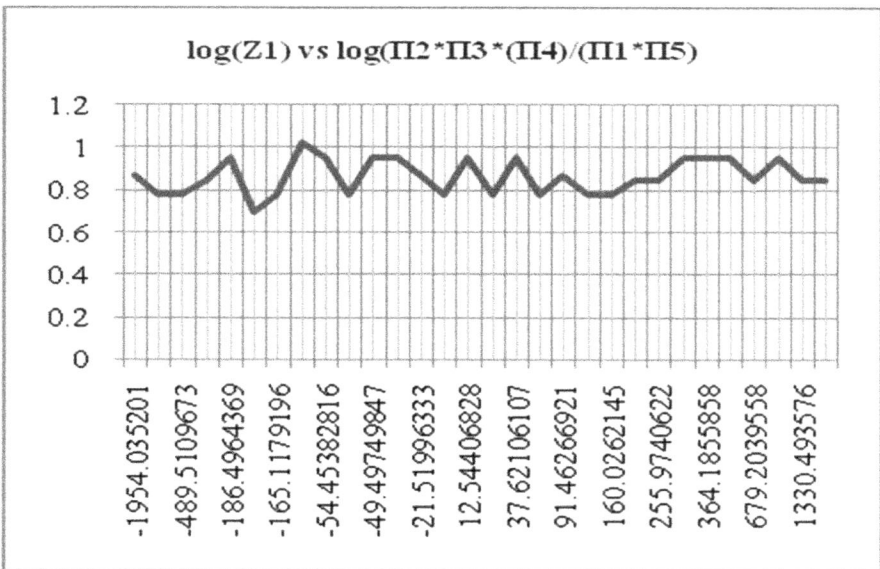

FIGURE 7.8   The variation indicates 10 peaks.

TABLE 7.5 Combination of Values in Eqns. (9)–(11)

SL. No.	9		10		11	
	Z1	Log $(\Pi_3 \times \Pi_4 \times \Pi_5)/$ $(\Pi_1 \times \Pi_2)$	Z1	Log $(\Pi4 \times \Pi5 \times \Pi1)/$ $(\Pi3 \times \Pi2)$	Z1	Log $(\Pi5 \times \Pi1 \times \Pi2)/$ $(\Pi3 \times \Pi4)$
1.	0.7782	−73.7706	0.7782	−0.0183	0.8751	−0.0160
2.	0.6990	−11.7985	0.7782	−0.0092	0.9542	−0.0073
3.	0.7782	−7.4780	0.8751	−0.0083	0.7782	−0.0068
4.	1.0212	−5.0111	0.9542	−0.0032	0.8451	−0.0064
5.	0.9542	−4.5263	0.9542	−0.0032	0.7782	−0.0062
6.	0.9542	−4.0386	1.0212	−0.0026	0.9542	−0.0053
7.	0.8751	−1.2348	0.9542	−0.0017	0.9542	−0.0018
8.	0.7782	−1.1411	0.9542	−0.0014	0.7782	−0.0015
9.	0.9542	−0.9782	0.6990	−0.0013	1.0212	−0.0014
10.	0.7782	−0.5795	0.7782	−0.0010	0.9542	−0.0013
11.	0.9542	−0.4519	0.8451	−0.0006	0.7782	−0.0010
12.	0.8751	−0.4446	0.7782	−0.0003	0.6990	−0.0005
13.	0.7782	−0.1262	0.7782	−0.0003	0.8751	−0.0005
14.	0.8451	−0.0073	0.8751	−0.0002	0.7782	−0.0003
15.	0.9542	0.0200	0.8451	0.0001	0.9542	0.0002
16.	0.9542	0.1391	0.8451	0.0002	0.8451	0.0002
17.	0.7782	0.1755	0.9542	0.0002	0.8451	0.0003
18.	0.8451	0.1892	0.8451	0.0003	0.8451	0.0003
19.	0.9542	0.2394	0.9542	0.0005	0.8451	0.0003
20.	0.9542	0.3443	0.9542	0.0005	0.7782	0.0008
21.	0.8751	0.3524	0.8451	0.0007	0.7782	0.0016
22.	0.9542	0.3583	0.9542	0.0012	0.9542	0.0017
23.	0.8451	0.3715	0.7782	0.0015	0.8451	0.0023
24.	0.7782	0.6079	0.8451	0.0015	0.9542	0.0025
25.	0.7782	1.2700	0.7782	0.0018	0.9542	0.0046
26.	0.8451	1.7904	0.8751	0.0035	0.8751	0.0057
27.	0.8451	2.5233	0.9542	0.0046	0.9542	0.0111
28.	0.7782	2.6972	0.7782	0.0104	0.9542	0.0116
29.	0.8451	2.7373	0.9542	0.0180	0.7782	0.0117
30.	0.9542	13.7239	0.7782	0.0184	0.7782	0.0135

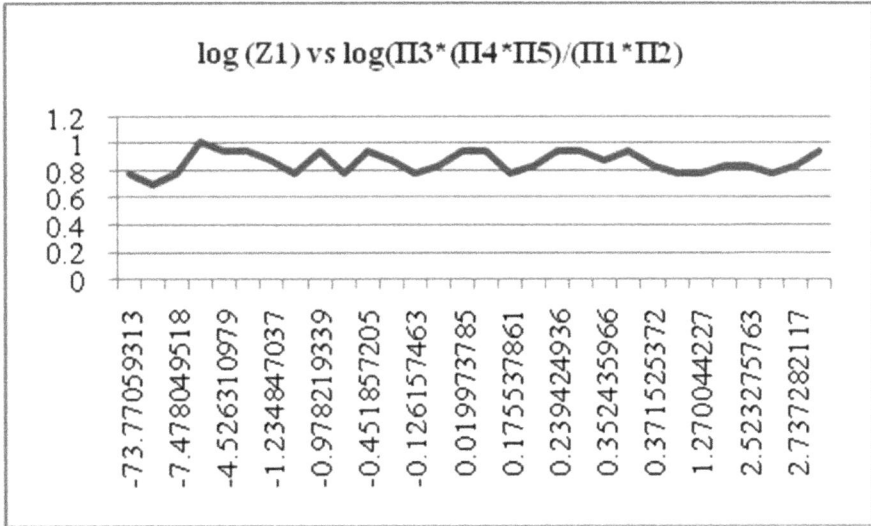

FIGURE 7.9   The variation indicates 12 peaks.

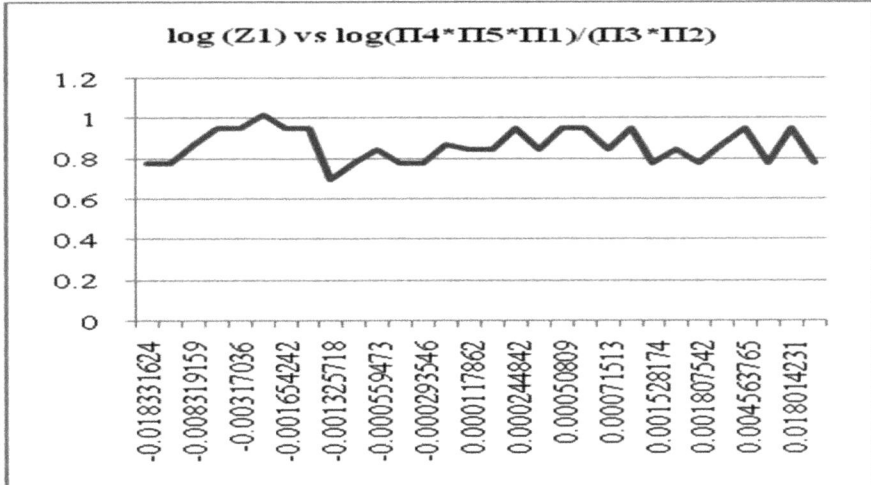

FIGURE 7.10   The variation indicates 12 peaks.

7.4   RESULT OF MODEL FOR Z1

Figure 7.12 is chosen for the model, because the graph chosen for the model should be with minimum number of peaks as per the theory of Modak. The graph has 10 numbers of peaks. The combination is Z1 Vs. log (($\Pi1 \times \Pi2 \times \Pi4 \times \Pi5$)/$\Pi3$). The total peaks in the graph were divided in 3 parts to get the model of the dependent Pi term Z1. It is very difficult to manipulate such a large number of peaks to present in one model, i.e., variation of Z1 vs combination of independent Pi terms. Hence this variation of the chosen graph is divided into three parts, as shown in Figure 7.12, to get a part wise model of Z1 vs log (($\Pi1 \times \Pi2 \times \Pi4 \times \Pi5$)/$\Pi3$).

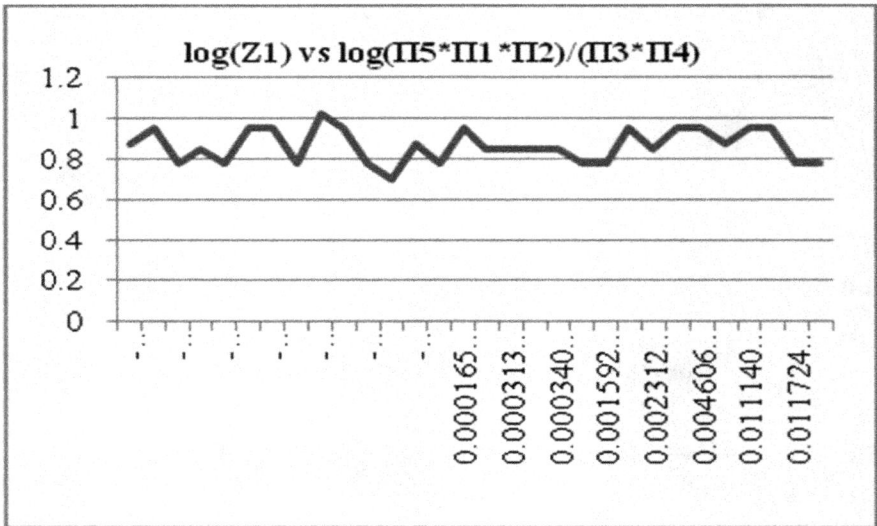

FIGURE 7.11 The variation indicates 13 peaks.

The three parts gave us three different models, shown in Table 7.6, but the model showing minimum error value and R^2 value nearest to 1 is chosen for the final model.

TABLE 7.6 Model and Range of 3 Parts of Z1

Range of Log $(\Pi_1 \times \Pi_2 \times \Pi_4 \times \Pi_5)/\Pi_3)$	Z1	Model	R^2
−0.0032 to 0.00004	0.7782 to 0.9542 0.954242509	$y = -3E\text{-}06x^6 + 0.000x^5 - 0.003x^4 + 0.032x^3 - 0.155x^2 + 0.363x + 0.551$	0.252
0.000074 to 0.00002	0.8451 to 0.7782	$y = -0.016x^2 + 0.063x + 0.794$	0.933
0.00032 to 0.00033	0.9542 to 0.7782	$y = 0.000x^6 - 0.004x^5 + 0.059x^4 - 0.361x^3 + 1.121x^2 - 1.676x + 1.812$	0.609

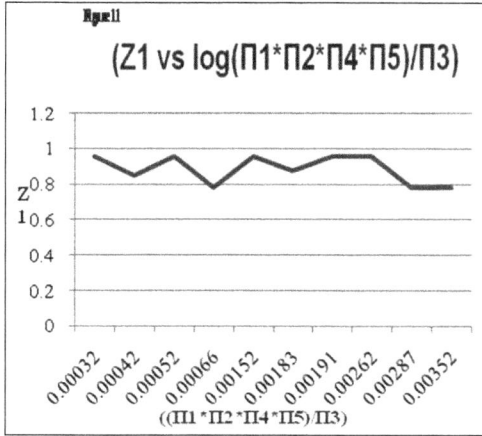

(a)

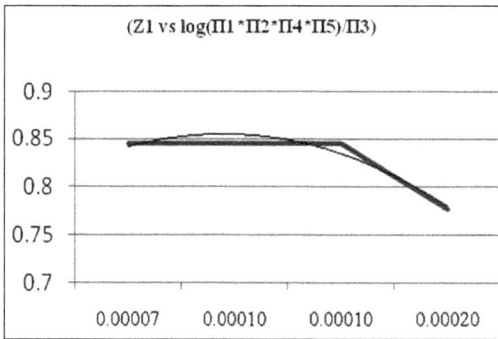

(b)

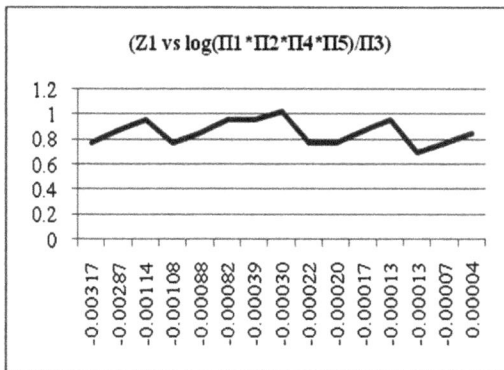

(c)

FIGURE 7.12   (a–c) Z1 model variation in 3 parts.

It may be noted that this graph is of transform Z1 and a combination of independent Pi terms in log form. As Process time Z1 is concerned, the best range of variations of Z1 is 0.8451 to 0.7782, with the model formed as $y = -0.016x^2 + 0.063x + 0.794$ having the reliability of 91%. Also, the value of $R^2 = 93.3$ with the only error of 7%.

7.5 CONCLUSION

1. The best range of variations of Z1 values of process time is with value R2 = 0.933 as the value of R2 is very much nearly 93%.
2. The combination of independent quantities for which range the reliability is highest should be the most appropriate for the purpose of design of future workstation.
3. The value of the process time range is between 0.8451 to 0.7782 and overall range value of equation is 0.000074 to 0.00002, which says that this value is of combination of all pi terms.
4. The complete pi term combination suggests changes in all aspects of workstation design, i.e., changes in seating stool, few changes in sewing machine workstation table inclination.
5. Also, changes in environmental aspects are expected, but these factors are not in the purview of the operator looking towards the overall workstation.
6. The above value helps us arriving at the best possible set of independent quantities in order to realize minimum process time for optimum production in a given time for the sewing machine operators.

KEYWORDS

- database modeling
- extraneous variables
- homogeneous groups
- machine system
- mathematical model
- musculoskeletal disorders
- sewing machine
- working system

REFERENCES

1. Modak, J. P., & Mishra, S. P., (2011). An approach to simulation of a complex field activity by a mathematical model. *Industrial Engineering Journal, 2*(20), 11–16.
2. Chopade, I. K., (2003). *Mathematical Simulation Modeling of Working of an Enterprise.* VNIT, Nagpur.
3. Bansod, S. V., (2005). *Evaluation of Existing Seats on the Basis of Ergonomic Criterion of Seat Design.* PhD Thesis S. G. Amaravati University, Amaravati.
4. Patil, S. G., (2005). *Assessment of Present Productivity and Suggesting Improvement in the Working of Manufacturing Operations in a Garment Manufacturing Unit.* PhD Thesis of S. S. G. Amaravati University, Amaravati.
5. Awate, A. U., (2007). *Ergonomics Study of Animal Drawn Weeders for Raw Crop Planting in Western Vidarbha.* PhD Thesis, S. G. Amaravati University, Amaravati.
6. Patil, C. R. *Prediction of Traffic Noise Pollution at Major Road Crossings and it's Effects on Community Health with Realistic Measures for Control: An Ergonomic Approach.* PhD Thesis, S.G. Amaravati University, Amaravati.
7. Junghare, A., (2007). *Analytical Models and Simulation for Dynamics & Vibration Response of Turbo Alternator Set.* PhD Thesis, VNIT, Nagpur.
8. Tatwawadi, V. H., (2007). *Mathematical Simulation of Working of an Enterprise Manufacturing Electric Motors.* PhD Thesis, R.T.M. Nagpur University.
9. Himte, R. L., (2010). *An Approach to Decide Prioritization of Mechanization of Mining Operations.* PhD Thesis, R.T.M. Nagpur University, Nagpur.
10. Vidhyasagar, V., (2014). *Modeling and Field Data Based Investigation of Man-Machine system for Ore Productivity.* PhD Thesis, R.T.M. Nagpur University.
11. Gupta, M. M., (2010). *Decision Support System for Behavioral Prediction of Industry Performance Indicators.* PhD Thesis, R.T.M. Nagpur University.
12. Bihade, O. S., (2011). *Ergonomics Evaluation of Construction Activity.* PhD Thesis, R.T.M. Nagpur University.
13. Mishra, S. P., (2011). *Mathematical Simulation of Man-Machine-System of some Construction Activities.* PhD Thesis, Sant Gadge Baba University, Amaravati.
14. Pote, G. P., (2012). *Design, Optimization & Development of Carding Machine.* PhD Thesis August, R.T.M. Nagpur University.
15. Dixit, K. S., (2012). *Formulation of Approximate Generalized Field Data Based Models of Automobile Service Function.* PhD Thesis Sept., R.T.M. Nagpur University.
16. Ikhar, S. R., (2014). *Field Data Based Mathematical Model and Simulation of a Stirrup Making Activity.* PhD Thesis Sept., R.T.M. Nagpur University, Nagpur.
17. Vikha, A. D., (2015). *Formulation of Approximate Generalized Field Data-Based Model and its Simulation, Optimization & Reliability Evaluation for some Operations of Manufacturing Enterprise.* PhD Thesis July, Sant Gadge baba University, Amaravati University.
18. Kadu, R. S. (2007). *Optimization of Tool Life for Cutting Tools for Boring Operation.* PhD Thesis R. T. M. Nagpur University.
19. Borkar, K. (2014). *Formulation of Approximate Generalized Field Data-Based Model for Overhauling and Maintenance operations of Diesel Loco Motive.* PhD Thesis R. T. M. Nagpur University Nagpur.
20. Agashe, A. A., (2013). *Formulation of Field Data-Based Model for Functioning of a Work Station of Flour Grinding System Emphasizing Ergonomics Considerations and Optimization.* PhD Thesis, R.T.M. Nagpur University.

21. Munje, S., (2016). *Formulation of Approximate Generalized Field Data Based Models for the Performance of Workers Influenced by Shift Work in Foundry Industries.* Master of Engineering (by Research) Degree of R.T.M. Nagpur University.
22. Shiwalkar, P. B., (2011). *Study of Kinematics & Kinetics of Biped Robot.* Master of Engineering (by Research) Degree of R.T.M. Nagpur University.
23. Askhedkar, R. R. (2015). *Estimation and Validation of Insertion Lost of Acoustic Enclosure Through Treatment of Field Data-Based Modeling.* PhD Thesis being submitted shortly to R.T.M. Nagpur University.
24. Khandelwal, N. S., (2012). *Formulation of Approximate Generalized Field Data-Based Model for Agglomerator.* Master of Engineering (by Research) Degree of R.T.M. Nagpur University.
25. Kulkarni, R. A., (2015). *Formulation of Approximate Generalized Filed Data-Based Modeling and Simulation for some Operations of Manufacturing Enterprise.* PhD Thesis, R.T.M. Nagpur University.
26. Chatpalliwar, A. S., (2012). *Formulation of Approximate Generalized Design Data-Based Model of a Plat Manufacturing Bio-diesel.* PhD Thesis, Nov., R.T.M. Nagpur University.
27. Agrawal, K. N., (2010). *Decision support system for Behavioral Prediction of Industry Performance Indicators.* PhD Thesis, submitted to R.T.M. Nagpur University.
28. Shank, Jr. H., (1967). *Theories of Engineering Experimentation.* McGraw Hill Inc., New York.
29. Modak, J. P., (2015). *Overview of Mathematical Modeling.* Lecture No. 1, Poornima University, Jaipur, India.
30. Murrell, K. F. H., (1956). *Nature of Ergonomics. Ergonomics (Man in His Working Environment).* Chapman and Hall, London, New York,
31. Eastman Kodak Co. Ltd., (1983). *Chapter V Environment Ergonomic Design for People at Work.* Van nustrand reinhold, New York,
32. Rao, S. S., (1984). *Optimization Theory & Applications* (2nd edn.). Wiley Eastern Ltd.

CHAPTER 8

Design of a 9-Level Cascaded H-Bridge Inverter-Driven Induction Motor with Fewer Harmonic

Y. S. BAIS,[1] S. B. DESHPANDE,[2] and K. K. HIREKAR[1]

[1]*Electrical Engineering Department, Govindrao Wanjari College of Engineering and Technology, R.T.M.N.U. Nagpur, Maharashtra, India, E-mail: yogeshbais@yahoo.co.in (Y. S. Bais)*

[2]*Electrical Engineering Department, Priyadarshini Institute of Engineering and Technology, R.T.M.N.U. Nagpur, Maharashtra, India*

ABSTRACT

An inverter structure with five or more levels can vary substantially lower down the harmonics in the inverter. The output voltage and power can be increased by increasing the number of levels. The voltage level can be increased appropriately in steps by the addition of a suitable switching device in each phase. This chapter elaborates a unique control scheme based on the phase-shifted carrier PWM method and its suitable implementation in 9-level cascaded inverters driving a 3-phase induction motor. This chapter evaluates the THD values of current and voltage waveforms of induction motors between various levels. It gives stress on enhancing the quality of output voltage waveform and thereby increasing the efficiency of multilevel inverter. Progressively new 5-level, 7-level and 9-level structures have been developed with reduced switches. The MATLAB simulations have been carried out, and circuitry is designed by making use of MOSFET's and IGBT's as the switching devices for 5 and 7-level inverter, respectively. Simulation is done for 9-level inverter and simulation results and THD is observed. The hardware of 9-level inverter is designed using IGBT's and an induction motor load is run on it. Using this scheme, the speed can be controlled desirably. Besides this, the vibration and noise in the induction motor can be reduced substantially.

8.1   INTRODUCTION

Converters using the multilevel concept have been suitably used for motor drives and utility interface systems. They give a good efficiency and high power factor. Multilevel converters fall into three main categories: (a) diode clamped, (b) flying capacitors, and (c) cascaded.

8.1.1   PROMINENT FEATURES OF MULTILEVEL CONVERTERS

1. Since it is possible to switch the devices at a lower frequency, they have higher efficiency.
2. They can be used for applications involving higher current and high-voltage.
3. Whether the converters are working in rectification (charge mode) or inversion (drive mode), no charge unbalance problem occurs.
4. No electromagnetic interference problem exists.
5. For multilevel converters, which are used as rectifiers for converting ac to dc, the power factor is very good nearing unity.

The proper voltage balancing of the dc bus capacitors which are connected in series is required in multilevel converters. Capacitors tend to completely discharge or overcharge, and under this situation, the multilevel converter comes down to a 3-level case unless some control mechanism is devised for balancing the charging of the capacitor. During the operations of the rectifier and the inverter, the balancing of the voltage needs to be done properly for the capacitor. In other words, the flow of real power out of the capacitor must be equal to the flow of real power into the capacitor, and the overall charge on the capacitor must remain the same for one complete cycle.

8.1.2   LEVEL INVERTER DESIGN

The inverter has an H-Bridge and a multi conversion cell consisting of 4 different voltage sources (V_{dc1}, V_{dc2}, V_{dc3} and V_{dc4}), 4 switches and 4 diodes.

Four multi-conversion cells are connected in cascade with each other. Each multi-conversion cell consists of one voltage source, one diode and one active switch that can make the output voltage only in positive polarity with different levels. To acquire both positive and negative polarity, only one H-bridge is present in the circuit along with the multi-conversion cell.

By turning on controlled switch S_a (S_b, S_c, and S_d turn off), output voltage $+1V_{dc}$ (first level) is obtained across the load. By turning on switches S_a, S_b (S_c and S_d turn off), $+2V_{dc}$ (second level) output voltage is obtained. Similarly, $+3V_{dc}$ level is obtained by turning on S_a, S_b, S_c switches (S_d turn off) and $+4V_{dc}$ level is obtained by turning on S_a, S_b, S_c, S_d as shown in Table 8.1.

From Table 8.1, it is observed that only one switch is switched ON for each voltage level, among the paralleled switches. By using the multi-conversion cell, the input DC voltage is converted into a stepped DC voltage. This is further given to the H-Bridge and output voltage as a stepped or nearly sinusoidal AC waveform is obtained.

TABLE 8.1   Switching Modes for 9-Levels MC-MLI

SL. No	Multi-Conversion Cells		H-Bridge		Voltage Levels
	On-Switches	Off-Switches	On-Switches	Off-Switches	
1.	S_a, S_b, S_c, S_d	D_1, D_2, D_3, D_4	S1, S2	S3, S4	$+4V_{dc}$
2.	S_a, S_b, S_c, D_4	S_d, D_1, D_2, D_3	S1, S2	S3, S4	$+3V_{dc}$
3.	S_a, S_b, D_3, D_4	S_c, S_d, D_1, D_2	S1, S2	S3, S4	$+2V_{dc}$
4.	S_a, D_2, D_3, D_4	S_b, S_c, Sd, D_1	S1, S2	S3, S4	$+1V_{dc}$
5.	D_1, D_2, D_3, D_4	S_a, S_b, S_c, S_d	S1, S2	S3, S4	0
6.	S_a, D_2, D_3, D_4	S_b, S_c, S_d, D_1	S3, S4	S1, S2	$-1V_{dc}$
7.	S_a, S_b, D_3, D_4	S_c, S_d, D_1, D_2	S3, S4	S1, S2	$-2V_{dc}$
8.	S_a, S_b, S_c, D_4	S_d, D_1, D_2, D_3	S3, S4	S1, S2	$-3V_{dc}$
9.	S_a, S_b, S_c, S_d	D_1, D_2, D_3, D_4	S3, S4	S1, S2	$-4V_{dc}$

During the positive cycle, only the switches S1 and S3 are switched on among the H-Bridge switches. While during the negative half cycle, only the other diagonal pair of switches S2 and S4 are made on. If S is the number of DC sources or stages, then the following equation can be used to find out the related number of output levels.

$$\text{Number of level, } N = 2S + 1 \tag{1}$$

For a sample case, if S = 3, the output voltage waveform will be having seven different levels ($\pm 3V_{dc}$, $\pm 2V_{dc}$, $\pm 1V_{dc}$ and 0).

8.1.3 PHASE SHIFT PWM STRATEGY (PSPWM)

Figure 8.1 shows the PSPWM technique wherein more than one carrier having equal amplitude and equal frequency are phase-shifted to one another by fixed degrees as decided by the number of levels. Thus, for a 9-level output, eight different triangular carrier waves, which are phase-shifted by 45 degrees have been used. There is only one sinusoidal reference waveform (i) the waveforms are having odd symmetry which results in even and odd harmonics for odd mf, and (ii) PSPWM waves are having quarter-wave symmetry which results in odd harmonics only for even mf.

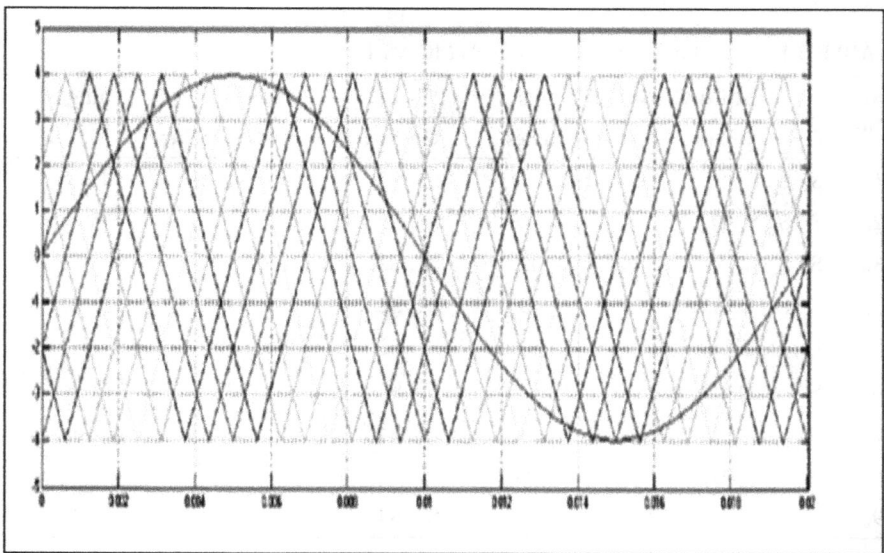

FIGURE 8.1   Carrier arrangement in phase shift pulse width modulation technique.

For PSPWM, amplitude of modulation index is:

$$Ma = Am/(Ac/2) \qquad (2)$$

8.1.4   SIMULATION RESULTS FOR 9-LEVEL INVERTER

The MATLAB Simulink model by using power system block set for the 9-level cascaded H-Bridge inverter is depicted in Figure 8.2. The parameter values taken for simulation are as follows:

$$V_1 = 100 \text{ V}, V_2 = 100 \text{ V}, V_3 = 100 \text{ V}, V_4 = 100 \text{ V};$$

$$f_c = 2000 \text{ Hz and fm} = 50 \text{ Hz}.$$

Gating signals for phase shifted carrier wave arrangement have been simulated for 9-level inverter (Figures 8.3–8.5).

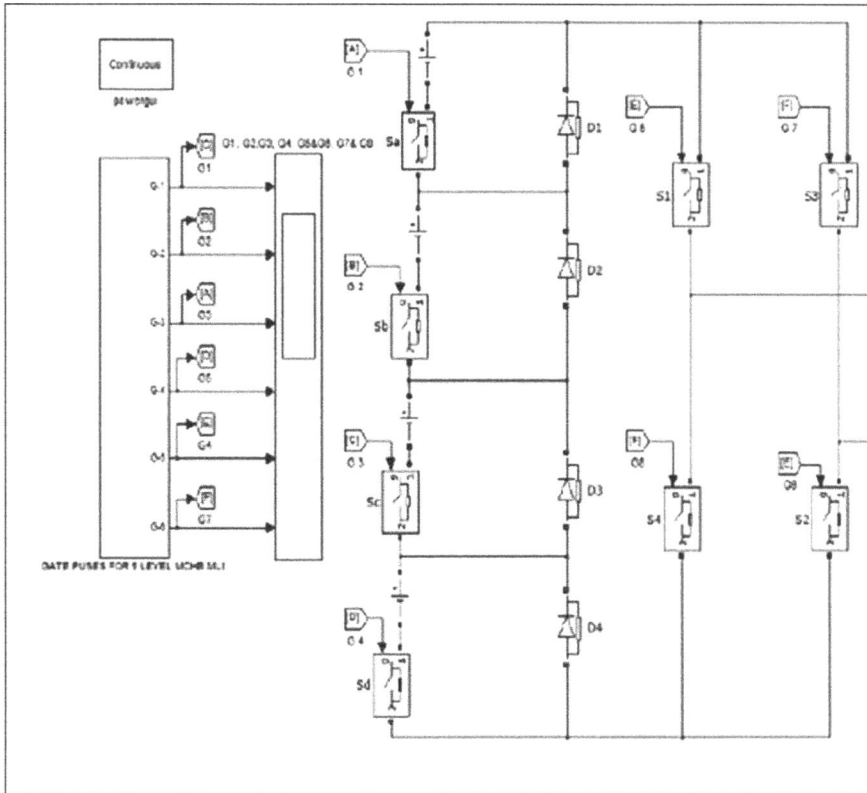

FIGURE 8.2   Modified cascaded nine-level inverter Simulink model.

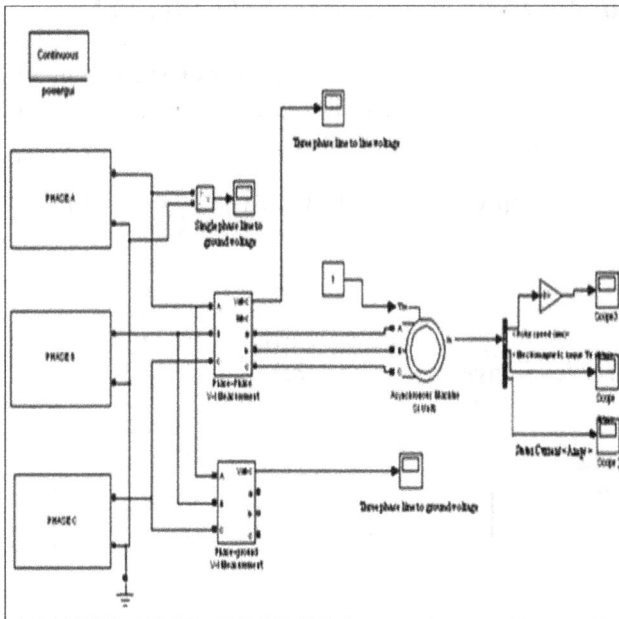

FIGURE 8.3   Simulation diagram for cascaded H-Bridge 9-level inverter driving an induction motor.

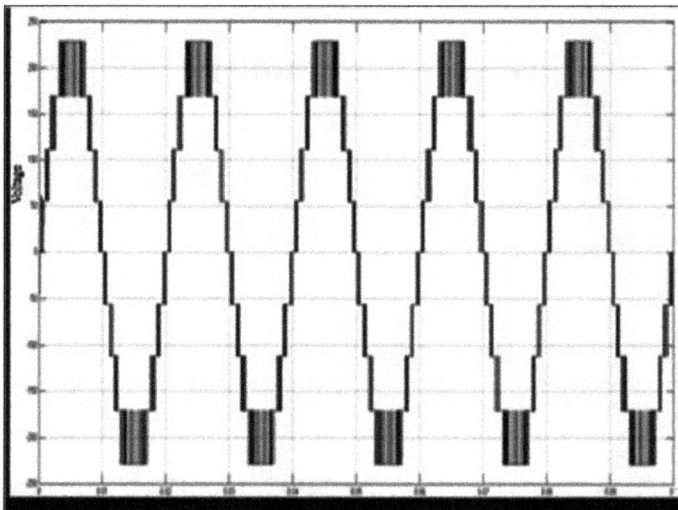

FIGURE 8.4   Simulated 9-level output voltage waveform by using phase shifted PWM technique.

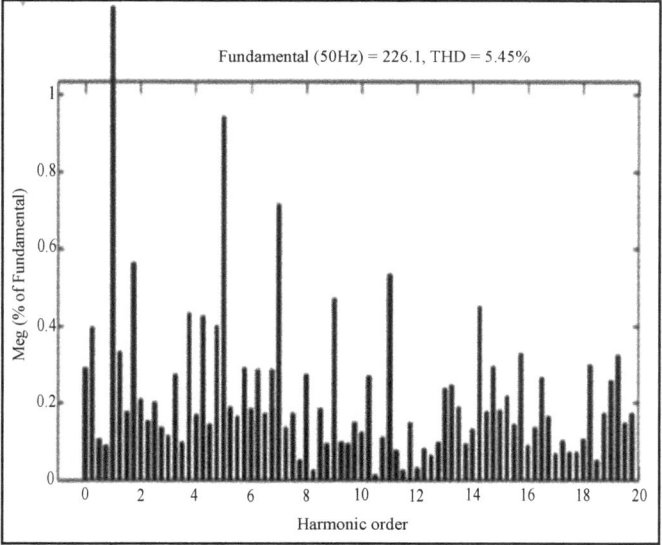

FIGURE 8.5   FFT plot of 9-level inverter by using phase shifted pulse width modulation technique.

8.1.5   3-PHASE 9-LEVEL INVERTER DESIGN (FIGURES 8.6–8.9)

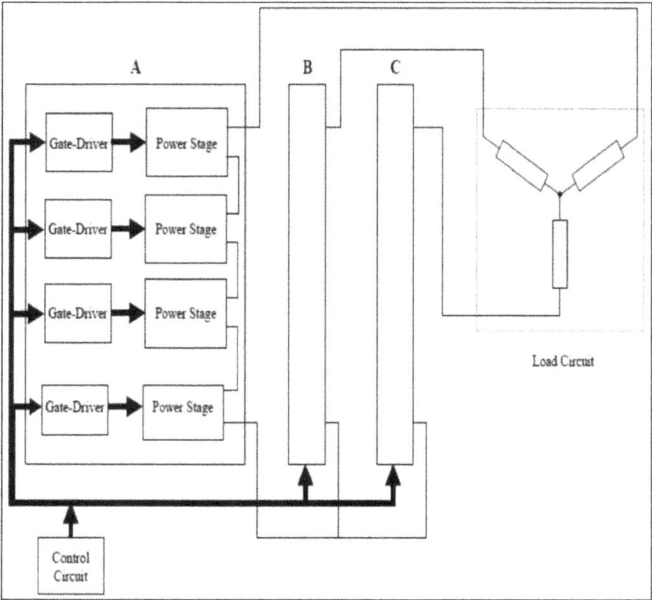

FIGURE 8.6   Hardware Design Construction of 9-level cascaded inverter for a 3-phase case.

FIGURE 8.7 Complete electronic circuitry of 9-level inverter using Microcontroller and variable frequency display.

FIGURE 8.8 Complete setup consisting of 9-level inverter, filter and 3-phase induction motor load.

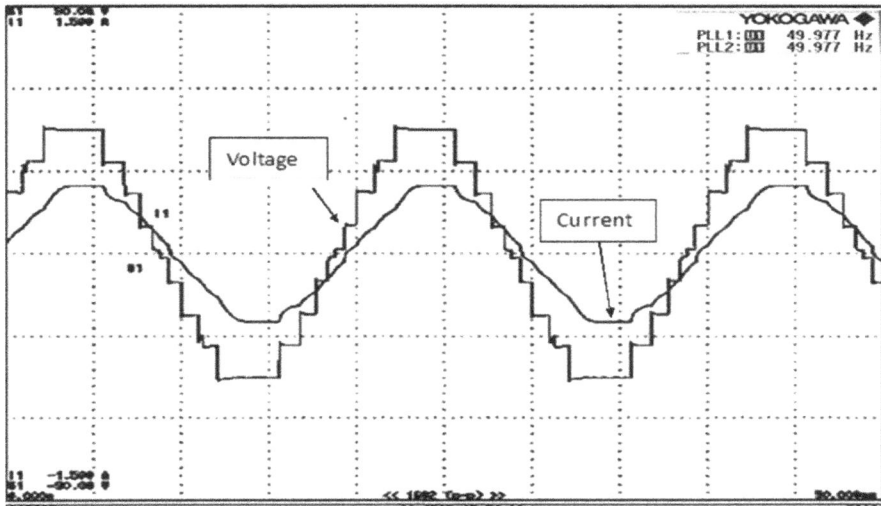

FIGURE 8.9   Experimental output voltage and current waveform.

8.2   CONCLUSION

The hardware design and implementation of a 9-level inverter for 3-phase case have been done and its output waveform shown. The 3-phase 9-levels cascaded multilevel inverter have been analyzed in detail by phase-shifted carrier sinusoidal Pulse Width Modulation technique.

The proposed configuration of 3-phase 9-level inverter has a lower total harmonic distortion. Also, it has reduced the number of switches. Demonstrated topology with carrier phase-shifted PWM technique is used to lower down the total harmonic distortion in comparison to other PWM strategies.

This 9-level inverter topology reduces voltage stress on semiconductors switches. It also improves output voltage and helps in reducing the higher number of semiconductor switches. Hence the size of the filter is reduced. One of the prominent advantages of this topology is that the THD is reduced, especially at higher level as the steps of voltage level rises. Simulation results demonstrate the performance of 9-level MLI with improved THD. This cascaded H-Bridge inverter is best suitable for grid-connected photovoltaic (PV) systems.

KEYWORDS

- **9-level inverter**
- **circuit arrangement**
- **electromagnetic interference**
- **induction motor**
- **multilevel converters**
- **phase-shifted carrier PWM**
- **voltage sources**

REFERENCES

1. Marquez, A., Leon, J. I., Vazquez, S., Portillo, R., Franquelo, L. G., Freire, E., & Kouro, S., (2017). Variable-angle phase-shifted PWM for multilevel three-cell cascaded h-bridge converters. *IEEE Transactions on Industrial Electronics., 64*(5), 3619–3628.
2. Ko, Y., Andresen, M., Buticchi, G., & Liserre, M., (2017). Power routing for cascaded H-bridge converters. *IEEE Transactions on Power Electronics, 32*(12), 9435–9446.
3. Suhalka, R., Sharma, G., Gupta, C., & Varshney, A., (2014). Reduction of total harmonic distortion in nine level cascaded multilevel inverter with different types of load. *IOSR Journal of Electrical & Electronics Engineering (IOSR-JEEE)., 9*(6), 12–20. www.iosrjournals.org (accessed on 30 June 2021).
4. Kumar, V., Mittal, A., & Abid, H. S., (2014). A new model of h-bridge multilevel inverter for reduced harmonics distortion. *International Journal of Engineering Research and Applications, 4*(11)(Part 5), 30–35. www.ijera.com (accessed on 30 June 2021).
5. Aravind, M. R., Shailaja, B., & Venkateshappa, (2015). Microcontroller based nine-level inverter for PV system with reduced switches topology. *International Journal on Advanced Electrical & Computer Engineering (IJAECE)., 2*(2).
6. Prabha, S. U., & Mahendran, K., (2016). An experimental investigation on implementation of advanced cascaded multilevel inverter for renewable energy applications. *International Journal of Innovations in Engineering & Technology (IJIET)., 7*(1).
7. Monopoli, V. G., Ko, Y., Buticchi, G., & Liserre, M., (2018). Performance comparison of variable-angle phase-shifting carrier PWM techniques. *IEEE Transactions on Industrial Electronics., 65*(7), 5272–5281.

CHAPTER 9

Comparative Study of Activity of Copper Ions onto Chitosan, Chloro-Chitosan, and Iodo-Chitosan

POONAM YEUL and SEEMA SHRIVASTAVA

Department of Applied Chemistry, Priyadarshini College of Engineering, Nagpur–440019, Maharashtra, India,
E-mail: seemashrivastava@rediffmail.com (*S. Shrivastava*)

ABSTRACT

Disposal of unsafe and dangerous metals by industries into the surroundings is leading to significant environmental concern due to its noxiousness, pertinacity, and accumulation. Although several researches are conducted using varied methods such as chemical precipitation, coagulation, adsorption, ion exchange, and membrane filtration for the removal of hazardous metals from wastewater, there are various restrictions associated with these methods. Of all the mentioned treatment methods, adsorption is the most efficient technique using adsorbents like activated carbon, but is less workable due to its relatively high cost. In this work, a natural biopolymer chitosan and its Chloro and Iodo derivative synthesized were used to separate Cu and Zn from wastewater. The affinity of chitosan and its by-product for Cu and Zn were studied using batch experiments on the artificial wastewater by UV-Visible spectrophotometer HACH DR-5000. The sorption of Cu and Zn ions was investigated under different conditions. The utmost sorption capacity was found to be 83% for Cu and 94% for Zn towards Iodo-chitosan.

9.1 INTRODUCTION

A toxic metal is a class of metals that exist naturally, and most of them may be found within the rocks and soil. Weathering of rocks and soil transports toxic, unsafe metals away into rivers and lakes. Though they occur naturally, they are available from various sources like burning of fossil fuels including coal, mining industries, burning garbage, forest fires, burning of hazardous remains of agricultural farms, burning of tobacco, etc. Toxic metals can be harmful to living beings at definite levels, and hence removal of harmful metal ions from wastewater is a crucial and widely studied research area. A number of technologies have been developed over the years in this field. However, the traditional technologies used so far are still not sufficient as they are costly. That is why, there is perpetual demand for new technologies which are cost-effective, energy-efficient and are low at maintenance [1, 2, 8].

In recent years, the use of natural biopolymer has increased enormously which bind the toxic metal ions into the environment. Chitosan is one such natural biopolymer which can be obtained from Chitin which is a natural polysaccharide. Chitosan can be shrimps obtained by deacetylation of Chitin. Chitin is found in the hard outer skeleton of some crabs, lobsters, or other shellfish [3, 4, 7].

Chitosan has the highest sorption capacity for various metal ions and is still low on cost compared to other alternatives. That makes it the best alternative absorbent. There are a number of different functional groups present in chitosan, such as amines hydroxyls, the metal ions bind to these functional groups by physical or chemical adsorption [9–11]. The purpose of this study is and to estimate the efficacy of chitosan and their Chloro and Iodo derivative [5, 6] for removal of metal ions like Copper and Zinc from wastewater.

9.2   MATERIALS AND METHODS

Chitosan, chloro-chitosan, iodo-chitosan, copper, and zinc solution and all the other reagents and chemicals are procured from Merck Co.

9.2.1   SYNTHESIS OF CHITOSAN

Chitosan was synthesized by deacetylation of chitin obtained from crab shell, by reacting 4 g of Chitin with 150 ml of 40% sodium hydroxide in a round bottom flask. This mixture was heated at 125°C for one hour. The residue obtained was chitosan which was cooled, filtered, washed thoroughly with distilled water, then dried at 60°C for 1 hour.

9.2.2   SYNTHESIS OF CHLORO-CHITOSAN

About 1 gm of synthesized chitosan with 15 gm of Triphenyl Phosphine (99.9% pure) and 100 ml of CCl_4 (9.37 M) were taken in a R.B flask, refluxed on a water bath for 5 h. After 5 h, the mixture was cooled at room temperature and 100 ml of n-pentane (99% pure) was added to it to precipitate Triphenyl phosphoxide. The formation of Triphenyl phosphoxide indicates incorporation of chloride group in chitosan, leading to the formation of Chloro-chitosan (0.8 g).

9.2.3   SYNTHESIS OF IODO-CHITOSAN

About 1 gm of synthesized chitosan with 1 gm of Tosyl chloride (Ts-Cl) in the presence of 50 ml alkaline chloroform ($\geq$ 99.8%) and 0.5 gm of sodium iodide yields Iodo-chitosan.

9.2.4   PREPARATION OF SYNTHETIC WASTE WATER

Copper solutions used in this study were obtained by dissolving weighed amount of $CuSO_4$ salts in distilled water, and different sets of experiment were set up for pH, time, doses, and the parameters studied, were optimized using a Jar test apparatus. The test was carried at 100 Rpm (revolution per minute) for rapid mixing, and 70 Rpm for flocculation process, and 2 minutes settling time for Cu removal. The filtrates were analyzed using UV-Visible spectrophotometer HACH DR-5000 to determine the quantity of copper adsorbed by Chitosan, Chloro-chitosan, and Iodo-chitosan.

9.2.5   ESTIMATION OF PERCENTAGE REMOVAL OF METAL IONS BY CHITOSAN, CHLORO-CHITOSAN, AND IODO-CHITOSAN

The following equation was used to calculate the percentage removal of copper ions from the solution by chitosan, Chloro-chitosan, and Iodo-chitosan during batch investigations;

$$Percentage\ removal = [(Co - Cf)/Co] \times 100$$

where; Co is the initial concentration; and Cf is the equilibrium concentration, both in ppm of metal ions in solution.

9.3 RESULTS AND DISCUSSION

9.3.1 EFFECT OF PH

The adsorption of Cu^{2+} ions were analyzed using atomic absorption spectrophotometer. It was found that the pH of the solution strongly affects the adsorption. Figure 9.1 shows, how in aqueous solutions, the pH influences the efficiency of removal of the copper ions. The results also indicate that at the pH 4, Cu (II) removal was maximum, and then decreased while pH, varied to 9 at 300 k. Agitation was constant at 100 rpm, t=10 min., initial Cu concentration = 1 mg/lt, and adsorbent dose = 50 mg/lt.

At pH 5; the maximum % removal of Cu (II) was found to be 52% for chitosan, 73% for Chloro-chitosan, and 74% for Iodo-chitosan. When the pH was lower than 5, free Cu (II) was mainly involved in the adsorption process, and it was the dominant species. When the pH surpasses 5, copper ions start precipitating as Cu $(OH)_2$ [3]. The adsorption of Cu ions decreases while pH > 5, it may be assigned to the formation of soluble hydroxyl complexes [4]. The adsorption of Cu ions was influenced by solution pH showing maximum % removal by Iodo-chitosan 74% followed by Chloro-chitosan 73% and 52% for chitosan.

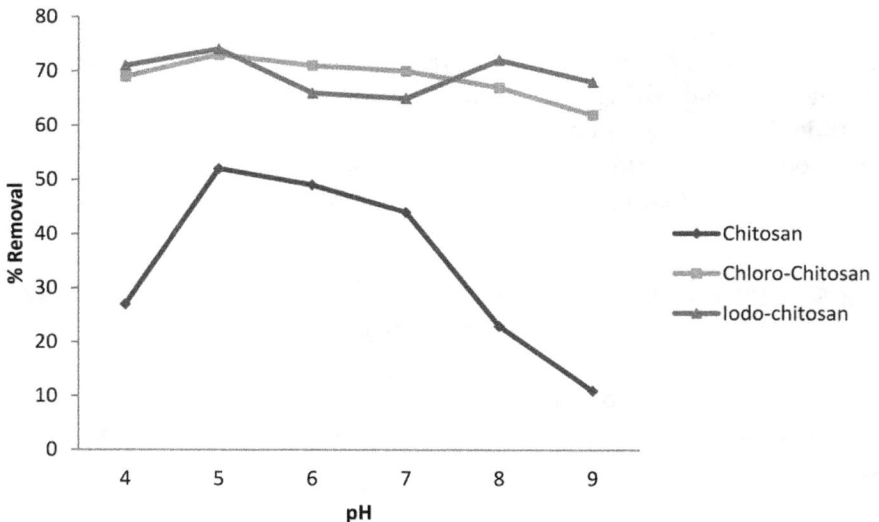

FIGURE 9.1   pH versus adsorption of Cu.

9.3.2 EFFECT OF CONTACT TIME

Removal of Copper ions was increased from 4% to 52% for chitosan with contact time variation from 2 to 12 minutes. After 12 minutes the % removal of copper remains constant (52%), which demonstrates that the equilibrium was attained at 12 minutes. Chloro-chitosan results from graph indicated that removal of copper increased from 8% to 88%, contact time varied from 2 to 12 minutes. Whereas for Iodo-chitosan, the removal of Cu increased from 7% to 87%, contact time varied from 2 to 12 minutes. After that, the percentage removal remains constant (87%). This showed that equilibrium was attained at 12 minutes (Figure 9.2).

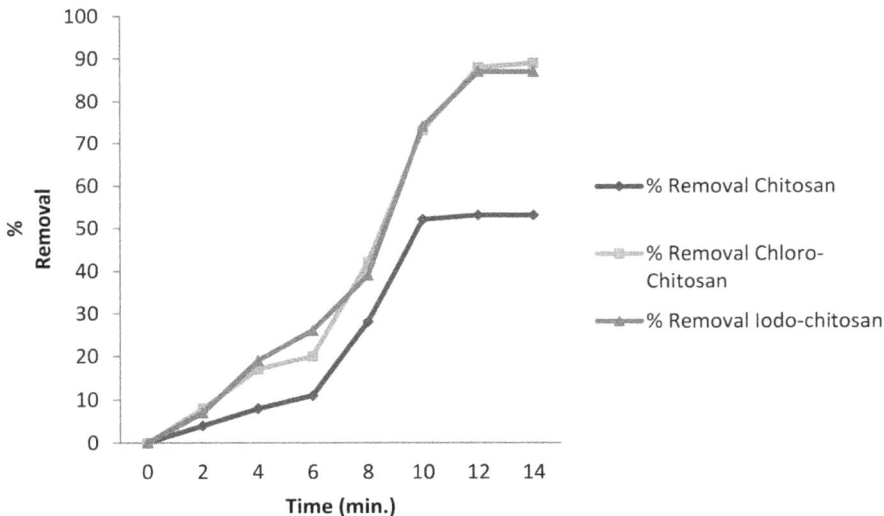

FIGURE 9.2 Contact time versus adsorption of Cu.

9.3.3 EFFECT OF INITIAL METAL ION CONCENTRATION

At a constant pH of 5, constant temperature 300°K, contact time of 12 minutes; the rate of absorption of copper was studied in the range of 1–100 mg/lt of initial metal ion concentration. It was found that the percentage of copper removal goes on decreasing while the initial copper concentration goes up.

For a given dose of adsorbent, i.e., 50 mg/lt, the number of adsorbent sites available remains constant. Hence while initial copper concentration increases, there is a decrease in removal of adsorbate (Cu ions) for chitosan,

Chloro-chitosan, and Iodo-chitosan. Thus, it is observed that the adsorption of copper was influenced by initial metal concentration (Figures 9.3 and 9.4).

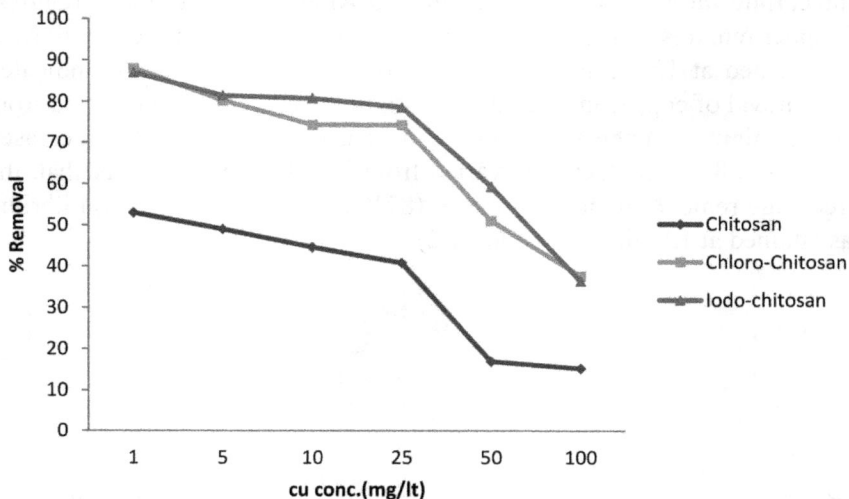

FIGURE 9.3   Metal ion concentration versus adsorption of Cu.

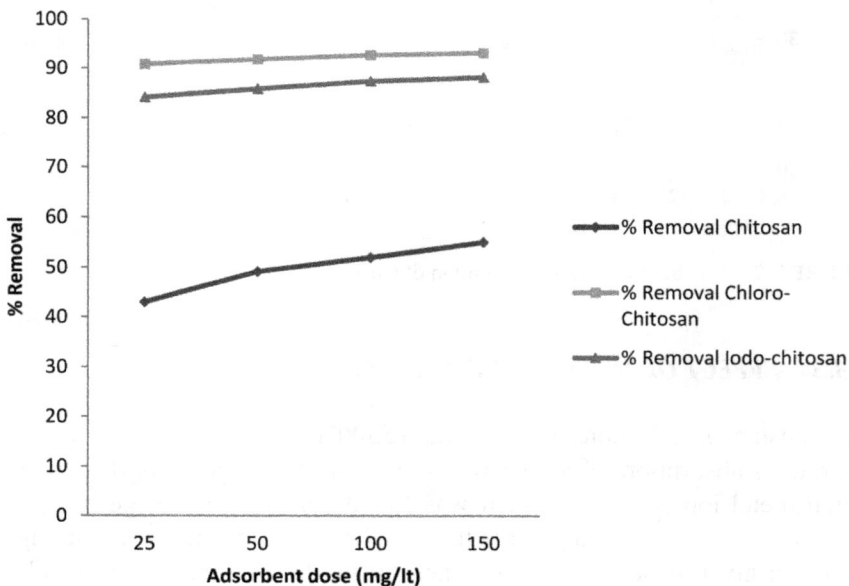

FIGURE 9.4   Adsorbent dose versus adsorption of Cu.

9.3.4   EFFECT OF DOSE OF ADSORBENT

The amount of adsorbents was varied from 25 to 150 mg/L while other parameters like pH = 5 and time of contact = 12 mins were kept constant to study the efficiency of copper adsorption. It is shown in Figure 9.4 that the efficiency of removal of copper increases as the doses of adsorbent are increased.

Although the removal of copper ions increases when the adsorbent dose is increases but the increase is substantial after 50 mg/lt of adsorbent dose. From the graph, it is clear that maximum copper ions are removed at an adsorbent dose of 50 mg/lt as compared to the other doses of adsorbent (25 mg/lt, 100 mg/lt and 150 mg/lt). From the graph, it is also clear that % removal of copper ions is more for Chloro-chitosan (91.81%) and Iodo-chitosan (85.85%) as compared to chitosan (49.16%) for the adsorbent dose of 50 mg/lt.

9.4   CONCLUSION

The result reveals that the maximum adsorption of copper by chitosan, Chloro-chitosan, Iodo-chitosan take place at optimized pH 5, adsorbent dose of 50 mg/lt, 1 mg/lt of Cu concentration at optimized time of 12 min. The obtained results indicated that the percentage of Cu ions removal was maximum by Chloro-chitosan and Iodo-chitosan. The maximum % removal of Cu (II) was found to be 52% for chitosan, 73% for Chloro-chitosan, and 74% for Iodo-chitosan. Thus, Chloro and Iodo chitosan are more efficient in removing copper ions as compared to chitosan and can be proved to be better alternates for heavy metal removal.

KEYWORDS

- **burning garbage**
- **chitosan**
- **chloro-chitosan**
- **crab shells**
- **iodo-chitosan**
- **metal removal**

REFERENCES

1. Duruibe, J. O., Ogwuegbu, M. O., & Egwurugwu, J. N., (2007). Heavy metal pollution and human biotoxic effects. *International Journal of Physical Sciences, 2*(5), 112–118.
2. Barakat, M. A., (2011). New trends in removing heavy metals from industrial wastewater. *Arabian Journal of Chemistry, 4*(4), 361–377.
3. Puvvada, Y. S., Vankayalapati, S., & Sudheshnababu, (2012). Extraction of chitin from chitosan from exoskeleton of shrimp for application in the pharmaceutical industry. *International Current Pharmaceutical Journal, 1*(9), 258–263.
4. Kumaria, S., & Rath, P., (2014). Extraction and characterization of chitin and chitosan from (Labeo Rohit) fish scales. *Procedia Materials Science, 6*, 482–489.
5. Yeul, P., & Shrivastava, S., (2016). Chloro and Iodo - stemmed products of chitosan. *Research Journal of Chemistry and Environment, 20*(9).
6. Labidi, A., Salaberria, A. M., Fernandes, S. C., Labidi, J., & Abderrabba, M., (2016). Adsorption of copper on chitin-based materials: Kinetic and thermodynamic studies. *J. Taiwan Inst. Chem. Eng., 65*, 140–148.
7. Mathur, N. K., & Narang, C. K., (1990). Chitin and chitosan, versatile polysaccharides from marine animals. *J. Chem. Edu., 67*, 938.
8. Demirbas, A., (2008). Heavy metal adsorption onto agro-based waste materials: A review. *Journal of Hazardous Materials, 15*(2, 3), 220–229.
9. Deans, Jr., & Dixon, B. G. (1992). Uptake of Pb^{2+} and Cu^{2+} by novel biopolymers. *Water Res. 26*(4), 469–472.
10. Findon, A., McKay, G., & Blair, H. S. (1993). Transport studies for the sorption of copper ions by chitosan. *J. Environ. Sci. and Health A 28*(1), 173–185.
11. Bailey, S. E., Olin, T. J., Bricka, R. M., & Adrian, D. D. (1999). *Water Research 33*(11), 2469–2479.

PART II
Sustainable Energy

CHAPTER 10

Choice of System Voltage (AC/DC) for a Standalone Nanogrid

SACHIN P. JOLHE,[1] G. A. DHOMANE,[2] and M. D. KARALKAR[3]

[1]*Department of Electrical Engineering, Government College of Engineering, Nagpur, Maharashtra, India*

[2]*Department of Electrical Engineering, Government College of Engineering, Amravati, Maharashtra, India*

[3]*Department of Electrical Engineering, Priyadarshini J.L. College of Engineering, Nagpur, Maharashtra, India,*
E-mail: minalkaralkar@gmail.com

ABSTRACT

"The war of current" is going to start for the nanogrid or niche operation with renewable power generation sources. For the higher and efficient operation of the system, which supply should be used?

This chapter outlines the different transmission frequency options characteristics for a distributed generation (DG) system, ranging from dc to high ac frequencies. Each option is evaluated to determine the most suitable transmission frequency for the system.

10.1 INTRODUCTION

The world is illuminated by the electricity since early 19th century (1802). The first carbon arc light was illuminated by Humphry Davy [1]. The lamp consists of the charcoal sticks, which are separated by 4-inch (100 mm), 2000-cell battery is used for producing an arc. But the arc lamp was used widely for the public use in the late 19th century. In year 1878, the world's first electric supply for the public was provided in Surrey town of Godalming in the UK. This is world's first hydropower plant which runs on water wheel on the River Wey. The generator is made by Siemens. This uses 10 horsepower (7.5 kW) generator, the world first Nanogrid [2–5], running 4 arc lamps and 27 incandescent Swan light bulbs [6, 7]. On September 4, 1882, Edison Illuminating Company was established to light the home with electricity at low DC power (Table 10.1) [1, 8–11].

TABLE 10.1   World First DC Power Grid

Particular	Specification
Location	255–257 Pearl Street in Manhattan
Area of power plant	50 by 100 feet (15 by 30 m)
Type	Coal-fired
Number of generators	six dynamos
Date of staring	September 4, 1882
Load	400 lamps
Number of customers	82 customers
Company	Edison illuminating company
Owner	Thomas Edison
Engine	high-speed steam engines (custom-made Porter-Allen)
Rating	175 Hp, 700 rpm
Voltage	110 Volt DC

Then Tesla invented the AC generator which overcome the limitation of the DC generator and motors. To transmit electricity up to long-distance, alternating Current (AC) is the best option. This is because of easy stepping up of voltage in alternating current, but because of high development in power electronics and communication system, direct current (DC) can be easily step up and down, and control. Now the same war is going to start for the nanogrid or niche operation with distributed generation (DG) with renewable power generation sources. For the higher and efficient operation of the system, which supply should be used? (AC or DC).

Renewable energy sources [12–14] are becoming the new ray of hope for many people around the world for the electricity supply. Globally more than 1 billion and In India near about half of people (47.5%) are facing the problem of n access to electricity or regular supply of electricity [15–17]. The central utility power supply grid can also serve the power to these, but most of the time it is uneconomical and unaffordable. Because of the remoteness, high dense forest, and hilly area, these places are still untouched by electrical power supply. The most advantage of the renewable energy sources are it is available from low to high power generation range, i.e., from few Watts to Gigawatt (by combination). As well as easy to install, low maintenance and easily available with government subsidy. The government mission "Power for all," can be fulfilled by this nanogrid [18]. With the increase in the nanogrid structures, all over the nation, will increase the living standard of the people as well as it will help in increasing in GDP of the nation.

- Nanogrid can operate in three modes [19–21]:
- Standalone;
- Grid-connected; and
- Combining with other neighbor nanogrid to for microgrid.
- According to supply:
- Alternating current (AC) nanogrid;
- Direct current (DC) nanogrid.

This chapter emphasis on the operation of the nanogrid in the standalone mode only. As this new emerging area of the power generation some problems and other things are coming up with regards to nanogrid [19, 22–24]:

- Energy management;
- Control of the nanogrid;
- Efficient operation;
- Different power generation sources and integration;
- Storage issue;
- Integration issues;
- Penetration in main grid issue;
- Stability issue;
- Optimization;
- Design of inverter.

In the above literature, different options of the transmission and utilization had not been studied. In this chapter, all options are studied for the nanogrid up to 5 km.

10.2   DISTRIBUTED GENERATION (DG)

As compared to the central ac power system DG has a different structure. In the conventional ac system, transformers and transmission lines are used for connecting a load to the generator. In the DG, the source may be static dc or variable dc/ac, which needs storage in batteries as well as it, requires power electronics converters between source and load for interfacing. Development in modern power electronics enables us to have variable output frequency which ranges from 0 to maximum frequency (dc to high frequency AC). Because of this advantage DG system is not restricted to traditional 50/60 Hz frequency. It may be advantageous to operate the DG system at different frequency than usual 50/60 Hz (may be higher or lower). The operating frequency should have the following characteristics:

- Efficient power transmission;
- Simple interface requirements between source and load;
- No major technical problems.

This chapter outlines the options for different transmission frequency characteristics for a DG system, ranging from dc to high ac frequencies. So that every option is assessed to find out the most appropriate transmission frequency for the system.

10.3   OPTIONS FOR TRANSMISSION FREQUENCY

10.3.1   DC

The first steam powered electric power station was developed by Thomas Edison and his company, The Edison Electric Light Company, in 1882, on Pearl Street in New York City. Initially the power station supplies power to 59 consumers for 3,000 lamps. The supply was given by DC and operated at a single voltage. Due to the drawbacks of the DC system, the voltage cannot change easily to up or down level, due to this distance was made limited to around 800 m between generators and load [19, 26].

After the invention of the AC supply system, the original dc systems were rapidly superseded. Because of transformer it is very much easy to step up and step-down voltage, this leads to high efficiency and transmission of electricity over a long distance is possible. Today dc transmission becomes practical with the use of modern power electronics technology, which finds niche applications. Simulation model for DC transmission is as shown in Figure 10.1.

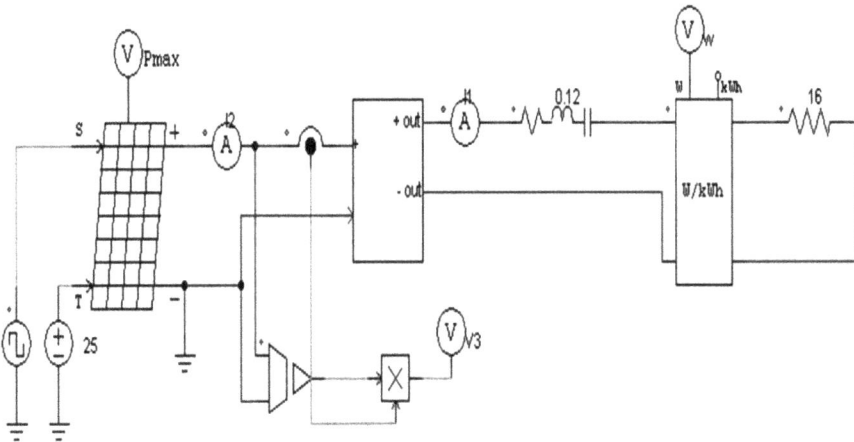

FIGURE 10.1   Solar standalone system with MPPT.

HVDC is the most common application of dc transmission. HVDC is economical for long-distancedistances (500 km and above) this is due to converter and inverter cost. For transferring the power at high voltage through the cable, DC is the only means. DC operation does not need generator synchronization. For connecting the asynchronous power system, the dc-link is very useful [27–29]. Simulation model for HVDC transmission is as shown in Figure 10.2.

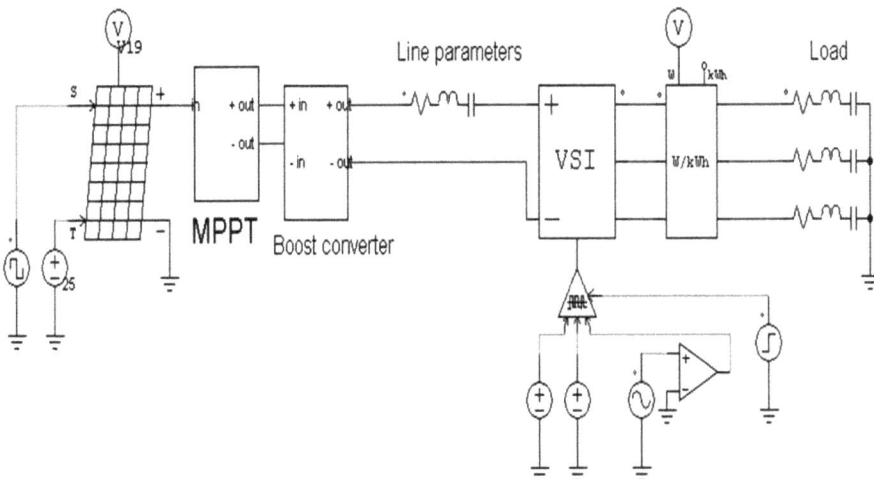

FIGURE 10.2   HVDC transmission with VSI.

In HVDC transmission, voltage source converters (VSC) is commonly used because of the progress in power electronic devices, for example, insulated gate bipolar transistors (IGBT). HVDC light is the HVDC scheme which is based on IGBT converters [30]. Since VSCs are well-suited for parallel operation, so offshore wind farms with multi-terminal systems based on VSCs have been proposed. The transmission line is needed, since the wind farms are 10–30 km offshore. DC power transmission is more efficient and reliable means for transmission of power to the shore via underwater cable than mains frequency ac. The application of the DC system with multi-terminal VSC-is proposed to distribute the power in large cities [31]. Under earth fault conditions, power electronic-based dc system has better performance with regard to power quality over the conventional ac system. The application of DC power is not restricted to HVDC system only. This is used by other systems also. However, these systems tend to be niche systems such as US Navy combat ships [32], telecommunications power supplies [33].

The dc power system offers a number of advantages like no need for intermediate distribution transformers which results in reduction in size and weight. Another major benefit is distribution frequency does not depend on the generator. Hence, generators can operate asynchronously, at their most efficient speed. With the use of power converters flexibility of the system enhanced to connect loads to the system. In the event of a fault, the converters limit current. In addition, the effect of current on the human body due to dc is significantly less than that of mains frequency ac. At 60 Hz, approximately 20 mA current gives severe shock and muscular contractions occur, while at dc, this threshold is three times greater [34].

The major disadvantage of dc is current interruption. The circuit breakers which are used for dc power systems having complex and expensive parts as compared to ac counterparts. To eliminate this problem, we can use hybrid circuit breakers. These circuit breakers comprise a solid-state switch and mechanical switch. To interrupt the fault solid-state switch are used and to provide electrical isolation, a mechanical switch is used in series [35].

10.3.2   16 2/3 HZ

In year 1878, two electricians at Godalming in England had developed the first-ever power plant and power system. The power system consists of two types of loads one is arc lamp operated at 250 V and seven in number, the other load is of incandescent lamp which operated at 40 V and 34 in number, which are supplied by power station which consist of two waterwheels. The supply was irregular to the lamps, so it was not useful for domestic application. In early 20th century, power is supplied at 25 Hz. This is especially for transmission at 6600 and 13200 volts for long-distance [36].

The transmission of electric power at a lower frequency than 50/60 Hz is called low frequency alternating current (LFAC). By using LFAC (16 2/3), the reach of ac transmission can be extended, and the investment cost that would have been necessary for VSC HVDC-based transmission reduced. Over the distance of 400–500 km, the power is transmission is 600 MW, and it is done with the help of 245 kV cable. So, it is observed that the whole power which is developed in the North Sea is transferred to the onshore grid with the help of 245 kV cables at the frequency of 16 2/3 Hz without HVDC line and off shore converter [37]. Simulation model for 16 2/3 Hz transmission is as shown in Figure 10.3.

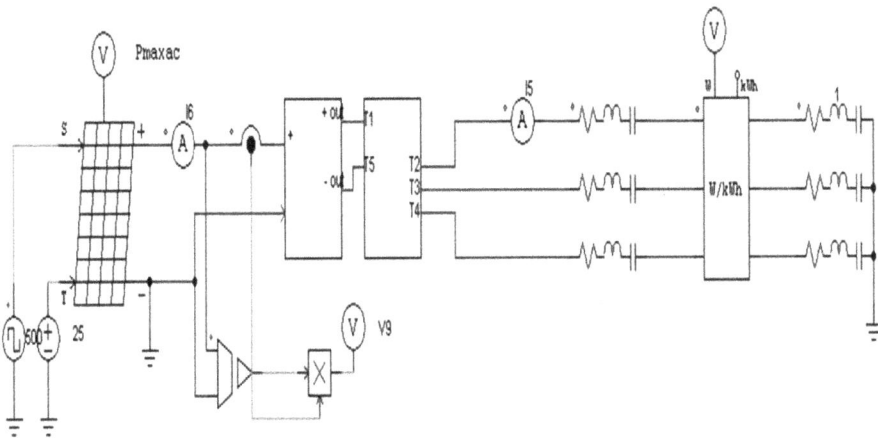

FIGURE 10.3   Standalone system with lower frequency.

10.3.3   50/60 HZ

Over the last 100 years, 50/60 Hz power system has been in existence. This is most advantageous system till date; as it has ability to transform voltages from one level to other in an efficient and reliable fashion with the use of a transformer is a major benefit even today. Power is easily transmitted at 50/60 Hz with an overhead transmission line, the switching operation and interruption of AC is easy. A major advantage is this system can directly connect to the grid [38].

Today, due to the availability of higher frequency supply, the size of magnetic components such as transformers and motors can be reduced. The main drawback of operating system at 50/60 Hz is that it increases the size and cost of the ac machines like transformers and motors as compared to the system which designed to operate at higher frequency.

10.3.4   400/500 HZ

There is one more option for frequency is of 400/500 Hz. This frequency is used in aircraft. This middle frequency is very much useful in reducing the size of the magnetic components [39]. With the use of this frequency in a DG, the size and weight of the magnetic components is reduced significantly. As the size of the magnetic component reduces, this also reduces the cost of the components, so that economically our system becomes viable. The major drawback in this system is line reactance. Line reactance is significantly high in 400/500 Hz system than that at 50/60 Hz. Another disadvantage of this system is a requirement of reactive power support. In addition, it is less efficient due to dielectric losses and skin effect.

10.3.5   HIGH-FREQUENCY AC

High-frequency ac is one of the options for the DG. But it is not usually employed as a transmission frequency of power systems. However, for distributing power in buildings, high frequency ac system has been organized with the frequency of 20 kHz [25]. The main advantage of the system is that it has a small size of magnetic components and inductive power transfer. The drawback of dielectric losses and skin effect is reduced by the use of Litz wire. The use of Litz wire renders this option impractical (economic viability) for longer distance transmission, while this transmission frequency may be a useful for small distances. Simulation model for 20 KHz frequency transmission is as shown in Figure 10.4.

10.4   TRANSMISSION LINE EFFICIENCY

After examining the qualities of various transmission frequencies, it is observed that DC and 50 Hz plays a very suitable and practical role in the DG system. The use of High frequency renders its option impractical (economic viability) for longer distance transmission, while this transmission frequency may be a useful for small distances (Figure 10.5). DC transmission is the best option for transmission in a DG system. With the use of modern electrical loads and lighting system such as CFL and LED, this option gives maximum reliability as well as efficiency. Common electrical loads are designed for operating on 50/60 Hz frequency, so it is always desirable to transmit power at this frequency from consumer point of view. For transmission with the lower (16 2/3) and higher frequencies (500, 20 kHz), additional equipment is required to convert frequency as per load specifications.

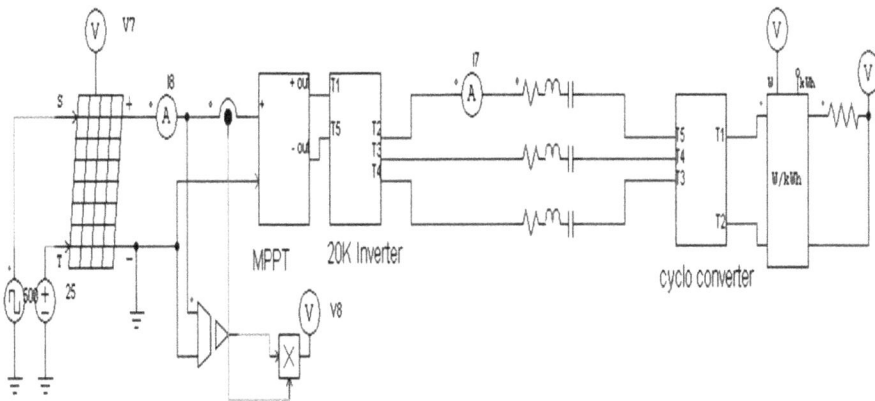

FIGURE 10.4   Standalone system with high frequency.

To find out the suitable transmission frequency among all these options, analysis is performed. The evaluation is also done for observing the effect of operating frequency on transmission efficiency.

10.5   CONCLUSION

The efficiency of the DC system is near about constant and highest up to 5 km, can be seen in Figure 10.5. So, for niche application DC is preferred. About 50/60 Hz is the common distribution frequency and mainly maximum load is operating on this frequency. The result shows the efficiency of this transmission is nearly the same as the DC and maximum. High frequency ac (20 kHz) is preferred for building with Litz cable.

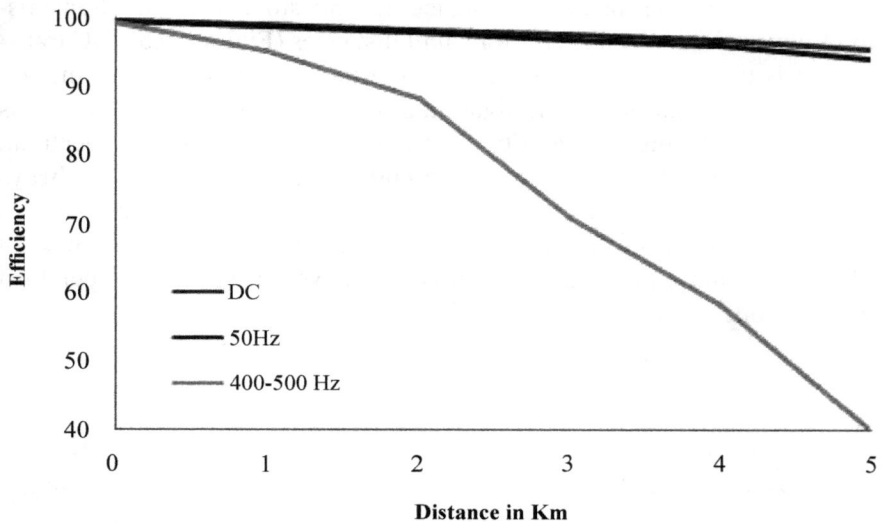

FIGURE 10.5   Graph of transmission line efficiency vs distance in km.

KEYWORDS

- direct current
- distributed generation
- high frequency
- low frequency
- renewable energy
- transmission frequency

REFERENCES

1. https://en.wikipedia.org/wiki/Electric_light (accessed on 30 June 2021).
2. Zhang, Z., Haoyue, T., Peng, W., Huang, Q., & Lee, W., (2020). Two-stage bidding strategy for peer-to-peer energy trading of nanogrid. *IEEE Transactions on Industry Applications, 52*, 1000–1008.
3. Bryan, J., Duke, R., & Round, S., (2004). Decentralized generator scheduling in a nanogrid using DC bus signaling. *IEEE Power Engineering Society General Meeting, 2*, 1–6.
4. Schonberger, J., Duke, R., & Round, S., (2006). DC-bus signaling: A distributed control strategy for a hybrid renewable nano-grid. *IEEE Trans. Ind. Electron., 53*(5), 1453–1460.

5. Burmester, D., Rayudu, R., Seah, R., & Akinyele, D., (2017). A review of nanogrid topologies and technologies. *Renewable and Sustainable Energy Reviews, 67*, 760–775.
6. https://en.wikipedia.org/wiki/Holborn_Viaduct_power_station (accessed on 30 June 2021).
7. https://en.wikipedia.org/wiki/Electric_power_industry (accessed on 30 June 2021).
8. https://en.wikipedia.org/wiki/Pearl_Street_Station (accessed on 30 June 2021).
9. Matthew, J. E., (1959). *Edison: A Biography*, McGraw Hill, New York, Chapter 13: Pearl Street.
10. Gene, A. (1997). *Thomas Alva Edison: Inventing the Electric Age*. Oxford University Press, New York, Oxford.
11. Keithley, J. F., (1999). *The Story of Electrical and Magnetic Measurements: From 500 BC to the 1940s*. Piscataway, Wiley-IEEE Press.
12. Park, M. S., Chun, Y. H., & Lee, S., (2019). Estimation of renewable energy volatility and required adjustable speed pumped storage power generator capacity considering frequency stability in Korean power system. *Journal of Electrical Engineering & Technology, 14*(3), 1109–1115.
13. Jolhe, S. P., Karalkar, M. D., & Dhomane, G. A., (2016). *Smart Grid IEEE Sponsored International Conference on Green Engineering and Technologies (IC-GET)*. pp. 1–5.
14. Sandeep, S. R., & Nandihalli, R., (2019). Optimal sizing in hybrid renewable energy system with the aid of opposition based social spider optimization. *Journal of Electrical Engineering & Technology*, 1–8.
15. Ministry of New and Renewable Energy (2020). http://mnre.gov.in/schemes/decentralized-systems/ (accessed on 31 July 2021).
16. United Nations Foundation. (2012). *Energy Access Practitioner Network—Towards Achieving Universal Energy Access by 2030*. p. 1.
17. Jolhe, S. P., Dhomane, G. A., & Karalkar, M. D., (2019). Solar home systems as the basis for bottom-up grids. *Helix, 9*(3), 5073–5080.
18. Power for All. Ministry of Power, Government of India. p. 13. https://powermin.gov.in/en/content/power-all (accessed on 31 July 2021).
19. Moussa, S., Ghorbal, M. J., & Ilhem, S. B., (2019). Bus voltage level choice for standalone residential DC nanogrid. *Sust. Cit. & Soc.*, 46.
20. Chauhan, A., & Saini, R. P., (2014). A review on integrated renewable energy system based power generation for stand-alone applications: Configurations, storage options, sizing methodologies and control. *Renewable and Sustainable Energy Reviews, 38*, 99–120.
21. Twaha, S., Makbul, A., & Ramli, M., (2018). A review of optimization approaches for hybrid distributed energy generation systems: Off-grid and grid-connected systems. *Sustainable Cities and Society, 41*, 320–331.
22. Vuarnoz, D., Cozza, S., Jusselme, T., Magnin, G., & Niederhauser, L., (2018). Integrating hourly life-cycle energy and carbon emissions of energy supply in buildings. *Suste. Cit. & Soc., 43*, 305–316.
23. Dahiru, A. T., & Tan, C. W., (2020). Optimal sizing and techno-economic analysis of grid-connected nanogrid for tropical climates of the Savannah. *Suste. Cit. & Soc., 52*.
24. Lee, S., Lee, L., Jung, H., Cho, J., Hong, J., Lee, S., & Har, D., (2019). Optimal power management for nanogrids based on technical information of electric appliances. *Eng. and Buildings, 191*, 174–186.
25. Ferreira, J. A., & Broeck, H. W., (1993). Alternative power distribution in residential and commercial buildings. *Fifth European Conference on Power Electronics and Applications, 7*, 188–193.

26. Bryan, J., Duke, R., & Round, S., (2003). Distributed generation - nanogrid transmission and control options. *International Power Engineering Conference*, 341–346.

27. Lei, L., Julong, W., Li, M., & Ming, C., (2018). A correction method for HVDC transmission plan considering the correlation between sending end new energy generation and receiving end loads. *Energy Procedia, 145*, 193–198.

28. Yihua, Z., Li, W., Qi, G., Dongxu, C., Jianbo, L., Yu, W., Xueming, L., et al., (2018). Research on security and stability characteristics and control strategies of power grid with VSC-HVDC. *Energy Procedia, 145*, 351–356.

29. Weimers, L., (1998). HVDC light: A new technology for a better environment. *IEEE Power Engineering Review, 18*(8), 19, 20.

30. Lu, W., & Ooi, B. T., (2003). Optimal acquisition and aggregation of offshore wind power by multiterminal voltage-source HVDC. *IEEE Transactions on Power Delivery, 18*(1), 201–206.

31. Jiang, H., & Ekstrom, A., (1998). Multiterminal HVDC systems in urban areas of large cities. *IEEE Transactions on Power Delivery, 13*(4), 1278–1284.

32. Ciezki, J. G., & Ashton, R. W., (2000). Selection and stability issues associated with a navy shipboard dc zonal electric distribution system. *IEEE Transactions on Power Delivery, 15*(2), 665–669.

33. Gruzs, T. M., & Hall, J., (2000). AC, DC or hybrid power solutions for today's telecommunications facilities. *International Telecommunications Energy Conference*, 361–368.

34. Walter, L., (1981). *Weeks Transmission and Distribution of Electrical Energy*. Harper and Row.

35. Meckler, P., & Ho, W., (2004). An electronic circuit breaker need electrical contacts? *IEEE Holm Conference on Electrical Contacts*, 480–487.

36. Steinmetz, C. P., (1910). Problems in design and operation of very large electrical generating systems. *IEEE Proceedings of the American Institute of Electrical Engineers, 29*(11), 7, 8.

37. Erlich, I., Shewarega, F., Wrede, H., & Fischer, W., (2015). Low frequency AC for offshore wind power transmission - prospects and challenges. *11th IET International Conference on AC and DC Power Transmission*, 1–7.

38. https://en.wikipedia.org/wiki/Deptford_Power_Station (accessed on 30 June 2021).

39. Takahashi, I., & Su, G. J., (1989). A 500 Hz power system - applications. *IEEE Industry Applications Society Annual Meeting, 1*, 996–1002.

CHAPTER 11

The Growth of the Indian Power Sector: Pre- and Post-Independence

SHUVAM SAHAY and NIRANJAN KUMAR

Department of Electrical Engineering, National Institute of Technology, Jamshedpur, Jharkhand, India, E-mail: shuvam.sahay90@gmail.com (S. Sahay)

ABSTRACT

Electricity played an essential role in the development and progress of the nations. The Indian power sector is one of the most variegated in the world. This chapter describes the growth of the Indian Power Sector from 1879 to till this date. It highlights the progress in generation, transmission, and distribution sector in terms of installed capacity, per capita consumption, losses, and actual generation. This chapter also presents comparison with some top countries like China, Japan, the USA, Brazil, Canada, and Germany in terms of energy production, energy consumption and transmission and distribution lose for showing the status of our country in the energy sector. This chapter also presents the scenario of energy production through renewable in India by comparing with other top countries of the world. Through this chapter, the authors want to present a picture in the minds of readers so that readers are able to know the progress and status of our Indian power sector and their shortcoming.

11.1 INTRODUCTION

Electrical power is an imperative component which is required for the economic development and welfare of a country. India also has an undesirable climate and geographical conditions. The electricity consumption of India is high due to its large population [1, 2]. Electricity supply act 1948 and creation of SEBs are the initial level of our Indian power sector. The situation of our power sector between 1956–1991 is known as the nationalization stage of the Indian power sector. Industry policy 1956, generation, and distribution of power under state control, grant, framework limitation, and resource compulsions are the parts of the nationalization stage of the Indian power system. The situation of our power sector between 1991–2003 is known liberalization Era of the Indian power sector. Legislative and policy action 1991, private sector involvement in generation, fast clearing of private sector participation projects, Electricity regulation act 1998 and establishment of Central Electricity Authority comes under this stage of growth. The stage beyond 2003 is called the growth era of the Indian power sector. Electricity Act 2003, National tariff policy 2006, exclusion of permission for generating power, giving permission to private entity in power generation to create competition in Indian power market, launch of UMPP scheme like JNNSM to promote energy production through solar, US nuclear agreement and give allowance to private investment comes undergrowth era of Indian power sector [3, 4].

11.2   HISTORICAL BACKGROUND

The first presentation of electric light in Calcutta was organized on 24 July 1879 by P.W. Fleury and Co. The first electric demonstration was on 1882 at Crawford Market in Mumbai. The first hydroelectric installation was installed in Sidrabong under Darjeeling Municipality Corporation on 1897. The first thermal power station was set up in Emambagh in 1899. Sivsamudaran Power House was set up in 1902, which supplies power to gold mines at 22 KV, to Bangalore and Mysore at 75 KV. The first generating power plant was operational in Mumbai in 1905 to provide electricity to tramways. The first electric street lighting in Asia is in Bangalore on 5 August 1905. Jammu power Plant was installed in 1909 for supplying electricity to Srinagar at 30 KV at 25 Hz. First-time tata trio of water plant utilized the monsoon precipitation for the generation in 1911. The first inland power station (Mettur power station) of capacity 2459 MW was installed for supplying electricity to Madras, Erode, and Moyer on 1925. Pykara power station of capacity 6.65 MW was installed in 1933 [5].

11.3   ELECTRICITY ACT

11.3.1   ELECTRICITY SUPPLY ACT 1910 [6]

1.   It gives basic architecture for electric supply industry in India;

2. It encourages the growth of different sectors like industrial, agriculture, and domestic through licensees permitted by state government;
3. This act starts giving licenses to supply electricity in a particular area;
4. It gives legal architecture for unwinding of wires and other related tasks;
5. It develops a relationship between licensee and consumer.

11.3.2   ELECTRICITY SUPPLY ACT 1948 [5, 6]

1. It authorized to create SEBs;
2. There is urgent requirement for the State to take action (through SEBs) to promote electrification (so far limited to cities) to the places which are lacking of it;
3. The main function of SEB's is the monitoring and controlling of the generation, transmission, and distribution of their states, respectively;
4. It constitutes the Central electricity Authority of India, which helps our governments and concerned authority on all technical issues related to different sectors of the power system. It also assists state governments, licensees, or power companies on subjects which encourage them to operate and maintain their system under their control in a revised manner.

11.3.3   ELECTRICITY ACT 2003 [7]

1. It removed the problem of taking license to generate power on a condition that they should follow the constraints given by the government;
2. Only authorized person can carry transmission and distribution business;
3. Central Government have started to monitor and regulate the different sector of the power system for the efficient flow of power from generation to user end;
4. It allows open access facility in transmission sector;
5. It creates a need of unwinding of SEB's but with old responsibilities;
6. It authorizes State Electricity Regulatory Commission (SERC);
7. An appellate tribunal is created to look after the issues related to CERC's and SERC's;

8.   It authorizes the metering of electrical energy supplied by them;
9.   It makes law stronger to tackle against thefts of electricity;
10. It mandates Regulatory commissions to fix a ceiling on trading margins;
11. It allows a stand-alone system for rural and remote areas;
12. It gives pace to the process of electrification;
13. It authorizes CEA to develop an electricity plan;
14. It authorizes CEA to develop an electricity policies and tariff policy.

11.4   POST INDEPENDENCE ERA

In 1950 installed capacity of India was 1713 MW and per capita consumption was 15 MW. About 63% of the generation and distribution was under private ownership. The first 220 KV line was laid down in 1950. The first SEB was set up in West Bengal in 1956. India began using grid management in 1961. The first 440 KV line was laid down in the year 1977. The first nuclear power plant was set up in TARAPUR in 1969. Other nuclear power plants were set up in Kota in 1973, Kalpakam in 1984, Narora in 1991, Kakrapara in 1992 and Kaiga in 2000. In mid-1970, due to lack of high-grade coal dependency over hydropower generation increases [5, 6]. After the 55-year plan, the Government of India (GOI) had assisted the state government in the transmission of electrical energy. NTPC, NPCIL, and NHPC were established in the year 1976. DNER was created in the year 1982, which incorporates CASE (1981) (Commission for additional source of energy) under its umbrella. The function of CASE was to formulate policies and implement them to develop new renewable sources and encourage R&D in this area after the energy crisis in the 1970's [7]. NTPC was later changed to PGCIL in 1992. In 1991 North Eastern grid and Eastern grid were interconnected. Some modification was carried out in electricity supply act 1910 and 1948; in 1991, to promote generating companies to generate and sell bulk power to the grid [5]. The first HVDC design (500 MW) was inaugurated in Vidhayanchal in the year 1989. PTC (Power Trading Corporation) was formed in the year 1995. It promotes efficient and smooth operation of power market in order to minimize the cost of energy for the industrial and commercial consumers and helps in power trading with the neighboring countries and also played a role of advisor in some national power trading. Currently, it guided a renewable energy of around 290 megawatts of wind energy in MP, Karnataka, and AP.

ERC act (Electricity regulatory Commission Act) was passed in the year 1998 under which CERC was set up at the national level and SERC was set up at the state level. CERC monitors and control the impost of generating companies governed by the central government. It regulates the interstate transmission of energy including tariff of the transmission utility. It regulates the interstate bulk sale of power. SERC monitors and control the charge of usage of transmission system for transmission of power. It plays an important role in the procurement process of transmission utilities and distribution utilities. It encourages competition, performance, and economy in the operation of the electricity companies and determines bulk and retail cost. Electricity Amendment Act was passed in the year 1998 which creates CTU at national Level and STU at the state level. It makes transmission as a different area to enhance contribution from the public and private sectors. CTU (PGCIL) discharges all function of planning and coordination relating to interstate transmission while STU do the same function relating to intrastate transmission. In 2003 Western grid were interconnected with the Eastern grid. In 2006 Northern grid were interconnected with the above and thus forming the central grid. Later, in 2013 southern grid interconnected with the central grid, forming the National grid of India [6].

Table 11.1 shows the trend of installed capacity and per capita consumption of India since independence [10, 11]. It indicates that the installed capacity of India is increasing year by year but the same as the rate of the increment of per capita consumption of India.

TABLE 11.1   Installed Capacity and per Capita Consumption: 1947–2019

Year	Installed Capacity (MW)	Per Capita Consumption (KWH)
1947	1362	16
1950	1713	18
1956	2886	31
1961	4653	46
1966	9027	74
1969	12957	98
1974	16664	126
1979	26680	172
1980	28448	172
1985	42585	229
1990	63636	329
1992	69065	348
1997	85795	465

TABLE 12.1   *(Continued)*

Year	Installed Capacity (MW)	Per Capita Consumption (KWH)
2002	105046	559
2007	132329	672
2012	199877	884
2013	223343	915
2014	245259	957
2015	271722	1010
2017	326833	1122
2018	344002	1149
2019	356100	1181

Table 11.2 displays the plan-wise growth of installed capacity in terms of different modes of generation. It indicates that the RES come into the picture after 1990. After 1974 transition occurs from non-renewable energy to renewable energy.

TABLE 11.2   Plan Wise Growth of Installed Capacity in India (MW)-Mode Wise

SL. No.	Year	Thermal	Hydro	Nuclear	RES	Total
1.	1947	854	508	0	0	1362
2.	1950	1153	560	0	0	1713
3.	1956	1825	1061	0	0	2886
4.	1961	2736	1917	0	0	4653
5.	1966	4903	4124	0	0	9027
6.	1969	7050	5907	0	0	12957
7.	1974	9058	6966	640	0	16664
8.	1979	15207	10833	640	0	26680
9.	1980	16424	11384	640	0	28448
10.	1985	27030	14460	1095	0	42585
11.	1990	43745	18308	1565	18	63636
12.	1992	48054	19194	1785	32	69065
13.	1997	61010	21658	2225	902	85795
14.	2002	74429	26269	2720	1628	105046
15.	2007	86015	34654	3900	7760	132329
16.	2012	131603	38990	4780	24594	199877
17.	2017	218330	44478	6780	57244	326833
18.	2018	222907	45293	6780	69022	344002
19.	2019	226279	45399	6780	77642	356100

Figure 11.1 shows that the rate of increment of installed capacity in the private sector is greater than the rate of increment of installed capacity in the state and central sectors. Table 11.3 shows the growth of generation and the contribution by the different modes of generation [8, 12]. In 1947, the total generation was 4073 gigawatt-hour in which the contribution by the thermal was 1877 gigawatt-hour, 2195 gigawatt-hour, and there was no contribution by the nuclear and RES. At the end of 2019, total generation of our country is 1371817 gigawatt-hour (Figure 11.2).

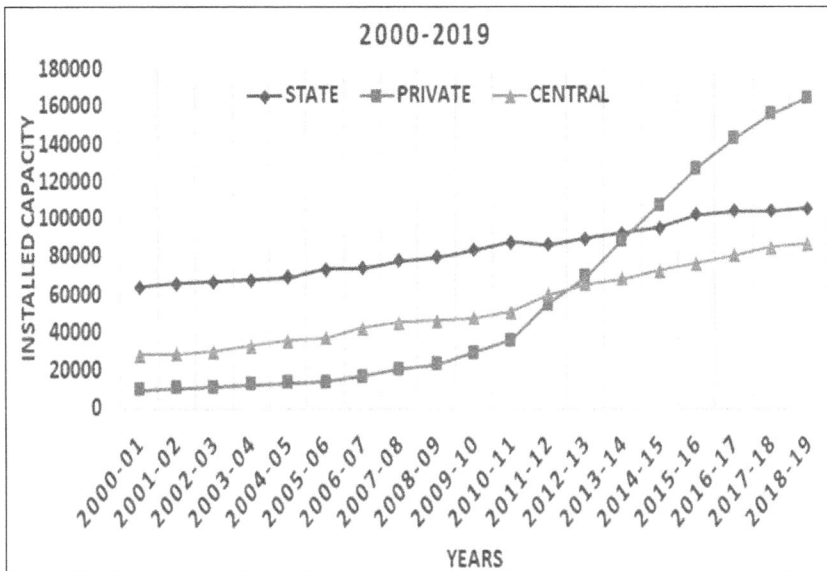

FIGURE 11.1   Sector-wise growth of installed capacity.

TABLE 11.3   Plan Wise Growth Electricity Generation in India (GWH)

SL. No.	Year	Thermal	Hydro	Nuclear	RES	Total
1.	1947	1877	2195	0	0	4073
2.	1950	2787	2519	0	0	5106
3.	1956	5600	4295	0	0	9662
4.	1961	9468	7837	0	0	16937
5.	1966	18158	15225	0	0	32990
6.	1969	27029	20723	0	0	47434
7.	1974	35321	28972	2396	0	66689
8.	1979	52594	47159	2270	0	102523

TABLE 12.3 *(Continued)*

SL. No.	Year	Thermal	Hydro	Nuclear	RES	Total
9.	1980	56273	45478	2876	0	104627
10.	1985	98836	53948	4075	0	156859
11.	1990	178690	62116	4625	6	245438
12.	1992	208708	72757	5525	38	287029
13.	1997	317042	68901	9071	876	395899
14.	2002	422300	73579	19475	2085	517439
15.	2007	528490	113502	18802	9860	670654
16.	2012	780427	130511	32287	51226	922451
17.	2017	993516	122378	37916	81548	1235358
18.	2018	1037146	126123	38346	101839	1303455
19.	2019	1072311	134894	37813	126800	1371817

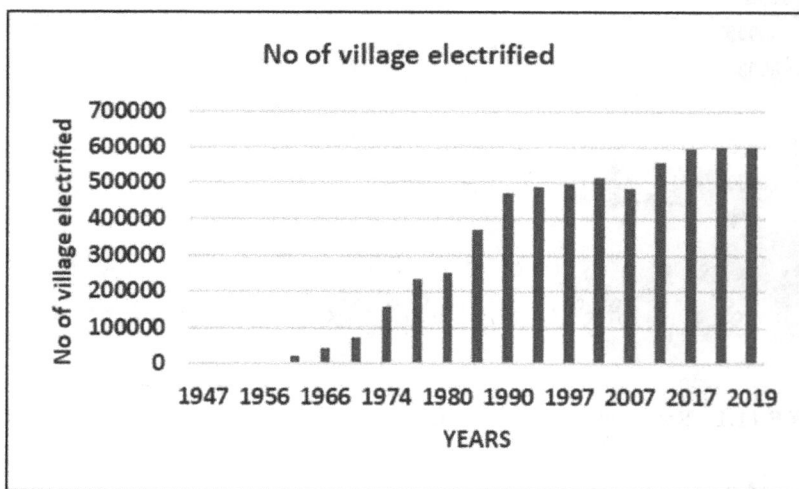

FIGURE 11.2   Number of villages electrified from 1947 to 2019.

When India got independence, no village in India was electrified. Indian villages got electricity after 1956. In 1950 around 3061 villages had electricity. Now around 597464 villages of India are illuminated by our government [13].

Figure 11.3 indicates the irregular variation of transmission and distribution losses. In 1947 transmission and distribution losses in India were 14.88%. In 1950 transmission and distribution losses were 15.2% but next year, it was 13.8%. Losses were highest in the year 2002 which was equal to 33.98% as shown in Figure 11.3 [9, 13].

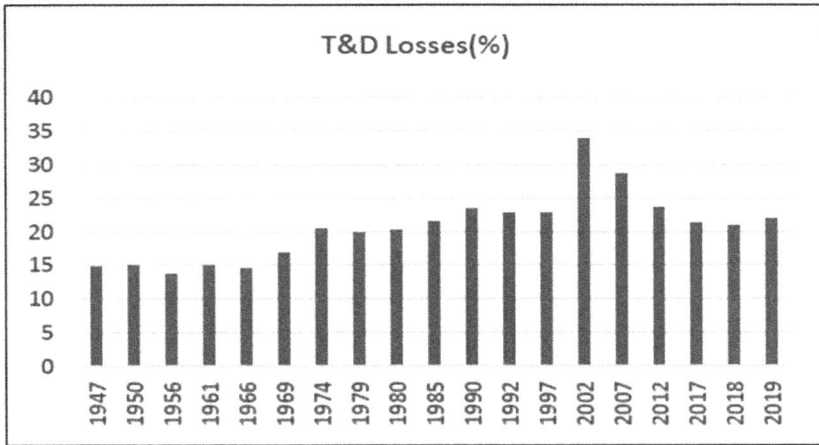

FIGURE 11.3   All India transmission and distribution losses.

11.5   GLOBAL SCENARIO

China has the largest generation of hydroelectricity, followed by Canada. USA has the largest generation through nuclear energy followed by European Union (EU). China has the largest generation through coal-based power plants, followed by the USA. India is 7th on the list in terms of hydroelectricity. India is 12th on the list in terms of nuclear power generation. India ranked 4th in terms of RES, leading by China, the USA, and Germany. India is 3rd on the list in terms of generation through coal-based power plants [13, 14].

Figure 11.5 gives you the comparison of India with the other countries in terms of energy production. China has the largest energy production equivalent to 2534 Mtoe, followed by the USA and Russia. USA and Russia have 2175 Mtoe and 1492 Mtoe, respectively. But India produces 441 million tons of oil equivalent. Figure 11.3 clearly presents you the status of energy production through renewables by comparing it with some top countries. In terms of production through renewable source, Brazil is at the top of the list. The share of the renewables in the energy production in Brazil is 82.46%, followed by Canada, in which the share is 65.86%. In other countries like the USA, China, Japan, Russia, France, and Germany, the share of renewables in energy production is 17.94%, 12.65%, 17.5%, 17.18%, 19.94%, and 36%, respectively. In India, the share of renewables in energy production share is 18.51% (Figures 11.4–11.6) [13, 14].

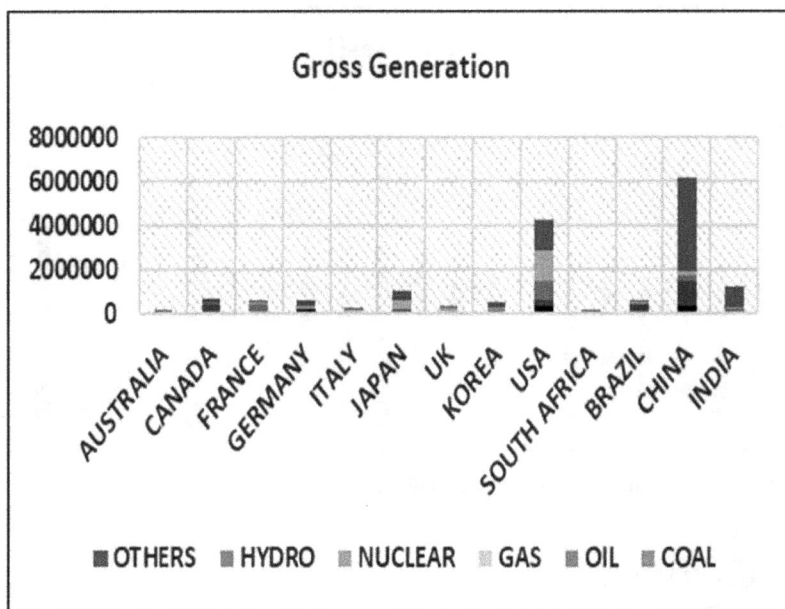

FIGURE 11.4 Gross electricity generation of various countries: Mode wise.

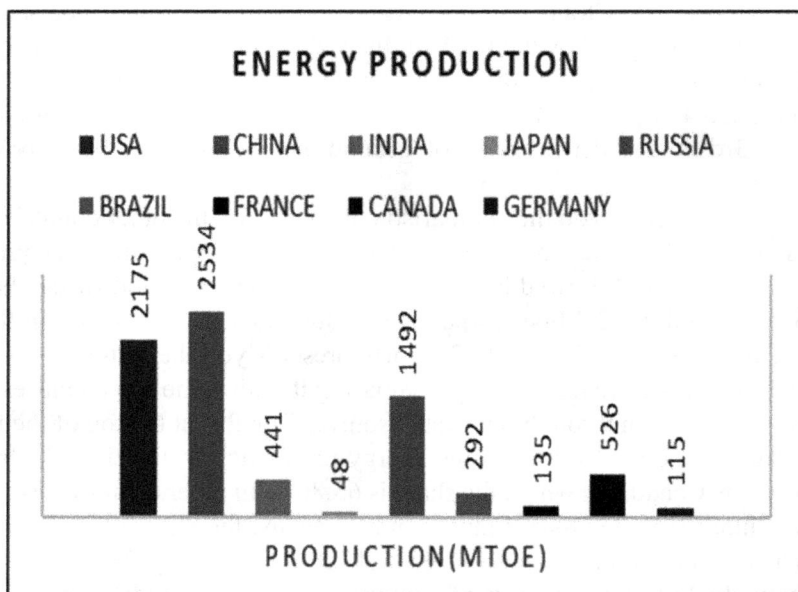

FIGURE 11.5 World-wide energy production.

PRODUCTION (RENEWABLES)%

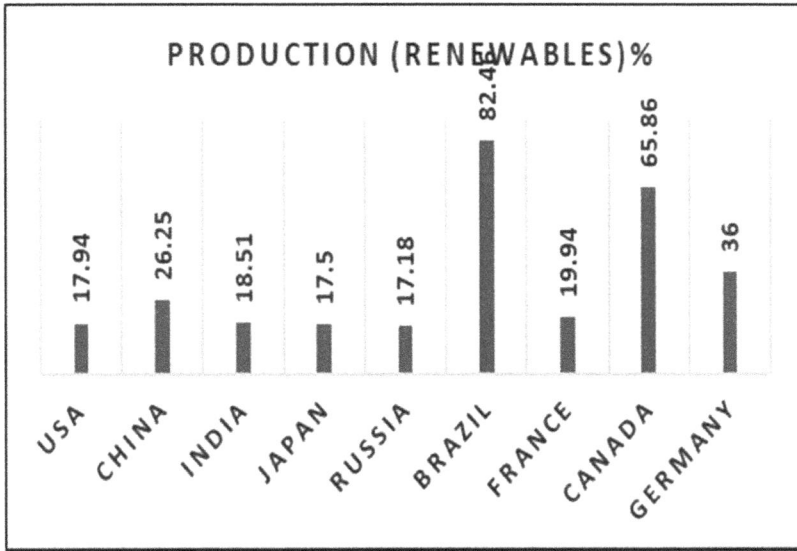

FIGURE 11.6   Share of energy production through renewables.

China has the highest energy consumption, which is equal to 3164 Mtoe, followed by the USA. The USA consumes 2258 Mtoe amount of energy. Russia consumes 800 Mtoe amount of energy. India's energy consumption is 929 Mtoe while other countries have energy consumption less than 300, which is shown in Figure 11.7 [13, 14].

Figure 11.8 displays percentage transmission and distribution losses of some top countries. India has the highest transmission and distribution losses, followed by Brazil and Russia, respectively. India has 21.34% transmission and distribution losses, whereas Brazil and Russia have 16.87% and 12.59%, respectively [13, 14].

11.6   CONCLUSION

Electricity is an imperative component without which no one either living or nonliving can do anything. Power sector should be restructured and deregulated at every stage of life. Every component of the power sector should be well designed and monitored so that our users can get supplies easily at a cheaper rate without hampering the transmission system. I have highlighted the current scenario of generation, transmission, and distribution so that researchers able to know the shortcoming of different sectors and give their

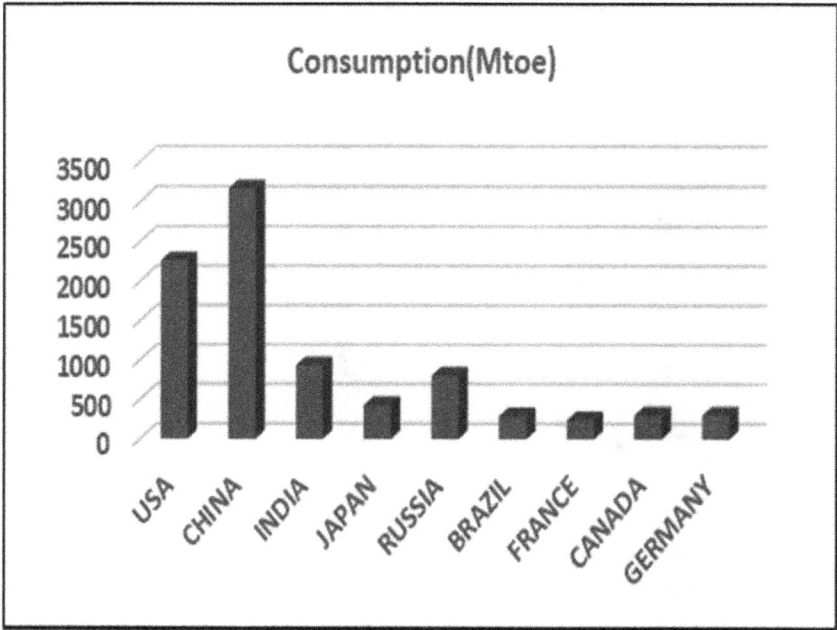

FIGURE 11.7   Worldwide energy consumption.

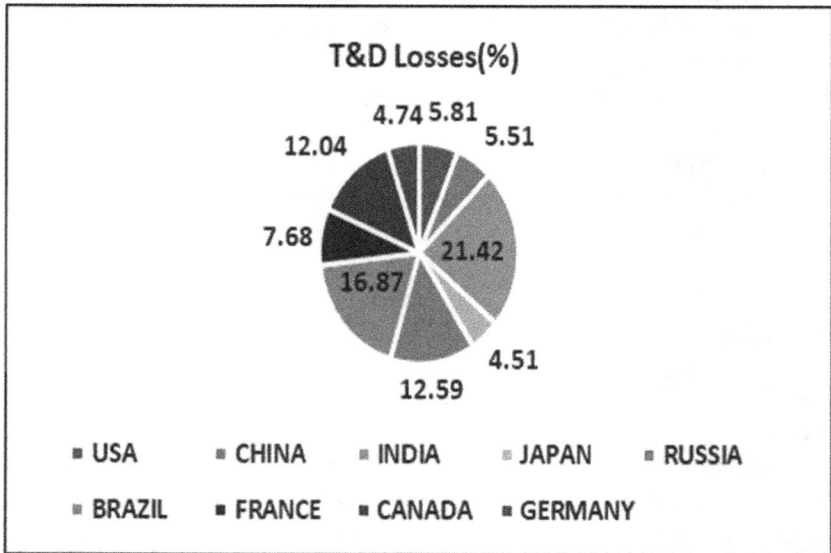

FIGURE 11.8   World-wide transmission and distribution losses.

contribution to resolve the issues that obstruct the smooth functioning of different sectors. I have compared our system with some top countries so that our concerned authority able to know which country has well-designed architecture for a particular sector and try to improve our system as per their architecture. China has the highest hydroelectricity generation. Japan has the lowest transmission and distribution losses. Brazil has the highest production through renewables. China has the highest consumption of electricity. At the time of independence installed capacity of India was 1362 MW which increases to 356100 MW in 2019. The generation of India was 4073 MW which increases to 1371817 MW in 2019. In India, variations of transmission and distribution losses are unexpected. In 1947 it was 15% and in 2019, it is 22%. The government is trying to reduce this level to 15%. Indian village got electricity after 1950 and around 597464 villages are getting electricity.

KEYWORDS

- **economic development**
- **electrical power**
- **electricity regulatory commission act**
- **geographical conditions**
- **installed capacity**
- **liberalization**
- **nationalization stage**
- **per capita consumption**
- **transmission**

REFERENCES

1. Sahoo, S. K., Sharma, K., Nayak, S., Karthikeyan, S. P., & Amalorpavaraj, A. J., (2019). Indian power sector-need of privatization and current status. *International Conference on Computer Communication and Informatics (ICCCI)* (pp. 1–7). Coimbatore, Tamil Nadu, India.
2. Sharma, K., Ramalingam, S. P., Karthikeyan, S. P., & Raglend, I. J., (2019). Indian power sector: Need of privatization and current status. In: *2019 Innovations in Power and Advanced Computing Technologies (i-PACT)* (pp. 1–7). Vellore, India.
3. Soonee, S. K., (2019). *Renewable Energy Integration in India: Present State and Long-Term Perspective* (pp. 1–6). IEEE Milan Power Tech, Milan, Italy.

4. IEEE Xplore. (2016). A history of automotive electric power supply systems. In: *Automotive 48-Volt Technology* (pp. 1–4). SAE Books.

5. Pahwa, A., & Li, Z., (2006). Electricity Industry restructuring in India. In: *38th North American Power Symposium (*pp. 99–105). Carbondale, IL.

6. Rahi, O. P., Thakur, H. K., & Chandel, A. K., (2008). Power sector reforms in India: A case study. *Joint International Conference on Power System Technology and IEEE Power India Conference* (pp. 1–4). New Delhi.

7. Kumar, Y. V., Padhy, N. P., & Gupta, H. O., (2010). *Assessment of Indian Power Sector Reform Through Productivity Analysis: Pre and Post Electricity Act-2003* (pp. 1–8). IEEE PES T&D, New Orleans, LA.

8. *Generation Report.* Available at: https://npp.gov.in/publishedReports/ (accessed on 30 June 2021).

9. *Transmission Report.* Available at: https://npp.gov.in/publishedReports/ (accessed on 30 June 2021).

10. *Installed Capacity Report.* Available at: https://npp.gov.in/publishedReports/ (accessed on 30 June 2021).

11. *Daily Renewable Generation Report.* Available at: http://cea.nic.in/reports/daily/renewable/2020/February/04.pdf (accessed on 30 June 2021).

12. *Power Sector at a Glance.* Available at: https://powermin.nic.in/en/content/power-sector-glance-all-India/ (accessed on 30 June 2021).

13. *Growth of Electricity Sector in India.* Available at: http://www.cea.nic.in/reports/others/planning/pdm/growth_2019.pdf/ (accessed on 30 June 2021).

14. *Global Energy Statistical Year Book*, (2019). Available at: https://yearbook.enerdata.net/electricity/world-electricity-production-statistics.html (accessed on 31 July 2021).

CHAPTER 12

Non-Edible *Cassia tora* Seed Oil as a New Source for Biodiesel Production

VIVEK P. BHANGE,[1] PRAVIN D. PATIL,[2] KIRAN D. BHUYAR,[3] and
MANJU A. SONI[1]

[1]Department of Biotechnology, Priyadarshini College of Engineering,
Nagpur–19, Maharashtra, India,
E-mail: vivekbhange@gmail.com (V. P. Bhange)

[2]Department of Basic Science and Humanities, NMIMS Mukesh Patel
School of Technology Management and Engineering, Mumbai–400056,
Maharashtra, India

[3]Department of Chemical Engineering, Priyadarshini College of
Engineering, Nagpur–19, Maharashtra, India

ABSTRACT

Diminishing supplies of fossil fuel and production of harmful emissions of dangerous constituents during burning processes led the scientific community to work towards renewable energy sources such as biodiesel. Non-edible *Cassia tora* seed oil is considered as second-generation biodiesel. In this work, we have mainly focused on finding a novel, non-edible oil source to produce biodiesel. The operation variables used as follows; methanol concentration (5–20% w/v), alkali catalyst concentration (0.1–0.9% w/v), and temperature (40–60°C). The concentration of methyl ester was determined by Gas Chromatography at different reaction conditions. Bio-diesel yield was measured while assessing several experimental parameters such as the concentration of catalyst, reaction temperature, concentration of methanol, and type of catalyst used. Pseudo-first order kinetics were performed for the conversion of *Cassia tora* seed oil to bio-diesel at varying temperatures. The activation energy and rate constant of reaction were calculated as 29.273 (kJ mol^{-1}) and 65.31 (min^{-1}), respectively.

12.1 INTRODUCTION

Biodiesel is considered as an example of the significant sources of renewable energy. American Society for Testing and Materials (ASTM) has characterized biodiesel as a blend of the long-chain monoalkylic ester from a fatty acid that can be utilized in diesel engines. Biodiesel is a fatty acid methyl ester (FAMEs) that is being generated by the transesterification process. Since it emits a low amount of carbon in the environment, it is a popular alternative to fossil fuel. Apart from possessing several benefits above conventional petroleum fuel such as, safer, and nontoxic, biodiesel also has low CO_2 emission and high fuel efficiency [1, 2]. Diesel fuel generated using non-edible oils has many advantages including easy transportability, renewability, biodegradability, ready availability, higher cetane number, greater heat content, greater flashpoint, and lower sulfur-aromatic content [3]. Biodiesel generation can be an energizing business and monitory advancement in rural area while developing replacement of fossil fuels for long term. Moreover, it can also help to lower the dependency on the crude oil import of developing countries such as India while availing enough supply to fulfill the demand [4]. A prior report demonstrated that the cost of biodiesel is around 0.5 \$ L^{-1} as matched to 0.35 \$ L^{-1} for diesel (petroleum) that makes it an efficient alternative to the conventional fuel [5].

Cassia tora plants are found in several Indian regions, and their seeds can be used as vegetable oil (non-edible) sources. *Cassia tora* is a potential source of triglycerides. It is legume in the subfamily *Caesalpinioideae* [6], widely found as a wild plant and is considered as a weed. Seeds of *Cassia tora* can produce oil that comprises several forms of unsaturated and saturated fatty acids. Several fatty acids were reported, including linoleic acid, palmitic acid, margaric acid, melissic acid, behenic acid, and linolenic acid [7]. *Cassia tora* oil can be employed as a substitute for diesel fuel. Along with this oil, methanol is used as an additive since it possesses higher efficiency. Synthesis of biodiesel using *Cassia tora* seed oil by trans-esterification with methanol in alkali presence, acting as a catalyst (KOH, NaOH), is possible.

The biodiesel generation has been widely studied using several types of oils such as rapeseed oil waste [8], frying oil waste, soap stocks, fats, greases [9], jatropha oil, rice bran oil [10, 11], etc. It was observed that oil contains free form of fatty acids that could not be successfully transformed into biodiesel by utilizing only an alkaline catalyst (NaOH, KOH). Acid catalyst can be directly used to produce biodiesel; however, it may end up having lower yield and longer reaction time. Hence, a two-step method has been employed to produce biodiesel where esterification (acid-catalyzed) pretreatment was needed to lower the free fatty acid (FFA) content followed by an alkali-catalyzed transesterification to complete the reaction [12, 13]. Demirbas examined several transesterification strategies for the synthesis of biodiesel from fats and vegetable oils [14]. Moreover, biodiesel synthesis by means of quick transesterification was also analyzed.

Biodiesel has numerous ecological points of interest where being a biodegradable and sustainable choice to traditional fuels; it offers lower emissions with the exception of nitrogen oxides (NOx), offering a more secure process while achieving complete combustion when compared to non-renewable energy sources [15–17]. In this study, the utilization of *Cassia tora* oil as biodiesel and the impact of several experimental conditions such as concentration and type of catalyst used, the molar ratio of methanol to oil, and temperature on the overall yield FAMEs were assessed.

12.2 MATERIALS AND METHODS

12.2.1 PLANT MATERIAL

Matured seeds of *Cassia tora* were gathered from a farmland area of Maharashtra state, India. The seeds were then oven-dried to remove moisture at 60°C for 8 h and was further crushed by employing a grinding machine. It was further air-dried and the oil was further extracted by employing the percolation method. The adequate amount of oil was extracted by using a solvent extraction process where *n*-hexane was utilized as a solvent. The seed powder to n-hexane ratio was 1 kg per 500 mL (w/v) in the percolator. It was then allowed to dissolve oil in the solvent for 24 hours. The liquid oil was skimmed after heating followed by filtration. The crude oil was separated using solvent extraction process, and it was further evaporated in order to recover the oil. Other impurities including soap stock, were separated from the extracted oil using centrifugation (at 4000 rpm for 30 min) after 24 hours. Residues were conditioned (dried) and reprocessed to produce by-products as animal feed. Finally, the extracted oil was used for the transesterification process in further studies (Figure 12.1).

12.2.2 PRE-TREATMENT AND ACID ESTERIFICATION

To remove the traces of water from highly dense crude oil, it was heated for 1 h at 60 °C followed by filtration to exclude dust particles. Removal of FFA from the oil was conducted in an acid pre-treatment (single-step) method. The reaction was carried out by using a magnetic stirrer in a 500 mL beaker at 500 rpm. Initially, methanol was charged in a beaker (20% w/w) followed by addition of sulfuric acid (0.6% w/w) while providing a constant stirring. Methanol and acid were mixed, followed by the addition of pre-heated oil, and this mixture was further stirred for 1 h at 50 °C. Finally, the reaction mixture was discharged in a separating funnel while allowing it to settle for 24 h. It was noted that the upper layer had water and unreacted methanol and while the bottom had FAMEs and oil [18].

FIGURE 12.1   Image of *Cassia tora* seeds.

12.2.3   *DETERMINATION OF ACID VALUE*

0.1 N KOH is used for titration with the acid esterified oil by applying a phenolphthalein indicator to determine the FFA's overall content. The process of acid transesterification was redone twice to obtain FFA content (up to 0.8%) that could be appropriate for transesterification reactions [19].

The blend comprised of acid treated oil with a lower percent of FFA was then preserved [20] and used for further investigations.

12.2.4   BIODIESEL PRODUCTION THROUGH ALKALI TRANSESTERIFICATION PROCESS

The reaction was conducted in a temperature-controlled condition while providing a constant stirring by using a magnetic stirrer for 180 min. The reaction temperature was monitored frequently, employing a thermometer. The pre-treated 150 g of *Cassia tora* oil was heated to 60°C and methanol with alkaline catalyst (NaOH, KOH) were appended in the amounts established for each experiment. Thereafter, a separating funnel was used where the mixture was allowed to set it down for 24 h. Two separate layers were obtained in the process where the upper layer constituted of methyl ester, a biodiesel product along with the bottom constituted of glycerol by-product.

The layer containing biodiesel was further separated by employing a separating funnel and was washed with an equal amount (1:1 v/v) of mild hot distilled water to separate and exclude the excess soap content and additional impurities. The water wash was extended unless clear biodiesel without any impurities were acquired. The biodiesel was then oven-dried for an hour (100°C) to remove the leftover traces of water. The overall transesterification reaction in catalyst presence can be represented by:

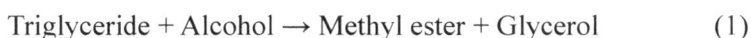

$$\text{Triglyceride} + \text{Alcohol} \rightarrow \text{Methyl ester} + \text{Glycerol} \tag{1}$$

12.2.5   UNIVARIATE OPTIMIZATION

One factor at a time approach was applied for primary optimization of varied operating conditions for the production of biodiesel. Two types of catalyst (NaOH and KOH), catalyst concentration (0.1–0.9% w/w), methanol amount (5–20% w/w) and temperature (40–60°C) were optimized to achieve the best yield of biodiesel.

12.2.6   GAS CHROMATOGRAPHY ANALYSIS

A gas chromatograph with a mass spectrometer (GC-MS) instrument (Bruker Scion, TQ MS System, California, USA) was used for the analysis of biodiesel samples. A Bruker Scion of TQ MS System was employed as a GC-MS system. A DB-Wax capillary column (30 m, 0.25 μm film thickness) of Agilent Technologies, USA was practiced for the analysis. Electron ionization framework (70 eV) was practiced for GC-MS recognition. A carrier gas (Helium, 99.99%) was employed at a constant flow rate (1.0 mL/min) with a split proportion of 10:1. The oven temperature was maintained as follows: 40°C held for 1 min, while raising to 150°C at the rate of 20°C min⁻¹ followed by inflating at the rate of 3°C min⁻¹, while holding for 0 min and ultimately increasing up to 300°C at the rate of 20°C min⁻¹, with 15 min hold time, while maintaining injector temperature and volume at 250°C and 3 μL, respectively. The total run time of GC was 55 min. The total GC running time was about 55 min. The MS was operated at an ionization voltage of 70 eV while keeping source temperature at 250°C with a total MS run-time of 51 min. The mass spectra of compounds were identified by examining the mass-spectra collected from their associated chromatographic peaks with NIST mass spectral libraries.

12.2.7 CHARACTERIZATION OF BIODIESEL

Several biodiesel characteristics including flash pour, cloud point along with viscosity, calorific value, iodine value, and ester content were identified with respect to the standard methods and matched with the traditional standards (Table 12.1).

TABLE 12.1   Biodiesel Properties

SL. No.	Parameters	Observation
1.	Density of oil at 15°C	0.92 g/cm^3
2.	Flashpoint	173°C
3.	Sulfated ash	0.01%
4.	Pour point	1°C
5.	Carbon residue	0.02%
6.	FFA	9%

12.2.8 KINETIC MODELING

The biodiesel is generated from a reaction among triglycerides existing in the vegetable oils and the alcohols (short-chain) with a catalyst. The stoichiometry equation of transesterification can be designed as follows:

$$CO + 3M \rightarrow G + 3ME \qquad (2)$$

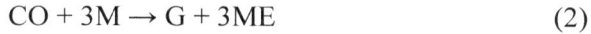

where; CO is the *Cassiatora* seed oil; M is methanol; G is glycerol; and ME is methyl esters.

The kinetic expression was assumed from an earlier study [19] and rate equation of transesterification reaction can be described as:

$$r = -\frac{d[CO]}{dt} = k1\,[CO]\,[M]^3 - k2[G][ME]^3 \qquad (3)$$

The formation of ME from TG can be assumed to be following first-order reaction while the equation of mass balance for *Cassia tora* seed oil [CO] can be written as:

$$r = -\frac{d[CO]}{dt} = K[CO] \qquad (4)$$

$$\frac{d[CO]}{[CO]} = -K\,dt \qquad (5)$$

The conversion CO to ME can be written as:

$$[CO] = [CO_0](1 - X) \qquad (6)$$

where; $[CO_0]$ is the initial concentration of CO and X is the conversion by substituting initial concentration in above governing equation we get as follow:

$$\frac{d[CO_0]\,[1 - X]}{[CO_0]\,[1 - X]} = -K\,dt \qquad (7)$$

$$\int^{X^d} {}_{0}\frac{[CO_0][1 - X]}{[CO_0][1 - X]} \, K\int_0^t dt \qquad (8)$$

$$\ln\,[CO_0][1 - X]_0^X = -K\int_0^t dt \qquad (9)$$

$$In\;f()\,[CO_0][1 - X] - In\;f()\,[CO_0] = -Kt \qquad (10)$$

$$\frac{In\;f()\,[CO_0]\,[1 - X]}{In\;f()\,[CO_0]} \qquad (11)$$

$$\ln[1{-}X] = -Kt \qquad (12)$$

$$K = -\frac{\ln\,[1 - X]}{t} \qquad (13)$$

The slope of the graph plotted between $-\ln(1{-}X)$ vs. t. can be used to determine the rate constant of the reaction.

12.3 RESULTS AND DISCUSSION

12.3.1 KINETIC STUDY

The kinetic studies were carried in a range (40, 50, and 60°C) of temperature to define the rate constant of the reaction. Figure 12.2 depicts the plot of –ln (1-X) vs. time for a range of temperature, implies that experimental data obtained can be fitted with the pseudo-first-order kinetics. The K (rate constant) was defined for a range of temperature provided in Table 12.2. Since the reaction temperature directs the rate constant, the activation energy needed for the Trans-esterification process can be defined by using the Arrhenius equation (Table 12.3):

$$K = Ae^{-Ea/RT} \tag{14}$$

Considering ln on both sides to the linear transformation of Eqn. (14) as follows:

$$lnK = lnA - (E_a/R_T) \tag{15}$$

where; k is the reaction rate constant; A is the Arrhenius constant or collision factor; E_a is the activation energy (J/mole); R is the gas constant (8.314 J/mole k); T is the absolute temperature (K).

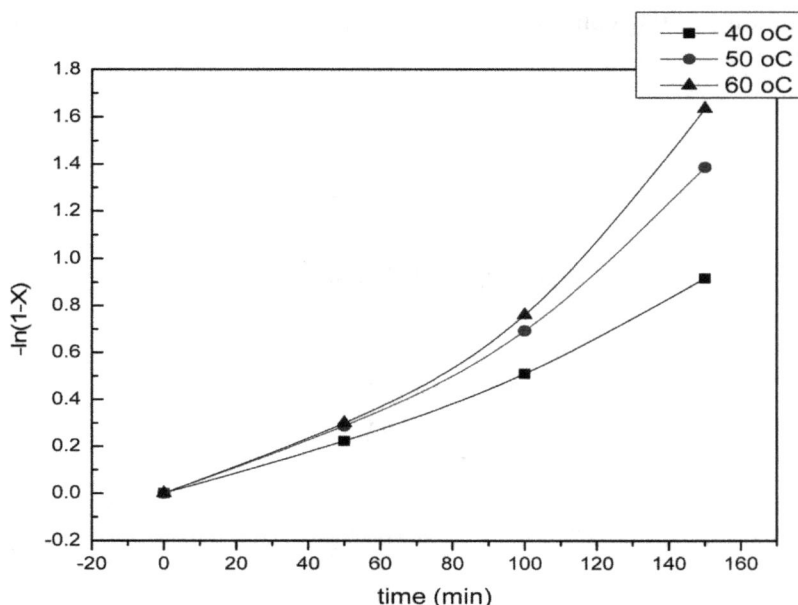

FIGURE 12.2   The kinetic studies of conversion of Cassia tora seed oil to bio-diesel at varying temperatures (40°C, 50°C, and 60°C). A plot of –ln (1–X) vs. time for different temperatures.

TABLE 12.2   Estimated Reaction Rate Constant

SL. No.	Temperature (°C)	K (min⁻¹)
1.	40	0.00574
2.	50	0.00812
3.	60	0.00922

TABLE 12.2   Estimated Reaction Rate Constant

SL. No.	Temperature (°C)	K (min^{-1})
1.	40	0.00574
2.	50	0.00812
3.	60	0.00922

TABLE 12.3   Calculated Values from Arrhenius Plot

SL. No.	lnK	1/T (K⁻)
1.	−5.16	0.003194
2.	−4.8134	0.003095
3.	−4.686	0.003

The E_a and A were defined from the slope and intercept of the graph plotted between lnK vs 1/T as depicted in Figure 12.3. In this investigation, E_a, and A values for transesterification of *Cassia tora* seed oil were defined as 29.273 kJ/mole and 65.31 (min^{-1}).

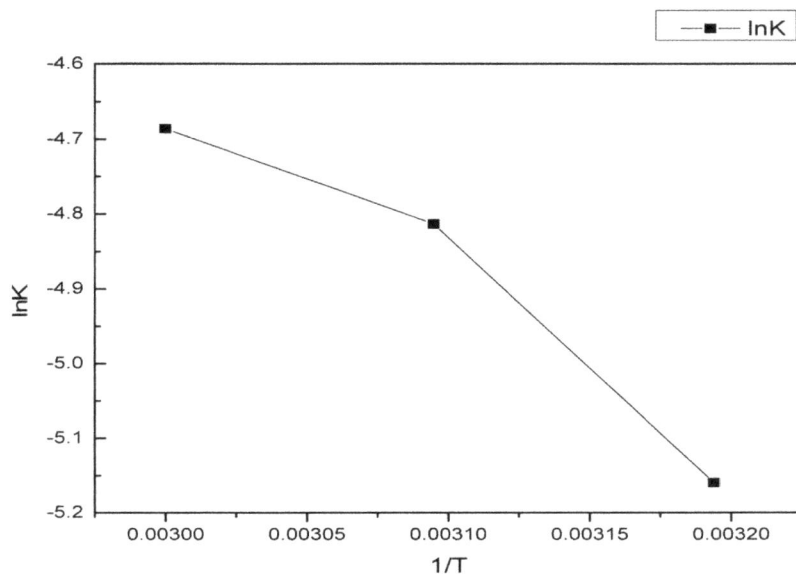

FIGURE 12.3   Arrhenius plot to determine the activation energy. Plot the graph between lnK vs. 1/T.

12.3.2   EFFECT OF CATALYST TYPE

The effect of various types of catalyst such as Na_2CO_3, KOH, NaOH on the yield of biodiesel was studied in different batches, the catalyst concentration was maintained at 0.9% by weight of oil, amount of methanol 20 wt.% of oil and temperature 60°C for 180 minutes. The results obtained are showed in Figure 12.4 with maximum conversion of *Cassia tora* seed oil was for NaOH catalyst 80.5%, KOH 15% and Na_2CO_3 3%. The result confirmed that the maximum yield of biodiesel produced by using NaOH as a catalyst.

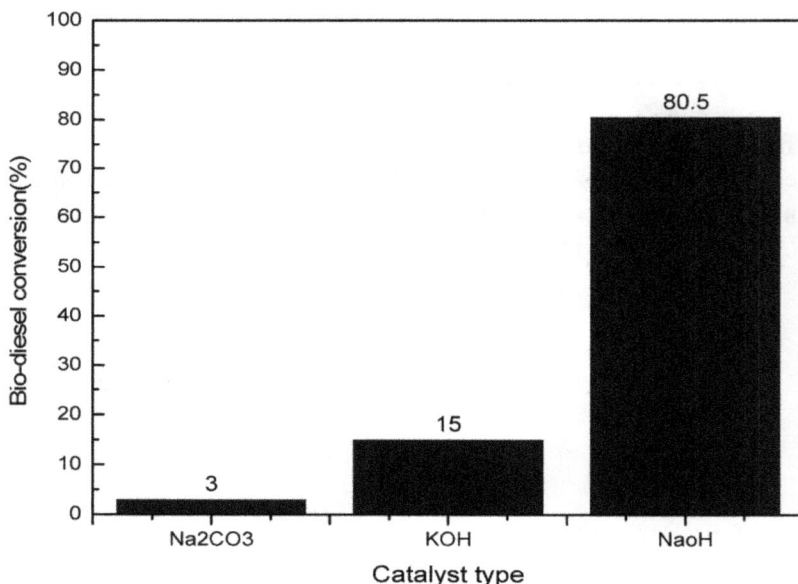

FIGURE 12.4   Bio-diesel conversion with different type of catalyst.

12.3.3   EFFECT OF METHANOL CONCENTRATION

The quantity of methanol applied for the transesterification process was studied with yield of biodiesel produced and it greatly affects the cost of production. The effect of methanol amount from 5–20 wt.% of oil on bio-diesel yield was examined. The reactions were conducted with NaOH as a catalyst, concentration of catalyst 0.9% by weight of oil and temperature 60°C for 180 min. Figure 12.5 shows that the highest yield of biodiesel 80.5% was received for 20% w/w of oil followed by 15, 10, and 5% w/w. This is possibly due to the fact that the higher amount of methanol favors the high yield of biodiesel due to increased catalyst solubility [21].

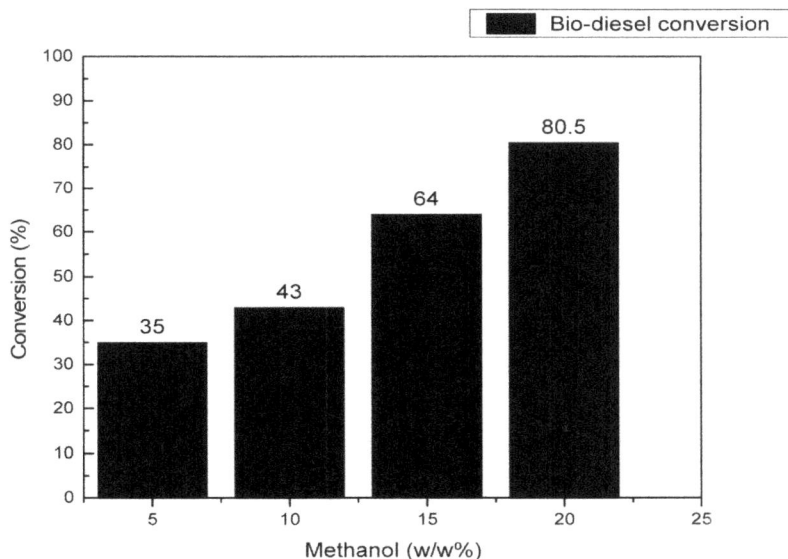

FIGURE 12.5   Effect of methanol amount on bio-diesel yield.

12.3.4   EFFECT OF CATALYST CONCENTRATION

The set of experiments were conducted with different concentrations of catalyst to monitor the effect on yield of biodiesel. The concentration varied from 0.1 to 0.9% (w/w) of oil and temperature was maintained at 60°C for 180 min. The maximum conversion of Cassia tora seed oil was obtained 80.6% at 0.9% (w/w) of NaOH followed by 63.2% for 0.7% (w/w), 42.6% for 0.5% (w/w), 31.2% for 0.3% (w/w) and 20.3% for 0.1% (w/w). Figure 12.6 shows that biodiesel yield could be enhanced with enhancement in concentration of catalyst up to 0.9% (w/w). The catalyst concentration higher than 0.9% (w/w) could lead to backward reaction and drop in the yield of biodiesel was observed. Similar results were reported in an earlier report [21].

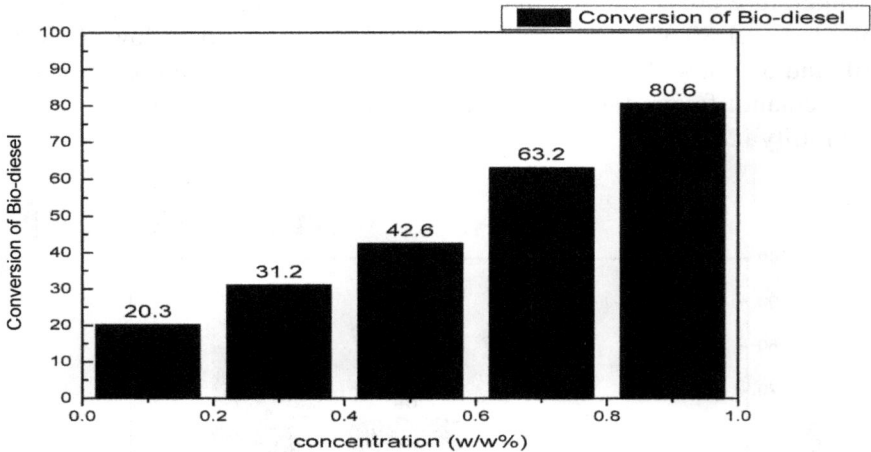

FIGURE 12.6   Effect of catalyst concentration on bio-diesel yield.

12.3.5 EFFECT OF TEMPERATURE

The temperature is an influential parameter for yield of biodiesel. Any rise in temperature enhances the rate of reaction rate since it elevates the collision between the reacting molecules that ultimately quicken the chemical reaction. Solubility of *Cassia tora* seed oil in methanol increases with increase in temperature. Due to solubility, contact between molecules increases, which tends to enhance in rate of reaction [18]. The effect of temperature on biodiesel yield was examined in the range of 40–60°C and methanol concentration was maintained at 20% (w/w) of oil, catalyst concentration at 0.9% (w/w). The obtained observations are depicted in Figure 12.7, where maximum biodiesel yield of 80.5% was obtained at 60 °C followed by 73% at 50 °C and 64% at 40 °C.

12.4 CONCLUSION

In the present work, the experimental conditions were optimized while employing conventional transesterification reaction of *Cassia tora* seed oil. Experiments were carried out at a varying range of experimental conditions such as temperature, catalyst concentration, methanol concentration and catalyst type to assess the final yield of biodiesel. The maximum yield of biodiesel was obtained 80.5% at 60 °C for 20% (w/w) of methanol and 0.9%

(w/w) of catalyst. The kinetic model of transesterification reaction was well studied with pseudo-first-order. Moreover, the activation energy of transesterification of *Cassia tora* seed oil was 29.273 (kJ mol^{-1}) with frequency factor of 65.31 (min^{-1}). Therefore, the biodiesel production from *Cassia tora* seed oil can be scaled up at industrial scale due to lower extraction process cost. Apart from being a cost-effective and economically efficient alternative to conventional fuels, *Cassia tora* biodiesel could also be proved as an excellent option to the petroleum fuel as it possesses related properties.

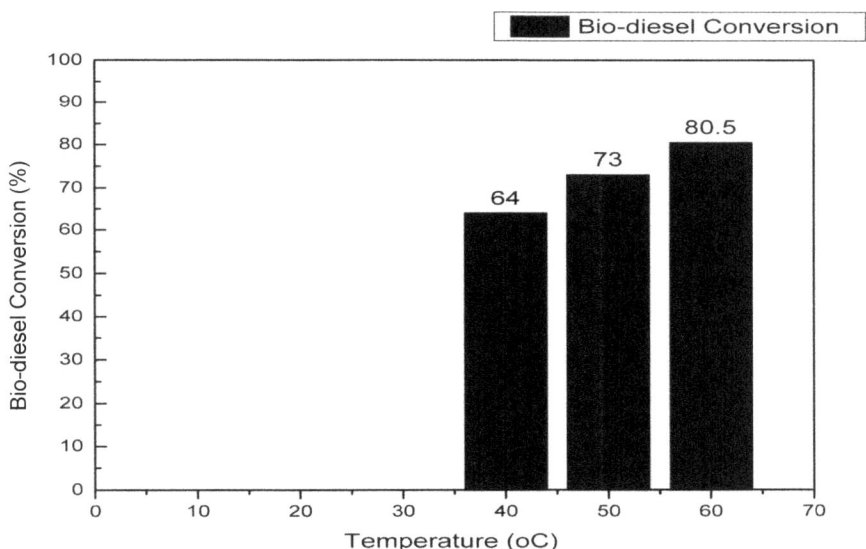

FIGURE 12.7 Effect of temperature on biodiesel yield.

KEYWORDS

- **biodiesel**
- **Cassia tora**
- **fatty acid methyl ester**
- **free fatty acid**
- **non-edible oils plants**
- **sustainability**
- **transesterification**

REFERENCES

1. Dincer, K., (2008). Lower emissions from biodiesel combustion. *Energy Sources, Part A Recover Util. Environ. Eff., 30*, 963–968.
2. Demirbas, A., & Dincer, K., (2008). Sustainable green diesel: A futuristic view. *Energy Sources, Part A Recover Util. Environ. Eff., 30*, 1233–1241.
3. No, S. Y., (2011). Inedible vegetable oils and their derivatives for alternative diesel fuels in CI engines: A review. *Renew. Sustain. Energy Rev., 15*, 131–149.
4. Moser, B. R., (2011). Biodiesel production, properties, and feedstocks. In: *Biofuels: Global Impact on Renewable Energy, Production Agriculture, and Technological Advancements* (Vol. 45, pp. 229–266).
5. Zhang, Y., Dubé, M. A., McLean, D. D., & Kates, M., (2003). Biodiesel production from waste cooking oil: 1. Process design and technological assessment. *Bioresour. Technol., 89*, 1–16.
6. Bruneau, A., Forest, F., & Herendeen, P. S., (2001). Phylogenetic relationships in the *Caesalpinioideae* (Leguminosae) as inferred from chloroplast trnL intron sequences. *Syst Bot., 26*(3), 487–514.
7. Shukla, S., Hegde, S., Kumar, A., Gaurav, C., Dalip, K. U., Mahesh, P., (2018). Fatty acid composition and antibacterial potential of *Cassia tora* (leaves and stem) collected from different geographic areas of India. *J. Food Drug Anal., 26*, 107–111.
8. Yuan, X., Liu, J., Zeng, G., Shi, J., Tong, J., & Huang, G. H., (2008). Optimization of conversion of waste rapeseed oil with high FFA to biodiesel using response surface methodology. *Renew Energy., 33*, 1678–1684.
9. Canakci, M., & Sanli, H., (2008). Biodiesel production from various feedstocks and their effects on the fuel properties. *J. Ind. Microbiol. Biotechnol., 35*, 431–441.
10. Krishnakumar, J., Karuppannan, V. S., & Elancheliyan, S., (2008). Technical aspects of biodiesel production from vegetable oils. *Thermal Sci., 12*, 159–169.
11. Sinha, S., Agarwal, A. K., & Garg, S., (2008). Biodiesel development from rice bran oil: Transesterification process optimization and fuel characterization. *Energy Convers Manag., 49*(5), 1248–1257.
12. Usta, N., Ozturk, E., Can, O., Conkur, E. S., Nas, S., & Con, A. H., (2005). Combustion of biodiesel fuel produced from hazelnut soap stock/waste sunflower oil mixture in a diesel engine. *Energy Convers Manag., 46*, 741–755.
13. Çetinkaya, M., Ulusoy, Y., Tekìn, Y., & Karaosmanoğlu, F., (2005). Engine and winter road test performances of used cooking oil originated biodiesel. *Energy Convers Manag., 46*, 1279–1291.
14. Demirbas, A., (2008). Comparison of transesterification methods for production of biodiesel from vegetable oils and fats. *Energy Convers Manag., 49*, 125–130.
15. Vasudevan, P. T., & Briggs, M., (2008). Biodiesel production - current state of the art and challenges. *J. Ind. Microbiol. Biotechnol., 35*, 421–430.
16. Demirbas, A., (2008). Biodiesel production via rapid transesterification. *Energy Sources, Part A Recover Util Environ Eff., 30*, 1830–1834.
17. Singh, S. P., & Singh, D., (2010). Biodiesel production through the use of different sources and characterization of oils and their esters as the substitute of diesel: A review. *Renew. Sustain. Energy Rev., 14*, 200–216.

18. Helwani, Z., Othman, M. R., Aziz, N., Kim, J., & Fernando, W. J. N., (2009). Solid heterogeneous catalysts for transesterification of triglycerides with methanol: A review. *Appl. Catal. A Gen., 363*, 1–10.

19. Muthukumaran, C., Praniesh, R., Navamani, P., Raghavan, S., Govindasamy, S., & Narasimhan, M. K., (2017). Process optimization and kinetic modeling of biodiesel production using non-edible *Madhuca indica* oil. *Fuel, 195*, 217–225.

20. Naik, M., Meher, L. C., Naik, S. N., & Das, L. M., (2008). Production of biodiesel from high free fatty acid Karanja (Pongamia pinnata) oil. *Biomass and Bioenergy., 32*, 354–357.

21. Hossain, A. B. M. S., & Mazen, M. A., (2010). Effects of catalyst types and concentrations on biodiesel production from waste soybean oil biomass as renewable energy and environmental recycling process. *Aust. J. Crop Sci., 4*(7), 550–555.

CHAPTER 13

Characterization of CdTe Film under Different Annealing Temperature for Solar Cell Application

RONEN P. DUTTA and NIMISHA NEOG

Department of Electronics, Sibsagar College Joysagar, Assam, India,
E-mail: nimi.neog2004@gmail.com (N. Neog)

ABSTRACT

CdTe thin film is prepared on a clean glass substrate by thermal evaporation (TE) technique at 10^{-5} mbar, then annealed for 1 h at 200°C, 250°C and 300°C. Characterization of the film such as optical characterization, structural characterization, and surface morphology is studied for different annealing temperatures using UV-Vis spectrophotometer, XRD, and SEM, respectively. As annealing temperature increases optical band gap decreases 1.57 eV to 1.47 eV and grain size increases from 18.83 nm to 33.92 nm. From optical spectra, it is observed that higher annealed film absorbs more compared to other two may be due to crystallinity improve at higher annealing temperature. Since the films absorb most of the incident light in visible range hence suitable material for solar cell application.

13.1 INTRODUCTION

Thin-film CdTe has great importance for solar cell application because of its ideal bandgap 1.45 eV and high absorption coefficient [1, 2]. It is one of the most promising candidates in the solar cell industry. Different technique has been adopted for the fabrication of the film such as thermal evaporation (TE) [3], chemical bath deposition (CBD) [4], closed space sublimation (CSS) [5], electron beam evaporation (EBE) [6], etc. But out of all TE method is cost-effective and set up is simple to handle. In this present work, CdTe thin film is prepared using TE technique and characterization of the film such as optical, surface morphology and crystal structure were studied under different annealing temperature.

13.2   MATERIAL AND METHOD

Very high purity (99.99% pure) CdTe thin films were deposited by TE in Vico Vacuum Coating Unit operating at about 10^{-5} mbar. Substrate temperature was kept at about 150°C during deposition of CdTe film. The films were deposited onto ultrasonically cleaned glass slide. As post-deposition processed individual CdTe films were annealed in vacuum at 200°C, 250°C and 300°C for an hour named as X1, X2, X3, respectively.

13.3   RESULTS AND DISCUSSION

13.3.1   OPTICAL CHARACTERIZATION

The absorption spectra (Figure 13.1), of all the CdTe films (X1, X2, and X3) shows good absorption in the visible range. Compared to other two higher annealed film X3 shows good absorption for longer wavelength region, this may be due to improvement in crystallinity at higher temperature. Again, from transmission spectra (Figure 13.2), transmission is higher above 800 nm and low in visible range that means it is a good absorber layer in visible range. Figure 13.2 also shows that a shift of transmittance peak towards the longer wavelength region as annealing temperature increases. Shifting of the peak occurs due to structural enhancement by crystalline grain size [7], which is proved by XRD results.

Figure 13.3 depicts the variation of bandgap with respect to annealing temperature. Bandgap decreases with increase of annealing temperature, that is maybe due to improvement in grain size with annealing. As annealing temperature increases grain size increases (from XRD results), which improves the crystallinity of the film and enhance the optical property [8, 9].

13.3.2   STRUCTURAL CHARACTERIZATION

The average size of CdTe grains has been obtained from X-ray diffraction pattern use Scherer's formula [10, 11].

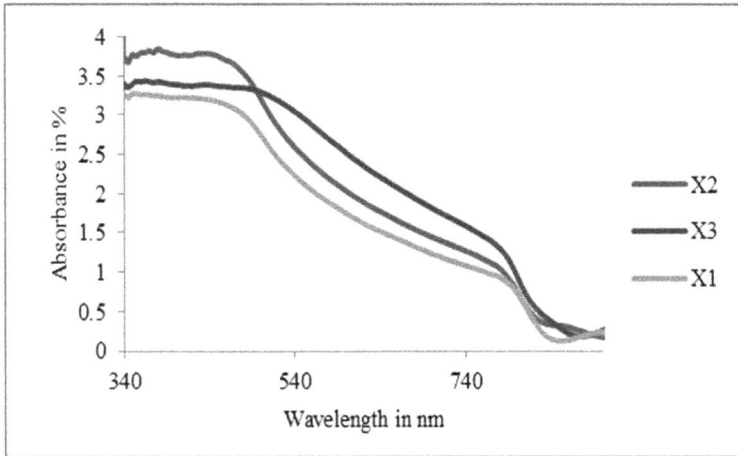

FIGURE 13.1   Absorption spectra of CdTe thin films for different annealing temperatures.

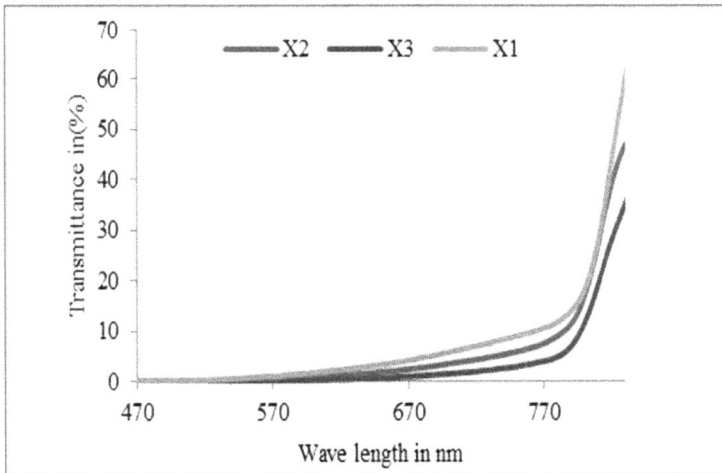

FIGURE 13.2   Transmission spectra of CdTe.

$$D = \frac{K\lambda}{\beta \cos\theta} \qquad (1)$$

The grain size of the films were determined using Scherer's Eqn. (1) from the XRD results of CdTe. As we can see from Figure 13.4, films with annealing temperature 250°C, i.e., X2 has higher intensity compared with the films annealed at 200°C (X1). Again, further annealed the film to a higher temperature leads to a decrease of the intensity of the film, this change may be due to a change in stoichiometry at higher annealing temperature [12].

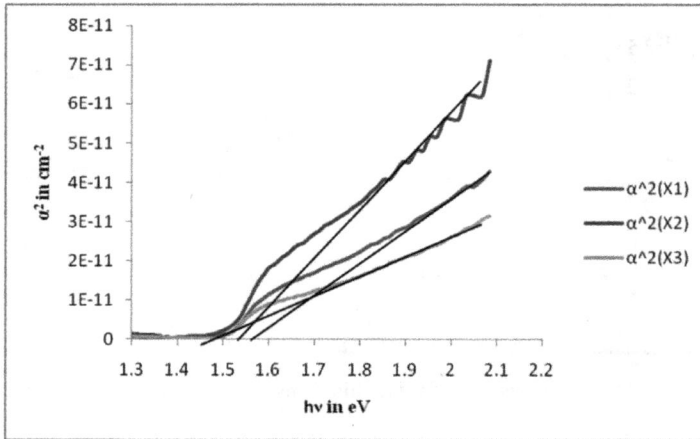

FIGURE 13.3   Plot of $(\alpha)2$ vs. hv plots of CdTe.

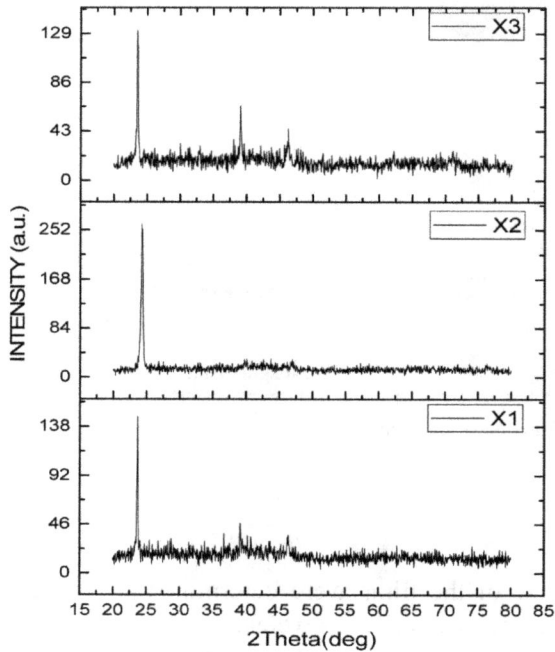

FIGURE 13.4   XRD result of CdTe.

The effect of grain size and bandgap to annealing temperature comparison is depicted in Figure 13.5 and Table 13.1. As annealing temperature increases grain size increases and band gap decreases.

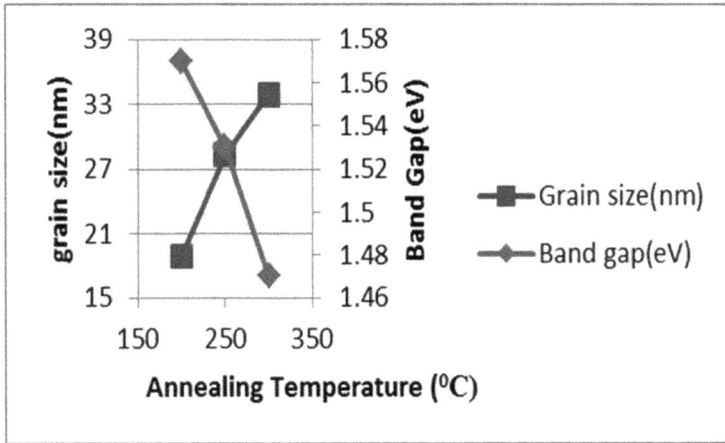

FIGURE 13.5   Bandgap and grain size variation with annealing temperature.

TABLE 13.1   Band Gap and Grain Size of CdTe Thin Film at Different Annealing Temperature

Sample Name	Annealing Temperature (°C)	Band Gap (eV)	Grain Size from XRD (nm)
X1	200	1.57	18.85
X2	250	1.53	28.30
X3	300	1.47	33.92

13.3.3   SURFACE MORPHOLOGY

Surface morphology is studied through SEM. SEM of higher annealed film X3 (Figure 13.6) reveals that the deposition of the film is smooth, uniform, and free from defect.

13.4   CONCLUSION

All the obtained results confirm that annealing temperature plays a significant role in the enhancement of different characterization properties of the films. Film annealed at higher temperature (X3) can be used in CdTe/CdS solar cell as an absorber layer.

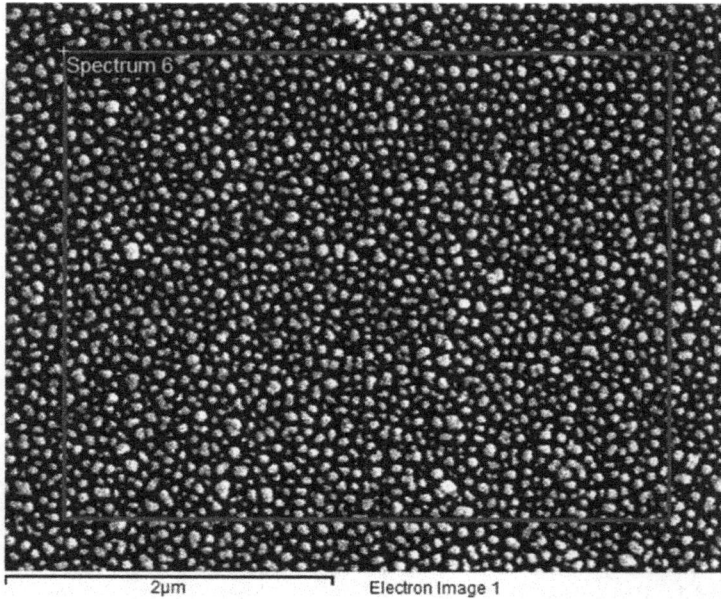

FIGURE 13.6   SEM image of CdTe annealed at higher temperature.

KEYWORDS

- **chemical bath deposition**
- **closed space sublimation**
- **electron beam evaporation**
- **optical band gap**
- **solar cell industry**
- **thermal evaporation annealing**

REFERENCES

1. Kastlander, J., Bargholtz, C., Batres-Estrada, G., & Behrooz, K., (2006). Development of methods to use CdTe detectors in field measurements. *Journal of Physics: Conference Series,*

[Online], *41*(1), 523–526. https://iopscience.iop.org/article/10.1088/1742-6596/41/1/060/pdf (accessed on 30 June 2021).

2. Lovergine, N., Prete, P., Tapfer, L., Mazro, F., & Mancini, A. M., (2005). Hydrogen transport vapor growth and properties of thick CdTe epilayers for RT X-ray detector applications. *Cryst. Res. Tech., [Online], 40*(10, 11), 1018–1022. https://onlinelibrary.wiley.com/doi/epdf/10.1002/crat.200410478 (accessed on 30 June 2021).

3. Singh, S., Kumar, R., & Sood, K. N., (2010). Structural and electrical studies of thermally evaporated nanostructured CdTe thin films. *Thin Solid Films, [Online], 519*(3), 1078–1081. https://doi.org/10.1016/j.tsf.2010.08.047.

4. Deivanayaki, S., Jayamurugan, R., Mariappan, R., & Ponnuswamy, V., (2010). Optical and structural characterization of CdTe thin films by chemical bath deposition technique. *Chalcogenide Letters [Online], 7*(3), 159–163. http://chalcogen.ro/159_Deivanayaki.pdf (accessed on 30 June 2021).

5. Ferekides, C. S., Marinskiv, D., Viswanathan, V., Tetali, B., Palekis, V., Selvaraj, P., & Morel, D. L., (2000). High efficiency CSS CdTe solar cells. *Thin Solid Films [Online], 361*, 520–526. https://doi.org/10.1016/S0040-6090(99)00824-X.

6. Begam, M. R., Rao, N. M., Kaleermulla, S., Shobana, M., Krishna, N. S., & Kuppan, M., (2013). Effect of substrate temperature on structural and optical properties of nanocrystalline CdTe thin films deposited by electron beam evaporation. *Nano and Electronics Phy. [Online], 5*(3), 03019. https://www.researchgate.net/publication/257410538_Effect_of_Substrate_Temperature_on_Structural_and_Optical_Properties_of_Nanocrystalline_CdTe_Thin_Films_Deposited_by_Electron_Beam_Evaporation (accessed on 30 June 2021).

7. Hussain, K. M. A., Faruqe, T., Parvin, J., Ahmed, S., Mahmood, Z. H., & Syed, I. M., (2015). Preparation of CdTe nuclear detector material in thin film form using thermal evaporation method. *Malay. J. Med. Biol. Res. [Online], 2*(3), 195–199. https://jmbr-my.weebly.com/uploads/1/3/4/5/13455174/6.4.pdf (accessed on 30 June 2021).

8. Geethalakshmi, D., & Muthukumarasamy, N., (2012). Effect of annealing on structural and optical properties of CdTe thin films. *J. Nanoscience and Nano Technology. [Online], 1*(1), 23–29. https://www.researchgate.net/publication/233978887_Effect_Of_Annealing_On_Structural_And_Optical_Properties_Of_CdTe_Thin_Films (accessed on 30 June 2021).

9. Mandal, M., Choudhury, S., Das, C., & Begum, T., (2014). Substrate temperature-dependent optical and structural properties of vacuum evaporated CdTe thin films. *European Sc. Journal. [Online], 10*(3), 442–455. https://www.researchgate.net/publication/263315620_Substrate_temperature_dependent_optical_and_structural_properties_of_vacuum_evaporated_CdTe_thin_films (accessed on 30 June 2021).

10. Caglar, M., Caglar, Y., & Ilican, S., (2006). The determination of the thickness and optical constants of ZnO crystalline thin film by using the envelope method. *J. Optoelectron Adv. Mater. [Online], 8*(4), 1410–1413. https://old.joam.inoe.ro/arhiva/pdf8_4/4Caglar.pdf (accessed on 30 June 2021).

11. Prabhar, S., & Dhanam, M., (2005). CdS thin films from two different chemical baths - structural and optical analysis. *J. Cryst. Growth. [Online], 285*(1, 2), 41–48. https://www.sciencedirect.com/science/article/pii/S0022024805009401 (accessed on 30 June 2021).

12. Bacaksiz, E. M. I. N., Basol, B. M., Altunbas, M., Novruzov, V., Yanmaz, E. K. R. E. M., & Nezir, S., (2007). Effects of substrate temperature and post–deposition anneal on properties of evaporated Cadmium Telluride films. *Thin Solid Films.[Online], 515*(5), 3079–3084. https://www.sciencedirect.com/science/article/abs/pii/S0040609006010212 (accessed on 30 June 2021).

CHAPTER 14

Solar Energy Conversion: The Need of an Era

A. M. BHAKE,[1] S. M. SAWDE,[1] MAITHILI PAIKANE,[2] and S. V. MOHARIL[3]

[1]Applied Physics Department, Priyadarshini Institute of Engineering and Technology, Nagpur, Maharashtra, India,
E-mail: aparna.dhond@gmail.com (A. M. Bhake)

[2]Humanities and Social Science Department, Visvesvaraya National Institute of Technology, Nagpur, Maharashtra, India

[3]Department of Physics, R.T.M. Nagpur University, Nagpur, Maharashtra, India

ABSTRACT

Solar energy is an inexhaustible energy source that is targeted by specially undeveloped and developing nations where solar energy conversion is the only task. Different technologies for solar energy conversion are developed, such as photovoltaic (PV) technologies. Also, solar cells of c-Si type are effectively commercialized and so very profitable because of established Si technology, but it is not affordable. There are three basic strategies that are to be considered for cost reduction. Some losses are found in the solar-cell-based energy exchange process, as a solution of it, special luminescent materials have been made and are used to curtail it. For changing the near UV-visible radiations into the solar spectrum regions, phosphors are quite beneficial. Over and above, mobile thin-film-based solar cells technologies such as copper indium gallium selenide solar cells, cadmium telluride solar cell, amorphous silicon solar cell: H, a perovskite (PSC); a unique technology is materialized. Predominantly, its great efficiency and its high bandgap makes PSC smarter. For improving the performance, i.e., efficiency, stability, manufacturability of PSC, environmental aspects number of laboratories are taking the great lead in exploring through an extensive range of alternatives.

This chapter pinnacles and compares the latest high-tech developments in the sphere of solar energy conversion technology.

14.1 INTRODUCTION

"We are star stuff harvesting sunlight" [1].

Energy is manifested in all the creatures existing in the universe. Every human activity is governed by the amount of energy it possesses and utilized in due course of time. Since times immemorial, Homo sapiens have not only explored the different sources of energy but, due to their impeccable mind and wisdom, have been able to convert them into different utility modes. This exploration exhausted as time passed by and the quest for a new and infinite source of energy became a new pursuit for the human race.

In this quest, man discovered both renewable and non-renewable energy sources. The permanence of renewable energy sources makes it an ever-lasting medium of fulfilling the needs of human beings in different ways. Energy received from the sun needs to be properly channelized through various technologies. Solar energy can truly prove to be beneficial due to its eco-friendly properties.

Countries of the world can develop themselves by emphasizing more on renewable energy sources. In this context, it is very aptly said that "If sunbeams were weapons of war, we would have had solar energy centuries" [1]. There has been a considerable and continuous rise in the generation of electricity from solar energy. Dependence on traditional energy generation is gradually decreasing and gaining secondary status as other renewable sources are coming in vogue. Solar energy conversion has gained momentum a few decades back and is fastly becoming a favored source of electricity. Though it is very expensive; the benefits of solar energy conversion cannot be undermined given its ecological advantage over other sources of electricity. No greenhouse gasses are released into the atmosphere in photovoltaic (PV) system used in solar panels.

Life exists on earth due to all the favorable conditions created by a supreme invisible energy. This energy is evident in different elements like water, sky/space, fire, earth, and air known as "PanchTatva" in Sanskrit language where Panch means five and Tatva means elements. The human body is as well composed of these five elements. It is by the virtue of their brains that they have made a commendable progress in almost all the fields ranging from agriculture to science and technology. Man knows how to exploit the natural resources to produce energy which can be utilized to fulfill the ever-increasing demands of people to live a luxurious life comprising of sophisticated electronic equipment and gadgets which run on electricity. Until recently, we were dependent on traditional methods of generating energy from non-renewable sources. But the unstoppable population in a developing country like ours gave rise to the usage of renewable energy sources to produce electricity. The "Sun" a part of space/sky is undeniably a divine natural source of heat and light energy; sky being one of the Panch Tattvas. Solar energy is generated by applying scientific methods and different technologies such as solar heating. The UV rays are trapped in PV panels to produce the solar energy.

Human nature is such that it tends to ignore or side track the natural resources and engross themselves completely in the pursuit of fuels which are harmful for the environment. Though our generation is witness to the production of solar cells in abundance, they have not been utilized in electricity production in its true sense. There has been a plethora of solar cells but electricity generation from fossil fuels still reigns supreme. Ample research in solar cells is carried out the world through to increase its efficiency and make it more economical. Ongoing research on different types of solar cells like (a-Si), (C-Si), (CdTe), etc., will definitely bring about a revolutionary change in the perception of human beings.

The field of science is enormously vast, and this is true in the case of solar energy conversion also. Unlimited invention can be made in solar cells by applying existing technology in its entirety.

Solar energy is unlimited heat and light energy from 'Sun' which is the cosmic natural source that is providing solar heating. The invention of solar cells uses this energy in an ideal way from long back, but until now, they have unable to fulfill a need of electrical energy out of a significant portion of overall electricity production. In recent years, even if there are shattering of solar farms but vestige fuels are still polluting. Even so, efforts are still in progress to explore in increasing productivity and dropping the cost of solar cells.

There are several kinds of solar cells such as a-Si, i.e., amorphous Silicon solar-cell, c-Si, i.e., crystalline silicon solar-cell, biohybrid type solar-cell, CdTe, i.e., cadmium telluride solar-cell, CVP, and HCVP, i.e., concentrated PV cell, [CI(G)S], i.e., copper indium gallium selenide solar-cells, float-zone Si, dye-sensitized solar cell (DSSC), perovskite solar-cell (PSC), etc. In this chapter, we shall discuss some of them.

14.2 PHOTOVOLTAIC (PV) CONVERSION

Out of all renewable energy sources, solar energy is having highest potential and by PV conversion with the help of solar cell, it is used to convert into electricity. In many countries, solar energy is converted into electrical power generation, which could start by PV industry. Above 85% global PV market is run on basis of c-silicon (c-Si) solar cells whose efficiency score is 26.7% against an intrinsic limit of ~29%.

Efficiency must enhance and cost of industrial modules must drop is today's requirement of current research and production trends. In this way, solar power based on PV is Worldwide initiating to give a significant contribution to the energy fusion; for example, solar energy expands about 4% of the averaged grid electricity in the Europe continent, and more than 7% in countries like Italy and Germany [2, 3].

Influence of PV technology has increased day by day at an optimum rate (~ 30–40% per year) in the last 15 years and is rising fast in many countries. Increased efficiency and reduced costs of commercial PV modules are the achievement at last. Along with the costs like financing, installation, maintenance, insurance, etc. The cost related to fabrication and segments of solar cells has declined to such a level that the solar cell effects for a fraction of the total cost, while more than 50% is as an outcome of electrical and other components. The conversion efficiency (which impacts the total cost/watt) must be increased is the primary requirement for reducing the total cost of PV energy. For that the production cost of the solar cell has a negligible impact [2, 3].

The wafer-based silicon solar cell covers the world PV market (above 90%) on the basis of several factors such as to acquire efficient PV conversion:

- Silicon has a bandgap within the ideal range;
- The earth's crust is rich with silicon as the second most abundant material;
- It is totally harmless and its overall usage is well understood by chemical as well as semiconductor industries;
- In 1999, the energy-conversion efficiency of silicon solar cells in the laboratory reached a highest value of 25%;
- The efficiency of wafer-based silicon modules has reached 24.4% and is always increasing in the laboratory as well as in the market;

- The power of silicon technology can reach the optimum value, which may be appreciated by looking at the international technology roadmap for PV [4].

A suitably designed solar cell must be electronically thin (i.e., to collect the photoexcited electron-hole recombination with minor losses) and optically thick (i.e., to absorb all or most of the incident rays of sunlight). The maximum efficiency can be achieved by fulfilling these two necessities which cause perfect thickness. Thus, the limiting efficiency can be known by deeply removing both the optical and the electronic issues [5].

14.3 C-SI SOLAR CELLS

Since 2011, the most corporate solar cells used are crystalline silicon solar cells. They are PV cells used in commercially available solar panels. They represent more than 90% of world PV cell market sales. Efficiencies factors of laboratory energy conversion of c-silicon PV solar cells for single-crystal cells is over 25% and for polycrystalline cells, it is over 20%. However, efficiencies ranging from 18%–22% can be achieved now by industrially manufactured solar modules by following routine tests. According to present DOE, i.e., dept of energy research, focus on innovative techniques for reduction in costs. Reduction in raw material supplies, with original ultra-thin crystalline silicon absorber layers, changing kerf-free wafer production techniques (kerf is silicon dust found wastage as blocks of silicon are cut into thin slices) improving growth processes are the jobs undertaken by Research and development cell [5].

Crystalline silicon solar cells are used mostly due to its following profits:

1. Great Quantity: Silicon is the second most abundant element found in Earth's crust (after oxygen).
2. Productivity: Silicon cell shows greater efficiencies than any other device made in industry. Less solar cells need to be fabricated and set up for a certain output as higher efficiencies lessens the cost of the final system.
3. Ripeness: There is a large amount of information on evaluating the steadiness and strength of the design, which is dynamic to obtain capital for arrangement projects.
4. Dependability: Crystalline silicon cells reach module lifetimes of more than 25 years and show little tough degradation.

It is as shown in Figure 14.1.

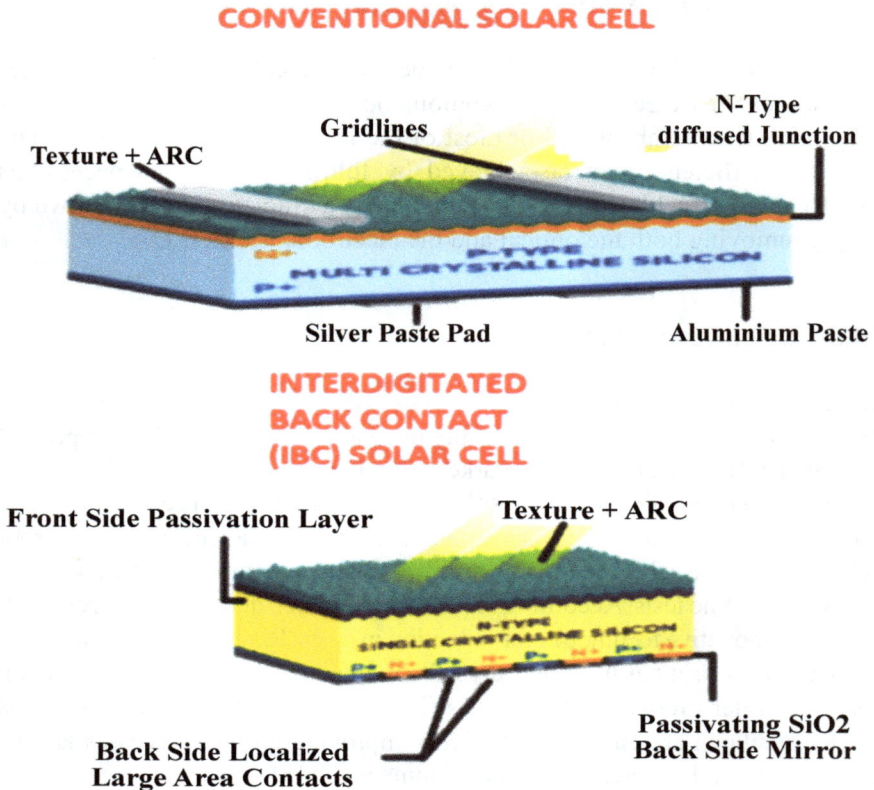

CONVENTIONAL SOLAR CELL

INTERDIGITATED BACK CONTACT (IBC) SOLAR CELL

FIGURE 14.1   Crystalline silicon photovoltaics.

In future, constant improvement in efficiency and reduction in cost of silicon solar cells will be needed so as to raise the contribution of PV power in the energy mix. In the laboratory and in industrial techniques, silicon solar cells are reaching the efficiency parameters, limitations of precision are already there and are certainly implemented. There is scope of tough lab research for revolutionary improvements with silicon phases which have bright prospective.

Merging of economics, physics, chemistry, engineering, and materials science as Inter-disciplinary subjects and related research will be very important for achieving short-as well as long-term advancement in the field.

14.4 AMORPHOUS SILICON SOLAR CELLS

NIR light shorter than 708 nm can be only absorbed by a-Si, i.e., amorphous Si solar cells which has a bandgap of approximately 1.75 eV. By using NaYF4:Er/Yb phosphors, the amorphous Si solar cells if upconversion of sub-bandgap NIR light (l > 700 nm) to visible emission can be combined and in this fashion the absorption limit can be extended [5].

14.5 DYE-SENSITIZED SOLAR CELLS

To construct low cost, flexible solar panels, next-generation PV cells that are Dye-sensitized solar cells (DSSCs) can be used. DSSCs mostly consist of photosensitive dyes and an electrolyte solution such as metal oxide nanoparticles which is not in case of conventional Si-based solar cells. Even though great efforts have been taken, the upgrading in the efficiency of state-of-the-art DSSCs remains a discouraging result, mainly due to the limited absorption spectrum of currently existing dyes. The ruthenium-based dyes are most normally used in recent DSSCs [5].

14.6 SOLAR CELLS WITH LUMINESCENT MATERIALS

Usually, semiconducting materials are successfully used which have special light-absorbing property called Solar cells [6].

The non-absorption losses and natural thermalization can be lowered by utilizing luminescent materials which will act as spectral converters. The emitted solar radiation extending to the earth is scattered over a broad range of the spectrum, considerable intensity being situated from about 350 nm–3000 nm. PV devices which are semiconductor-based do not utilize the whole of the spectrum. Suppose the emitted light radiations which are unutilized are converted, for example: making use of luminescence, to a particular region of the spectrum, then taking into account the whole performance will apparently enhance [7].

Light emitted by these solar radiations is scattered above a very extensive spectral range starting from 300 nm and extending into far infrared, with a maximum emission around 500 nm. The region 400–500 nm has to be targeted as unused radiations for solar spectrum modification which are important in the context of improving the solar cell efficiencies. The phosphors to be used must have strong excitation which will cover this region. In enthusiasm for obtaining down conversion, this has not been achieved. It was decided to tackle the problem from the opposite end and looked for the solids which can absorb 400–500 nm radiations effectively. The next step is to find which of these can re-emit in the near-infrared (NIR) region. Semiconductors with direct bandgap around 2.5 eV answer to this description. Hence, one of the ways for obtaining suitable phosphors for solar spectrum modification is to investigate luminescence in such semiconductors [8].

Research efforts on 'luminescence solar concentrators' are carried out during earlier phase. Then it is extended to 'quantum cutting phosphors.' Reviews to the greatest extent are accomplished associated with c-Si solar cells [9].

Adapting the solar spectrum causes improvement in silicon solar cells, as these are most effectively commercialized. Thus, using these phosphors, it is very interesting that a photon of high energy (350–500 nm) is inefficiently converted by c-Si. It is divided into two NIR photons. These photos are almost utilized in PV methods. Different fruitful prospects are tried to enhance the solar cell productivity with the help of this event [10–15].

Nevertheless, at the commercial scale, the predicted benefits have not been reached even after nearly 15–20 years. On the theory of 'spectral matching' or 'solar spectrum modification' a huge amount of papers has given the idea which is based on demonstration, but a practical device has not been produced yet [16].

By entirely overlooking the excitation spectrum's spectral overlap, the greater number of the conflicts are fixated for obtaining 'quantum cutting' [17–19]. By utilizing the theory of spectrum modification for better organizations, for different solar cells like Cadmiumd Sulfide-Cadmium Telluride, CIS, CIGS, c-Si, etc., needs to convert spectral region (350 nm–500 nm) to NIR is needed. During this transformation, the significance should be given to the longer wavelength region of this spectrum.

There is a need to show an exponential growth of the fraction of the solar spectrum with the wavelength as shown in Figure 14.2. Therefore, only 4% of the entire solar emission occurs in the interval 300–400 nm, whereas more than 15% for the range 400–500 nm. By using an energy gap of semiconductor with E_g about 2.5 eV efficient absorption can be ensured in 400–500 nm regions. Suppose the energy absorbed energy is well shifted to the stimulators like Yb^{3+} or Nd^{3+}, then proper phosphor can be understood for solar spectrum variation. e. g. For the first time, NIR emission in the host Bi_2MoO_6 from Nd^{3+} and Yb^{3+} stimulators are detected. The host makes Nd^{3+} emission sensitive, and thereby the comparable excitation is broadened band which covers Ultraviolet-blue region. Line excitation occurring within f-f transitions of Nd^{3+} is observed. Instead, it is Nd^{3+} which makes sensitive the emission of Yb^{3+}. NIR emissions are detected because of Host $\rightarrow$ Nd^{3+} and Nd^{3+} $\rightarrow$ Yb^{3+} energy transfers. There is not two-step energy transfer Host $\rightarrow$ Nd^{3+} $\rightarrow Yb^{3+}$. Energy transfer from the Host $\rightarrow$ Nd^{3+} occurs at Bi_1 sites, whereas transfer of energy from $Nd^{3+} \rightarrow Yb^{3+}$ occurs at the Bi_2 sites engaged by adding impurity Nd^{3+}/Yb^{3+} ions. Some of these conclusions are quite remarkable for solar PVs in the background of alteration of solar spectrum. Broadband excitation overlaps superbly with the solar spectrum is remarkably significant. Considering vast volume of earlier work reporting NIR emission for enhancing presentation of c-Si solar cells, based on f-f excitations in Ultraviolet region. $Bi_2MoO_6:Nd^{3+}$ phosphor would be appropriate for several applications which needs NIR emission [7].

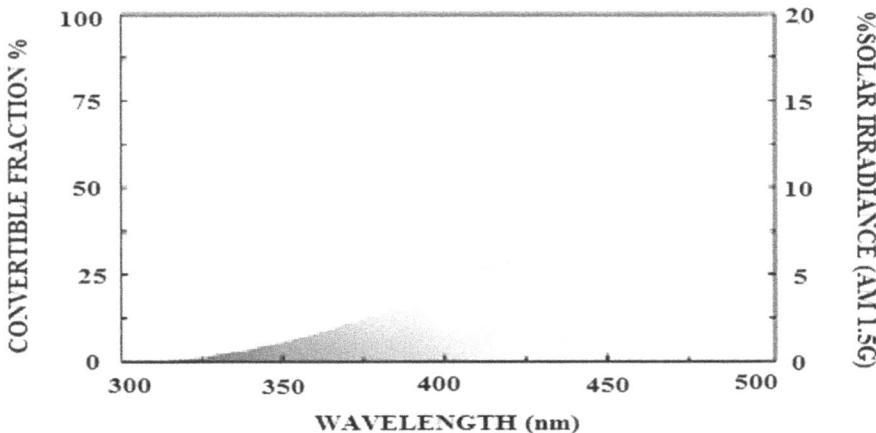

FIGURE 14.2   Part of solar light emitted wavelength: 300 nm–500 nm.

14.7   A-SI:H, CD TE AND CIGS BASED ON THIN-FILM

There is a global attention towards solar cells like a-Si:H thin films, CIGS, and CdTe, which are flexible in nature. Until now, the PV market is ruled by Si cells in order to do cost-cutting of module production, cheap flexible substrates are preferred. This decreases the charges for setting up and transfer, which in turn decreases the system price also. Other variable substrate options for flexible solar cells are banknotes, papers used in security, papers made from cellulose. At times, plain white papers are also used. It is not necessary that films made of plastic, metallic foils, and flexible glass will be used every time.

14.7.1　CDTE

Figure 14.3(a–c) demonstrates the graphical representation of how particular configurations of substrate will change the look of solar films which are thin films in nature. Generally, the sequence where high efficiency process is used decides the selection of a configuration. Many a times there can be an option of varied design factors. The manufacturing of CdTe-based modules takes place in superstrate level where front glass layers are in front contact with transparent conductive oxide. In substrate configurations are possible in PSC technology as it functions only in low temperature processes. In this, the most common set up is of superstrate composition. A glass pane/a protective foil are required to screen up the stopped components. Fully supple segments are probable along with a supple surface and a supple front careful front layering or sheet metal.

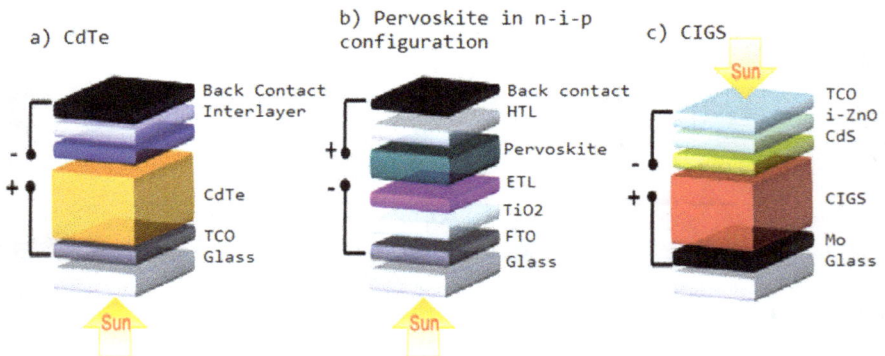

FIGURE 14.3　(a) Superstrate arrangement for CdTe chambers; and for (b) perovskite chambers in the n-i-playout; and (c) layer of CIGS cell design.

Although the market of PV module is overlooked by new silicon technology, i.e., wafer-based technology we witness the increase in volume of PV markets. There has been a significant contribution of the thin film manufacture in the field of CdTe and CIGS technology. These are becoming larger in volume but are lagging in due proportion or greater quantity.

Thin-film PV technologies provide ample scope for opportunities in designing field. Their appearance is identical, and hence it becomes a smart option for integration of building if it is combined with design elements which are specifically frontal in facing due to huge integration process, there is no restriction on the cell size of the modules. They can undergo any shape as required. There is a great chance to change the parameters electrical output by changing the number of cells and then width. There has been a rise of efficiency of CdTe and CIGS as against the lower efficiency as silicon modules. Perovskite solar cells reveal cell efficiencies which are high small area. Solar cells which are perovskite-based are a good option due to their wide-bandgap. An additional feature is that it is a thin-film tandem device in top cells. To fulfill this objective, there are simplicity constraints for front and back contacts. Perovskite based solar cells have a configuration which is semitransparent. It has an elevated transmittance where photons are transmitted with energy which is lower than the energy bandgap suitable for use in a bottom cell in a tandem device.

One of the salient features of the compound semiconductor system is that it can be possible to create and alter the absorber layer due to good efficiency in conversion technique. Such type of solutions facilitates band gap ingredients which are present in the absorber. There is a chance to adjust the bond formatting in a junction that is heterogeneous. There are certain specific variations which affects properties like concentration of charge carrier and stability of phase.

Combination of Zn and Cd with S, Se are used in II-VI compounds which are the main components of the chalcogenide alloy system. Chalcopyrite-III-V-12 system consists of Cu and Ag combination with In and Ga. It also comprises of Cu and Ag with S, Se, and Te. In the course of solid solution bandgap, corresponding energy levels can be regulated. There is an inadequate concentration in III to V Compounds due to lattice matching like chalcogenide II-VI and I-III V-12 compounds. This can even consist to structural defects.

In the beginning stage of exploration, the PSC material system is static. At this stage, bandgap, band alignment, etc., can be tuned due to variations in composition. Efficient heterostructures can be created by engineering of the alloys and graded barrier layers. Selective contacts with low recombination losses can be formed by graded barrier layers.

A barrier of minority charge is formed as a strategy. This reduces the recombination of back contact to increase the bandgap. This further increases the voltage which is already built up. We can make improvements in the carrier collection by introduction of drift fields. In Figure 14.4, we can witness the advancement of how thin-film solar cells perform. Graph drawn as per the details accumulated by NREL accessible online [20].

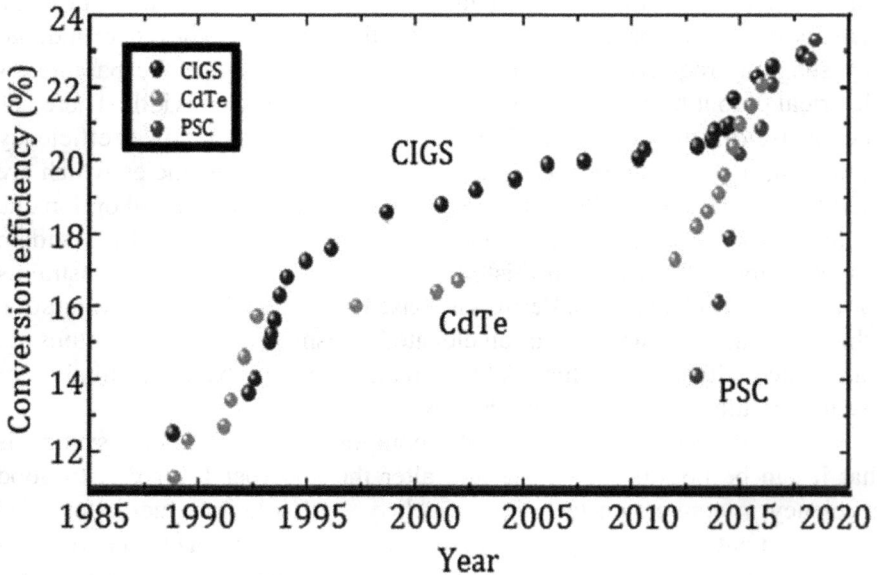

FIGURE 14.4   Report of effectiveness of thin-film solar chambers CIGS, CdTe, and PSC (from 1987).

Figure 14.4 states the record of efficiencies of different thin-film solar cells. Then graph is plotted taking into consideration data that is collected by NREL and is available online. Due to the commercial values of PV technologies; CdTe and CIGS are produced in great quantities. In the year 2017, 4 GW of thin films modules were carried equivalent to around 4% of the existing modules [21, 22]. As compared to CIGS, the market share of CdTe is thrice than that of CIGS. In the near future, online platforms will crop up for new thin films which are in the production phase.

Market for thin film and the entire PV market will go hand in hand. In due course of time, as production capacity increases, there will be cost reduction due to untapped potential of thin-film technologies. The cost cut potential increases assimilated production lines that blend far reaching areas and cheap processes which are highly effective and low energy consumption. Whenever we are thinking of commercialization of the product, we have to think about the modules long term stability and lifetime potentiality. In the 1970's, when Cu_2S based Solar cells were developed, they came in the way of stabilizing the character of thin-film solar cells. The material which is present in the modern thin-film solar cells have proved that they are stable for the long term. Under favorable operating conditions CdTe and CIGS are thermodynamically stable [23, 24].

In fact, met stabilities which are visible under few operational conditions along with under light and forward bias can lead to better efficiency. The stable nature of perovskite cells is an ongoing research topic where improvements are reported from time to time [25]. Investigations and assessments are in progress to make information available to investors regarding the expected performance of perovskite-based cells as compared to the other modules.

14.7.1.1 MODULES BASED ON CDTE

14.7.1.1.1 CONCEPT AND CHARACTERISTICS OF DEVICE

One of the materials which is investigated for the application of solar cells is CdTe. Almost ultimate features for PV cells are available in CdTe. According to Shockley-Queisser, CdTe has a bandgap of 1.5 eV. This bandgap occurs near to prime requirement as conversion of energy trapped from the sun [26]. Later than the first experimentation, it needed some time and more experiments to reach the required material properties. Since 1950's, fairly widespread in-depth study and research states the progress of CdTe [27].

Several valuable characteristics of CdTe for applications of solar cells are:

- The chapter states dominance of thin-film mechanics for purpose of creating large-area PV modules;
- Bandgap of 1.5 eV is supposed to be near perfection;
- Due to its recombination property which is very easy in application and bear-ability to electronic defects;
- During the process of thin-film deposition film properties can be easily controlled;

- CdTe can function at a moderate substrate temperature below 600°C which leads to the synthesis of high-quality films. It is available in abundance and is ecofriendly. Hence drawbacks are limited. CdTe-based modules can be easily produced on a huge scale [28].

Several characteristics of the materials are predominantly favorable for the manufacturing process in industry. At moderate substrate temperatures, we can see the deposition of high-quality semiconducting films.

The progress of the control of composition is smoothened by the congruent evaporation and condensation. Individually cadmium sulfide and Te (Tellurium) are having excessive vapor force as compared to the resulting Compound Cadmium Telluride. Hence it becomes very easy to achieve stoichiometric composition. This will result in re-evaporation of at least one component after getting separated from stoichiometric composition. Because of these beneficial characteristic features various thin film deposition methods like thin-film evaporation, sputtering, electrodeposition, etc., are allowed [27]. At present evaporation is bench mark. Due to this, solar evaporation procedure is applied on a gigawatt scale [29]. Figure 14.5 demonstrates closed space sublimation (CSS) and Vapor Transport deposition which are the two most usual vaporization methods for thin-film accumulation of Cadmium Telluride.

These two techniques occupy the evaporation of Cadmium Telluride and its precipitation on a prime cooler Substrate. In the case of Confined space sublimation, the origin and substrate both should be placed in akin cells. After this placing, the substrate is mounted on the fount.

One of Vapor transfer accumulation works as a carrier gas to transit the vapor from the source to the substrate, thereby a Controlled configuration is permitted which is having an advantage for mechanical necessity on the substratum. It is even fruitful for movement within the machine where it is deposited.

At length the composition of CdTe comprises of Tin oxide (SnO_2) coated glass along Cadmium sulfide window layer and Cadmium Telluride soaker. It can be regarding the current generation.

Effectiveness could be only up to 16% for those cells which are laboratory-based. Major failure was because of the parasitic absorption which took place in the CDS window layer. One more reason of major shortfall was the open-circuit voltage of 0.6–0.8 V. Advancements and high efficiency experiments for 10 years in succession were due to features loss analysis. It can be in respect to current generation. The assembly, improvement, and open-circuit voltage charge factor led to further progress [30]. Whatever learnings were done through the advances resulted in the development of new technologies in solar cells. An example of this is, when composition gradients were employed, the CdTe technology could accomplish an improvement in the doping control and reduced reorganization of combinations at interface and contacts [30]. If we thoroughly analyze the material, interface, and other required properties, then we can achieve efficiency above 25%. Suitable material and effective preparation technique will help us attain quality polycrystalline thin-layer solar panel [31].

CdTe Deposition
a) Close space sublimation

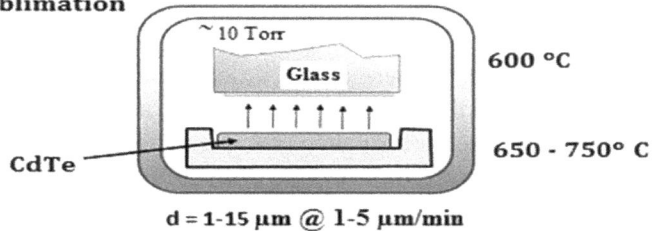

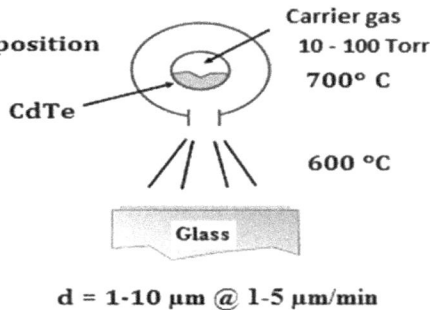

FIGURE 14.5   Demonstrates closed space sublimation and vapor transport deposition which are the two most usual vaporization methods for thin-film accumulation of cadmium telluride.

Betterment in the control of impure materials and doping has led to the enhancement of basic material properties in the recent times. In the past, p-type carrier concentration was increased by the word of Cu, but then Cu doping was known to be a source of defects and instabilities [32, 33] Competencies in doping is achieved through making use of P and As group V elements. To easily handle the production environment by an optional replacement is a matter of experimental research [34]. Similarly, there is a need to find out a kindling step of an absorber layer through cadmium chloride treatment.

For a long time, the efficiency of a bandgap of CdTe was restricted because of open-circuit voltage; but CdTe band gap can tap solar energy conversion to the maximum level. The best possible method was to reduce the shortfall in voltage [35]. Recently the prospective that was displayed was of Te material. The ideal structure comprises of crystalline materials where single crystals are in abundance or epitaxial films [36, 37].

Alloys can be formed by combining cations and anions due to the element compounds of II-VI group. Table 14.1 states the summary of band gaps. Earlier the alloys (Zn, Cd) Te did not result in improved devices.

The alloy of anion shows important bending between addition bandgap in Te/Se or Te/S ratio by least possible bandgap fixed in the endpoints. Superior photocurrents are a result of shrinked bandgap. Even after the establishment of sturdy gradients, there is still little deficiency in Cd(Se, Te) alloy. It can be compared to the novel CdS/CdTe structures which are heterogeneous in nature and where performance was adequate. This was due to the imperfections which occurred during the interdiffusion region of two compounds. CdS-CdTe lead to the notifications in the band structure of thin-film cells by suitable alloying. A breakthrough in efficiency was possible because of Cd (Se, Te) alloy system [38, 39].

Proper grading of absorber offers perfections in open-circuit voltage and photocurrent. Bandgap is decreased because of bowing, which leads to a lower wavelength in the absorption process. There is an alternative to it where extrinsic semiconductor Cadmium selenide offers a good junction between two different types of semiconductors, i.e., heterojunction [40]. In comparison to the binary Compound in Cadmium Telluride Cd (Se, Te) has a lower GB reconnection velocity.

TABLE 14.1   Energy Gaps of II-VI Compounds Applied in CdTe-based Solar Cells

Compound Technique	Least E_g (eV)	Small E_g (eV)	Big E_g (eV)
Zinc (Selenium, Tellurium)	2.1	2.26	2.8
(Zinc, Cadmium) Tellurium	–	1.6	2.26
Cadmium (Sulfur, Selenium)	–	2.0	2.5
(Magnesium, Cadmium) Tellurium	–	1.6	3.0
Cadmium (Selenium, Tellurium)	1.34	1.34	1.8
Cadmium (Sulfur, Tellurium)	–	1.6	2.6

Even after having a low bandgap of 1.7 eV in n-type still a good blue response is achieved in CdSe-CdTe heterojunction. Excellence of materials and amazing movable properties existing in Cd (Se, Te) helps in the current collection from the n-type heterojunction. Recombination velocities of less than 100 cm/sec at polycrystalline Al_2O_3/Cd(Se, Te) interfaces [41] are demonstrated by favorable electronic properties possessed by the alloys. By increasing p-type doping in the absorber additional optimization of the absorber is obtained. Cu has the property to be a lifetime killer hence group V elements are now considered for this purpose of absorption.

More developments are possible to increase the efficiency of device only because of optimization of contacts. For these changes must be done in heterojunction and the contact. Table 14.2 comprises of the list of materials used in the creation and steady and well-organized equipment. Also, the good tolerance with respect to defects, the option of passivation layer charge carrier of recombination which takes place at the grain boundaries and interfaces looks controllable. After an explanation of several blends for passivation and counter interfaces, wide range of options for materials optimization were revealed. In a recent work, the pliable nature of the material combinations, which discloses the notable inherent properties of the material for higher productivity is shown using amorphous-Silicon as a p-type contact [42].

TABLE 14.2 Metals Contacts and Impurity Added to Cadmium Tellurium Solar Compartments

Back Contact	Impurity Added	Heterojunction Barrier Layers	Remark
Carbon	Phosphor	amorphous-Silicon	–
Zinc Tellurium	Arsenic	(Magnesium, Cadmium) Tellurium	–
Tellurium	Cadmium Chloride/ Magnesium Chloride	(Zinc, Manganese) Oxygen	–
Copper Tellurium	Copper	–	Copper should be ignored

As measured by photoluminescence (PL) decay, the lifetime is increased to over 100 ns. Exceedingly low recombination at Cadmium Tellurium/ Magnesium, Cadmium) Te heterointerfaces are also depicted [43]. Coating by Al_2O_3 pilots the amalgamation of velocities below 100 cm/sec [41]. Efficiencies around 20% can be attained by the connection of appropriate window layers (Zinc, Manganese) Oxygen and (Magnesium, Cadmium) Tellurium [42]. Other advancements are connected to the back contact. To obtain a good ohmic contact, Cu containing compounds were applied extensively. We can obtain control over back contact stability by optimizing alloys with ZnTe. Hole can be conducted by using this contact.

New expansion too benefitted from standard measurement phenomenon like lifetime tomography [44]. This assigns the consequence treatment of $CdCl_2$ on a microscopic scale. When space wavelength and Spectroscopic technique obtained at time intervals are applied, issues related to dopant separation and grain boundary reformation are disclosed. Analysis of this outcome should take into account the organization of the entire device. Solar cell performance is influenced by internal fields and patent barriers.

14.7.1.1.2   UNIT CHARACTERISTICS, SOLAR CELL, AND COMPONENT PRODUCTION

Manufacturing of CdTe modules is possible by use of monolithic integration technique. This can be compared to the technique elaborated for CIGS technology in the part Sec. III E while actual execution [29]. Table 14.3 shows few selected data related to the laboratory cell and module performance.

TABLE 14.3   Chosen Details of CdTe Report Lab Chambers and Modules

Cell/Module	Sector (cm²)	V_{oc} (V)	j_{sc} (mA/cm²)	FF (%)	Efficiency (%)
CdTe x-Tal	0.3	1.007	22.0	62	14
Module	7039	111.1	1.54	74	19 [45]
Cd (Se, Te) cell	0.5124	0.8964	32.01	79	22 [45]

The efficiency of 22.1% exhibits the completion level that can be expected for the development of future modules. First Solar, the most modern creation comprises of modules of the size 2.0–1.2 m² where the nominal power is 440 Wp has dominated the CdTe module market. So far around 17 GW modules are sold in the world. The latest improvement consists of increasing the size of products three times more than the original size, i.e., from 0.72 m² to 2.4 m². This new huge size has an advantage of cost in manufacturing and fixing. This lessens the overall expenses for generation of electricity. "Topaz Farm" built from 2001 to 2014, located in California; USA is regarded as the largest power plant with a capacity of 550 MWp. This plant connects around 9–10[6] of the erstwhile 80-W-Sized panels [46].

Figure 14.6 states the satellite image of the Topaz Farm from Google earth. In the past there was an issue because of loss of concentration of p-type carrier and solidity of back contact [47]. Meanwhile, the stability and lasting conduct of first solar CdTe modules proved elongated testing phase in practical execution and at particular test sites. New modules show high solidity and execution ratio [48]. There is a recurring stability issue in almost all PV modules which is due to partial shading. An exception to this is Si technology. Those people who install solar set ups and end users of these set ups, must be careful while installing the solar panels and even during the cleaning process. This is so because mitigation planning for thin-film modules and partial shading of these modules will result in breakage of the module due to localized area of reverse bias. This feature is already being dealt with in Section III E [29] in the description of CIGS technology. Partial shading means when the individual cells generate minimum Current as Compared to the rest of the module.

FIGURE 14.6   Space station profile of useful-scale insertion "topaz farm."

There is a vital argument related to the possible production scale which may rise to terawatts for CdTe modules centers; this is due to the limited availability of Te. If we consider the aspects of recycling and study the market dynamics [49], we know that installation will be about 2 TWp until 2050. This states that before Te an important share in the market is possible. Cost is driven by availability of the product. Sinha and Wade have published a study wherein they claim that environmental footmark for CdTe is the lowest among all PV technologies. They have done an in-depth study, looking into the environmental concerns arising due to development of new technologies [28].

14.8   PEROVSKITE SOLAR CELLS (PSC)

PSC consists of a structured perovskite structured compound. It is made of a material which is based on tin halide or a fusion of organic-inorganic lead. This is the dynamic layer which can gather light. Methylammonium lead halides are inexpensive perovskite materials. Because of this, they are very simple to manufacture. Absorption of spectrum, charge separation, lengthy passage distance of electrons and holes and elongated carrier separation are few characteristics of perovskite as shown in Figure 14.7 [49, 50]. Calcium Titanite and perovskite share the same group of materials. At the time of installation, perovskite solid is located on the upper surface of a silicon solar cell [51]. This raises the efficiency of generating electricity as it minimizes thermal loss in the silicon. Perovskite act as an additional support to silicon solar cells at minimum costs of installation, which boosts the efficiency of these cells. Outstanding electronic properties like Si or GaAs, affordable cost of material and fabrication of solar cell, thin-film skill manufacturing process on flexible substrate are some of the outstanding features of perovskite. Perovskite solar cells give efficient electrical performance, which is equivalent to those solar cells technologies that are firmly established.

A spin-coating technique is used by researchers. The liquid perovskite material is to be deposited on a substrate which is one of several promising routes. The processes like dot-slide coating, blade coating, and ink-jet printing are to be recycled wherever required for better results [52].

A four-terminal solar cell constructed by the researchers at Imec is as shown in Figure 14.8. The perovskite solar cell is mounted at the top. It is optically attached to the lower silicon cell. Control of power reaching to these two cells is separate. Overall efficiency is always levitated by combining perovskite and silicon. The tandem solar cell is also a two-terminal type, where the upper and lowest cells are electrically fused. With alignment of four-terminal, there is only optical link between both solar cells. From these and previous results, it seems that improvement in the efficiency which will be well-adjusted the extra industrial costs of solar cells with their electronics. Imec researchers tested the cell using the outline in Figure 14.9.

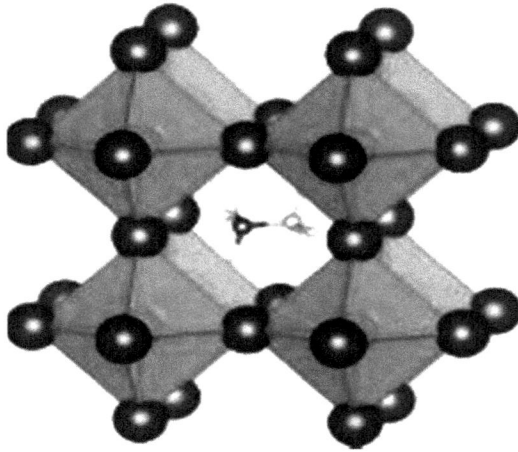

FIGURE 14.7   Perovskite solar cell.

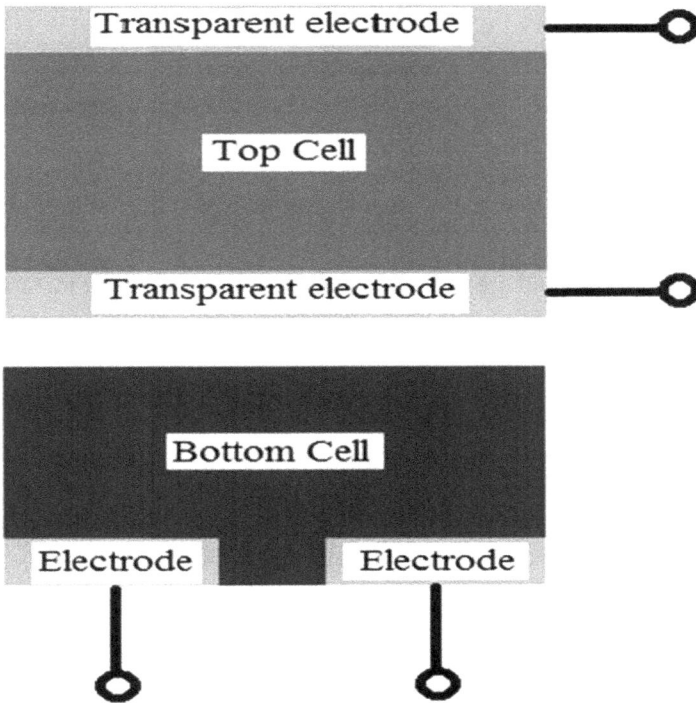

FIGURE 14.8   A two-part solar compartment consisting top portion of perovskite and bottom portion of Si used by Imec scientists. Optically linked two cells.

The measurement system is made up of two things such as a regulated 'Sun simulator' and a Keithley's Source Meter. With an optical power of 1000 W/m², the solar cell illuminates with the simulator and the cell's electrical current as well as voltage is measured by the Source Meter here which is used as a measuring instrument only. Communication with the Source Meter, displaying the output and calculating efficiency is possible with the help of Commercial software.

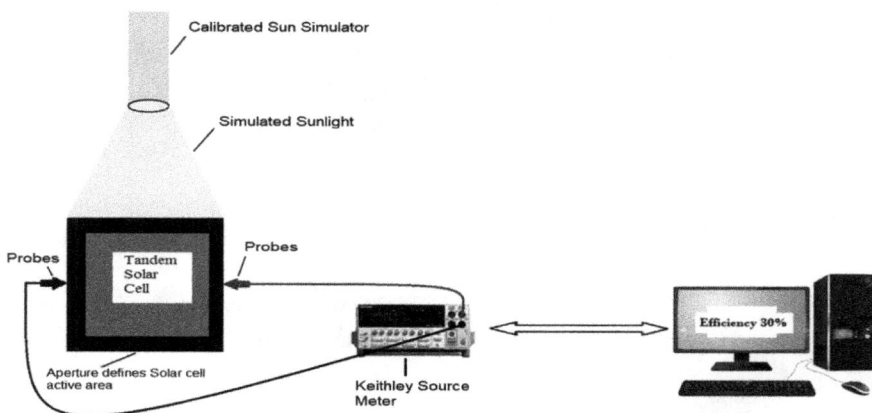

FIGURE 14.9   In order to illuminate the chamber under study, Imec's test set-up utilizes a calibrated sun simulator which leads to greater power conversion efficiency than Si alone.

The presence of lead creates perovskite issues which is the most alarming issue. Primary perovskite materials vanished their crystalline structure in minutes, i.e., short lifetime was another issue. To overcome it, today, the permanency of the structures can be maintained for more than thousands of hours as scientists have designed new but better material formulas. These materials which have improved steadiness while the absorption needed to increase adeptness of power is also maintained [53]. Once the lead issue is resolved, lifetimes, and efficiencies both will likely improve. In this way, perovskite materials could offer the help that solar cells necessity to grow our use of inexhaustible energy like solar energy ultimately.

14.9   CONCLUSION

From last two decades, different technologies for solar energy conversion are used such as PV technologies. c-Si solar cells are effectively commercialized due to established Si technology, but it is too expensive. Using a small amount of material-thin film solar cells, low-cost materials other than Si, increasing efficiency so as to reduce the cost/Watt are the solutions found. Luminescent materials, i.e., phosphors have been synthesized and used to reduce the losses in the energy conversion process, which are based on solar cell. It rises the productivity of solar cells which is significant for an ecological point of view. In addition to supply thin-film solar cells technologies such as copper indium gallium selenide solar cells, cadmium telluride solar cell, amorphous silicon solar cell: H; a novel technology is emerged, i.e., known as a perovskite (PSC). Its great efficiency and large bandgap make PSC predominantly smart. For improving the performance, i.e., efficiency, stability, manufacturability of PSC, ecological aspects, etc., a number of laboratories are taking great lead in investigating through a wide range of routes. In this way, all undeveloped and developing countries are on the path of development by using solar energy as renewable energy.

KEYWORDS

- **environmental aspects**
- **luminescent materials**
- **phosphors**
- **renewable energy**
- **solar cell efficiency**
- **solar energy**

REFERENCES

1. https://www.goodreads.com/quotes/tag/solar-power (accessed on 30 June 2021).
2. *Fraunhofer ISE Photovoltaics Report,* (2017). Available at: https://www.ise.fraunhofer. de/en/publications/studies/photovoltaics-report.html (accessed on 30 June 2021).
3. European Commission-JRC, (2017). *PV Status Report-2017.* Available at: https:// ec.europa.eu/jrc/en/publication/eur-scientific-and-technical-research-reports/pv-status-report-2017 (accessed on 30 June 2021).
4. *International Technology Roadmap for Photovoltaic.* Available at: http://www.itrpv.net/ (accessed on 30 June 2021).
5. Lucio, C. A., Angelo, B., Piotr, K., Marco, L., & Lisa, R., (2019). Silicon solar cells: Toward the efficiency limits. *Adv. Phys., 4,* 1548305. https://doi.org/10.1080/23746149.2 018.1548305.

6. Nelson, J., (2002). *The Physics of Solar Cells*, Imperial College Press, London.

7. Tumrama, P. V., Kautkara, P. R., Wankhede, S. P., & Moharil, S. V., (2019). NIR emitting Bi_2MoO_6:Nd^{3+}/Yb^{3+} phosphor as a spectral converter for solar cells. *Journal of Luminescence, 206*, 39–45.

8. Tawalare, P. K., Belsare, P. D., & Moharil, S. V., (2020). Semiconductor host for designing phosphors for modification of solar spectrum. *Opt. Mater., 100*, 109668.

9. Strumpel, C., McCann, M., Beaucarne, G., Arkhipov, V., Slaoui, A., Svrcek, V., Del, C. C., & Tobias, I., (2007). Modifying the solar spectrum to enhance silicon solar cell efficiency—An overview of available materials. *Sol. Energy Mater. Sol. Cells., 91*, 238–249.

10. Ten, K. O. M., Jong De, M., Hintzen, H. T., & Van, D. K. E., (2013). Efficiency enhancement calculations of state-of-the-art solar cells by luminescent layers with spectral shifting, quantum cutting, and quantum tripling function. *J. Appl. Phys., 114*, 084502.

11. Badescu, V., & Badescu, A. M., (2009). Improved model for solar cells with up-conversion of low-energy photons. *Renew. Energy, 34*, 1538–1544.

12. De Vos, A., Szymanska, A., & Badescu, V., (2009). Modeling of solar cells with down-conversion of high energy photons, anti-reflection coatings and light trapping. *Energy Convers. Manag., 50*, 328–336.

13. Badescu, V., Vos, A. D., Badescu, A. M., & Szymanska, A., (2007). Improved model for solar cells with down- conversion and down-shifting of high-energy photons. *J. Phys. D: Appl. Phys., 40*, 341–352.

14. Badescu, V., & Vos, A. D., (2007). Influence of some design parameters on the efficiency of solar cells with down-conversion and downshifting of high-energy photons. *J. Appl. Phys., 102*, 073102.

15. Van, S. W. G. J. H. M., (2008). Simulating performance of solar cells with spectral downshifting layers. *Thin Solid Films, 516*, 6808–6812.

16. Bryan, M. V. D. E., Linda, A., & Andries, M., (2009). Lanthanide ions as spectral converters for solar cells. *Phys. Chem. Chem. Phys., 11*, 11081–11095.

17. Klampaftis, E., Ross, D., & McIntosh, B. S., (2009). Richards Enhancing the performance of solar cells via luminescent down-shifting of the incident spectrum: A review. *Sol. Energy Mater. Sol. Cells, 93*, 1182–1194.

18. Richards, B. S., (2006). Luminescent layers for enhanced silicon solar cell performance: Down-conversion. *Sol. Energy Mater. Sol. Cells., 90*, 1189–1207.

19. Vanm, S. W. G. J. H. M., Meijerink, A., Schropp, R. E. I., Van, R. J. A. M., & Lysen, E. H., (2005). Enhancing solar cell efficiency by using spectral converters. *Sol. Energy Sol. Energy Mater. Sol. Cells., 87*, 395–409.

20. NREL Research Cell Record Efficiency Chart, (2018). https://www.nrel.gov/pv/cell-efficiency.html (accessed on 30 June 2021).

21. Mints, P., (2018). *20 Years PV from 1997–2017 Changes, Challenges, Growth Other Topics.* SPV market research. https://www.renewableenergyworld.com/2018/05/03/20-years-of-pv-changes-challenges-and-growth/ (accessed on 30 June 2021).

22. Burger, B., Kiefer, K., Kost, C., Nold, S., Philipps, S., Preu, R., Rentsch, J., et al., (2018). *Photovoltaics Report.* Fraunhofer Institute for Solar Energy Systems (ISE); PSE Conferences & Consulting GmbH. https://www.ise.fraunhofer.de/content/dam/ise/de/documents/publications/studies/Photovoltaics-Report.pdf (accessed on 30 June 2021).

23. Dobson, K. D., Visoly-Fisher, I., Hodes, G., & Cahen, D., (2001). Doping and inter-mixing in CdS/CdTe solar cells fabricated under different conditions. *J. Appl. Phys., 90*, 2553–2558.

24. Guillemoles, J. F., Rau, U., Kronik, L., Schock, H. W., & Cahen, D., (1999). Cu (In,Ga) Se₂ solar cells: Device stability based on chemical flexibility. *Adv. Mater., 11*, 957–961.

25. Fu, Q., Tang, X., Huang, B., Hu, T., Tan, L., Chen, L., & Chen, Y., (2018). Recent progress on the long-term stability of perovskite solar cells. *Adv. Sci., 5*, 1700387.

26. Shockley, W., & Queisser, H. J., (1961). Slip patterns on boron-doped silicon surfaces. *J. Appl. Phys., 32*, 510–519.

27. Basol, B. M., & McCandless, B., (2014). Brief review of cadmium telluride-based photovoltaic technologies. *J. Photonics Energy., 4*, 040996.

28. Sinha, P., & Wade, A., (2018). Addressing hotspots in the product environmental footprint of CdTe photovoltaics. *IEEE J. Photovoltaics., 8*, 793–797.

29. Powalla, M., Paetel, S., Ahlswede, E., Wuerz, R., Wessendorf, C. D., & Friedlmeier, T. M. (2018). Thin-film solar cells exceeding 22% solar cell efficiency: An overview on CdTe-, Cu(In,Ga)Se₂-, and perovskite-based materials. *Appl. Phys. Rev., 5*, 041602.

30. Gloeckler, M., Sankin, I., & Zhao, Z., (2013). CdTe solar cells at the threshold to 20% efficiency. *IEEE J. Photovoltaics, 3*, 1389–1393.

31. Kanevce, A., Reese, M. O., Barnes, T. M., Jensen, S. A., & Metzger, W. K., (2017). The roles of carrier concentration and interface, bulk, and grain-boundary recombination for 25% efficient CdTe solar cells. *J. Appl. Phys., 121*, 214506.

32. Yang, J. H., Metzger, W. K., & Wei, S. H., (2017). Carrier providers or killers: The case of Cu defects in CdTe. *Appl. Phys. Lett., 111*, 042106.

33. Guo, D., Brinkman, D., Shaik, A. R., Ringhofer, C., & Vasileska, D., (2018). Metastability and reliability of CdTe solar cells. *J. Phys. D: Appl. Phys., 51*, 153002.

34. Major, J. D., Turkestani, M. A., Bowen, L., Brossard, M., Li, C., Lagoudakis, P., Pennycook, S. J., et al., (2016). In-depth analysis of chloride treatments for thin-film CdTe solar cells. *Nature Communication, 7*, 13231.

35. Burst, J. M., Duenov, J. N., Albin, D. S., Colegrove, E., Reese, M. O., Aguiar, J. A., Jiang, C. S., et al., (2016). *Nat. Energy, 1*, 16015.

36. Zhao, Y., Boccard, M., Liu, S., Becker, J., Zhao, X. H., Campbell, C. M., et al., (2016). Monocrystalline CdTe solar cells with open-circuit voltage over 1 v and efficiency of 17%. *Nat. Energy., 1*, 16067.

37. Becker, J. J., Boccard, M., Campbell, C. M., Zhao, Y., Lassise, M., Holman, Z. C., & Zhang, Y. H., (2017). Loss analysis of monocrystalline CdTe solar cells with 20% active-area efficiency. *IEEE J. Photovoltaics, 7*, 900–905.

38. Munshi, A. H., Kephart, J., Abbas, A., Raguse, J., Beaudry, J. N., Barth, K., Sites, J., Walls, J., & Sampath, W., (2018). Polycrystalline CdSeTe/CdTe Absorber cells with 28 mA/cm² short-circuit current. *IEEE J. Photovoltaics, 8*, 310–314.

39. Paudel, N. R., & Yan, Y., (2014). Enhancing the photo-currents of CdTe thin-film solar cells in both short and long-wavelength regions. *Appl. Phys. Lett., 105*, 183510.

40. Baines, T., Zoppi, G., Bowen, L., Shalvey, T. P., Mariooti, S., Durose, K., & Major, J. D., (2018). Incorporation of CdSe layers into CdTe thin-film solar cells. *Sol. Energy Mater. Sol. Cells, 180*, 196–204.

41. Kuciauskas, D., Kephart, J. M., Moseley, J., Metzger, W. K., Sampath, W. S., & Dippo, P., (2018). Recombination velocity less than 100 cm/s at polycrystalline Al₂O₃/CdSeTe interfaces. *Appl. Phys. Lett., 112*, 263901.

42. Burst, J. M., Duenow, J. N., Kanevce, A., Moutinho, H. R., Jiang, C. S., Al-Jassim, M. M., Reese, M. O., et al., (2016). Interface characterization of single-crystal CdTe solar cells with Voc> 950 mV. *IEEE J. Photovoltaics., 6*, 1650–1653.

43. Zhao, X. H., Liu, S., Campbell, C. M., Yuan, Z., Lassise, M. B., & Zhang, Y. H., (2017). Ultralow interface recombination velocity (~1 cm/s) at CdTe/MgxCd xte heterointerface. *IEEE J. Photovoltaics., 7*, 913–918.

44. Barnard, E. S., Ursprung, B., Colegrove, E., Moutinho, H. R., Borys, N. J., Hardin, B. E., Peters, C. H., et al., (2017). 3D lifetime tomography reveals how $CdCl_2$ improves recombination throughout CdTe solar cells. *Adv. Mater., 29*, 1603801.

45. Green, M. A., Hishikawa, Y., Dunlop, E. D., Levi, D. H., Hohl-Ebinger, J., & Ho-Baillie, A. W. Y., (2018). *Prog. Photovoltaics Res. Appl., 26*, 427.

46. https://iopscience.iop.org/article/10.1088/2053-1591/ab3089/meta (accessed on 31 July 2021).

47. Gretener, C., Perrenoud, J., Kranz, L., Cheah, E., Dietrich, M., Buecheler, S., & Tiwari, A. N., (2016). Solar energy materials and solar cells. *Sol. Energy Mater. Sol. Cells, 146*, 51–57.

48. Ngan, L., Strevel, N., Passow, K., Panchula, A. F., & Jordan, D., (2014). Performance characterization of cadmium telluride modules validated by utility-scale and test systems. In: *Conference on 40th IEEE PVSC* (pp. 1957–1962).

49. Houari, Y., Speirs, J., Candelise, C., & Gross, R., (2014). Prog. A system dynamics model of tellurium availability for CdTe PV photovoltaics. *Res. Appl., 22*, 129–146.

50. *Imec Beats Silicon PV with 27.1% Perovskite-Silicon Tandem*, (2018). Press release. https://optics.org/news/9/7/40 (accessed on 30 June 2021).

51. Snaith, H. J., (2013). Perovskites: The emergence of a new era for low-cost, high-efficiency solar cells. *J. Phys. Chem., 4*, 3623–3630.

52. Di Zhou, Tiantian Zhou, Yu Tian, Xiaolong Zhu, & Yafang Tu (2018). *News Release: Perovskite Technology Scalable, But Questions Remain about the Best Methods*. National Renewable Energy Laboratory, U.S. Dept. of Energy. https://www.nrel.gov/news/press/2018/perovskite-technology-is-scalable-but-questions-remain-about-the-best-methods.html (accessed on 30 June 2021).

53. Perovskite improves solar-cell efficiency, Article by: Martin Rowe, 21 Jan 2019. *Video: The IV Curve*. Technical University of Denmark. https://www.coursera.org/lecture/solar-cells/the-iv-curve-8YtK8 (accessed on 30 June 2021).

CHAPTER 15

Futuristic Scope for Biofuels and Their Production

U. V. GAIKWAD,[1] A. R. CHAUDHARI,[1] and S. V. GAIKWAD[2]

[1]*Priyadarshini Bhagwati College of Engineering, Nagpur, Maharashtra, India, E-mail: umagaikwad651@gmail.com (U. V. Gaikwad)*

[2]*Dr. Babasaheb Ambedkar College of Engineering, Nagpur, Maharashtra, India*

ABSTRACT

Rapid growth in the use of fossil fuel and the consequent CO_2 emission is demanding the need for new alternative energy sources which are renewable and ecofriendly. The disadvantage of fossil fuel is that burning of it has increased the atmospheric concentration of some greenhouse gases (GHG), which are responsible for global warming.

Biofuel is the fuel obtained from biological sources like crop wastes, animal wastes, exhausted mineral oils or fats and algae, etc. Biofuels generation, driven with the potential to enhance the energy security and climate issues. A few nations have started approaches to help biofuel improvement, creation, and use in transportation area. Biofuels acting like an important source to stimulate rural development and create employment opportunities. In this chapter, the challenges and future prospects of ecologically and economically viable biofuel product is outlined.

15.1 INTRODUCTION

Bioenergy actually characterized as a sustainable source of energy which can be obtained from biomass. Natural waste materials, goes under the wide area of biomass. Biofuels are an option in contrast to petroleum products, which are fluid or vaporous powers which are generated from biomass sources. These fuels can be utilized alone or blend another non-renewable source, for example, petroleum. Biofuels arranged into three era biofuels [1].

Origin of biofuels are starch sugars and seed oils. Since oils from vegetables and non-edible oils like lubricants has high viscosity, density which can be harmful for the diesel engines [1]. Sometimes neem oil was taken as the parent oil, ethanol/methanol as solvent and KOH as catalyst [2]. Though the neem oil has more fat content, owing to the availability of neem oil and its cost, it has been chosen as parent oil for this work.

Demand of energy increases slowly which basically obtained from the non-renewable resources. There are limited resources which go on decreasing because of large uses of energy. More use of energy leads to environment pollution. So, it is important to find some renewable as well as ecofriendly resources of energy which fulfill the energy demand by taking care of the surrounding atmosphere. A Researcher has taken or rather taking many efforts for alternative energy sources. While doing literature survey, it is observed that microalgae are one of the sources. This resource appears as low power product but availability as well as estimate of production which are the main reasons due to which scientists are giving so much attention. By photosynthetic process Algae convert CO_2 into the oil form [3].

As indicated by fundamental figuring, petroleum derivatives will have depleted in around 200 years. Utilization of Fuel causes discharge of carbon dioxide, bringing about breakdown of tuning between liberated carbon dioxide and gas that can be consumed by plants. It can be evaluated, that if there should arise an occurrence of proceeded with utilization of conventional fuel sources till 2030, carbon dioxide levels will increase to 39 billion Mg for every year [5].

Worldwide emission of CO_2 in various nations, will keep on developing, due to increasing creation in non-industrial nations, for which clean advancements and putting resources into environmentally friendly power sources are as well costly. Presently, such a pattern can be watched, on the grounds that in the created nations, carbon dioxide discharges was decreased by 6.5% [5]. Then again, the agricultural nations expanded the discharges up to 3%. Thus, the economy of the 21st century will be rely upon non-renewable energy source resources [5].

While doing overview, it is seen that India is inadequate in consumable oils, non-eatable oil, and the primary decision for creating biodiesel. As indicated by Indian government strategy and Indian innovation impacts, some advancement works have been completed concerning the creation of transesterified non-palatable oil and its utilization in biodiesel by units, for example, Indian Organization of Science, Bangalore, Tamil Nadu Farming College Coimbatore, and Kumaraguru School of Innovation [6].

So much study has done that a neural network model was framed to predict the performance of the engine. Backpropagation neural network with gradient descent algorithm was used for the modeling in MATLAB [2].

15.2   BIOFUEL PRODUCTION

There is a various aspect for production of biofuels. Biodiesel creation used for generating the biofuel, biodiesel, by the synthetic retorts of transesterification and esterification. It includes vegetable or animals' fats as well as oils being retorted with short-chain alcohols. The condition is that alcohols must be having low atomic weight [1].

A biofuel is the fuel which is produced through modern processes from biomass, rather than a fuel produced by the very slow ecological processes involved in the formation of fossil fuels, such as oil. It can be produced from organic matter, or biomass, such as corn or sugar, vegetable oils or waste feedstocks. As biofuels emit less carbon dioxide (CO_2) than conventional fuels, they can be blended with 90 existing fuels as an effective way of reducing CO_2 emissions in the transport sector. The word biofuel is typically saved for fluid or vaporous energizes, utilized for transportation. On the off chance that the biomass utilized in the creation of biofuel can regrow rapidly, the fuel is commonly viewed as a type of sustainable power [5].

Biofuels can be created from plants crops from agrarian, business, homegrown, and additionally modern squanders (if the waste has a natural origin). Renewable biofuels for the most part include contemporary carbon fixing, for example, those that happen in plants and microalgae through the cycle of photosynthesis [5]. There are different bioenergy sources, biomass alludes to natural material obtained from living or late living beings which can charred to create Bioenergy. Advantageous wellsprings of bioenergy incorporate vegetable plants, green growth, monocot plants, vegetable oils and fats of creature [1]. Monocots plants are the type of plants which have a solitary cotyledon. These plants utilized as a wellspring of bioenergy which include corn, wheat, sugarcane, etc., [1]. Ethanol production from plant which is around 0.9 L around 2.6 kg of corn grains. As a cellulosic or second-age feedstock and corn stover utilized in various energy applications not affecting food production [10].

Nowadays metabolic engineering [13] have increased the production of various liquid biofuels like fatty acid derivatives of fatty acids and many more.

15.3 PRODUCTS OF BIOFUELS

There are various products which were obtained while the generation of Biofuels. For example, Biodiesel, Biogas, Butanol, Ethanol, crude oil, Diesel, gasoline, and many more.

15.3.1 BIODIESEL

Biodiesel is the fuel obtained from the esterification process occurs in feedstocks, for example, animals' fat, and vegetable oil. This refining strategy utilizes ethanol or methanol and a catalyst to change over the oil into an unsaturated fatty fuel. Biodiesels are utilized in diesel with next to zero changes and ignites with less sulfur and sweet-smelling emanations [8].

15.3.2 BIOGAS

Biogas produced from sewage, biomass, and solid waste. It used an anaerobic process in which few microbes help to fragment organic matter in sealed containers by converting it into the hydrocarbons. The sealed containers act like a digester in which biogas is produced, which is actually a combination of gases like methane and maximum percent of carbon dioxide, which is greater than 80% out of the overall percentage. Quantity of other gases like hydrogen sulfide, nitrogen, hydrogen as well as oxygen observed smaller [8].

Extra hydrocarbons, for example, propane, can delivered utilizing digesters that warm water to very high temperatures and pressure, which is necessary for the fermentation process of sugars found in corn as well as in sugarcane. Biogases and biopropane have various uses like residential boiler to industrial uses and as a transportation fuel [8].

15.3.3 BUTANOL

Butanol properties offer all the more inspiring outcomes contrasted with those of lower chain alcohol, for example, methanol or ethanol. However, commercially production of butanol as a biofuel has is not so easy due to its high estimate. Butanol is formed by the process of acetone-butanol-ethanol (ABE) fermentation. Though a lot of efforts are taken for obtaining higher butanol concentration from ABE fermentation but still relatively expensive and challenging [9].

Butanol can burn without alterations in a current gasoline engine which is less corrosive than ethanol. Also, it does not emit carbon monoxide [8].

15.3.4   ETHANOL

Ethanol which is basically alcohol produced by fermentation of the sugar as well as starch components in the crops. While its production enzyme digestions, fermentation, and distillation are the various steps. Nowadays, ethanol is the most utilized biofuel in the world, like in Brazil as well as in the United States. It is used as an alternative fuel used in fuel vehicles [8].

15.3.5   RENEWABLE CRUDE OIL

Revolutionary innovations, for example, green growth, intend for the scope of green crudes from woody biomass as well as green growth. These crudes can be handled in usual purifying innovation which promptly accessible to change over inexhaustible into sustainable gas, fly rather sometimes diesel also [8].

15.3.6   RENEWABLE DIESEL

Today maximum Progressed biofuels advancements plan to make inexhaustible diesel is equivalent to oil diesel fuel. Petro-diesels are regularly characterized as one of the fluid fuels used in diesel motors, with the most well-known being partial purified for generation of the diesel. Diesel fuel is delivered from oil and different sources. The subsequent items compatible in many utilities. Diesel-controlled vehicles for the most part have a superior mileage than equal gas motors and produce less ozone depleting substance outflow. Their more prominent economy is because of the larger energy per-liter substance of diesel fuel and characteristic proficiency of the diesel motor [8] (see Figure 15.1).

15.3.7   RENEWABLE GASOLINE

Inexhaustible fuel or manufactured gas is on the plan of a few progressed biofuels organizations. Some of them will change sugars over to gas, whereas others utilize cellulosic materials, and gasification of Current gas, additionally refer as petroleum, is an oil determined fluid blend which most generally utilized as fuel of decision in interior ignition motors. It comprises generally of aliphatic hydrocarbons obtained through fragmentary refining of oil, improved by iso-octane or by sweet-smelling hydrocarbons toluene, benzene to expand its rating of octane. Little amounts of different added substances are normal, for purposes, for example, standardization motor execution, lessening hurtful fumes outflows. Some gas combinations accessible in the market now a day additionally contain low mixes of ethanol as a halfway elective fuel (see Figure 15.1) [8].

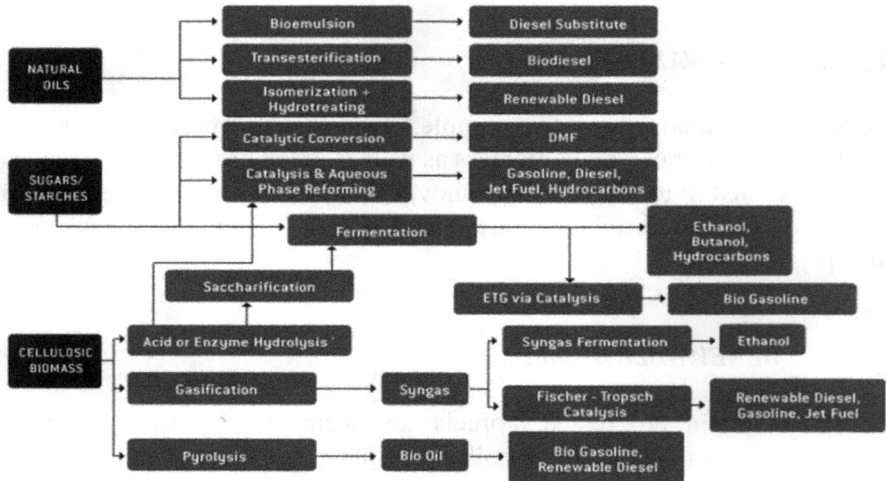

FIGURE 15.1 Formation of renewable fuels.

Source: Adapted from Ref. [8].

15.4 TECHNOLOGY BEHIND THE GENERATION OF BIOFUELS

15.4.1 *CATALYSIS*

Catalysis is the process in which molecules undergo a reaction made possible due to the help of catalysts. Catalysts which reduced the activation energy which is used for a chemical reaction and increased its rate also.

Due to faster chemical reactions, there is an increase in productivity in less amount of energy. Catalysts utilizes different feedstocks and produces a range of various molecules [8].

15.4.2   CELLULOSIC

Generation of biofuels from lignocellulose has a lot of benefits of large amount and various raw material which requires a huge amount of processing to make available the sugar monomers that are typically used to generate ethanol by fermentation. There is minimum two different methods of delivering ethanol from cellulose, both of which utilize refining to detach unadulterated ethanol, gasification, and cellulosic processes [8].

Gasification is the process which transforms the lignocellulosic material into carbon monoxide and hydrogen. All these gases easily converted to biofuels by fermentation or compound catalysis. The gasification depends upon the breaking of cellulose into sugar molecules. By partial combustion, the carbon is converted easily into synthesis gas. The carbon monoxide, carbon dioxide, and hydrogen may then be taken care of into an exceptional sort of fermenter. This cycle utilizes a microorganism as opposed to yeast which devours carbon monoxide, carbon dioxide and hydrogen and produces ethanol and water [8].

The cellulose particles are made out of long chains of sugar atoms. In the hydrolysis cycle, these chains are separated to free the sugar, before the fermentation for alcohol production. There are two significant cellulose hydrolysis measures: a synthetic response utilizing acids, or an enzymatic response. With the fast advancement of protein innovations over the most recent 20 years, the corrosive hydrolysis measure has slowly been supplanted by enzymatic hydrolysis [8].

15.4.3   FERMENTATION

Mainly the alcohol fermentation occurs due to sugar, for which renewable materials like sugar canes, sugar beets are used. Fermentation happens after the pretreatment processes which separates cellulose, lignin, and pectin. Once the sugars are removed, they can be easily fermented into biofuels. A number of new cellulosic companies plan to turn syngas into ethanol by the same process. Alcohol is a short-chain maturation item created by microbial digestion, so making recuperation of the alcohol is energy concentrated and costly with the necessity of complex distillation [8].

15.4.4   GASIFICATION

Gasification basically is the cycle by which strong biomass is divide using a high temperature, high weight measure and changed over in to more modest particles. The subsequent gas blend, combination gas, or syngas, is an ignitable fuel with various valuable applications including the creation of sustainable biofuels [8]. Nowadays, advanced gasification processes expel toxic impurities carbon dioxide, and various impurities are removed or captured. The result is a clearer and more efficient than conventional production methods. Some gasification processes become more efficient by using a very high amount of solar energy to achieve very high temperature, which is very necessary for the gasification process. Thus, this technology can produce syngas without affecting another material which create necessary heat for the proper process functioning [8].

15.4.5   HYDROLYSIS

Hydrolysis literally water-based chemical reaction which converts the complex polysaccharides in the renewable materials to simple sugars. In biomass-to-biofuels transformation process lot of acids and enzymes are act as a catalyst. In the synthetic cycle, an atom is divided into two sections by the increase in a particle of water. One section of the parent atom increases a hydrogen particle (H +) from the extra water particle. The other gathering gathers the rest of the hydroxyl gathering (OH⁻) [8].

15.4.6   HYDROPROCESSING

Hydroprocessing is not only a one process, but for any chemical process, it breaks heavy hydrocarbons into lighter amount by the addition of hydrogen. It is the common process in the traditional petroleum industry [8]. In the biofuels area, hydroprocessing has very vast applications which transform waste animal fats into a wide variety of fuel like bio-diesel, and jet fuels.

15.4.7   SYNTHETIC BIOLOGY

Synthetic Biology is actually an area of biological research which combine science and engineering promotion. Synthetic consist of both the design as well as fabrication of biological components and systems which does not exist in the natural world [8].

15.5   BIOFUEL POLICIES

Biofuels shares a miserable part in absolute fuel utilization, and its use is estimated to rise consistently in the future. One of the key drivers for the improvement of biofuel creation and concurrent weakening of biofuels in the energy market which is the nation's approach. Thus, this basic survey centers around the development of biofuel approaches in the BRICS (Brazil, Russia, India, China, and South Africa) countries and effective execution of the arrangements by the separate nations. In detail, this chapter gives legitimate structure and approaches provoked the development of biofuel exploration and businesses over the BRICS countries. It widely presents important laws, mixing targets, current guidelines, public activity plan, government uphold component (sponsorships), yield targets, and significant obstacles and extension in biofuel arrangements of BRICS countries for productive presentation of biofuel to the energy market [7].

15.6   LIMITATION

Biofuels obtained from plants as well as harvests. Biofuels like biodiesel and bioethanol are very much efficient one. These Plants' powers used as an inexhaustible source and easily developed at anyplace. Contrasted with petroleum products, biofuels have low carbon emissions. Biofuels even assistance in diminishing discharge of ozone-depleting substances. So, biofuels can give a valuable method to diminish the reliance on non-sustainable non-renewable energy sources too as they can be useful to the climate around [1].

Biofuels have so many points of interest still at the same time had few drawbacks moreover. It discharges huge amounts of carbon while the burning process, an incredibly important harvest land utilized in biofuel creation. There is the possibility of deficiencies of food if the crops are developed for the creation of biofuels. The best possible maintenance of biofuel crops, water needed in huge amounts which can be put a strain on accessible water assets. The energy yield of biofuels is lesser in comparison to that of oil based energizes, so as to achieve a similar energy stage, biofuels should utilize in enormous amounts [1].

15.7 FUTURE SCOPE (ENTREPRENEUR OPPORTUNITIES)

Biofuels will not just be an answer for a decent ecological quality, however may likewise get an expansion food creation. This situation, which alludes to adequately immense attraction of biofuels, competent to bring better natural quality opposite food security. Biofuels can possibly offer a success to improve natural quality, whereby better ecological quality may advance a sizeable gain in food creation [12].

The experimental analysis and simulation model has proved vegetables oil fatty acid alkyl esters to be a suitable alternative for petroleum diesel. However, there are some difficulties to predict performance of the engine. Therefore, the following are the recommendations for extending this work further. Investigations could be carried out to store biodiesel efficiently and use it after a long time as diesel. Efforts could be taken to perform experimental study to substitute pure biodiesel for petroleum diesel. There is a scope of modeling also in viewpoint to analyzes performance of biodiesel engine. There is a lot of scope in entrepreneur opportunities like energy crop farming for which biofuel feedstock is done, manage industrial waste for production of biofuel, as well as opportunities in storage and distribution of biofuel systems.

KEYWORDS

- acetone-butanol-ethanol
- algae
- bioenergy
- biofuels
- biomass
- carbon dioxide
- ecofriendly
- fossi

REFERENCES

1. Rasool, U., & Hemalatha, S., (2016). A review on bioenergy and biofuels: Sources and their production. *Brazilian Journal of Biological Sciences, 3,* 3–21.
2. Keith Krueger, (2016). *Chapter 7: Conclusion and Future Scope,* Anna University, Chennai.

3. Hossain, A. B. M. S., Salleh, A., Boyce, A. N., Chowdhury, P., & Naqiuddin, M., (2008). Biodiesel fuel production from algae as renewable energy. *American Journal of Biochemistry & Biotechnology, 4,* 250–254.
4. Rasul, M., Masud, K. K. M., & Sharma, S., (2014). *Review of Biodiesel Production From Microalgae: A Novel Source Of Green Energy.* Central Queensland University, Conference Paper.
5. Biernat, K., Malinowski, A., & Gnat, M. (2013). *Chapter 5: The Possibility of Future Biofuels Production using Waste Carbon Dioxide and Solar Energy.* DOI: 10.5772/53831.
6. Magdum, S., (2015). Future Biofuel Potential and Scope for Lipid-Based Biodiesel. Researchgate Technical Report.
7. Saravanana, A. P., Pugazhendhib, A., & Mathimanic, T., (2020). Review Article Comprehensive Assessment of Biofuel Policies in the BRICS Nations: Implementation, *Blending Target and Gaps Journal, 272.* Article 117635.
8. http://advancedbiofuelsassociation.com/page.php?sid=2&id=7   [Online   Article], Advanced Biofuel Association.
9. Ibham, V., Mohd, F. M. S., & Zulkarnain, A. L., (2019). *Progress of acetone-butanol- ethanol (ABE) as biofuel in gasoline and diesel engine: A review. Fuel Processing Technology,* 196.
10. Klingenfeld, D., (2008). Corn Stover as a Bioenergy Feedstock: Identifying and Overcoming Barriers for Corn Stover Harvest, Storage, and Transport. Washington, DC: National Commission on Energy Policy.
11. Shafaque, F., (2017). A review: Advantages and disadvantages of biodiesel. *International Research Journal of Engineering and Technology, 4,* 11.
12. Subramaniam, Y., Masron, T. A., & Azman, N. H. N., (2020). Biofuels, environmental sustainability, and food security: A review of 51 countries. *Energy Research & Social Science., 68,* 101549.
13. Choi, K. R., Jang, W. D., Yang, D., Cho, J. S., Park, D., & Lee, S. Y., (2019). Systems metabolic engineering strategies: Integrating systems and synthetic biology with metabolic engineering. *Trends Biotechnol., 37,* 817–837.

CHAPTER 16

Selection of Fall Protection Equipment for Windmill Tower

SHIKHAR CHAKRAVORTY,[1] VINAYAK SOOD,[1] and VISHAL SHUKLA[2]

[1]Research Scholar, Shri Ramdeobaba College of Engineering and Management, Nagpur, Maharashtra, India, E-mail: shikharchakravorty@gmail.com (S. Chakravorty)

[2]Associate Professor, Department of Mechanical Engineering, Shri Ramdeobaba College of Engineering and Management, Nagpur, Maharashtra, India

ABSTRACT

Wind energy contributes substantially to renewable energy. With growing demand, it only follows that a large number of workers will be employed for the service and maintenance of the windmill shafts. Therefore, the workers' safety must be prioritized. Industrial facilities face accidents which are due to falling from a height, being struck by a moving vehicle, struck by a moving object, and other kinds of accidents.

Thus, the selection of proper safety technology should be of prime importance within the wind power industry. Hence in this chapter, an attempt is made to compare different technologies to arrest the accidental fall while climbing up a windmill tower. The comparison is made on the basis of cost, safety, reliability, space requirement, work fatigue, life of system, power required, etc., for rail type fall arrester (RTFA), rope type fall arrester (ROTFA), climb assist system (CAS) and service lift used for windmill towers.

16.1 INTRODUCTION

Energy is a most required input in all sectors of any country's economy [1]. It is crucial for human development index as human development is positively correlated to energy consumption [2]. Till the late-1980s, energy has been generated largely by burning coal, hydrocarbon oil, and natural gas, leading to huge carbon emissions. Hence, environmental crisis has become a critical concern for the world today [3]. Emission of greenhouse gases (GHG), limited coal availability, environment distortion, rising prices of fossil fuels, and pressure on foreign exchange reserves have created hindrance in the prolongation of these resources [4]. In India, more than 70% of all electrical energy is produced using thermal power plants, i.e., coal and petroleum [5]. These percentages are need to be replaced by the renewable sources of energy. Due to this, a number of researches have been carried out in the area of renewable sources of energy. Due to the geographic conditions of India, there is plenty of renewable energy sources available such as solar, wind, biomass, hydro, and tidal [6]. India is one of the fastest-growing, developing country in the world and is considered as a favorable investment destination.

Wind energy policies issued by the Indian government are very investor-friendly and offer attractive tariff and regulation that provides healthy growth to this sector [3]. MNRE has introduced a generation-based incentive scheme to provide financial incentive for every unit of generation up to 10 years [7]. Due to all this support and subsidies from government in India wind power accounts for nearly 10% of India's total installed power generation capacity and generated 62.03 TWh in the fiscal year 2018–2019, which is nearly 4% of total electricity generation and 48.54% of the total renewable energy capacity. As of 31st March 2019, the total installed wind power capacity was 36.625 GW [8], the 4th largest installed wind power capacity in the world. The capacity of the all the wind turbines installed across the world has reached to 597 GW by the end of year 2018 [9]. The Potential for wind power generation for grid interaction has been estimated at about 1,02,788 MW taking sites having wind power density greater than 200 W/sq. m at 80 m hub-height with 2% land availability in potential areas for setting up wind farms at the rate of 9 MW/sq. km (Figure 16.1) [10].

Working as a technician in a windmill is severely demanding on the body. The technicians working at a windmill for maintenance and repairs have a highly taxing job description. They have to climb up and down more than 70 m high ladders to carry out repairs along with heavy equipment. On an average, a windmill technician spends about 4–5 hours inside the nacelle or windmill tower climbing more than 6–7 times on the windmill ladders. It would be very helpful to use a device that could act as a fall protection in case of fall (Figures 16.2 and 16.3).

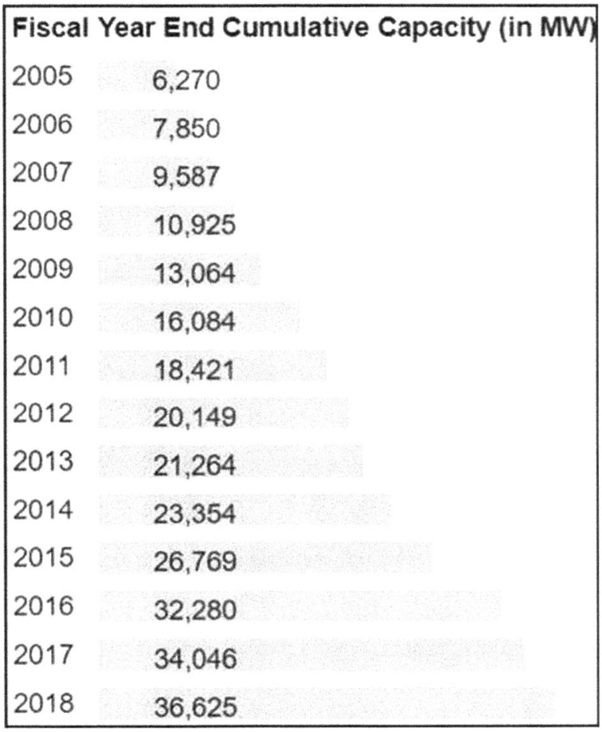

Fiscal Year End Cumulative Capacity (in MW)	
2005	6,270
2006	7,850
2007	9,587
2008	10,925
2009	13,064
2010	16,084
2011	18,421
2012	20,149
2013	21,264
2014	23,354
2015	26,769
2016	32,280
2017	34,046
2018	36,625

FIGURE 16.1   Installed wind power capacity in India [11].

The protection devices used in a windmill industry are rail type fall arrester (RTFA), ROTFA, climb assist system (CAS) and Service Lift. In RTFA, a C section is used as a guide rail and a fall arrester attached to harness of person runs in the section and during fall a pawl comes out and lock within the slots provided in C section to arrest the fall. In ROTFA, a metallic rope is used as a guiding rope and a fall arrester attached to harness of person runs around the rope and during fall a pawl comes out and due to the friction between pawl and rope the fall is arrested. In CAS, a mechanical motorized system is used as an assisting system to support the climber. It provides an upward force and provides a continuous pull which is equal to some percentage of the weight of the person. In CAS the basis aim is to reduce the efforts required by the person to climb up the tower. This reduces the climber's fatigue. In service lift, a small size lift is used to carry the person inside the windmill tower. They all perform the same function, but they differ in terms of cost, safety of person, reliability, space requirement, installation, worker fatigue, life of system, external power required for operation, present, and future demand. In this chapter, the comparison of these three-fall protection equipment is done on the basis of differing factors and a conclusion is drawn for the same.

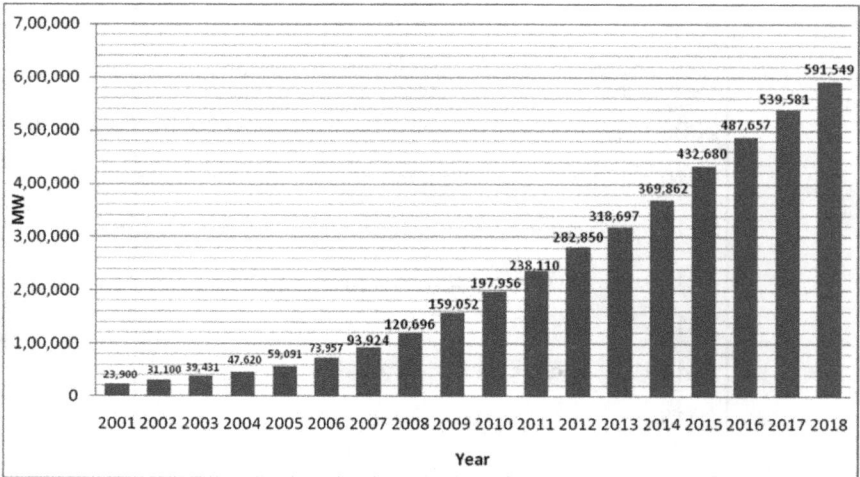

FIGURE 16.2   Total installed capacity across the world [12].

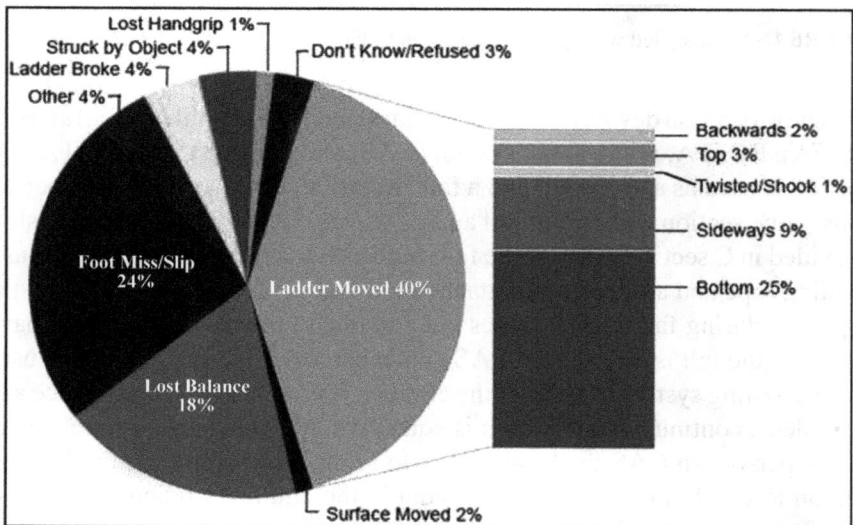

FIGURE 16.3   Causes of fall from ladder.

16.2   COMPARISON OF DIFFERENT FALL PROTECTION DEVICES

In this chapter, literature related to various devices used as fall arresting devices. While writing this paper, market was surveyed and various products were brought to the fore and surveyed. This information is compared and put before people through this paper.

16.2.1   COST

After comparing the cost of different products from different manufacturers it is found that the cost of ROTFA is least among these four fall protection devices, which is followed by RTFA, CAS, and Service Lift. The cost per piece is Rs. 5,000, Rs. 15,000, Rs. 1,50,000, and Rs. 5,00,000 for ROTFA, RTFA, CAS, and service lift, respectively. The cost of installation is also maximum for service lift and comparatively less for CAS. The cost of installation is for RTFA and ROTFA is involved in windmill tower installation. The cost of operation is least for RTFA and ROTFA because they do not require any kind of external power source, but on the other hand, CAS and service lift require external power to drive the motor. The CAS and service lift have more mechanical and electrical components which increase the maintenance frequency and hence increase the cost of maintenance. ROTFA requires more maintenance than RTFA due to the use of rope due to this, proper tension is required to be maintained in the rope, and since it arrests the fall using friction so proper cleaning of the rope to remove oil and grease is also required. On the other hand, RTFA uses rigid rail which requires least maintenance hence least maintenance cost. Also, after the case of accident or fall if the rail gets damaged due to indentation, then that part of rail can easily be replaced because the generally the rail is manufactured in pieces.

16.2.2   CAPACITY

The capacity means the number of person and the range of weight which is allowed by the manufacturer, including all tools for safe operation of the fall protection devices. The load-carrying capacity of service lift is around 250 kg and most of the lifts are designed for 2 person and minimum load required for the working of lift is zero because the lift is operated by means of electric motor. The maximum load-carrying capacity for ROTFA and RTFA is around 140–150 kg with 1-person minimum weight required is 50 kg since the working of fall arrester is based on gravity so some minimum weight is required to push the pawl out. The capacity of the CAS is measured in terms of the upward force provided by the motorized system. The upward pull reduces the effort of the person to climb up the windmill tower and hence reduces the climber fatigue. The upward pull is only some fraction of the weight of the person. It does not completely lift the climber. The upward pull provided by the CAS is ranged from 50 kg to 70 kg, and this upward pull is also adjustable in most of the CAS so that it can be adjusted according to the climber's weight.

16.2.3 RELIABILITY

The service lift is the least reliable because it has more electrical and mechanical components, so there are more chances of failure. Hence it is always required to have a reserve system of ladder and fall arrester. The CAS is competitively more reliable due to less electro-mechanical components, but it also required a backup system. On the other hand, ROTFA, and RTFA do not require any reserve system. The RTFA has protection for backward fall but the ROTFA does not have any such arrangement. But the ROTFA is less reliable than RTFA since in RTFA the fall is arrested by locking the pawl in slots provided in the rail but in ROTFA the fall is arrested by the friction between the rope and pawl of the ROTFA, but in case if there is some oil and grease is present on the rope which is the most possible case, then the friction is considerably reduced and then there is the chance that the fall may not get arrested. The CAS also has inbuilt fall protection system which also works on the same principle of ROTFA. The service lift does not require any such arrangement. So, the RTFA is most reliable, followed by ROTFA and service lift.

16.2.4 WORKER FATIGUE

The worker fatigue is negligible in the service lift because the work required to be done to climb the tower is done by electric motor. In CAS the worker fatigue is not considerably less than RTFA and ROTFA but more than a service lift because in CAS the fraction of weight of person is pulled up by the CAS so a part of work is done by CAS and other part is done by the person himself to climb the tower. RTFA and ROTFA the complete work is done by the person himself so fatigue is maximum. The ease of use is also high for service lift followed by the CAS. The technician does not require any special training to use the service lift but for ROTFA, RTFA, and CAS it is required (Table 16.1).

TABLE 16.1 Comparison between Fall Protection Device for Windmill

Type of Device Parameter	Rail Type Fall Arrester	Rope Type Fall Arrester	Climb Assist System	Service Lift for Wind Mills
Cost/Piece*	Rs. 15,000/-	Rs. 5,000/-	Rs. 1,50,000/-	Rs. 5,00,000/-
Weight carrying capacity*	50–140 kg	50–140 kg	50 kg**	0–250 kg
Number of persons at a time*	1	1	1	2
Backward fall protection	Available in some products	Not available	Not required	Not required
Reliability	High	High	Low	Low
Worker fatigue	High	High	Low	Negligible
External power required/power consumption*	Nil	Nil	0.4 to 0.6 kW	2 to 3 kW
Installation	Easy	Easy	Slightly difficult	Very difficult
Ease of use	Less	Less	High	High
Maintenance frequency	Regular maintenance not required	Regular for tightening the rope	Regular for checking the machineries	Regular for checking the machineries
Standards followed	EN 353:2014 EN 353:2002	EN 353:2014 EN 353:2002	EN 610000:2008 EN 610000:2007	EN 1808:2015 EN 795:1997
Present and future markets	Highly demanded fall arrester due to highest reliability	Highest used fall arrester due to least price	Currently preferred over lift due to less cost	Currently not in much use due to cost but demand is continuously increasing

*This is the approximate value of the device based on the market price of different manufacturers.

**It is the upward pull which is provided by climb assist.

16.3 CONCLUSION

As of now due to high cost and less reliability, currently the service lift and CAS is not in much use in India. But as the technology is advancing cost of manufacturing and installment is continuously decreasing so, while considering the worker fatigue, ease of operation and speed of climb there are high chances that in future each tower will have at least have a CAS. But considering the reliability of electrical components, there will always be a reserve system in terms of ROTFA and RTFA to use in case of emergency.

KEYWORDS

- climb assist system
- fall protection device
- rail type fall arrester
- rope type fall arrester
- service lift
- technician safety

REFERENCES

1. Singh, A. K., & Parida, S. K., (2013). National electricity planner and use of distributed energy sources in India. *Sustainable Energy Technologies and Assessments, 2,* 42–54.
2. Dwivedi, G., Sharma, M. P., & Kumar, M., (2014). Status and policy of biodiesel development in India. *International Journal of Renewable Energy Research, 4*(3), 246–254.
3. Sangroya, D., & Nayak, J. K., (2015). Development of wind energy in India. International *Journal of Renewable Energy Research, 5*(1).
4. Mani, S., & Dhingra, T., (2013). Policies to accelerate the growth of offshore wind energy sector in India. *Renewable and Sustainable Energy Reviews, 24*(c), 473–482.
5. January 2019 Report. Central Electricity Authority, Ministry of Power, Government of India.
6. Ashwani, K., Kapil, K., Kaushik, N., Sharma, S., & Mishra, S., (2010). Renewable energy in India: Current status and future potentials. *Renewable and Sustainable Energy Reviews, 14*(8), 2434–2442.
7. Khare, V., Nema, S., & Baredar, P., (2013). Status of solar-wind renewable energy in India. *Renewable and Sustainable Energy Reviews, 27,* 1–10.
8. https://mnre.gov.in/physical-progress-achievements (accessed on 30 June 2021).
9. https://wwindea.org/information-2/information/ (accessed on 30 June 2021).
10. http://www.mitrasias.com/wind-energy/ (accessed on 30 June 2021).
11. All India Installed Capacity of Power Stations, (2019). https://www.cea.nic.in/old/reports/monthly/installedcapacity/2019/installed_capacity-12.pdf.
12. http://www.gwec.net/wp-content/uploads/2012/06/Global-Annual-Installed-Wind-Capacity-2001-2016.jpg (accessed on 30 June 2021).

CHAPTER 17

Integrated Architecture Pedagogy for Energy Efficient Design

APARNA TARAR and SHRUTEE DHANORKAR YEOLEKAR

Priyadarshini Institute of Architecture and Design Studies, Nagpur, Maharashtra, India, E-mail: tararaparna7@gmail.com (A. Tarar)

ABSTRACT

Architecture Design is an important subject in the curriculum of Architecture Education. Architectural education should be more responsive to environmental issues. The academic community should be involved in providing opportunities for future architects to develop more socially and culturally responsible and environmentally responsive architecture. Environmental and Sustainable architecture should be considered as one of the most important parameters for developing architecture. It is therefore imperative that we produce design solutions that are eco-sensitive, energy-efficient, and climate-responsive, user-friendly, and cost-effective too. In architecture design studios, at the stage of designing of a form or building envelope, many times students face the difficulties in finding solutions for the factors such as heat, light, ventilation, sound, etc. All such factors and decisions, which directly affects the energy consumption of a building needs to be taken care at the schematic design stage in the overall design process. In order to achieve this at each level of curriculum, a scientific method needs to be followed with a balanced comprehension between energy efficient approach and creativity in design. This chapter is an attempt to develop systematic design processes for design studios of different levels with different pedagogical approaches for energy efficient building design.

17.1 INTRODUCTION

Incorporation of the sustainability principles in architecture is a need of the 21st century. The basic understanding of the sustainability, and its importance in architecture design, has to be taught in the graduation level curriculum, in order to achieve a better sustainable built environment in the future. This will also require the fundamental reframing of the architecture curriculum to emphasize on the energy efficiency and use of natural resources in building design. In architecture sustainability denotes the terms green, ecological, environmentally sensitive, climate-responsive and energy-efficient. Traditional and vernacular architecture are the best references for passive heating and cooling techniques to be learned from, also since many years, some architects are practicing these strategies to reduce the energy needs of the building. Now a day's many passive techniques are used in buildings to increase the ability to capture energy or generate their own energy.

17.2   ENERGY EFFICIENT DESIGN

In today's scenario, for comfortable and modern living style, people are opting for new building materials and advanced building services, which automatically leads to consumption of natural resources and excessive energy, again indirectly leads to a serious environmental issue. The approach towards sustainable architecture design helps to establish positive and appropriate contributions to the social environment. The energy efficiency in buildings is affected by decisions to be taken at all the stages, i.e., Design, construction (materials and techniques) and maintenance/management. Design is all about optimum resource usage. Sustainability need not have some extra efforts to be put in, but it has to be intrinsic in designs. Solar radiation and wind are two important aspects that could be favorable with respect to the context. Use of appropriate building design and building techniques, as well as external inputs, such as the local environmental conditions, solar energy, wind, and water, for improvising the indoor climate of buildings, without using mechanical means, is a real challenge in architecture design.

There are multiple approaches that can be adopted to achieve the energy efficiency in the buildings:

- Climatic study and application of bioclimatic architectural principles, a climate-responsive approach in building design.
- Building planning and design level considerations to meet the need of thermal and visual comfort of the occupant by providing optimum natural lighting and ventilation.
- Integration of, passive heating and cooling techniques in building design to minimize the load.

- Use of low embodied building material with efficient structural system, and reduction of transportation energy.
- Designing energy efficient lighting, heating, and ventilation/air conditioning systems of building.
- Use of renewable energy systems to meet some part of total building load such as, use of solar energy.

17.3   ARCHITECTURE PEDAGOGY

For a better future of the sustainable development, it is utmost necessary to provide knowledge to the budding architects through the architecture curriculum at every stage of learning. Architecture education is one of the most distinctive branches of education, which requires various capabilities and ways for the transmission of knowledge. Along with the creative thinking abilities, some technical skills and new design methodologies can be taught to deal with the environmental issues. The subjects included in the architecture curriculum can be broadly divided into three types, i.e., Skill related subjects; Knowledge related subjects and the application-based subjects. All these three types of subjects need to adopt different methods of teaching. In the early years of the UG course, sustainability-related topics could be covered in knowledge-based subjects like Environmental architecture, regional architecture, Vernacular Architecture, Climatology, etc. In support of the theory, creative hands-on activities also could be taken as an experiential learning assignment for better and fast learning (Figure 17.1). For application-based subjects, such as in Architecture design, the rigid format that has been established and followed till date will have to be loosened up. Dealing with various issues and integrating the knowledge acquired in various subjects of architecture curriculum in design studio must deal systematically at appropriate stages. Design is not to be seen as a 'subject' of learning but an application of what is learnt from support subjects to design. The concept of an integrated studio is not necessarily new, but innovative approaches need to be adopted in its implementation.

17.4   DESIGN STUDIO

A design studio is an environment that enables both the teacher and the learner to explore the domains of cognitive, affective, and psychomotor skills. Design is not to be seen as a subject of learning but an application of what is learnt from support subjects to design. The studio is like a crucible where all applied knowledge melts into one another to obtain a unified whole. This is what we call as the 'total integration,'-integration of applied knowledge. In the conventional approach of architectural design teaching, the emphasis is often given to form-based solutions. Recent researches and studies state that the architectural design pedagogy majorly focuses on the product-based approaches, rather study of context, climate, and environment. The concept of an integrated studio is not necessarily new, but innovative approaches need to be adopted in its implementation. The rigid format that has been established and followed till date will have to be loosened up. Integration of acquired knowledge from allied subjects, practical and mental abilities, and creative skills are the central tools which can be used in the architectural design studios.

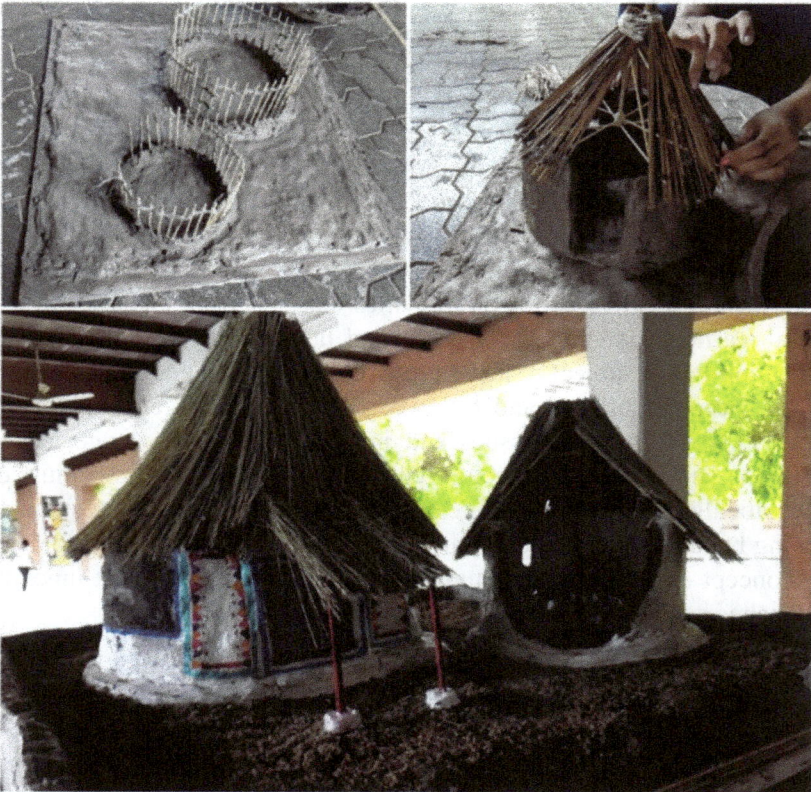

FIGURE 17.1 Miniature model of traditional Bhunga house made (Source: Second year students of PIADS, Nagpur under Author's instructions) in the subject Vernacular Architecture.

17.5   DESIGN PROCESS

Design is an analytical process, which involves the critical thinking skills and decision-making abilities at each given design situation. Design is a continuous process, where at every subsequent stage, one needs to look back for the previous decisions made. The design process has been studied for quite a long time. Many theories have expressed their views on the design process. Many conferences and societies were formed for research work. The consensus was on design process to be a problem-solving procedure, with a three-phase sequence of:

$$Analysis \rightarrow Synthesis \rightarrow Evaluation$$

The final product of the design is too assumed by the designer before the means of achieving it can be explored. There are certain common steps necessarily involved in any type of design, though the emphasis may differ:

- Acquiring knowledge;
- Setting objectives;
- Counter checking;
- Implementing.

An educator could develop his own design process for the integrated studio, to deal with many issues related to sustainability in architecture. A systematic approach is necessary in design studio working, which we could achieve through a design process. The incorporation of theory with design, results in better understanding of the issues, than the simple theory teaching. The design processes could be framed as per the learning ability of the students; the complexity in design could be increased, based on different design approaches or design theories for 1st year level to higher semesters.

17.5.1   SUSTAINABLE ARCHITECTURE PRINCIPLES

Vernacular architecture is strongly associated with sustainable architecture principles, as they share a common goal of developing environmental friendly building solutions, considering the local climate and surrounding conditions, utilizing the local materials and resources, use of green architectural principles of energy efficiency, which makes that construction lasts for many years.

Exercise for first-year (Figure 17.2):

FIGURE 17.2 Documenting the vernacular tribal habitats at Manav Sangrahalay, Bhopal (Source: First year students of PIADS, Nagpur under Author's instruction).

- Introduction to traditional vernacular architecture of different regions;
- Study tour to vernacular sites/villages/tribal habitat museum;
- Documentation of vernacular houses.

17.5.2 CLIMATE RESPONSIVE DESIGN PROCESS

Climate and built form are two key issues directly related to the environmental set-up. In climate-responsive approach, at every stage of design, one needs to consider climate as a basic parameter for the application of passive design strategies. These various aspects or parameters should follow a logical sequence. The sequence should proceed from macro-level details, i.e., from site level considerations to micro-level details, i.e., building-level strategies [2]. In architecture design studio, implementation of these climatic criteria and passive techniques could be achieved by systematic approach. Stage-wise inputs in the site planning stage, building planning stage and building detailing could help students in better understanding. Precedent studies, and other climatology-related inputs made their ideas clear. Hence, the design considerations for climate-responsive architecture should follow majorly three levels study and its integration in the design process (Figure 17.3). As 2nd-year students are aware of climatic factors, integration of Climatology in 2nd-year design could result in better climate responsive design solutions.

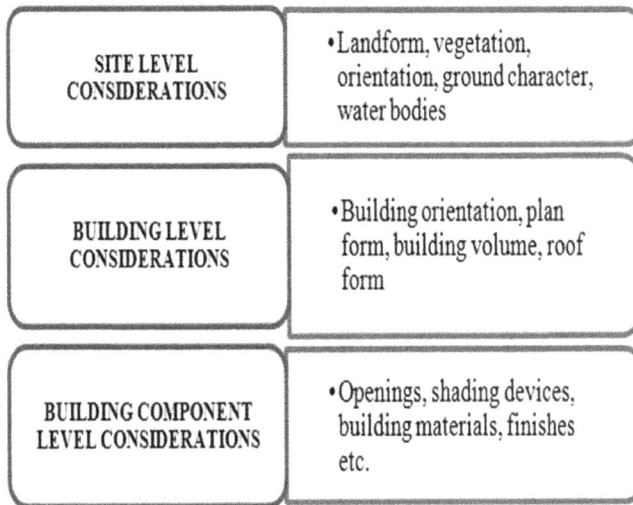

SITE LEVEL CONSIDERATIONS	•Landform, vegetation, orientation, ground character, water bodies
BUILDING LEVEL CONSIDERATIONS	•Building orientation, plan form, building volume, roof form
BUILDING COMPONENT LEVEL CONSIDERATIONS	•Openings, shading devices, building materials, finishes etc.

FIGURE 17.3 Climate responsive scientific process of design.

Source: By Arvind Krishnan, Climate Responsive Architecture, A design handbook for energy-efficient buildings.

17.5.3 ENERGY CONSCIOUS DESIGN MODEL

In the book "New Trends in Architectural Education-Designing the Design Studio," Prof. Salama [1], begins with an introspective view of the current scenario of the architecture profession. In his book, he has presented innovative and practical methods of teaching design in the studio. The systematic stage-wise design model has been proposed by the author named as Energy Conscious Model. As this model is based on climatology and energy-related issues, and have to follow steps in process, therefore can be integrated in third year level of architecture design.

17.5.3.1   ENERGY CONSCIOUS MODEL

This model has been developed by Raymond Cole in the mid 1970's [1]. Cole argues that experience develops when theory is incorporated with design. The main focus of this model is to examine energy issues. Students are encouraged to think within energy terms to influence the manner in which various design solutions are directly and indirectly explored. The model attempts to combine a specified theory with design solutions. The studio work goes concurrently with a directed study, since ideas, issues, and processes are discussed in the context of the evolving design solutions. The process of energy conscious design involves parallelism between the studio exercise and the directed theoretical course. This model focuses on three general phases, each of which has underlying steps that govern the process (Figure 17.4). The phases are separate; after each one, students are asked to respond to the information and ideas within their design proposals.

17.5.4   ENERGY SIMULATION TOOLS

In the 4th year level of architecture graduation, students are more aware of the computer software. Integration of energy modeling or energy simulation software's in design studio is a new experimental approach adopted nowadays [7]. There are many simulation tools available such as 'ECOTECT,' 'RADIANCE,' and 'CONTAM,' etc., to analyze the quantity of energy requirement, based on which, students can evolve design forms and propose the energy solutions.

Leadership in Energy and Environmental Design, 'LEED,' is the widely used rating system used for calculating the sustainability achieved in building. For calculating these ratings, one has to study a number of attributes in depth. With the use of 'BIM,' Building Information Modeling as a design tool, the LEED rating credit points can be achieved in very less time. Hence BIM and other simulation tools can be integrated in the design process as an architecture study for better understanding and awareness of sustainability from their education level.

Phase 1 :   Inputs about energy and architecture
             Implication of energy issues on site and form
Output: Students conduct a site visit, Develop and present solutions.
Building typology: Simple Residence

Phase 2: Application of same issues that have been handled before
Output: Site visit, sketch design & Slide presentation
Building typology: More complex building.

Phase 3:   Inputs about fabric sensitivities, daylight and model studies.
Output: Steps includes site visit, design proposal, model presentation.
Building typology: More complex

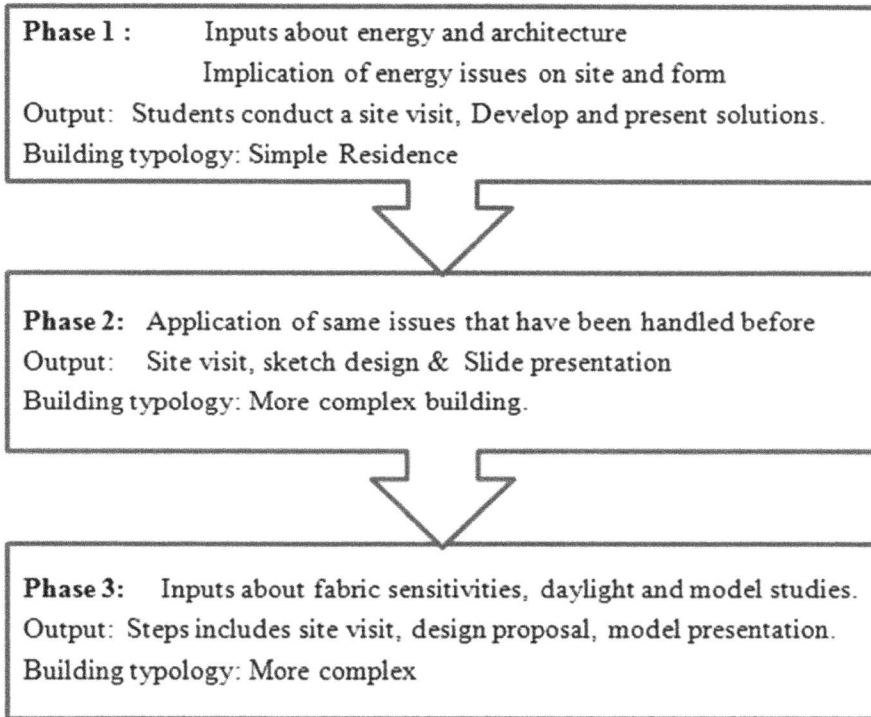

FIGURE 17.4   Phases of energy-conscious model.

Source: By Raymond Cole.

17.6   CONCLUSION

Adopting the conventional design processes in design studios, results in functional, esthetical, and form-oriented outcomes. Integration of the different theories, inputs about sustainability and energy efficiency is a very important ingredient for a successful design studio. This chapter tries to propose, integrated approaches at the different levels of architecture education, which would discuss about the energy efficiency, climatology, or sustainability issues. The chapter also wants to discuss and promote the architecture design studio pedagogy with integrated design processes to meet the current challenges and future requirements for energy-efficient architecture.

KEYWORDS

- **architecture pedagogy**
- **architecture sustainability**
- **design process**
- **design studio**
- **energy-efficient design**
- **solar energy**

REFERENCES

1. Ashraf, S., (1995). New Trends in Architectural Education-Designing the Design Studio. *Tailored Text & Unlimited Potential Publishing*, USA, 0-9647950-0-0.
2. Arvind, K., (2001). *Climate Responsive Architecture: A Design Handbook for Energy Efficient Buildings.* Tata McGraw-Hill Education., 0074632183, 9780074632185.
3. Kolhatkar, S. L. (2015). *Solar Efficient Building Design.* Course material compiled by Urja Mitra-Solar passive Design Consultancy, Pune, India.
4. Priya, C., Madhura, R., & Sonal, F., (2009). *Energy Efficient Buildings: A Comprehensive & Integrated Design Approach.* Time, Space & People; COA Publication.
5. Alpana, R. D., (2011). Climate responsive approach to architecture design. *J. Design Research, 9*(4), 2011–2321.
6. Aparna, T., & Shrutee, D., (2017). An approach towards energy-efficient building through integrated design studio. *IJERT, 10*(1).
7. Khaled, T., Mohamed, A., & Maya, K., (2018). Teaching energy simulation in the architectural design studio: An experimental approach. *ESMT, 2*(2).

CHAPTER 18

Energy Efficient Systems in Traditional Indian Cities

RASHMI DANDE

Priyadarshini Institute of Architecture and Design Studies, Nagpur, Maharashtra, India, E-mail: rashmi.piads@gmail.com

ABSTRACT

Energy-efficient is the buzzword in the present scenario and is often discussed in every aspect of day-to-day life and even in respect to the building industry. This industry plays a key role in the growth of India's economy by creating investment opportunities in the related sectors. The Government's emphasis today is to create necessary physical infrastructure in India and has planned for a massive investment in the building construction sector. In the 12th Five Year Plan, the investment up to 9% as estimated of the GDP. In construction industry, energy consumption is immense and needs to be reduced by substantial efforts for a sustainable future. So, by conserving energy or reducing its use, eco-efficiency has to be achieved. It also achieves improved environment quality. Energy also can be conserved by reducing the wastages, leakages, and losses caused by negligence, by technological advancements and by using improved machines and equipment and undergoing maintenance. Energy efficiency can be very well achieved through efficient building Designs. Energy-efficient Measures can be well achieved in the case of current building designs, but they were prevalent in the ancient building designs and in the design of traditional Indian Cities. The main aim of this chapter is to investigate the energy-efficient systems adopted in ancient Indian cities in water management supply systems, sewage systems, water harvesting systems that were integrated in the building design and housing. This will help to understand the aspect of energy efficiency for a sustainable future.

18.1   INTRODUCTION

Energy efficiency means optimum use of energy, it is the goal to reduce the amount of energy required to provide products and services as stated by Wikipedia, Free encyclopedia. Energy Efficiency means using energy efficiently or effectively, but before we understand the term energy efficiency, we should find an answer to the question why we should save energy. Energy can be categorized as-Kinetic energy and Potential energy, and the different types of energy we use are heat energy, light energy, mechanical energy, chemical energy, electric energy, nuclear energy, etc. And the primary sources of energy are the fossil fuels in the form of oil, natural gas, and coal.

The renewable energy is available in the form of solar energy, wind, and water energy, nuclear energy. With rapid urbanization, the consumption of these various forms of energy has increased tremendously, causing exhaustion of these recourses. So, the efficient use of energy is to be achieved. Secondly, the gap between the supply and demand of energy in every sector is increasing day by day and this gap can be reduced by conserving energy [1].

The construction industry is the biggest consumer of energy, here energy is required in the production of the raw materials, procuring the raw materials from the source, transporting it to construction sites, here energy is required in the implementation of the construction techniques to shape the final product-Building, this energy is called initial embodied energy, After the completion of the building, it will require the energy to function-operational energy, the building in its life span will require maintenance, this will be called recurring embodied energy, and the energy that will be required for the disposal of the building called demolition energy. And if we look into energy consumption in the building industry as compared to other sectors, it is the highest consumer of energy and so we should achieve energy efficiency. Amongst the above discussed four types of energy the maximum consumption in a building is operational energy, which is required for the functioning of the building which can be reduced by efficient HVAC systems, effective water management systems, rainwater harvesting systems, effective gray water recycling and management, effective reduction in the consumption of electricity by using solar energy.

18.1.1   ENERGY EFFICIENCY AND ITS CHALLENGES

The demand for energy is increasing day-by-day and with the exhausting resources, the available energy resources need to be utilized efficiently till we find an alternative resource. Energy Conservation should form a crucial part of the Government of India's (GOI) policies. A comprehensive conservation policy should be formulated to bridge the gap between the supply and demand of energy by reducing the consumption, by spreading awareness amongst the masses, by shaping finances to support energy efficiency, by structuring the city to achieve energy efficiency, by promoting the use of renewable energy resources so that this precious resource is available for future generations. Researchers and scientists are finding a solution to this, and alternative energy solutions are explored as they were the primary source of energy in the past.

18.1.2   ENERGY EFFICIENCY AND CITIES

Buildings collectively with its inhabitants' form cities; they are one of complex systems and the highest consumer of energy. Cities function efficiently with an efficient road network, adequate water supply network, efficient sewage disposal systems, efficient refuge disposal systems, efficient transport system, affordable housing. Apart from this, the city cannot stand or function with its hinterland, its region. A successful city must balance its social, economic, and environmental needs, its recourses. A city should also strive to achieve a sustainable future for its citizens. Urban population is expected to grow by another 2 billion people over the next three decades, [2] and with this rapid expansion of cities managing the energy footprint will be a major challenge for cities. And hence energy management through conservation has become an integral part of urban planning [3].

Cities have built environment, network of roads, and services, and while operating these, large energy is consumed and managing its carbon foot print is its biggest challenge so an integrated approach to energy management will help the Urban Planners in managing this carbon footprint. So, policies need to be made to manage this increasing energy demand, future carbon emissions to achieve sustainable goals, the energy crises have raised due to lack of adequate infrastructure to meet the growing energy demand [3].

By the year 2050, India will have 2.7 million population which accounts to 18% of world's population and at present is India consuming 6% of the world's energy. The Urban areas are contributing to 50% to 60% to the greenhouse gases (GHG), making the task difficult and increasing the pressure on the non-renewable energy recourses, which has further aggravated climate change and global warming [3].

18.1.3   PLANNING ENERGY EFFICIENT AND LIVABLE CITIES

The various examples of cities around the world indicate that, efficient work-ability of a city depends upon its physical form, i.e., the spatial distribution and its dimensions, streets, open spaces, apart from this spatial organization of residential spaces, jobs, services, and environmental amenities. These livable cities follow:

- Compact densities in residential areas, strategically distributed with social and environmental amenities.
- Proper hierarchy of roads interconnected to all the functional sectors of the city, variety of sizes to cater to all types of users like pedestrians, cyclists, etc.
- Strategically located public transport which is easily accessible to the residents.
- Civic facilities and amenities like schools, hospitals, open spaces, etc., should be easily accessible.
- Mixed land uses to be planned in the neighborhoods.
- Climate responsive building designs and neighborhoods should be promoted for high energy efficiency and liveability, which requires consistent urban planning policies, capable local institutions, and able citizen support [4].

18.1.4   INDUS VALLEY CIVILIZATION

Many traditional Indian Cities like Amravati, Mohenjo-Daro, Harrappa, Hampi, Lothal, Madhurai, Nalanda, Vaishali, Takshashila, Prayag, etc., all functioned very efficiently during their times. They were traditionally planned and organically grown; this chapter explores the various systems that achieved energy efficiency in these traditional cities. Among the historic cities that developed under Indus valley civilization were advanced in terms of city planning, they followed a disciplined road pattern, used oven-baked bricks of regular size, the citizens were given the high level of safety, their cities were guarded with fortified walls, they enjoyed a healthy social life and highly sophisticated amenities in terms of great bath as compared to the people who resided in west in Egypt or Mesopotamia at that time. The cities that developed at that time were Kalibangan, Mohenjo-Daro, Harrappa, Dholavira, and Lothal and all has a very strong local government. The Indus

Valley Civilization can be best explained by the structure of its two towns-Mohenjo-Daro as seen in Figure 18.1 and Harrappa, as seen in Figure 18.2 which are located in the Sindh region 40 Km away from the Larkhan Town and Harrappa is located 644 km North-East of Mohenjo-Daro in Punjab province and several other mounds are also there which are under excavation indicating other cities of the civilization.

FIGURE 18.1   The ruins of Mohenjo-Daro.

18.1.5 TOWN PLANNING

The people of Indus valley had planned cities that were an example of advanced knowledge and technical know-how. The City of Mohenjo-Daro was designed on two levels, as seen in Figure 18.1; on the higher level was located the citadel, which had all the administrative areas, and on the lower level was located the residential areas. This was done so that city could be administered well. The cities were designed holistically with a gridiron pattern of road network; it had houses for commons and for elites, Sarai's and inns, granaries, community facilities as great bath, government houses, the city had a centralized administrative system, the central authorities maintained effective drainage system, encouraged rainwater harvesting, they had developed an advanced system for drinking water supply, they constructed wells, they used even-sized bricks, stone, and wood for construction of their houses, In Mohenjo-Daro the houses were planned on both sides of the roads and also in lanes. The residences had doors that opened on the internal streets and not on roads and had high plinths, the design of the residential areas was such that the houses to open on internal lanes and not on the main streets, ensuring the safety of the residents. The design of the houses was climate-responsive as it had courtyards and rooms designed around them. In Mohenjo-Daro, people practiced community living, they have public baths, Figure 18.3 shows the great bath, which was as a place of socializing and granaries to store food grains, local authorities maintained proper cleanliness in the city, and the city dwellers lead a healthy clean and a luxurious life.

FIGURE 18.2   Ruins of Harrapa.

The Great Bath was a water tank for community bathing. It was used for leisure activities and for religious activities also, it reflects unique construction features with the locally available materials. In Figure 18.3, the tank itself measures approximately 12 m north-south and 7 m wide, with a maximum depth of 2.4 m. Two wide staircases lead down into the tank from the north and south, and the floor of the tank was made watertight with gypsum plaster, and with a thick layer of bitumen.

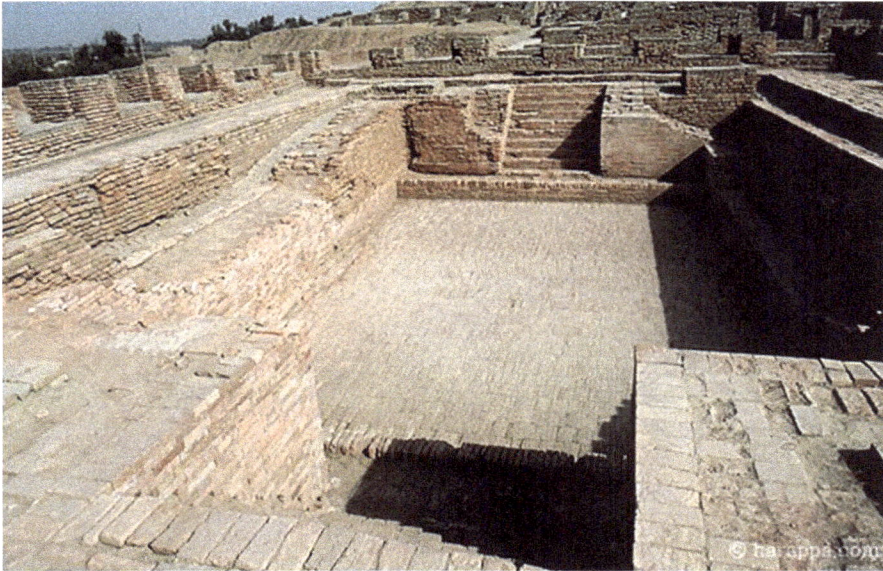

FIGURE 18.3   Great bath of Mohenjo-Daro.

18.2   DRINKING WATER SUPPLY

There were 700 wells in their city, one well for every third house. Figure 18.4 shows a well. These wells were constructed with tapering bricks. The city of Mohen-Jo-Daro received very little rainfall and was situated far away. Hence it was necessary to collect and store water for various purposes. In addition to wells; they constructed reservoir for water storage. These reservoirs were fortified with stonewalls. Figure 18.5 shows the well near the reservoir.

In every house the disposal of wastewater happened through streets, the toilets were placed near the streets for easily disposing the wastewater, ablution taps were placed very close to the toilets, if the toilets were placed on upper floors, they were drained with the help of terracotta pipes, the material used for all types of bends, turns, and spigot ends were also made in terracotta, which was made from naturally available material mud that was baked, and it lasted for more than 5000 years. People of Mohenjo-Daro were well received with water from well and rivers and their dirty water was drained off through sewers [5].

FIGURE 18.4   Well in Harrappa.

FIGURE 18.5   Wells in Mohenjo-Daro.

18.3   SANITATION

The Indus valley civilization had a very efficient sanitary and drainage system, the local authorities of Mohen-Jo-Daro maintained a highly efficient drainage system (Figure 18.6). Each and every house was connected to the main drain. The sewage system had inspection chambers for maintenance. The sewer pipes were connected to the main drains was concealed below the pavement level and was covered with flat stones and sturdy tile bricks. The drain pipes were connected to the larger sewerage outlets which finally took the gray water outside the residential areas. They had earthenware waste pipes to carry sewage from each home to the nearby agricultural fields, rivers, or streams. The waste from kitchens, bathrooms, and indoor toilets was collected; the main drains even had inspection points, as seen in Figure 18.7. The houses had excellent plumbing facilities for provision of water; Toilets had brick seats and were flushed with water from jars.

FIGURE 18.6   Showing the drainage system.

FIGURE 18.7 A drain.

The waste flowed out through clay pipes into a drain in the street. Waste was carried away along the drains to 'soak pits' (cesspits), the sweepers use to take the waste away. They also took the refuge from bins on the side of houses. The drains were usually ranged from 46 to 61 cm below the street level, (Figures 18.8 and 18.9) and varied in dimensions from 30.50 cm deep and 23.00 cm wide when the drain could not be covered by flat bricks, or stone slabs, the roof of the drain was corbeled [5].

The Indus valley civilization had an undivided culture and decentralized form of government. The Indus civilization was spread widely different regions and terrain, yet these regions were knit together by common bonds of religion and culture. The cities of Indus valley civilization had well planned drainage, sanitation system, dockyards, and hydraulic engineering. Their residences had their own wells, bathrooms, and toilets. Every bathroom in the house had a fine sawn burnt brick pavement, often with a surrounding curb. The pipes that started from the bathrooms of the houses were connected to the main sewer on the street. They had equal distribution of resources, and effective utilization amongst the masses. Effective use of the locally available building material (Clay used to manufacture sun-dried bricks) and construction technique, and efficient watersheds and drainage systems and water supply and showcased a holistic approach for the appropriate development [5].

FIGURE 18.8   Showing a drain.

FIGURE 18.9   A drain.

18.4   CONCLUSION

The cites of Indus Valley indicated efficient workability, and it entirely depended upon its physical form, i.e., the spatial distribution and its dimensions, streets, open spaces, apart from this spatial organization of residential areas, and environmental amenities. Apart from this compact density in residential areas, strategically distributed with social and environmental amenities.

Proper hierarchy of roads, interconnected to all the functional sectors of the city, variety of sizes to cater to all types of users like a pedestrian, etc., strategically located public transport easily accessible to the residents. Climate responsive building designs and neighborhoods that be promoted High energy efficiency and livability, which requires consistent urban planning policies, capable local institutions, and able citizen support. This all existed in the planning of the Indus Valley Civilization.

Energy efficiency was very well achieved through efficient building Designs in traditional Indian Cities, and thus this explains us the aspect of energy efficiency for a sustainable future.

KEYWORDS

- **buildings**
- **civilization**
- **embodied energy**
- **energy efficiency**
- **environmental amenities**
- **low energy building materials**
- **sustainable future**

REFERENCES

1. Bhansali, K. V., (1995). Energy Conservation in India: Challenges and Achievements (pp. 365–372). International Conference on Industrial Automation and Control Hyderabad, India, *IEEE*.
2. Khalil, H. A. E. E., (2009). Energy efficiency strategies in urban planning of cities. In: *7th Annual International Energy Conversion Engineering* Conference, 45th AIAA/ASME/SAE/ASEE Joint Propulsion Conference & Exhibit. Denver, Co, USA.
3. Rahiman, R., Yenneti, K., & Panda, A., (2019). *Making Indian Cities Energy Smart,* TERI-UNSW Policy Brief (New Delhi: The Energy and Resources Institute).
4. Serge Salat, Mansha Chen, & Feng Liu, (2014). *Planning Energy Efficient and Livable Cities.* Energy Efficient Cities. Mayoral Guidance Note # 6. ESAMP, Knowledge Series 022/14, 1–2.
5. Khan, S., (2014). Sanitation and wastewater technologies in Harappa/Indus valley civilization (ca. 2600-1900 BC). In: *Evolution of Sanitation and Wastewater Technologies Through the Centuries* (p. 25). IWA Publishing.

CHAPTER 19

Study of Energy-Efficient Building Envelope Design for Climate Change

RUPALI THOKAL

Priyadarshini Institute of Architecture and Design Studies, Nagpur–440019, Maharashtra, India, E-mail: rupali.piads@gmail.com

ABSTRACT

This chapter presents a study of the energy-efficient building envelope design for future climate change. The extremes of climate change are the challenge to the building element and building envelope to better the thermal performance. It was initiated in building design that various external envelope energy-efficient strategies can be implemented depending on the climatic zones. As the climatic zone changes, human body react it as per the change. Same way building envelops need to customize thermal comfort according to climate change. The concept of natural ability of thermoregulation to maintain human body temperature can be similarly applied to energy-efficient building envelope design. Different strategies need study and implement on building envelop design rather than placing high-value insulation material on the exterior of the building to improve energy efficiency and thermal comfort. Climate change was majorly inclined towards the increase and shifts this demand to cooling conquered. In better-insulated buildings to reduce cooling load are more critical. The aim is to identify optimal refurbishment strategies for energy-efficient building envelope design like the concept of thermoregulation of the human body. It was concluding that external envelope energy-efficient strategies are limited and can be implemented to different climatic zones.

19.1 INTRODUCTION

Climate change is for an undefined period of time, changes in the average weather patterns which define "Earth's local, regional, and global climate" [1].

Building "Envelope" usually refers to those building elements that enclosed air-conditioned spaces and through thermal energy is transferred to one space of the external environment, which can be affects the indoor environment [2]. Building envelope consists of all exterior components of the building like roof, skylight, external wall of façade, windows, doors, etc. Building envelope influences building energy consumption and maintain indoor environment of the building, which is important factor consideration for building design. There are different parameters which affects building envelope energy performances by the climate change (e.g., configuration of the exterior fenestration). Constant or sudden change of outdoor temperature of the site may affect the building envelope. Ample amount of work is done related to active and passive building envelope designs in cold-dry climates, hot, and dry climates, warm, and humid climates with a focus on super insulation, extreme air rigidity, high performance fenestration, the probability of passive envelope design in warm and humid climate. Now "aim" is to investigate about the energy-efficient building envelope design for climate change for any zone all over the world. In this study, trying to evaluate which design element are the most sensitive and appropriate to climate change also prove as a universal design strategy for climate change.

It studies that the envelope design can be achieve by the process of conception of the solution in climate change, identifying potential of envelop design measures for energy efficiency building envelope, setting up the envelope models, heating, and cooling load analysis for indoor environment, and making envelope design decisions. Studying and conceptualizing the effect of climate change on energy efficiency solutions may provide acquaintance to the proposal of energy-efficient building envelopes in context with climate change.

The concept for framing methodology is the human body which is the best example for maintains of internal body temperature according to the climate change to some extent.

The deflection of outdoor temperature does not affect the core body temperature. Because of the phenomenon of thermoregulation which helps to maintain core body temperature. If this phenomenon of thermoregulation is incorporate to design the building envelop to achieve energy efficiency during the temperature variation due to climate change (see Figure 19.1). Firstly, need to study different types of passive building envelope system for the different climatic condition. Secondly, study of energy efficiency through building envelope.

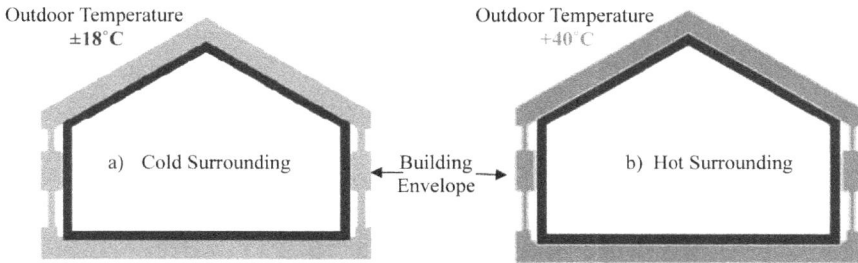

FIGURE 19.1   Building envelope for different outdoor temperature.

19.1.1   LITERATURE

In the past decade, there is insubstantial paper related to energy-efficient building envelope. As per study reported by Bojić in 2006 [3], it "examined the effectiveness of shading device and side fins act as envelope helps: in the energy savings of multistoried residential buildings. Also, the application of fenestration overhangs helped to reduce the electricity consumption up to 5.3% on topmost floors and other floors are in shadows of surrounding buildings. The conclusion specified that insulating the building envelope, it would be "effective in reducing the annual space cooling load on sources up to 38%," That means up to 6.6% can be achieved. The "simulation" results recommend that the strategies on recuperating the thermal performance of external barrier are more "effective" than those for fenestration [4]. This will analyze the energy efficiency of particular envelope elements only rather than in the case of climate change. Nevertheless, considering the warm and humid climate context, how, why, and what would be primary design measures and envelope design are identified was unknown for the climate change. The sources of energy use in buildings are simulated only towards the outdoor temperature. All "studies" are pointing towards the energy consequences of buildings because of the "increased urban temperatures caused by the urban-heat-island and climate change" [5–9].

Majorly climate change and the increase of the "air temperature" are responsible for cooling requirements in the world. Demand of indoor cooling will be larger over the world except northern part of the world. Prediction of climatic conditions to each zone is necessary for application to the building model.

19.1.2   BUILDING ENVELOPE

According to study building envelope design can be design according to regional climatic conditions. Now a day it is difficult to study the pattern of climate in a particular region because of climate change. The amount of heat flows through a building envelope due to outdoor temperature divergence on inside and outside of the building is a function of the magnitude of that difference, the resistance to heat flow by the envelope materials, and the area of the envelope design [10]. As heat flows from hot to cold, if the indoor temperature is more than the outdoor environment, the heat flow will be outward direction to the building. If the inside of the building is cooler than the outside environment, the heat flow will be inward direction to the building.

To design building envelope, broadly, there are two types of building envelop system. One is dual-stage system and single-stage system. A dual-stage system considers as initial barrier for the building with secondary water proofing system for the protection of a building. We can say that the brick masonry veneer wall is one of the examples of dual-stage system [11]. A single-stage system depends on the exterior skin or envelope which is leakage proof. We can say roof membranes and insulated metal panels are the example of single-stage system.

19.1.3   CLIMATE CHANGE AND BUILDING ENVELOPE

To deal with the climate change, some energy efficient techniques are needed to study for change in climatic seasons. The energy-efficient strategies need to study for all types of building. Few principles are utilized to deal with the consumption of energy increase and overheating in case of any commercial building: 'switch off' or/and 'absorb' or/and 'reflect' or/and 'blow away' and 'convict.' The principle 'switch off' is realized with the optimization of solar heat gains in the inside of the building; with the application of shading on exterior and glazing will be energy efficient. The 'switch off and absorption' principle is approached the non-insulated fabric for external insulation. Energy-efficient glazing for the fenestration incorporates both "principles 'switch off' and reflect by removing and 'reflecting' the unwanted solar gains." Instead glazing of fenestration applied as an envelope to the building, the 'reflect' option can be demonstrating by increasing the reflectance of the outer elements of the building and relieving indoor spaces as of extreme high temperatures hours. Automatic sensor ventilation system with mechanized control in the morning hours as well as night hour's ventilation according to the outdoor temperature and the indoor temperature of all zones proves to be more efficient.

The experts supervise all the data for the period 1970–2010, because of climate change, the simulation results identify that cooling load exceed by 33% and 22% reduces heating demand. The performance of insulating material for building envelope is suggested simply for the optimum building extended in the year 2080 [12].

19.1.4   ENERGY-EFFICIENT ENVELOPE DESIGN STUDY

As per building typology consider for envelope, different designs of envelopes are necessary for diverse climatic zones. An Energy-efficiency for building-envelope planning should more concentrate on structure response to the outdoor environment, (i.e., which will benefit maximum from the regional climatic conditions).

In warm and humid climate, the properties of excessive performances building envelopes which allow daylight inside the building from fenestration, preventing unnecessary solar heat gain, better thermal resistance and maximizing natural ventilation [13]. In the energy efficiency, daylight harvesting aims to increase natural lighting, and decrease artificial-lighting utilization and enhance indoor-environment excellence. Heat evaporation includes solar and thermal control during adequately designed shading device; balance glazing, efficient insulation so on. As we know, natural light and ventilation is a successful measure to save the source of energy consumed in buildings, but with climate change it is quite difficult to balance indoor ventilation and to recover indoor air-quality.

The amount of required "energy vanished through the envelope" which influenced by planning strategies and materials used for envelope during the climate change (see Table 19.1) [14]. For climate change design strategies affect the placement of fenestration, the location and size of fenestration which can be optimized to reduce energy convention [15]. The appropriate building envelope materials need to determine the energy performance for the climate change.

19.2   CONCLUSIONS

In this study, the energy-efficient building envelope designs inculcate in the broad level. The impact of climate change on the relative significance of

TABLE 19.1   Basic Performance Requirements for Climate Change-Function and Performance of Building Elements

Systems		Walls	Glazing	Roof	Climate Change Effect	Comments
Principal performance requirements	Thermal	Major determinant	Major determinant of influence	Influence	Influence	Relative quantities largely determine thermal performance of wall and glazing insulation
	Moisture protection	Major determinant	Influence	Major determinant	Minor determinant	All systems important, particularly glazing interface with walls, roof interface with skylights.
	Light transmission	Influence	Minor determinant	Minor determinant	Minor determinant	Glazing determinant, quantity, and location: roof-important if skylights are provided.
	Natural ventilation	—	Minor determinant	—	—	Coordination with HVAC system-essential to ensure energy-efficient system.
	Durability	Minor determinant	Minor determinant	Minor determinant	—	All systems contribute to overall durability of the building.

energy-efficient design characteristics of basic system has been studied by considering increasing outdoor temperature, indoor heating requirement and indoor cooling demand are currently both significant. It was noticed that external building-envelope energy-efficient strategies can be beneficial to different climatic zone which are studied but need to study about climate change in every zone in terms of rising outdoor temperature for all type and functional buildings. There are different perspectives to extensively optimize cooling energy consumption with envelope design strategies and materials for the building envelope. Also, the chapter emphasizes the consequence of incorporating the study of energy modeling towards the design of building envelopes in constricting further energy-efficient buildings.

KEYWORDS

- **building envelope**
- **climate change**
- **energy efficiency**
- **global climate**
- **indoor environment**
- **thermal comfort**

REFERENCES

1. Garcia, R., Cabeza, M., Rahbek, C., & Araújo, M., (2014). Multiple dimensions of climate change and their implications for biodiversity. *American Association for the Advancement of Science, 344,* 1, 2.
2. Li, D. H. W., Yang, L., & Lam, J., (2012). Impact of climate change on energy use in the built environment in different climate zones: A review. *Energy, 42*(1), 103–112.
3. Bojić, M., (2006). Application of overhangs and side fins to high-rise residential buildings in Hong Kong. Civil Engineering and Environmental Systems, 23(4), 271–285.
4. Bojić, M., & Yik, F., (2005). Cooling energy evaluation for high-rise residential buildings in Hong Kong. *Energy and Buildings, 37*(4), 345–351.
5. Kolokotroni, M., Zhang, Y., & Watkins, R., (2007). The London heat island and building cooling design. *Solar Energy, 81,* 102–110.
6. Kapsomenakis, Kolokotsa, D., Nikolaou, T., Santamouris, M., & Zerefos, S. C., (2013). Forty years increase of the air ambient temperature in Greece: The impact on buildings.' *Energy Conversion and Management, 74,* 353–365.
7. Santamouris, M., Synnefa, A., & Karlessi, T., (2011). Using advanced cool materials in the urban built environment to mitigate heat islands and improve thermal comfort conditions.' *Solar Energy, 85,* 3085–3102.

8. Hassid, Santamouris, M., Papanikolaou, N., Linardi, A., Klitsikas, N., Georgakis, C., & Assimakopoulos, D. N., (2000). The effect of the Athens heat island on air conditioning load. *J. Energy and Buildings, 32*(2), 131–141.

9. Kolokotroni; Gowreesunker, L., & Giridharan, R., (2013). Cool roof technology in London: An experimental and Modeling study. *Original Research Article Energy and Buildings, 67,* 658–667.

10. Saladin, K., (2018). *Anatomy & Physiology—The Unity of Form and Function* (8th edn.). McGraw-Hill.

11. Arnold, C., (2016). FAIA, RIBA (pp. 12, 13). Building Systems Development Inc.

12. Attia, S. A., (2012). *Tool for Design Decision Making Zero Energy Residential Buildings in Hot-Humid Climates.* Presses Univ. de Louvain.

13. Scott, M., Wrench, L., & Hadley, D., (1994). Effect of climate change on commercial building energy demand. *Energy Source, 16*(3), 317–332.

14. Wan, K. S. Y., & Yik, W., (2004). Representative building design and internal load patterns for modeling energy use in residential buildings in Hong Kong. *Applied Energy, 77*(10), 69–85.

15. Blezer, D., Scott, M., & Sands, R., (1996). Climate change impacts on U.S. commercial building energy consumption: An analysis using sample survey data. *Energy Source, 18*(2), 177–201.

Part III
Sustainable Environment

CHAPTER 20

Perceptions of Waterfront Open Spaces: Case Studies of Lakefront Spaces of Nagpur City

ARCHANA BELE[1] and UJWALA CHAKRADEO[2]

[1]*Priyadarshini Institute of Architecture and Design Studies, Nagpur, Maharashtra, India, E-mail: archana.piads@gmail.com*

[2]*Smt. Manoramabai Mundle College of Architecture, Nagpur, Maharashtra, India*

ABSTRACT

While urban space planning is often concerned with open space provision and access, measuring the perceived recreation quality of urban space, particularly waterfront, is rarely carried out since users' perceptions are not directly visible. As a result, space might not suit the users' needs and requirements. However, knowledge of recreation quality indicators, such as public perception, behavior, experience, and satisfaction at urban waterfronts can provide useful information for sustainable urban open space development. Hence the study aims at analyzing the recreational quality indicators at waterfront environments so as to investigate the significance of these spaces in enhancing the quality of urban life.

The methodology is based on comparative analysis of identified five lakefronts in Nagpur through six dimensions of urban design, namely morphological, perceptual, social, visual, functional, and temporal. The analysis incorporates qualitative and quantitative approach through the study of morphological maps, activity mapping, mind mapping, observations, questionnaires, walking interviews and photo documentation. The study concludes that the design of lakefront development should cater to the end users' needs and expectations and should encourage a variety of activities and events which will help in offering opportunities for recreation and social activities for wider age group and all class of people.

20.1   INTRODUCTION

For the sustainable development of cities, it is necessary to establish a balance between nature and social life. A natural element like water body plays an important role in the establishment of this balance, particularly in urban areas. Water acts as an important urban planning element for gaining social, physical, and psychological comfort [1] thus increasing the livability of the residents.

While sustainable city and green space planning are often concerned with open space provision and access, measuring the perceived recreation quality of urban space, particularly waterfronts is rarely carried out since users' perceptions are not directly visible. As a result, space might not suit the users' needs and requirements. This leads to a lack of understanding of the relationship between the intended design of such spaces and its experience, use, and perception by the public. Hence investigation of recreational quality indicators, such as public perception, behavior, use, and experience is essential for spatial planning of waterfront spaces which will facilitate in enhancing the quality of urban life.

20.2   DIMENSIONS OF URBAN DESIGN

In the book' Public places, urban spaces,' Matthew Carmona et al. [2] define urban design as the process of making better places for the people that would otherwise be produced. It emphasizes the value of place-making concerning both local and global contexts. They identified six different dimensions of urban design, namely morphological, perceptual, social, visual, functional, and temporal, which are linked and related by the conception of design as a process of problem-solving. Urban design can be holistic if all these overlapping dimensions are considered simultaneously. Hence in this chapter, urban open space in the form of lakefront is studied through parameters of these six dimensions.

20.3   AIM AND OBJECTIVES

The chapter aims at evaluating lakefront spaces in the city of Nagpur based on user perception, experience, and use so as to frame the guidelines regarding the design of such open spaces. The objectives are i) to identify the existing lakefront spaces in the city. ii) to study the identified lakefronts through six dimensions of urban design namely morphological, perceptual, social, visual, functional, and temporal iii) to analyze the lakefronts with a view of understanding people's satisfaction based on their perception, experiences, and use of these lakefronts iv) to formulate directions and guidelines regarding design and development of such lakefronts.

20.4   RESEARCH METHODOLOGY

This study adopts multiple case studies (5 nos.) in the form of lakefronts in Nagpur city. The methodology incorporates relevant literature study and comparative analysis of the identified lakefronts through six dimensions incorporating qualitative and quantitative approaches through the study of morphological maps, activity mapping, mind mapping, observations, questionnaires, walking interviews and photo documentation. Finally, the directions are formulated for the development of lakefront spaces.

20.5   RESEARCH CONTEXT

Nagpur is the Tier-II city of Central India rapidly developing and chosen for round two of the Smart Cities Initiative. The large metro region of the city has a population of about 3.6 million. It is considered as one of the most livable cities with adequate healthcare, green spaces, and public transportation facilities. The city witnesses all three seasons with hot summers, cold winters, and plentiful rains during monsoons. Nagpur is considered as a Lake City with a large number of lakes (about 11 nos.) located in and around it (Figure 20.1).

20.6   ABOUT THE LAKEFRONTS

The chapter incorporates the study of five major lakes in Nagpur. Ambazari Lake, situated in the southwest part of the city was built in the year 1870 under Bhonsla rule. It is the biggest lake under consideration and spread over 380 acres of land. Futala lake situated in the western part of the city was built under the Bhonsla regime. Ancient Futala lake exists for 200 years. Gandhisagar lake historically known as Shukrawari lake is said to be existing for more than 275 years and is situated almost in the center of the city. Chand

Sultan, the then ruler of Nagpur, established it as a source of water supply to the city. Gorewada lake, located at the outskirts is at the north-west corner of the city. Waterworks department developed it in 1912 for water supply to the city that is still serving as the city's primary source of drinking water. Sonegaon lake is situated in the southern part of the city. This 300-year-old heritage structure acting as a significant water body of the city was also built by the Bhonslas. It is the smallest of all the lakes under study spread in about 35 acres.

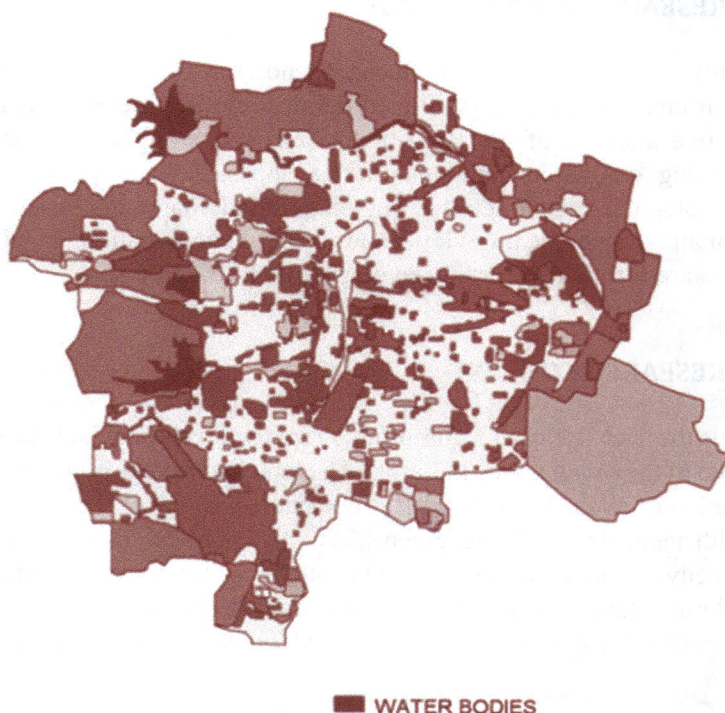

■ WATER BODIES

FIGURE 20.1   Water bodies of Nagpur city.
Source: Nagpur Improvement Trust, Nagpur.

20.6.1   LAND USE AROUND THE LAKEFRONTS

Each lakefront is placed in a unique kind of surrounding development. Ambazari lake is surrounded by garden and no development or agricultural zone. Futala Lake is bordered by the garden on one side, institutional zone, and no development zone on the remaining sides. Gandhisagar lake is placed within a highly congested commercial zone on all sides at the rear side of which lies the residential zone. Gorewada lake is surrounded by park area on one side and no development zone in the form of the forest on the rest of the part. Sonegaon lake is positioned in the residential zone, whereas one side of it abuts the boundary of the transport sector in the form of airport authority land.

20.7   PARAMETERS OF IDENTIFIED DIMENSIONS

20.7.1   MORPHOLOGICAL DIMENSION PARAMETERS

The study of urban morphology helps in understanding the effects and outcomes of thoughts and intentions behind the shaping and molding of our cities. It provides tools to assess and classify the urban form and shape of developments [3]. The major elements of morphological analysis are buildings, gardens, streets, parks, water bodies, etc. These elements are constantly used and hence are liable to transform through time. In the case of urban water bodies, potential mechanisms responsible for fetching morphological changes include intentional removal or addition of water bodies, physical reshaping of water bodies, and selection of it as development sites.

20.7.1.1   SIZE AND GEOMETRICAL FORM

It is a well-established fact that the size, shape, and connectivity of water bodies (lakes, ponds, and wetlands) have important effects on ecological communities and ecosystem processes. Exhaustive modifications of water bodies in the cities during the process of urban development, which include reshaping, construction, drainage, burial, etc., may influence the water bodies. Alterations specifically like construction and reshaping result in a change in the shape and size of water bodies [4]. The same is the case with Nagpur city as some of the lakes were used to be bigger in size, but the haphazard and unplanned development has resulted in the reduction of the size. This holds true for Gandhisagar lake which was about 1.5 times its existing size. The western part was filled up as the city spread towards Sitabuldi fort in the early part of the 20th century [5]. Presently it is spread over 44 acres of area with a rectilinear geometry.

Ambazari Lake is the largest of all spread over 380 acres of land. It is having almost triangular geometry along with few branches rising from one of its edges. Since there is no much development in its catchment area, it is not much modified. Futala Lake is also having triangular geometry with an area of about 99 acres. Today, it is in the midst of the city and its catchment area has very little open space. Gorewada lake, the second largest with an area of 343 acres is having an irregular shaped geometry. Alike Ambazari lake, Gorewada lake being surrounded by forest on all sides is successful in retaining the size due to no development in the catchment area. Sonegaon Lake, smallest of all the lakes, having 35 acres of land is having a rather pentagonal shape.

20.7.1.2 MICRO-CLIMATE EXPERIENCE

Waterfront acts as microclimate modifier and hence help create appropriate living conditions if utilized in a particularly favorable climate. It can provide effective cooling for particular climatic conditions so as to create comfortable living environments [6]. Outdoor space quality is generally determined by the perceived thermal comfort. Thermal comfort is a situation in which people prefer neither cooler nor warmer temperatures, i.e., the ideal [7].

The morphological changes in urban water bodies may also affect the conditions in adjacent terrestrial environments, which includes the effect on climatic comfort. According to Sun and Chen [8], small water bodies cool the surrounding landscape more efficiently than large water bodies per unit area, whereas large water bodies help in greater cooling intensities. They also found that more compactly shaped water bodies (square or round) intensify the cooling effect. While comparing the microclimatic experience, it was found that Ambazari lake being the largest and surrounded by garden on one side with dense plantation felt to be cooler by most of the users. The cool breeze flowing from the lake made the surrounding microclimate felt soothing, leading to thermal comfort. Thus, it was found that the visitation frequency at Ambazari lake during all the times of the day was very high. Futala lake was felt moderately cool, being small in size with not much plantation around it. Hence people mostly preferred to visit it during cool morning and evening times, avoiding the harsh afternoon sun, particular during summers. Gandhisagar lake though compactly shaped was felt very humid being surrounded by heavy traffic roads on all its sides and hence was not much preferred. Gorewada lake, with its huge size but irregular shape and being surrounded by dense forest on one side, felt to be coolest and soothing. It was found to be the most comfortable in terms of thermal comfort. Sonegaon lake being the smallest one with less quantity of water was not found to be much effective in enhancing the surrounding conditions, which resulted in making the surrounding microclimate felt humid (Table 20.1). Thus, in contrast with what Sun and Chen [8] uttered, it was realized that it is not just the compact shape of the water body, but the surrounding development also influence the microclimatic conditions at waterfronts. It was also revealed that the thermal environment of a place affects the decision making of people regarding the time of visit as well as the duration of visit. Consequently, visitors' experience of comfort is greatly associated with the heat and temperature at the lakefront and surrounding spaces particularly during summers [9].

TABLE 20.1   Perception of Microclimate at the Lakefronts

SL. No.	Name of the Lake	Morphology of the Lake	Perception of Surrounding Microclimate
1.	Ambazari Lake		Cool
2.	Gorewada Lake		Cool
3.	Futala Lake		Moderately cool

TABLE 21.1   *(Continued)*

SL. No.	Name of the Lake	Morphology of the Lake	Perception of Surrounding Microclimate
4.	Gandhisagar Lake		Humid
5.	Sonegaon Lake		Humid

20.7.1.3   *PERCEPTUAL DIMENSION PARAMETERS*

The perceptual dimension of urban design emphasizes on the people and how they perceive and value the urban environment. We affect the environment and, in turn, get affected by it. For this interaction to happen we must perceive-that is, we must get stimulated by sight, sound, smell, or touch, which offer clues about the world around us [10]. Usually gathering the environmental information and organizing and interpreting the environmental stimuli are considered two distinct processes which are called 'sensation' and 'perception.'

Though perception is simply a biological process, it is to be socially and culturally learnt. Most of the times sensation is experienced similarly by everyone however, perception might differ from individual to individual depending upon factors such as age, gender, ethnicity, lifestyle, length of residence in an area and on the physical, social, and cultural environment in which a person lives and was raised [2]. Perceptual parameters at lakefront open spaces specifically in urban areas can incorporate visual, auditory, tactile, and psychological effects.

20.7.2   VISUAL PERCEPTION

Water is considered as the most fascinating element on the earth. Various studies related to senses and psychological healing have shown positive effects of water on human beings [6]. Viewing and experiencing the water body and its surrounding areas can activate all our senses and create opportunities for relaxation and recreation. This proved true at all the lakefronts as people rated water as the most visually impressive element at all the lakefronts when they were asked to rate the most visually impressive elements (natural or human-made) on to a scale of 1 to 5. The second most impressive element people rated was the historic basalt stone viewing deck provided at the revetment, particularly at Futala and Sonegaon lake reason being that the decks brought the people in close vicinity with the water. The landscaped garden at the periphery of the Ambazari lake was rated at the third position. In the case of Futala lake, the wide promenade with an interesting paving pattern was rated higher than the landscape (Figure 20.2). At Gandhisagar lake, after water, the landscaped garden was found to be impressive at second position, whereas the bridge that connected the island with the street was at the third position (Figure 20.3). As there was no development carried out along Gorewada lake, people mostly found water as most impressive element in its natural setting. The lake being surrounded by natural landscape in the form of forest, the landscape got second preference after water (Figure 20.4). At all the lakefronts pathways were rated as the least visually impressive element (Figure 20.5).

The findings unveil that though natural elements like water body and green cover enhance the visual quality of space and create psychological satisfaction, human-made elements like stone viewing decks or paved promenades, if designed in harmony with natural elements offer pleasing views and comfort resulting in greater opportunities to refresh, relax, and achieve a sense of balance. The perceived visual quality is positively influenced by natural elements like water bodies and landscape as well as human-made elements like type of revetment and the promenades provided.

20.7.2.1   AUDITORY AND TACTILE PERCEPTION

Here the auditory perception is interpreted in terms of the ability to identify, interpret, and attach meaning to sound heard at the waterfronts. In order to understand the auditory perceptions, the perception of noise levels at all the lakefronts was studied taking into account specifically the audibility of water

FIGURE 20.2   Promenade at Futala lakefront.

FIGURE 20.3   Connecting bridge at Gandhisagar Island Garden.

FIGURE 20.4   Gorewada lake with forest at the backdrop.

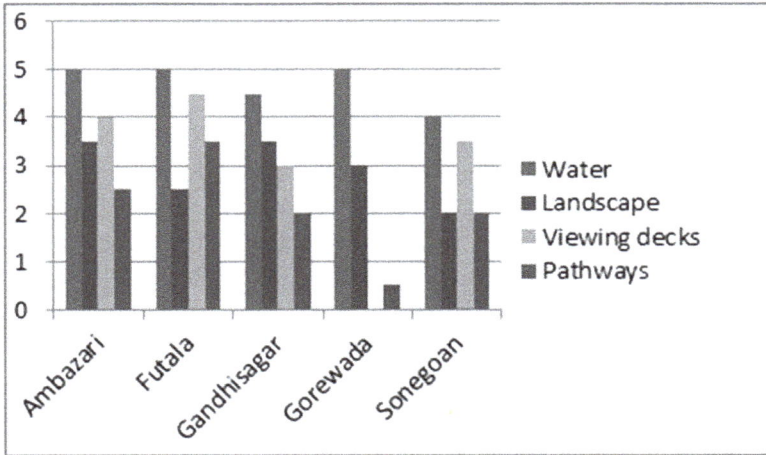

FIGURE 20.5   Rating for visually impressive element.

and chirping of birds. Gorewada lake being located away from the city and heavy traffic road and being surrounded by forest, the noise level was almost nil. As a result, the sound of water or chirping of birds could be clearly heard, making it one of the most preferred bird watching sites as well. Also, Ambazari lakefront, being located away from noisy and heavy traffic roads, the noise level was less and the sound of water and chirping of birds could be easily heard especially from the garden side. Futala lakefront being near to heavy traffic road was having a high noise level. The highest noise was perceived at Gandhisagar lake being surrounded by commercial zone with heavy traffic roads as a result of which, the sound of water and chirping of birds could hardly be heard except at the island garden. Sonegaon lake being located in a residential colony and surrounded by fewer traffic roads, the noise level was moderate. Consequently, people preferred those lakefronts where noise level was the least resulting into calm and quiet milieu (e.g., Gorewada and Ambazari lakefront). The low noise level and high vegetation density which attracted greater biodiversity made these lakefronts the most preferred bird watching sites.

For tactile perception, the surface underfoot was studied and found that almost all the lakefronts except Gorewada were having properly paved pathways that were found comfortable. Instead of paved pathways, Gorewada lake was having nature's trail passing through the forested area in its most natural form without any pavements. Although the trail was uncomfortable, people enjoyed it because of its wild natural appearance and surrounding forest wilderness.

Gorewada, being located in its most natural and quiet setting, was found to be most tranquil as against Gandhisagar lakefront being located in the noisiest locality was found to be most disturbing (Figure 20.6). Thus, as affirmed by Tse et al. [11], besides visual perception, auditory perception of lakefront spaces plays a significant role in the acceptability of urban spaces, particularly lakefronts. Hearing sounds of water and chirping of birds increase the likelihood of offering a higher degree of acoustical comfort to the visitors. On the contrary, hearing sounds of vehicles from heavy traffic considerably reduce the probability of extending high acoustical comfort. Wilderness though found physically uncomfortable, but visually perceived positively by the people and thus appreciated and enjoyed.

20.7.2.2   PERCEPTION OF SAFETY FACTOR

Users' perception of safety is ultimately the result of interaction between the users' characteristics (social, psychological, cultural) and the characteristics of the physical environment in terms of its physical and social context [12]. At lakefronts, safety could be perceived in terms of two aspects, safety from water and safety from crime and antisocial elements. Safety from crime and antisocial activities was perceived high at Ambazari lakefront especially along the garden side as most of the spaces were accessible and filled with people which was evident from the presence of a higher number of families particularly, women, and children whereas it was low at Futala lakefront because of some inaccessible and visually hidden spaces along the promenade which invited some antisocial activities. Noticeably, the most frequent user group at Futala lake was found to be the younger people, particularly during evening hours as compared to the elderly and women. Gandhisagar lake due to its location along the busy street was also perceived safe. On the contrary, Gorewada lake being surrounded by dense forest and difficult in terms of wayfinding, was perceived to be most unsafe. Sonegoan lake being located in a quiet residential area was also perceived safe.

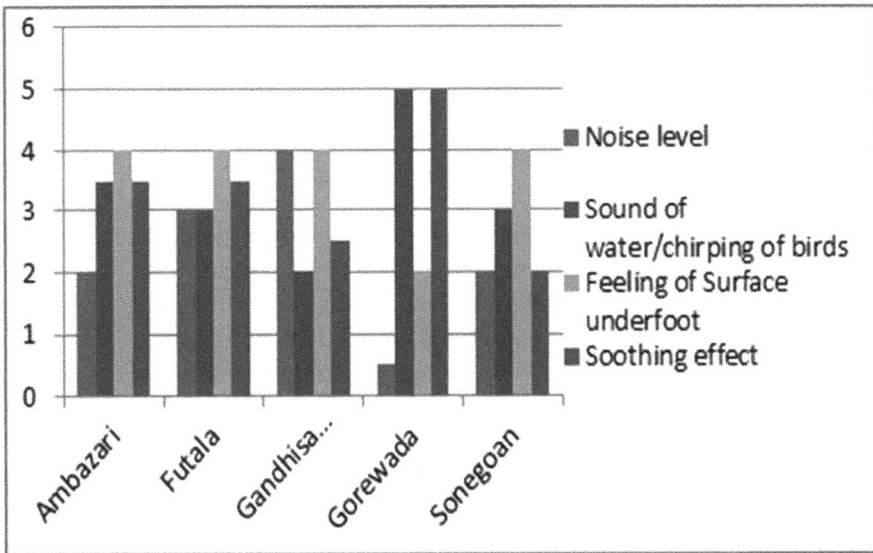

FIGURE 20.6   Rating of auditory and tactile perception.

Talking in terms of safety from water, Futala lake was perceived most safe as the stone viewing decks of revetment were provided with safety walls at the periphery, whereas viewing decks at Sonegaon lake did not have any safety walls which rendered it most unsafe. Proper railings were provided along the water body at Ambazari and Gandhisagar lake for safety purposes. Though Gorewada lake did not have any barrier from water still it was perceived safe as the water level gradually decreased towards the bank and people can safely come in contact with water (Figure 20.7).

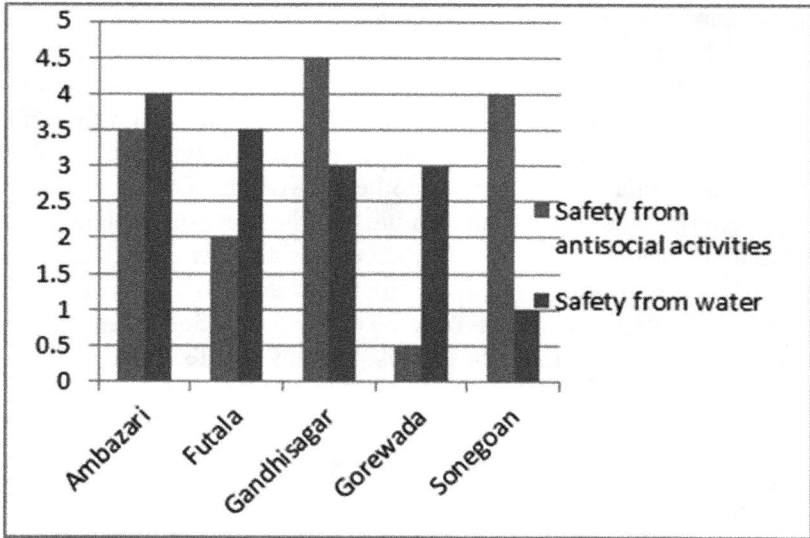

FIGURE 20.7   Rating of safety factor.

Thus, as asserted by İlknur and Rengin [13], a lower perception of safety could discourage users from enjoying the public spaces. Places that are accessible, open, having highway finding ability, and filled with people are perceived to be safe, whereas places that are visually and physically inaccessible are not considered safe. Also, different demographic groups of people perceive safety factors differently where younger people value it lesser as compared to elderly and women. Thus socio-demographic factors were also found to influence safety perceptions greatly. Moreover, it was discovered that even if the lakefront is visually impressive, if it is not perceived safe, it will not be used. Lakefronts which are located near residential areas are preferred more by the public because of easy accessibility, proximity, and connectivity with their residences. Despite the lakefronts being esthetically pleasing, if located in a congested and crowded place or if not easily accessible, it will not be much favored by the people.

20.7.3   SOCIAL DIMENSION PARAMETERS

As the environment influences and changes people likewise people also influence and change the environment. Urban waterfronts projects throughout the world mostly focus on creating places with public amenities that offer activities for leisure, culture, commerce, and hospitality [14]. The social aspects of urban waterfront development incorporate different ways of experiencing and using the edges of seas, lakes, or rivers and understanding their qualities for the community. It should focus on increasing the awareness of planners, decision-makers, and the community about the social and cultural aspects of waterfront areas. Based on the analysis of social impact assessment studies and urban waterfront studies, the social dimension of urban waterfronts can be studied through parameters like access and connectedness, interactivity, and user group and satisfaction level [15].

20.7.3.1 ACCESSIBILITY

Every public space is required to be open and accessible to all regardless of age, gender, ethnicity, and socio-economic status. While analyzing accessibility, it was found that the Ambazari lakefront was majorly accessible from the garden side. Recent development around the lakefront indicated that the top of retaining wall of the lake was also paved so as to make it accessible for walking and jogging, leading to increased accessibility at the lakefront. The part of lakefront behind the Maharashtra Forest Development Corporation office though accessible was not much used by the public since it was mostly perceived unsafe by the public. In the case of Futala lake, the promenade was connected with the street through steps making it easily accessible for the public. It is also the only lake provided with ramps at regular intervals making it barrier-free. Gandhisagar lake was accessible from all its sides as well as from the island garden being surrounded by roads all around. Gorewada lake being surrounded by forest on all three sides, only the part of the lakefront along nature's trail was accessible for the public. In the case of Sonegaon lake, the side of the lakefront where the temples are located was accessible. Rest all the sides being surrounded by retaining walls were inaccessible for the public (Figure 20.8).

Thus, it is important to note that public spaces like lakefronts should provide accessibility to all groups and classes of people in the society and hence should have provisions for making it universally accessible. It was also noted that increase in accessibility results in increased perception of safety because people find places occupied with the public as the safest as that are monitored by the public themselves, which ultimately results in high use frequency. Lakefronts with continuous public access are much anticipated than those which are having discontinuity in accessibility. Stretches of water that are inaccessible to the public deprive the people of experiencing the water.

20.7.3.2   CONNECTEDNESS

In terms of connectedness, except Gorewada, all the lakes were well connected with the major parts of the city by private as well as public transport. Ambazari lakefront was found to be most popular and visited by the people residing in a radius of about 15 km or even more. School trips were also found to be organized for the children from nearby villages as it was well developed and had more recreational facilities for kids. Futala was also found to be popular where people visited from a radius of about 12 km or more. Gandhisagar lake though being very well connected with the city and located in the center of the city was not much preferred because of the surrounding congested commercial development and heavy traffic roads. Gorewada lake being located far away from the city and not being connected properly by means of public transportation was least visited, and mostly, people from a radius of 5 km only visited the lake. Sonegaon lake was located within the city limits and well connected. But since it was not much developed, it was preferred and visited only by the people living in the close vicinity with the lake, i.e., residing within a radius of 3–5 km (Figure 20.8). It was found that lakefronts that are nearer to people's residences were frequently visited, even if underdeveloped. Thus, public transits as well as residential proximity both are found to affect the social connectedness of lakefronts.

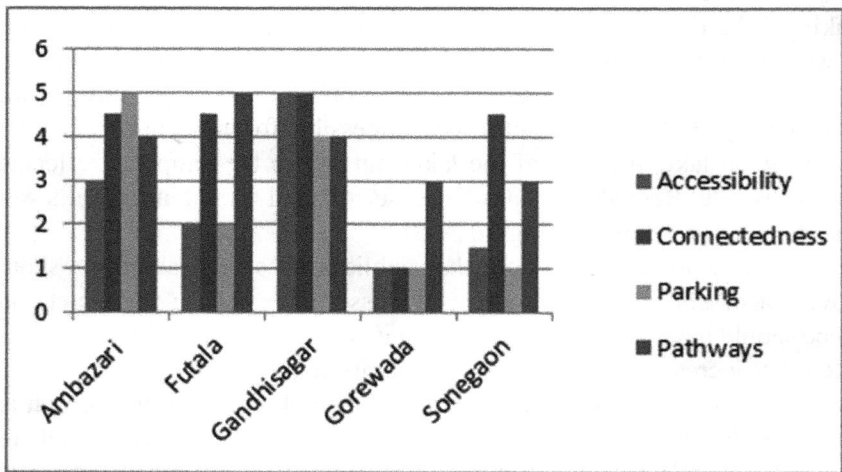

FIGURE 20.8   Rating of accessibility, connectivity, parking, and pathways parameter.

20.7.3.3   PARKING

Ambazari lakefront was having proper parking facility, especially at the garden side. At Futala lake, there was no special provision for parking and people had to park their vehicles by the side of heavy traffic road along the promenade which invited safety issues as well. At Gandhisagar lake, the side of the lake where the island garden was located was having demarcated parking provision along the roadside. In the case of Gorewada and Sonegaon lake also, there was no provision for parking and people had to park the vehicle along the street (Figure 20.8). The most appreciable case was with the Ambazari lakeside garden where proper parking provision was given whereas, for the rest of the lakefronts, people had to park along the roadside, which was not considered as a desirable situation. Thus, it could be inferred those basic facilities like parking provisions are utmost importance which influence the decision making of the people regarding visitation.

20.7.3.4   PATHWAYS

Except Gorewada, all the lakes had well-paved pathways with curbs that were used by the people for walking as well as jogging (Figure 20.8). Though people found paved pathways comfortable for walking and jogging, nature's trail at Gorewada lake in its natural form was also appreciated because of its wild and rustic appearance. At all the lakefronts, there were certain areas where the network of pathways was not reachable, and those areas were considered to be unsafe. So, providing continuous pathways at all places will ensure that the place is visited by the people frequently, leading to improved safety factor.

20.7.3.5   INTERACTIVITY

When the space encourages people to meet their friends, greet the neighbors and feel comfortable to interact with the strangers, a strong sense of place and attachment is felt by the people, which help in fostering social activities [16]. In affirmation to this, Rad et al. [17], point out that interacting with each other establishes a strong bond between their society and the space. Ambazari lakefront garden was rated highest in terms of interactivity as it provided facilities that catered to all age groups and classes of people. Also, the sitting arrangements provided was such that it encouraged interaction. Similarly, Gandhisagar island garden was also having high interactivity amongst all age groups of people due to facilities like kids play zone and outdoor gymnasium. Futala lakefront was also having high interactivity level, but it catered mostly to younger people due to lack of facilities and provisions for other age groups. The safety wall at stone decks played a major role in increasing interactivity. Sonegaon lakefront was having a moderate level of interactivity, whereas Gorewada lakefront was having very little interactivity as it was mostly visited by the people coming for walking or jogging only (Figure 20.9 and Table 20.2). Accordingly, it is evident that the type and arrangement of sitting elements affect the interaction level. Thus, interactivity is directly related to the kind of facilities provided and will be high if it satisfies the needs of all age groups of people. Consequently, it could be measured and evaluated by the presence of different social groups and types of facilities provided.

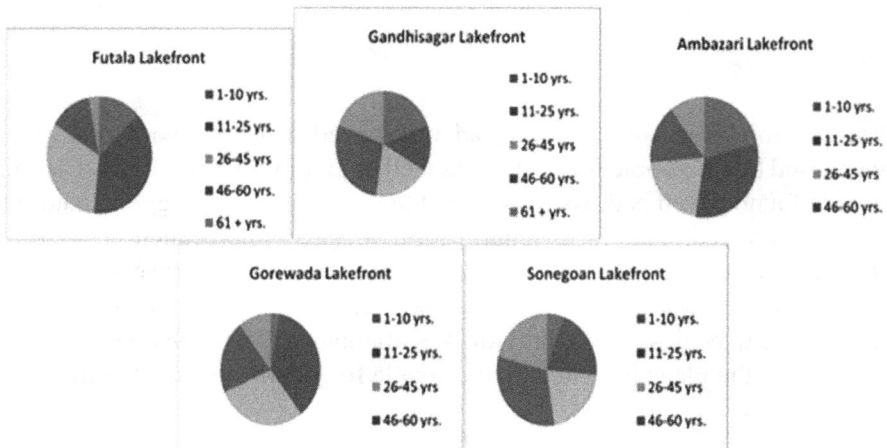

FIGURE 20.9   Age wise percentage of users at the lakefronts.

20.7.3.6   PRIME USER GROUP

At Ambazari lake, since the lakefront development catered to all the age groups, it was found to be used extensively by all age groups of people, including children, youngsters, and the elderly. Futala lakefront was mostly used by the younger people, particularly along the wider safety walls, which could be used for sitting near the water while interacting. At Gandhisagar lake, only the island garden was used majorly by the children and elderly people. The rest part which was abutting heavy traffic roads was left unused. Gorewada lakefront, being away from the city and not much developed and families did not prefer to visit. Hence it was used mostly by the youngster and elderly for jogging, walking, and bird watching. Sonegaon lakefront also did not have provision for children's activities and was primarily used by the elderly because of the presence of the temple complex along the lakefront (Figure 20.9 and Table 20.2). So, it could be analyzed that providing various types of facilities is of paramount importance not only for attracting users from various demographic groups but also for increased interactivity amongst them.

20.7.3.7 SATISFACTION LEVEL

The satisfaction level was highest at Ambazari lakefront as it catered to all age groups of people whereas it was lowest at Sonegaon lake as it was catering mostly to the elderly people. At Futala lake because of the viewing decks and surrounding eateries, satisfaction level was high, particularly amongst youngsters. Gandhisagar lake though located at the center of the city was not much preferred by the people, and satisfaction level was high only at the island garden situated inside the lake. Gorewada lake though not much visited by the people being away from the city, undeveloped lakefront in its natural wilderness was greatly appreciated by the people in terms of its scenic beauty (Table 20.2).

TABLE 20.2 Rating of Sociability Parameter

Lakefront	Sociability Parameter			
	Interactivity	Prime User Group	Spaces for Socialization and Celebrations	Satisfaction Level
Ambazari	Very high	Children, youngsters, and elderly	✓	Very high
Futala	High	Mostly youngsters	✓	High
Gandhisagar	Moderate	Children and elderly	✓	Moderate
Gorewada	Very low	Youngsters and mostly elderly	×	High
Sonegaon	Moderate	Youngsters and elderly	✓	Low

20.7.4   VISUAL DIMENSION PARAMETERS

The visual dimension of the urban environment is the outcome of human perception and cognition. This means how the environment appeals to our mind and emotion involves perceiving the stimuli along with processing, interpreting, and judging the information gathered. Beyond the expression of simple taste, esthetic appreciation also has to be socially and culturally learnt. Gordon Cullen [18] envisaged the concept of 'serial vision' to explain the visual aspect of the townscape. Accordingly, to generate drama and visual interest and to strengthen or develop the sense of place, it is necessary to consider buildings, streets, hard and soft landscaping, and street furniture collectively.

Most waterfront spaces attract users because of the impressive views they present. Architectural considerations and landscape design can enhance the visual quality of any waterfront space. Various physical features like water, vegetation in the form of trees and shrubs, street furniture, and pathways are highly valued as they attract the public towards waterfront spaces. Visual contact with the natural environment can relieve people from the stresses of daily life [19].

Although the visual effect is a subjective concept, there are some general esthetic standards and common sense that can be quantified. At urban waterfronts, natural elements are dominant and important for the visual effect. To get a good visual effect, skyline, and subjective experience are the key features. At Ambazari lake, attractiveness in terms of the skyline was high since the skyline did not show much development in terms of built form. It was also having a considerable green cover which complemented the vast amount of water in the lake rendering the sight appealing. Though Futala lake lacked in terms of green cover, particularly along the promenade, extensively paved pathways in gray color with interesting patterns complemented with the massive historic decks built out of black basalt stone, forming interesting foreground for the lake, and enhancing the attractiveness of the lakefront. In the case of Gandhisagar lake, the green cover was limited only at the island garden and the rest all the lakefront space was having sparse vegetation in the form of roadside tree plantation. Only the island garden formed a major attraction along with the bridge which connected the garden with the street (Figures 20.10 and 20.11). But the haphazard development at the periphery revealed the dominance of human-made features over natural features in the skyline.

Gorewada lake topped in terms of attractiveness since the forest present on three sides formed the background to the huge amount of water present in the lake, adding to the scenic beauty of the lake in the absence of manmade features around. Sonegaon lake though situated in a residential zone did not appear much attractive because of less quantity of water, moderate vegetation, and low-rise buildings at the periphery. Only the massive historic black basalt stone viewing decks served as a major attraction (Figures 20.10 and 20.12). Interestingly, all the users acknowledged that watching the water and the scenic beauty around the lakefronts helped them in achieving psychological comfort and relaxation.

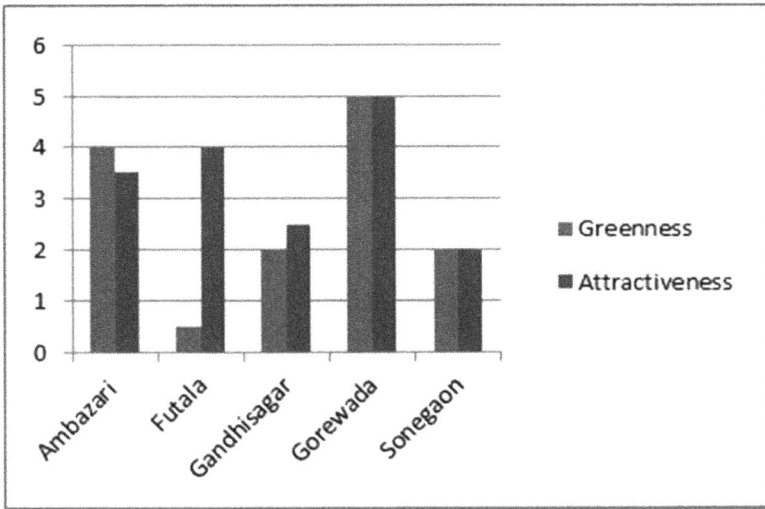

FIGURE 20.10   Rating of greenness and attractiveness parameter.

Thus, as analyzed by Zhao Ye [20], a good skyline needs appropriate fluctuation, landmarks at the exact position, and natural link between city and nature. The variety and color coordination are equally important, the higher the better, while the proportion of manmade and natural areas is correlated inversely, the less the better. Having said that, it is also found that, the human-made elements at waterfronts, if designed in harmony with the natural features, could heighten the allure of the lakes.

20.7.5   FUNCTIONAL DIMENSION PARAMETERS

Carr et al. [21] assert that apart from being meaningful and democratic, public spaces should also be responsive, that means designed and managed to serve the needs of the users. There are five primary needs that people

FIGURE 20.11   Island garden in Gandhisagar lake.

FIGURE 20.12   Stone decks along with Ganesh and hanuman temples forming the backdrop to Sonegaon lake.

seek to satisfy in public space, which include comfort, relaxation, passive engagement and active engagement with the environment, and discovery. Successful places often serve more than one purpose.

Functionality and vitality of waterfront can be determined by public activities, the intensity of use, and accessibility by all user groups. Yamashita and Hirano [22] recorded waterfront activities into three types: activities on the water, activities on the bank, and group activities. Some design features support user's activities, whereas some features serve as constraints [23]. Engaging physically with water through activities like swimming, splashing, walking in shallow waters, etc., help in providing various sensory and psychological experiences.

20.7.5.1   USE FREQUENCY ALONG THE LAKEFRONTS

In order to investigate the spaces which people prefer to use, onsite mapping was carried out at the lakefronts by demarcating the high, moderate, and low frequency or inaccessible areas around the lakefronts. At Ambazari lake, the garden side was the high-frequency use area whereas the retaining wall used mostly for jogging and walking was moderately used. The rest of the areas around the lakefront were rarely used being inaccessible. Futala lake was having the highest frequency along the promenade side and along the food kiosks, whereas the rest of the part being difficult to accessible was having the lowest use frequency. At Gandhisagar lake, the highest frequency use area was the island garden located inside the lake, and the street along that side was widely used due to the presence of informal sectors in the form of tea stalls, etc. The rest of the area was moderately used. There were no areas inaccessible at Gandhisagar lake being surrounded by roads on all sides. At Gorewada lake, the area along nature's trail was only used extensively. Other parts of the lake abutted the forest safari area hence entry to the lake was prohibited from those sides. Thus, accessibility of the space determined the frequency of use. Easily accessible areas were found to be frequently visited and used by the people.

20.7.5.2   STREET FURNITURE

Ambazari lakefront was having proper provision for benches below the trees that were facing the lake. It also had few raised gazebos for people to relax and at the same time enjoy the view of the lake. Sitting provision near kids' play zone was highly appreciated by the people as it helped them in keeping an eye on the kids while relaxing and interacting. Futala lake had the provision of benches, but people preferred to sit on the stone safety wall by the side of the water. Gandhisagar lake had provision for benches below the trees at the island garden as well as along the roadside footpath facing the water. Gorewada lake did not have many provisions for benches. It had only one steel gazebo at the lakefront. Hence people did not prefer to visit the lake during day time due to a lack of protection from the harsh sun. Sonegaon lake also had the provision for benches, but due to lack of shady trees, those were also not much used (Figure 20.13).

Thus, position of sitting arrangement is important for enhancing the comfort level. People seek shady areas for sitting and relaxing. Also, the benches that impart a view of the lake are significant as people like to see the water body while sitting and relaxing.

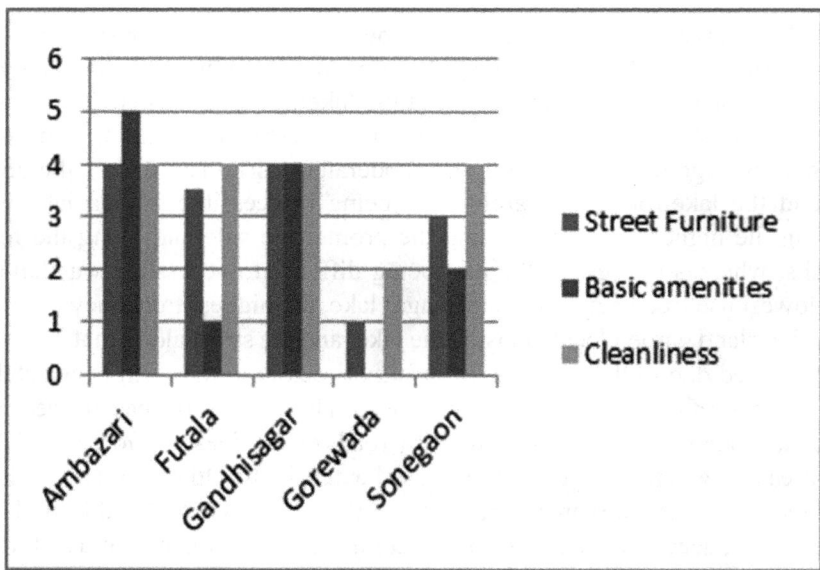

FIGURE 20.13   Rating of street furniture, basic amenities, and cleanliness parameter.

Apart from seating, signage, and lighting are crucial elements which facilitate and enhance the safety and comfort level of the visitors. It was also observed that the position of lighting and signage was not appropriate and sufficient at all the lakefronts, which led to security issues, particularly during evening times. The absence of such elements resulted in inconvenience and discomfort to the uses.

20.7.5.3   BASIC AMENITIES

Ambazari, Futala, and Gandhisagar lakefront had proper provisions of toilets, dustbins, etc., whereas Gorewada and Sonegaon lake lacked these basic amenities and hence were not much preferred as people seek such provisions at public spaces (Figure 20.13).

20.7.5.4   CLEANLINESS

All lakefronts like Ambazari, Futala, Gandhisagar, and Sonegaon were found properly maintained and clean except Gorewada lakefront which was not much maintained and hence lacked cleanliness. But a lot of displeasure was observed in the case of physical quality of water at all the lakefronts as the water surface was covered with garbage and trash, particularly along the water's edge leading to environmental as well as visual pollution. Gandhisagar and Gorewada lake also showed the growth of algae on its surface (Figure 20.13).

20.7.5.5   PASSIVE ENGAGEMENTS

As described by Project for Public Spaces [24], comfort not only includes perceptions about cleanliness and safety but also about the availability of places to sit. The significance of giving people the options to sit where they want is generally underestimated. This leads to the fact that people need to be given a number of choices in terms of sitting provisions where they can decide their favorite places to relax and interact. As discussed earlier, almost all the lakefronts were having proper sitting provisions in the form of benches, gazebos, etc. At Futala people also used the stone safety wall along the side of water and wide steps leading towards the lake for sitting and relaxing. It was also found that people also enjoyed sitting on the lawn, which served as greenery for people to sit and rest. Thus, more choices for sitting should be provided in order to improve the comfort level (Table 20.3).

20.7.5.6   ACTIVE ENGAGEMENTS

20.7.5.6.1   WALKABILITY

Walkability was also good at all the lakefronts since promenades and a proper network of pathways was provided at all lakefronts except at Gorewada. But nature's trail along the side of the lake passing through the forest though uncomfortable, added thrill to the walk due to its wild natural form. Thus, good quality paving, pathways, and promenade placed at appropriate locations imparting attractive views contribute to the pleasant walking experience (Table 20.3).

TABLE 20.3   Uses and Activity Parameters

Lakefront	Uses and Activity Parameters			
	Provision for Sitting	**Walkability**	**Activities**	**Food and Other Facilities**
Ambazari	✓✓✓	✓✓✓	Both leisure and sports (active and passive engagement)	✓
Futala	✓✓✓	✓✓✓	Mostly leisure (passive engagement)	✓✓✓
Gandhisagar	✓✓✓	✓✓✓	Both leisure and sports (active and passive engagement)	×
Gorewada	✓	✓	Mostly leisure (passive engagement)	×
Sonegaon	✓✓	✓✓	Mostly leisure (passive engagement)	×

20.7.5.6.2   ACTIVITIES

As asserted by Efroymson et al. [25], a variety of activities should occur in public spaces as it acts as an indicator that the place has much to offer in terms of spontaneity, energy, creativity, and livability. Accordingly, Ambazari lake involved the people in both active and leisure activities in terms of jogging, walking, playing, exercising, yoga, fishing, boating, etc., since it had the provision for children's play equipment, paved pathways, landscaped gardens, outdoor gymnasium, etc., which attracted all age group of people. This resulted in the formation of a lively atmosphere at the lakefront (Figure 20.14). At Futala lakefront people were mostly involved in leisure activities of observation of the lake and other people's activities apart from walking, jogging, and sometimes fishing also. Another major activity along the lakefront was eating at the food stalls and street vendors. At Gandhisagar lake also, people were mostly involved in passive activities of observing the natural environment and the people around them apart from jogging and walking. At the island garden, facilities like children's play equipment, outdoor gymnasium were provided, which were extensively used by the people. At Gorewada and Sonegaon lakefront people preferred to involve in passive activities of observation and active engagements in terms of walking and jogging as there was no provision for play equipment, landscaped garden, or open gymnasium (Table 20.3).

FIGURE 20.14   Lively atmosphere at Ambazari lakefront.

Lakes like Gandhisagar and Sonegaon were having historic temple complexes situated at the banks. Thus, these lakes proved to be of historical significance and development around it contributed to the important heritage structures of the city. The temple complexes not only served as worship places for local residents but also as meeting places for young and elderly people (Figure 20.15). They also act as platforms for festive celebrations during various religious and cultural events. Such social activities promote local heritage and concerns for sustainable development and assist in keeping the local legacy alive [26].

FIGURE 20.15   Historic Mahadev temple at Futala lake serving as interaction space for the elderly.

20.7.5.7   FOOD FACILITIES

The food facility was available at Ambazari lake only in the form of a small cafeteria and a few street vendors. Futala lakefront was most developed in terms of food facility with a series of food outlets present along the side of the road as well as street vendors which also served as the main attraction and fascinated a large number of people. The rest of the parks did not have any food facility, which was not considered as much preferable situation as people showed the desire to relish food while relaxing and enjoying at recreational spaces (Table 20.3).

20.7.6   TEMPORAL DIMENSION PARAMETERS

Time has an important role to play while transforming spaces into places. Time makes the space more meaningful. Urban designers need to understand activity patterns and design so as to encourage activities through different time periods. According to Lynch [27], although activity timing is as important as activity spacing, it is less often consciously manipulated. It is generally observed that many spaces are used greatly for a certain period of time and then remain empty for longer periods. This holds true for lakefronts like Sonegaon and Gorewada as people visited those only during morning and evening hours. As such, it was observed that all the lakes were used extensively in the morning and evening hours for carrying out activities like walking, jogging, exercising, and yoga. But lakefronts like Ambazari and Gandhisagar were used during the afternoon period as well because of the gardens placed along the lakefronts, which were having provisions for children's play equipment, gazeboes, benches under shady trees, etc. Futala lakefront was used during night time also because of the food stalls and recreational facilities along the promenade.

Nevertheless, successful public places are those that are full of people day and night. Apart from providing opportunities for carrying out day to day activities like jogging, strolling, playing, viewing, they also serve as a stage for the celebration of life like festivals, public performances, market-places, and other energetic gatherings. Environmental factors like sun, rains, or cold should not let those remain empty. Unfortunately, no lakefront was developed keeping in mind the seasonal changes that occurred throughout the year. As a result, during summers, the lakefronts were mostly used during morning and evening hours. During winters, because of getting colder, lake-fronts were not used very often. During rains also, no lakefronts were used much since there was no provision for seeking protection from rains. Hence while considering the temporal dimension of urban design, it is important to understand the impact of time on places.

20.8 DISCUSSION

Based on the study, it is found that the size of the water body does not matter when it comes to the use of waterfront by the public. What matters is the location of the water body, its accessibility, and connectivity with the rest of the city areas. Apart from the size and shape of the water body, the surrounding development affects the microclimate greatly. Water bodies surrounded by high development around cannot effectively modify the microclimate in terms of cooling the adjoining areas. On the contrary, dense vegetation at the periphery of the water body helps in creating comfortable micro-climatic conditions, particularly in cities with hotter summers.

People find natural elements like water as the most visually impressive element. Nevertheless, human-made element like stone viewing deck is also cherished by the people if designed in harmony with nature as it gives people the opportunity to observe and enjoy water closely. Thus, revetments should be designed and built with natural or semi-natural materials instead of concrete which will create a harmonious and pleasing visual association with natural elements present at the waterfronts.

Noise level which ultimately determines the serenity of the lakefront plays a major role. Calm and quiet lakefronts are preferred by the people. Safety is of utmost importance in public places. Hence the lakefront design should aim at making the spaces open and easily accessible, physically as well as visually, through suitable design strategies and techniques which will ensure high visitation frequency amongst all social groups. Safety from water is equally important and safety elements should be carefully located and designed as those also serve as elements that bring people in close vicinity with the water. Even if it is not possible or feasible to give people the opportunity to actually touch the water, the design of safety elements should ensure the proximity of people with the water.

A well-designed network of continuous pathways should run throughout the site making all the places easily accessible that will result in achieving higher perception of safety. Inclusive and universal design approaches need to be taken into account, particularly while designing public spaces like lakefronts. Properly developed public transport infrastructure should be facilitated, which will assist in increasing social connectedness and reduce social segregation resulting in optimum utilization of public spaces.

Lakefronts are highly appreciated for the interesting views and scenic beauty they offer which affect the psychology and emotional state of the visitors. Hence a systematic approach towards the design and development of waterfronts could assist in reconstructing the presently lost relation between humans and water, thereby bringing people back towards water. Activities like fishing and boating are still enjoyed by the people at lakefronts. Hence high-frequency use zones are those areas where various recreational facilities along with basic amenities are provided. Thus, the design of lakefront spaces should make proper provisions for basic amenities like restrooms, parking, proper sitting and lighting provisions, food facilities, etc., at appropriate places. Design should also encourage use of the lakefronts during all seasons and all times of the day.

A variety of facilities that involve people in active as well as passive engagements should be provided at lakefronts as its presence is directly related to the usability and interactivity amongst various user groups. Therefore, the new trend of outdoor fitness zones established by the development authorities at few lakefronts, which involved people in active engagements, is welcomed heartily by the people as it takes care of fitness as well as recreation in an open natural environment. Thus, the satisfaction level with those waterfronts which catered to higher facilities and diversity of functions for a variety of user groups is greater amongst the people and hence such facilities should be taken care of and maintained.

Lakefront spaces should facilitate social interactions amongst people by accommodating voices of people from all walks of life, removing barriers that discourage interactions and serve as stages for exhibitions, performances, and recreational activities, thereby encouraging sociability. Making provisions for a variety of sitting facilities like benches, gazebos, decks, lawns, etc., will give people a choice for selecting their desired destination for interaction with others as well as with nature. It was evident from the presence of a large number of youngsters at all lakefronts that water bodies attract the younger generation greatly whether it is developed or not whereas children and elderly people are sensitive towards the development of the lakefront and use only if it is properly developed and found safe. The absence of certain facilities and physical features and lack of management contribute towards public dissatisfaction and fail in attracting visitors.

Cultural heritage at waterfronts should be valued and conserved as it has an important role to play in terms of keeping the city inheritance alive, thereby helping in the maintenance of such structure and serving as gathering and meeting spaces for the locals. They also facilitate in raising awareness regarding the cultural heritage of the city among the residents.

Thus, it is important to acknowledge waterfronts as a landscape that comprises natural and cultural components which help in reinforcing social and cultural connection amongst the citizens [28]. According to Roberts [29], waterfronts, instead of being a mere place for tourist attraction, should function as a diverse and culturally vibrant urban landscape reflecting a variety of social and spatial practices, thereby assuring its sustainability. The creation of a city's image can be a planned process [30]. Apart from the formation of a friendly urban place for residents, city planners should strive for the development of the appropriate image of a city. This will not only contribute in fetching profit by attracting tourists but such development will also be enjoyed by generations to come.

20.9   CONCLUSIONS AND RECOMMENDATIONS

Based on the above study of perceptions of the people at all five lakefronts, the following recommendations are suggested for the development of each lakefront from the user's point of view.

At the lakefront garden of Ambazari, for the safety of the people, steel railings are provided by the edge of the lake, which could be replaced by wider safety walls in a few places where people can sit comfortably while enjoying the view of the lake. Care should be taken those railings are retained, and no walls are provided in front of the benches facing the lake in such a way that will hamper view of the lake while sitting on the benches. For enhancing accessibility for all social groups, provision of ramps at appropriate locations is sought for. Also, inaccessible part of the lakefront behind Maharashtra Forest Development Board could be developed and made accessible so that a large number of people could be benefitted by the facilities along the lakefront.

In the case of Futala lakefront, vegetation cover at the promenade could be improved by planting shady trees which will solve the dual purpose of creating a comfortable micro-climate and providing shade so that the lakefront could be utilized during all times of the day. This will aid in addressing the issue of a lack of green cover. Sheltered spaces (e.g., Gazeboes) could be provided with sitting provisions so that the lakefront could be utilized during all seasons and different times of the day. Design of sheltered spaces should be such that it would not interfere with the view of the lake. Heritage elements like basalt stone viewing decks and revetments should be retained and conserved. The existing encroachment on the south side could be removed, and the area could be developed and made accessible both physically as well as visually. Futala lake being used extensively during evening hours, inaccessible, and visually hidden narrow spaces along the promenade should be widened to enhance physical and visual accessibility. Adequate lighting provisions should be made to reinforce the sense of safety. Development of a kids play zone would help in attracting more number of children, women, and families towards the lakefront.

At Gandhisagar lake, though the island garden is having a considerable vegetation cover, the rest of the areas surrounding the lakefront are having sparse vegetation, which could be enhanced so that the surrounding areas could be felt thermally comfortable. The vegetation cover will also help in achieving the noise barrier from surrounding heavy traffic roads and assist in elevating the esthetic value. Except island garden there are very few places from where people can comfortably come in close contact with the water.

To address this issue, the existing heritage elements in the form of basalt stone viewing decks which are right now in poor condition and neglected, could be restored and maintained so that people can come in close vicinity with the water. The island garden needs to be made barrier free by making provisions for ramps at appropriate locations. Improving lighting conditions could aid in maximizing the use of space during evening times as well. The historic Ganesh temple and Mahadev temple on the banks of the lake should be preserved as these places not only promote local history, but also serve as stages for festive celebrations and social events.

In case of Gorewada lake, the edge of the lake along the garden is undeveloped and undulating. It could be provided with a narrow promenade and viewing decks with safety walls made out of natural material like stone which will blend with the surrounding natural environment and help people enjoy the view of the lake and forest while being close to water. As people like the wilderness of the nature's trail, the surrounding dense, wild natural landscape should be retained. To provide ease of walking for the joggers, it could only be provided with narrow pathways made out of natural stone without disturbing the naturalness of the trail. Being located at the outskirts of the city, public transport needs to be improved so that people from all parts of the city could visit the place frequently. Apart from this, maintenance needs to be improved which is lacking at present. No evening activity should be encouraged that may adversely affect and disturb the biodiversity, i.e., flora, and fauna surrounding the lake and create safety and security concerns. To address the issue of lack of sitting provisions, benches could be provided below the existing trees arranged in such a way that will encourage interaction amongst the visitors. Sitting arrangements facing the lake would also be highly appreciated so that people can interact passively with the nature while enjoying the scenic beauty. The materials used for making sitting arrangements should be made out of natural materials like stone, timber, or bamboo. Facilities like outdoor exercise equipment or kids play equipment should not be encouraged as those might disturb the peaceful and serene ambience of the forest. Thus, utmost care should be taken while developing the lakefront, keeping in view the fact that people appreciate its high degree of calmness, naturalness, and scenic beauty, which need to be kept undisturbed and preserved.

Sonegoan lakefront also has a lot of scope for improving the vegetation cover, particularly along the temple complex side. The abruptly ending jogger's track should be redesigned to maintain its continuity for the comfort of users and should run through inaccessible places that will enhance the sense of safety at all places. It also needs to conserve the historic basalt stone viewing decks.

They should be provided with safety walls made out of the same material alike Futala lake, which would increase the safety factor at the same time could be utilized for sitting, interacting, and enjoying the view of the lake. Better provisions of lighting could encourage the use of the lakefront during evening hours. Looking at the dearth of sitting provisions, more sitting arrangements could be provided with shady trees around for physical comfort and use of the space during different times of the day. Facilities like kids play zone and outdoor fitness zone should be provided so that people of all age groups could enjoy a variety of activities. The historic temple complex on the banks of the lake should be preserved as it not only serves as a place of worship, but also as a meeting place while promoting festive events and celebrations.

It is observed that there is an urgent need for the development of separate parking zones at all lakefronts except Ambazari lakefront garden since people need to park their vehicles along the side of heavy traffic roads in the absence of better parking facility. Similarly, all lakefronts need better restroom facilities. Poor quality of lake water is also a major concern at all the lakefronts. Proper measures should be taken to ensure better water quality to check the physical and visual pollution of water. Apart from this, all the lakefronts need better food facilities which is highly appreciated and desired by the users of all age and social groups. Organizing social events frequently at all lakefronts could heighten the use of lakefronts during different seasons and times of the day and will enhance the sense of cohesion and community amongst the residents of the city.

The study concludes that the design of urban lakefronts should be honest to its local history and location and should respect the culture and existing heritage character of the site. The visual character should be developed such that it strengthens the natural beauty of the waterfronts and the existing character of the city and surrounding development. At waterfronts, the interface of the public with water is of utmost importance. The elements which bring people closer to the water should be developed sensitively keeping in mind the psychological effects water creates on the human beings and safety factor as well. The design should appreciate the events and activities occurring in the spaces, celebrate it and generate revenue also for its maintenance. Waterfronts should celebrate water by offering greater opportunities for carrying out diverse activities like festive celebrations, street plays, concerts, and other social events in all seasons and all times of the day, which will encourage community interaction and socialization. The lakefront development demands a more sensitive approach towards its design. A deeper insight into the perception, use, and experience of the public at urban lakefronts can help enhance the quality of urban life.

KEYWORDS

- **dimensions of urban design**
- **public perception**
- **quality of life**
- **recreational quality indicators**
- **spatial planning**
- **waterfront urban spaces**

REFERENCES

1. Timur, U. P., (2013). Urban waterfront regenerations. *Advances in Landscape Architecture*. Murat Özyavuz, IntechOpen. doi: 10.5772/55759. https://www.intechopen.com/books/advances-in-landscape-architecture/urban-waterfront-regenerations (accessed on 30 June 2021).

2. Carmona, M., Heath, T., Taner, O., & Tiesdell, S., (2003). *Public Places Urban Spaces - The dimension of Urban Design*. Architectural Press Publications.

3. Song, Y., & Knapp, G. J., (2004). Measuring urban form - is Portland winning the war on sprawl. *Journal of the American Planning Association, 70*(2), 210–225.

4. Steele, H., (2014). Morphological characteristics of urban water bodies: Mechanisms of change and implications for ecosystem function. *Ecological Applications, 24*(5), 1070–1084.

5. Roy, A., (2013). *Celebrating Nagpur City embellished by Lakes, But They Need Conservation*. https://timesofindia.indiatimes.com/city/nagpur/Celebrating-Nagpur-City-embellished-by-lakes-but-they-need conservation/articleshow/18840695.cms (accessed on 30 June 2021).

6. Sourabh, N., & Srinivas, D., (2016). Water as an element in architecture. *Best: International Journal of Management, Information Technology and Engineering, 4*(1), 49–60.

7. Nasir, R. A., Ahmad, S. S., & Ahmed, A. Z., (2013). Physical activity and human comfort correlation in an urban park in hot and humid conditions. *Procedia Soc. Behav. Sci., 105*, 598–609.

8. Sun, R., & Chen, L., (2012). How can urban water bodies be designed for climate adaptation? *Landscape and Urban Planning*, 27–33.

9. AlMohannadi, M., Zaina, S., Zaina, S., & Raffaello, F., (2015). Integrated approach for the improvement of human comfort in the public realm: The case of the corniche, the linear urban link of Doha. *American Journal of Sociological Research., 5*(4), 89–100. doi: 10.5923/j.sociology.20150504.01.

10. Bell, P. A., Fisher, J. D., Baum, A., & Greene, T. C., (1990). *Environmental Psychology* (3rd edn.). Holt, Rinehart Winston, Inc., London.

11. Tse, M. S., Chau, C. K., Choy, Y. S., Tsui, W. K., Chan, C. N., & Tang, S. K., (2012). Perception of urban park soundscape. *J. Acoust. Soc. Am., 131*(4), 2762–2771.

12. Shehayeb, D., (2016). Safety and security in public space. *Crime Prevention and Community Safety* (pp. 107–112). International Report.

13. Türkseven, D. I., & Zengel, R., (2017). Analysis of perceived safety in urban parks: A field study in Büyükpark and hasanağa park, Metu. *Journal of the Faculty of Architecture, 34*(1), http://dx.doi.org/10.4305/metu.jfa.2017.1.7.

14. Doucet, B., (2010). *Rich Cities with Poor People: Waterfront Regeneration in the Netherlands and Scotland.* PhD Dissertation, Utrecht University, Utrecht. https://dspace.library.uu.nl/bitstream/handle/1874/42400/doucet.pdf?sequence=1 (accessed on 30 June 2021).

15. Rauno, S., & Satu, K., (2006). Assessing social impacts in urban waterfront regeneration. *Environmental Impact Assessment Review, 26,* 124, 125.

16. Madden, K., & Schwartz, A., (2000). *How to Turn a Place Around: A Handbook of Creating Successful Public Spaces.* New York, NY: Project for Public Spaces.

17. Rad, V. B., & Ngah, I., (2013). The role of public spaces in promoting social interactions. *International Journal of Current Engineering and Technology, 3*(1), 184–188.

18. Cullen, G., (1961). *Townscape.* Architectural Press, London.

19. Rahman, M. A., (2010). *Development Opportunities for the New Waterfront in South Side of Kungsholmen in Terms of Tourism and Recreation: An Urban Design Approach to Vibrant Urban Waterfront Development in Stockholm.* Degree Project, SoM EX 2010, master program urban planning and design, Stockholm. http://kth.diva-portal.org/smash/get/diva2:386869/FULLTEXT01.pdf (accessed on 30 June 2021).

20. Ye, Z., (2013). *The Evaluation and Improvement Method of Waterfront Urban Landscape* (pp. 3–5). 49th ISOCARP Congress.

21. Carr, S., Francis, M., Rivlin, L. G., & Stone, A. M., (1992). *Public Space.* Cambridge University Press, Cambridge.

22. Yamashita, S., & Hirano, M., (1995). Residents' evaluation and the recreational uses of urban rivers. In: Craig-Smith, S. J., &. Fagence, M., (eds.), *Recreation and Tourism as a Catalyst for Urban Waterfront Redevelopment: An International Survey.* Westport, Connecticut and London: Praeger Publishers.

23. Shang, W., (2017). *Role of Waterfront in Shaping City Center Landscape: Perception of Tianjin Haihe Riverfront Landscape.* BiblioBazaar.

24. Project for Public Spaces, (2007). *What is Placemaking?* https://www.pps.org/article/what-is-placemaking (accessed on 30 June 2021).

25. Efroymson, D., Ha, T. T. K. T., & Ha, P. T., (2009). *Public Spaces: How They Humanize Cities.* Health Bridge - WBB Trust Dhaka.

26. Agnieszka, K., (2018). The revitalization of the old harbor in Reykjavik by a cultural economy. *Space and Culture, 21*(4), 424–438.

27. Lynch, K., (1981). *A Theory of Good City Form.* MIT Press, Cambridge, Mass.

28. André, F., Figueira, D. S. J., & Regina, S., (2017). The cultural heritage in the postindustrial waterfront: A case study of the south bank of the Tagus estuary, Portugal. *Space and Culture,* 1–22.

29. Les, R., (2010). Dis/embedded geographies of film: Virtual panoramas and the touristic consumption of Liverpool waterfront. *Space and Culture, 13*(1), 54–74.

30. Kowalewski, M., (2018). Images and spaces of port cities in transition. *Space and Culture,* 1–13.

CHAPTER 21

Zero Liquid Discharge: Water Recycling in Industries Towards Sustainability

N. N. RAO[1] and SUMITA N. RAO[2]

[1]*Former Chief Scientist and HOD, Wastewater Technology Division, CSIR-National Environmental Engineering Research Institute (NEERI), Nagpur–440020, Maharashtra, India, E-mail: nnrao.neeri@gmail.com*

[2]*Department of Applied Chemistry, Priyadarshini College of Engineering, Hingna Road, Nagpur–440019, Maharashtra, India*

ABSTRACT

Water is an essential raw material for various industrial operations and domestic uses. In recent years our country has witnessed deteriorating water quality in several rivers and other water bodies like lakes and ponds. Well water and groundwater too got substantially impacted with pollution. Uncontrolled discharge of domestic and industrial effluents may have led to pollution of these water bodies. The pollutant levels have increased so high in these water bodies that they are no longer fit for drinking without further treatment. In some cases, the available water requires to be pre-treated adequately even for use in industries. Therefore, impending water quality issues coupled with growing water shortage in various parts of India force us to consider reuse/recycling of treated industrial and domestic effluents. In this chapter, we endeavor to discuss current water scenario and water policy, status of industrial effluent treatment, status of domestic waste treatment, introduce zero liquid discharge (ZLD) technology for obtaining recyclable treated water from raw effluents, and highlight its benefits through a few case studies taken from the authors own experience in the field.

21.1 INTRODUCTION

21.1.1 CURRENT WATER SCENARIO, WATER POLICY, INDUSTRIAL, AND DOMESTIC EFFLUENTS, COMPLIANCE STATUS AND AVAILABLE TECHNOLOGIES

Water is a natural resource that supports sustainable development. With only 4% of the world's renewable water resources, India population is more than 18% of the world's population. Further, the availability of water across the country differs widely due to differences in precipitation and climatic conditions; some parts of the country are dry while other parts receive high precipitation resulting in floods and crop damage. India being a fast-developing nation its demand for water will increase in the future while unfavorable climatic conditions in some parts will lower the water availability there and create an imbalance in water availability across the country. Water stress scenario is already provoking conflicts among different user groups and water-sharing States. Due to water stress conditions, there is already wide parity in water pricing across the country, in respect to drinking as well as industrial uses. Apart from this; inefficient use of water up to 40–50% in some cities, and pollution by non-point discharges into water bodies is an economic and health concern. Many rivers dry up seasonally, and perennial rivers experience a shortage of water flows leading to ponding in the river beds and basin. Thus, with due concern for sustainable development, GOI in its National Water Policy proposed a set of rules and guidelines, institutional revamping, and action plans [1].

The NWP 2012 is the third in series since 1987, and it is currently in force. The major policy innovations are:

- Planning, development, and management of water resources to consider the "river basin/sub-basin" as a unit;
- To restrict water levels in reservoirs and dams and ensure sufficient water flow in the Ganga as well as in other rivers throughout the year;
- To maintain health and hygiene of people by ensuring supply of a minimum quantity of potable water to households;
- A complete evaluation of environmental, economic, and social impacts is critical in case any decision regarding inter-basin transfers of water is necessitated.

Currently, water stress is experienced by 600 million in India, and about 2.0 lakh people die every year due to inadequate access to safe drinking water. By 2030, the country's water demand will sharply increase and get doubled over the available supply. This means that severe water scarcity is looming, and hundreds of millions of people are likely to be affected. Further, the per capita availability of water in India has decreased from 2209 m^3/year in 1991 to 1545 m^3/year in 2011 and it is estimated to decline further up to 1140 m^3/year in the year 2050 [2].

21.1.2   STATUS OF INDUSTRIAL EFFLUENT TREATMENT

Pharma, Pulp, and Paper, Tanneries, Textile Dyeing, Chemicals, Power Plants and many more, generate wastewater having high salinity/total dissolved solids (TDS). The concentration of TDS in these effluents is above the statutory discharge limit of 2100 mg/l. If the discharge of treated wastewater having high TDS is continued, both ground and surface waters get polluted.

Conventional physicochemical-biological treatment cycle, which is good for removing organic pollutants is grossly ineffective for the removal of TDS from saline effluents. To restrict or prevent water pollution from these Industries, Central Pollution Control Board brought out guidelines in 2015 encouraging industries to adopt ZLD-a regulatory tool that prohibits discharge of effluents into inland surface waters. Tamil Nadu compulsively took the lead on ZLD as it has no fully flowing perennial rivers. This state shares water from neighboring states and it is fraught with socio-legal complications. Industries in Gujarat and Karnataka are also inclined towards ZLD. Industries in 'landlocked areas and issues related to sea discharge of 'treated' wastewater have also indirectly pointed to the need of ZLD. Further, water scarcity, water economics, and regulatory pressure are the initial motivators; the higher water pricing greater than Rs. 50–100 per cubic meter has acted as a revenue driver for ZLD. The GOI intended to extend subsidies and incentives to implement and to encourage reclamation of water from industrial effluents for recycling/reuse [1].

21.1.3   STATUS OF DOMESTIC WASTEWATER TREATMENT

The importance of public health in building sustainable society need not be exaggerated. The first step towards protecting public health and the environment is to properly collect, treat, and dispose of human wastes. Allowed methods of disposal per se surface water discharge and land application dictate the degree of wastewater treatment needed. Disposal to surface waterbodies, such as rivers, lakes, estuaries, and oceans, is by far the most common approach in the world. A land application system is also referred to in some select cases, but there is fear of contaminating groundwater or surface waters.

It has been observed that both ground and surface water bodies are progressively getting contaminated with discharges from urban, semi-urban, and industrial sectors. Only one-fifth of sewage from total generation is currently getting treated in urban/semi-urban sectors. The major gap is the lack of adequate infrastructure for treatment of generated wastewater (domestic wastewater/sewage). The cities/towns should build more numbers of functional sewage treatment plants/Decentralized treatment facilities to address the problem of water pollution and related health issues. Similarly, the industries should also exercise their Corporate Responsibility and build necessary effluent treatment plants to ensure a high degree of treatment to their effluents. Recycle and reuse of adequately treated domestic sewage needs to be promoted in a large way.

Class I cities (500) with a population over 1,00,000 accounts for 93% of the urban sewage generated in the country. But only 32% of the volume of domestic effluents they generate are treated. Similarly, Class II towns (410) with population between 50,000 and 1,00,000 have a combined treatment capacity of only 8% of total liquid waste generated by them. While the treatment infrastructure set up can treat about 30% of total sewage generated in the country, the actual and effective treatment capacity is only 19–22% [3]. A total of 61754 MLD sewage is generated in the country while the sewage treatment capacity exists only for 22963 MLD. The treatment status of sewage treatment plants is also appalling; out of the 152 STPs, 9 were under construction, 30 are non-operational, and the concentration of biochemical oxygen demand (BOD) in treated effluent from 49 STPs is above the BOD standards. The two conclusions drawn from the study were (i) there is need for lessening the gap between sewage generation and installed treatment capacity and (ii) 24/7 availability of electric power and deployment of skilled manpower for effective operation and maintenance of STPs. Lack of the O&M capacity among the municipal bodies also resulted in poor performances of the existing STPs which often fail to match the statutory discharge norms. Therefore, operation, and maintenance lacunae should be given full consideration before treatment technology is recommended as SOP for all to follow.

It is imperative that more infrastructure needs to be set up and operated to remove pollutants from sewage before they are discharged into rivers. A better scenario would be to treat adequately to enable recycling and reuse of treated domestic wastewater in some suitable applications. Under National River Conservation Program, the GOI took the initiative to finance the setting up of sewage treatment plants in cities along riverbanks. There is still a large gap in sewage generation and sewage treatment capacity in India which needs to be addressed comprehensively by state governments as well.

Multiple reports and publications over rivers and lakes pollution due to discharge of untreated and treated sewage has led to the belief that a uniform standard for a country as vast as India with varying geographic and climatic features is an inadequate regulation. A classic case is pollution of Ballandur and Vartur lakes in Bangalore, which often froth out onto streets due to excessive pollutant loading. The Bangalore authorities are planning to pump out entire water, desilt lakes and then revive them.

21.1.4 COMPLIANCE STATUS AND SHORTCOMINGS

The objective of a wastewater treatment scheme (Figure 21.1) is to reduce the concentration of polluting parameters to the stipulated standards laid down by MoEFCC, Government of India (GOI) [4]. Discharge Standards for treated sewage have been set as per this act with respect to the following major parameters (Table 21.1).

TABLE 21.1 Discharge Standards for Treated Sewage

SL. No.	Parameters	Concentration* (mg/L)
1.	Suspended solids	100
2.	BOD (3 d, 27°C)	30
3.	COD	250

*General standards for discharge of environmental pollutants, Part-A: Effluents, The environmental rules (1986).

Source: Environmental Protection Act (1986).

The standard clearly states, "*While permitting the discharge of effluent and emission into the environment, State Boards have to consider the assimilative capacities of the receiving bodies, especially water bodies so that quality of the intended use of the receiving waters is not affected. Where such quality is likely to be affected, discharges should not be allowed into water bodies.*" Recently the discharge standards have been made more stringent with a view to achieve improved water quality of surface water bodies (Table 21.2) as per MOEF&CC [5].

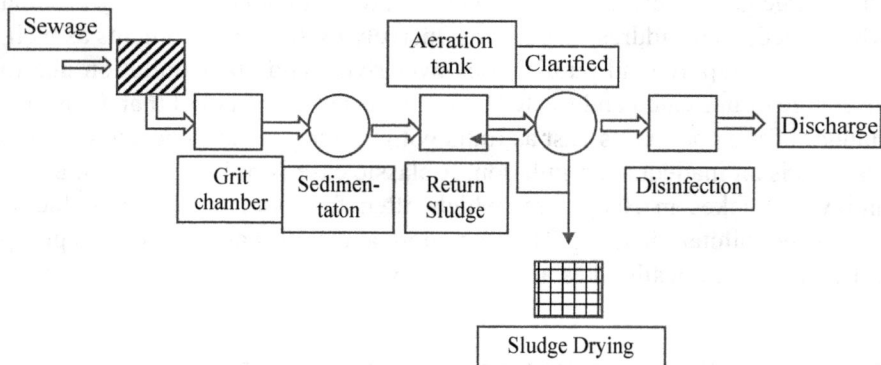

FIGURE 21.1   Schematics of physicochemical, biological based sewage treatment plant.

TABLE 21.2   Revised Standards for Discharge of Treated Sewage as per Gazette Notification (2017)

SL. No.	Parameters	Concentration (mg/L)
1.	Suspended solids	<10
2.	BOD (3 d, 27°C)	20
3.	COD	50
4.	NH_4-N	5
5.	N	10
6.	Fecal coliform, MPN/100 mL	<100

To achieve these limits, additional tertiary units need to be implemented in the existing units, and proposed STPs shall need to be designed for meeting these stringent norms. This will have cost implications as the capital cost will almost double and the O&M cost shall also increase by 25–35% depending on the treatment technology adopted. The currently applied regulatory limits for discharge of treated domestic effluents call for choosing technologies for meeting specified limits, e.g., achieving nutrients limits demand use of sequential bioreactor (SBR) or membrane bioreactor (MBR) technologies. This implies that while regulation does not require use of any specific technology but the treatment performance to achieve effluent discharge limits are based on performance of specific technologies. It is to be noted that no one-treatment methodology will be able to achieve the desired quality of treated effluent conforming to the discharge standards, which are already in place.

Different technologies as given below find applicability based on raw sewage characteristics and desired quality of treated effluent:

- Waste stabilization pond systems (WSPS);
- Duckweed pond systems (DPS);
- Facultative aerated lagoon (FAL);
- Trickling filter (TF);
- Activated sludge process (ASP);
- BioFOR technology (biological filtration and oxygenated reactor);
- High rate activated sludge BioFOR-F technology (ASF);
- Fluidized aerated bed (FAB) technology;
- Submerged aeration fixed film (SAFF) technology;
- Cyclic activated sludge process (CASP);
- Up-flow anaerobic sludge blanket (UASB) process;
- Decentralized water treatment systems (DeWATS);
- Moving bed biological reactor (MBBR);
- Membrane bioreactor (MBR);
- Sequential batch reactor (SBR);
- Natural wastewater treatment systems and engineered natural sewage treatment system.

DeWATS has been considered suitable for the treatment of sewage collected from small communities, apartments, and colonies. MBBR, MBR, SBR are finding increasing application for sewage treatment to meet the stringent standards for discharge. The last item, Natural Wastewater Treatment Systems and Engineered Natural Sewage treat systems is upcoming, and several small capacity plants (10–1000 m^3/day) have come up in some places. Natural wetlands which host a variety of flora, fauna, and biota also contribute to the treatment of domestic effluents through diverse mechanisms [6–8]. It is widely accepted that a variety of bacteria grow on the submerged roots and stems of aquatic plants in the wetlands and contribute immensely to the removal of BOD. Other ways removals of pollutants by sedimentation, adsorption/filtration, ion exchange/adsorption capacity involving the plants' roots and stems, natural sediments are also possible, with different significance levels. The climatic conditions such as wind, sunlight, and temperature appear to favor the removal mechanisms.

Scientists have also devised artificial systems that mimic the functions of wetland in the wastewater treatment. These are called 'constructed wetlands.' They are the over the ground constructed facilities with specific water flow conditions such as free water surface flow systems (FWS) with shallow water depths, subsurface flow systems (SFS) with water flowing laterally through the sand or gravel beneath a bed of plants. Both horizontal and vertical FWS can be designed as per the site requirements. The removals of BOD and TSS are often greater than 70–80%, while nutrients such as N&P also get removed with >50–60% [8]. In the constructed wetlands, the aquatic vegetation such as cattails (Typha), rushes (Scirpus), and reeds (Phragmites) are planted, while the natural wetlands can have much wider species depending upon the growth-promoting conditions.

21.2 ZERO LIQUID DISCHARGE (ZLD) TECHNOLOGY FOR OBTAINING RECYCLABLE TREATED WATER FROM RAW EFFLUENTS

21.2.1 WHAT IS ZLD?

ZLD essentially means that an industry should treat and recover all its effluents and not release any liquid effluents into the water bodies or neighboring lands. As per MoEF&CC and CPCB (R11), '*ZLD refers to installation of facilities and systems to enable the industrial effluents for recycling and converting solute into residue into solid by adopting method of concentration and thermal evaporation.*' ZLD is a solution for hard-to-treat or dangerously contaminated wastewater streams [9–21]. The term "Zero-discharge" is used to promote conservation of the quality of the environment. The concept received force because of i) reduced availability of freshwater for process use, ii) difficulty to comply with the regulatory standards, iii) regional considerations banning the priority pollutants iv) economics due to unaffordable costs of freshwater v) "good neighbor" policy (sensitivity to nature and community).

The concept of zero discharge can be understood clearly with the help of Figure 21.2. Zero discharge implies that all reactant materials are converted into products, and all processing aids are re-used, and no waste is produced. This refers to an ideal process. However, in practice, a fraction of raw materials escapes as waste; unusable processing aids such as energy, fuels, salts, catalysts, reagents, solvents, including water, can become a part of the process waste stream (real process).

Thus, a 'Zero Discharge' industry implies that air, liquid, solid wastes are completely contained and appropriately treated, used, and managed within the premises of the industry.

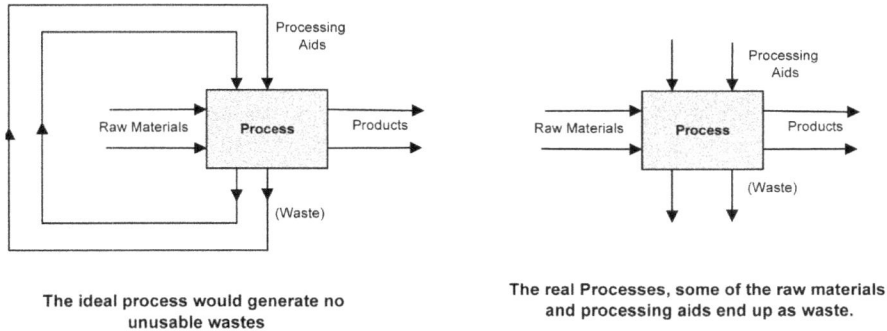

The ideal process would generate no unusable wastes

The real Processes, some of the raw materials and processing aids end up as waste.

FIGURE 21.2   Ideal and real processes explain the ZLD concept.

21.2.2   *DRIVERS FOR WASTEWATER RECYCLING, APPROACHES, AND STRATEGIES*

The following policies, regulatory, and sustainable growth-related issues act as drivers for ZLD implementation [9]:

- National water policy;
- National water mission;
- Non-availability of quantity and quality of water for specific application;
- Improving water efficiency;
- Cost reduction and resource conservation;
- Sustained growth and environmental considerations;
- Statutory requirements;
- Plant operational requirements;
- Availability of suitable technologies;
- Reducing energy consumption in water systems.

21.2.3   *HOW TO APPROACH ZLD?*

The strategy for zero discharge facility involves (Figure 21.3):

- Minimization of waste and wastewater generation by choosing:
  - Alternative process method;
  - Alternative raw materials;
  - Purer feedstocks;
  - Alternative operating conditions.
- Segregation and reuse of wastewater streams: Wastewater streams having low concentration of pollutants may be segregated and reused to clean up progressively more contaminated streams.
- Treatment and reuse of final effluent wastewater streams. Stringent discharge limits can necessitate wastewater clean-up so that it may be reused. Common reuse options are as cooling tower make up water and boiler water make up.

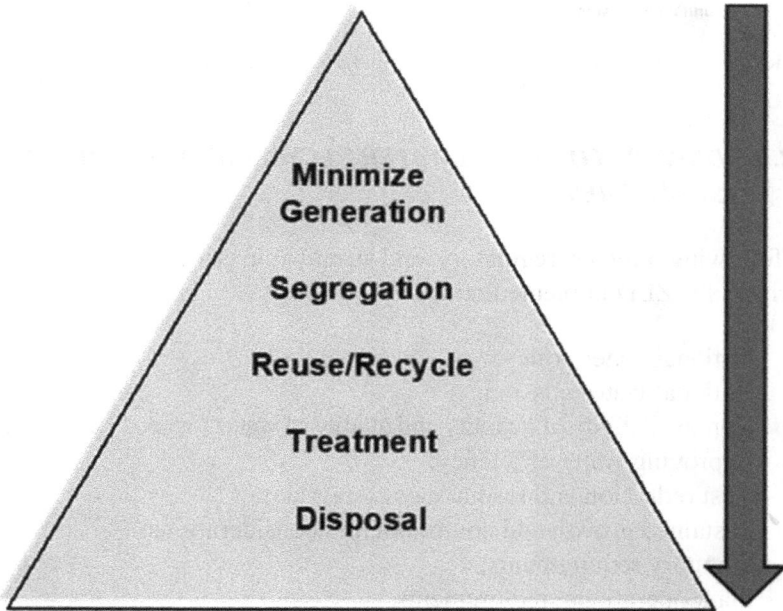

FIGURE 21.3   Strategy for prioritizing actions towards zero liquid discharge.

21.2.4   WASTE MINIMIZATION APPROACHES

The following waste minimization steps may be useful prior to implementation of ZLD in an industry:

1. Inventorization-Management and Improved Operations:
   i.   List and quantify all raw materials;
   ii. Choose materials that have fewer toxic impacts;
   iii. Ensure safety in material storage and handling.

2. Upgrade and Retrofit:
   i.   Older equipment which has less performance efficiency may have to replace with better efficiency equipment (pumps, fans, aerators, evaporators, mixers, compressors, etc.), to minimize waste generation;
   ii. Operating efficiency of equipment tools to be improved;
   iii. Adopt preventive maintenance program periodically to avoid breakdown of operations.

3. Green Production Process:
   i.   Wherever feasible substitute hazardous raw materials, reagents, solvents with less or non-hazardous materials;
   ii. Practice segregation of waste streams. This may facilitate recovery of certain raw materials reuse or upcycling;
   iii. Eliminate sources of leaks/spills;
   iv. Optimize raw material use in reactions (atom efficiency).

4. Recycling and Reuse:
   i.   Evaluate feasibility of installing closed-loop systems that supplement material and energy needs;
   ii. Promote recycling either on-site or off-site (water, salts, etc.);
   iii. Exchange wastes-a few wastes may find application in some other industry. Keep exploring alternative uses.

There are at least three stages for strategizing zero liquid discharge (see Figure 21.4).

- Baselining of water and wastewater footprint, risk profiling and sustainability assessment;
- Response formulations; and
- Integration into long-term planning.

21.3 ROLE OF REGULATORS AND LACK OF UNIFORM POLICY

The Pollution Control Board (PCB) has been essentially a regulator. Often, it cannot offer any "Technological Solutions," but can issue advisory to protect environment. In some instances, the discharge standards set are stricter and harder to achieve through best available and techno-economically feasible technologies. In the current scenario, the PCB has daunting task of safeguarding quality of environment by battling against pollution so much so that it is a respondent in related Court and NGT cases. It is also a herculean task for them to ensure 'round-the-clock' monitoring of industrial discharges, which is also not practical due to high staff requirement. Moreover, several states and even districts interpret the law differently and allow special provisions (water pricing, energy cess, relaxed norms, etc.), to promote industrialization and encourage businesses. This, in effect, results in 'shifting of pollution' to neighboring states and districts. A classic recent case is plastic ban by Maharashtra State. Many Plastic manufacturing industries moved to Gujarat as there was no ban on plastic manufacturing in this State.

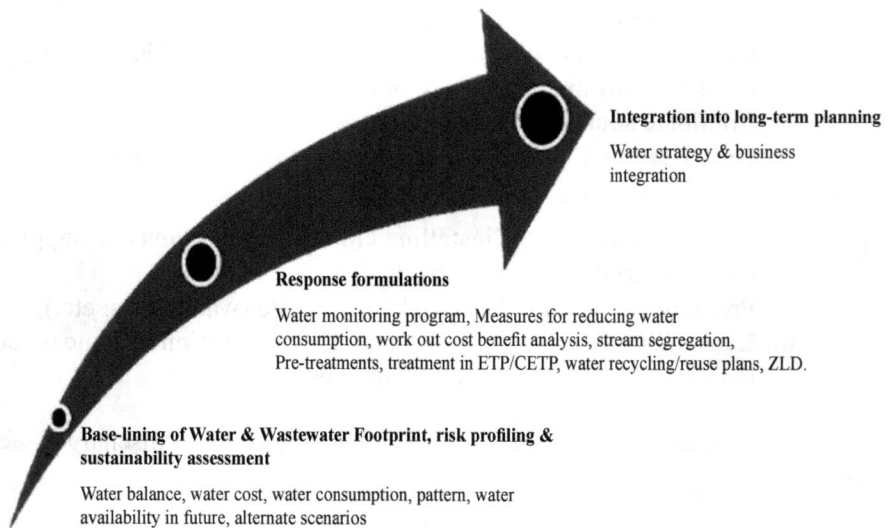

Integration into long-term planning

Water strategy & business integration

Response formulations

Water monitoring program, Measures for reducing water consumption, work out cost benefit analysis, stream segregation, Pre-treatments, treatment in ETP/CETP, water recycling/reuse plans, ZLD.

Base-lining of Water & Wastewater Footprint, risk profiling & sustainability assessment

Water balance, water cost, water consumption, pattern, water availability in future, alternate scenarios

FIGURE 21.4   Stages in strategizing ZLD for an industry.

Currently, four industrial sectors, namely textile (wet processing), distilleries, pulp, and paper and tanneries have been advised to apply ZLD guidelines. Several states sharing the Ganga basin area have been asked to submit ZLD action plans. The ZLD guidelines have compelled industry to value water, an exhaustible natural resource. Industry in water stressed areas have already understood the importance of water reclamation from liquid effluents. However, implementation of policy uniformly in the country is the key to success of water reclamation.

21.4 TECHNOLOGIES FOR ZLD

The following treatment technologies are common within a ZLD scheme as highlighted in Figure 21.5:

- Pre-treatment in ETP or CETP (physicochemical and biological, polishing);
- Membrane processes (UF, NF, RO);
- Thermal processes (evaporators and crystallizers).

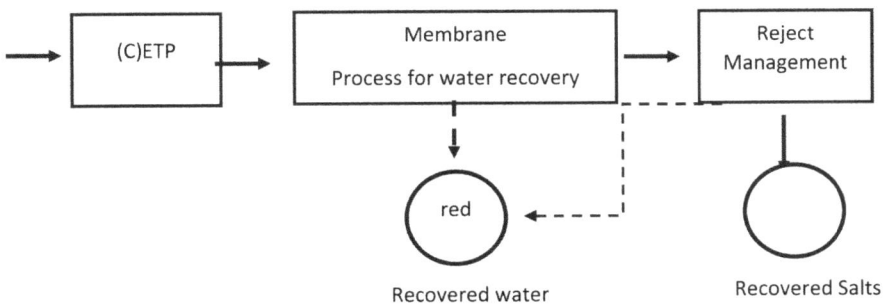

FIGURE 21.5 Process components in a typical ZLD system.

21.4.1 EARLIEST AND RECENT ZLDS

Power plant discharges from wet gas scrubbing, runoff from coal storage area, and leachate from gypsum sheds typically contain dissolved salts. The salinity of rivers increases, which often receive discharges from the power plants. Its discharge into rivers cause an increase in salinity. The increase in salinity of Colorado river in USA in early 70s due to Power Plant discharges acted as driver for considering regulatory method to control increase in salinity and this served as beginning for ZLD concept. Proposals for setting up industries incorporating ZLD approaches were given preference and given early clearance. The ZLD systems produce clean water that is recycled into the plant. The solid waste/product is sent for landfill disposal [22–34]. This policy helped to spread ZLD application in USA and Germany. In India, some tanneries in TN started ZLD activities already in 2006, mostly driven by freshwater necessities and regulatory constraints.

21.4.2 TEXTILE SECTOR: A BIG POLLUTER

Textile industry is water-intensive and releases large volume of wastewater. The industry set up effluent treatment facilities a couple of decades ago, however, the treatment was less satisfactory. As a result, partially treated wastewater used to be disposed of into rivers and drains. Water pollution was so severe that some of the rivers and drains in south India near Tirupur and Erode in TN and in some other cities in Gujarat turned colored, inviting public outcry and filing of court cases. The honorable courts had ordered the closure of some of the factories in Gujarat and Tamil Nadu in 2011. For example, 22 textile units in Vapi (Gujarat) were shut down after it was noticed that these industries have immensely damaged the ecology through water pollution. The textile industry was directed to adhere to ZLD norms via a draft notification. This is a move to curb the massive pollution caused by the textile industry. The guideline [9] says that '*all the textile units; be it dyeing units, cotton or wool processing units and integrated factories generating over 25 kiloliter of effluents daily must install Zero Liquid Discharge effluent treatment plants.* The policy is facing strong criticism from several sections due to strict ZLD norms and other clauses.

As a spin-off, about 200 CETPs were established in textile dyeing, leather tanneries, chemicals, electroplating, and pharmaceutical industries. These CETPs employed physicochemical, biological treatment technologies, which helped to remove the contaminants such as organics, heavy metals, dissolved, and suspended solids. However, these technologies are not suitable for removing salinity from these wastewaters which contained the inorganic TDS in the range from 5,000 to 20,000 mg/l or even more. The TDS in the treated effluent is ranging from 5,000–7,000 mg/L in case of textile effluent and 10,000–15,000 mg/L in case of tannery effluent. The process could also not achieve a reduction in salinity even after segregation of saline liquors, like soak and pickle liquors in tanneries or dye bath liquor in textiles. The TDS constitutes chlorides, sulfates, and other salts which increase the salinity of the receiving water bodies such as rivers, lakes, and groundwater and affected large tracts of soil severely impacting agriculture.

21.4.3 TIRUPUR SECTOR

Tirupur is textile manufacturing hub of India, was directed by the Hon'ble Supreme Court in October 2010 to complete and operationalize the ZLD system within a period of 3 months. The sector differed the directives and did not escape the ire of Hon'ble court, which then ordered closure of several dyeing and bleaching units in the region and held that no unit should be allowed to reopen and operate unless it achieves ZLD. Initially, Tamil Nadu Pollution Control Board (TNPCB) permitted Arulpuram CETP to demonstrate ZLD. After witnessing the successful operations at Arulpuram CETP, the other CETPs too decided to carry out the modifications required for ZLD. TNPCB gave permission to these CETPs for trial operations with restrictions in the effluent flow, while the Tamil Nadu government extended financial assistance to the industry to set up 20 CETPs with ZLD facilities in Tirupur [25].

The aim of 'zero liquid discharge (ZLD)' plants is to prevent discharge of wastewater across industry's boundary and to recover and reuse of wastewater for industrial purposes. The regulators insist that they reject from RO operations shall be managed through TE and shall not discharge rejects onto land or into deep wells [9]. Small ETPs may use solar evaporation pans for evaporator blowdowns.

The typical concentrations of different pollutants in textile raw effluent are given in Table 21.3. The TDS concentration gets concentrated by 4–5 times after RO operations and finally recovered as salts through concentration drying process (Figure 21.6). The recovered water has <200 mg/L TDS and other pollutants well under control. This recovered water is fit for reuse in the process (Table 21.4).

TABLE 21.3 Raw Effluent Characteristics of Textile Wastewater

SL. No.	Parameter	Range*
1.	pH	8.5–10.0
2.	Biochemical oxygen demand (BOD)	400–500
3.	Chemical oxygen demand (COD)	1000–1200
4.	Total suspended solids (TSS)	200–300
5.	Total dissolved solids (TDS)	6000–7000
6.	Chloride (Cl⁻)	400–700
7.	Sulfates (SO_4^{2-})	2500–3100

*Except pH, all other values in mg/L.

TABLE 21.4   Quality of Recovered Water from ZLD Treatment Textile Wastewater

SL. No.	Parameter	Influent (After Secondary Treatment)	Recovered Water
1.	pH	9.0	7.0
2.	Biochemical oxygen demand (BOD)	251	BDL
3.	Chemical oxygen demand (COD)	1034	BDL
4.	Total suspended solids (TSS)	200–300	BDL
5.	Total dissolved solids (TDS)	6744	170
6.	Chloride (Cl-)	734	34
7.	Sulfates (SO_4^{2-})	3142	19

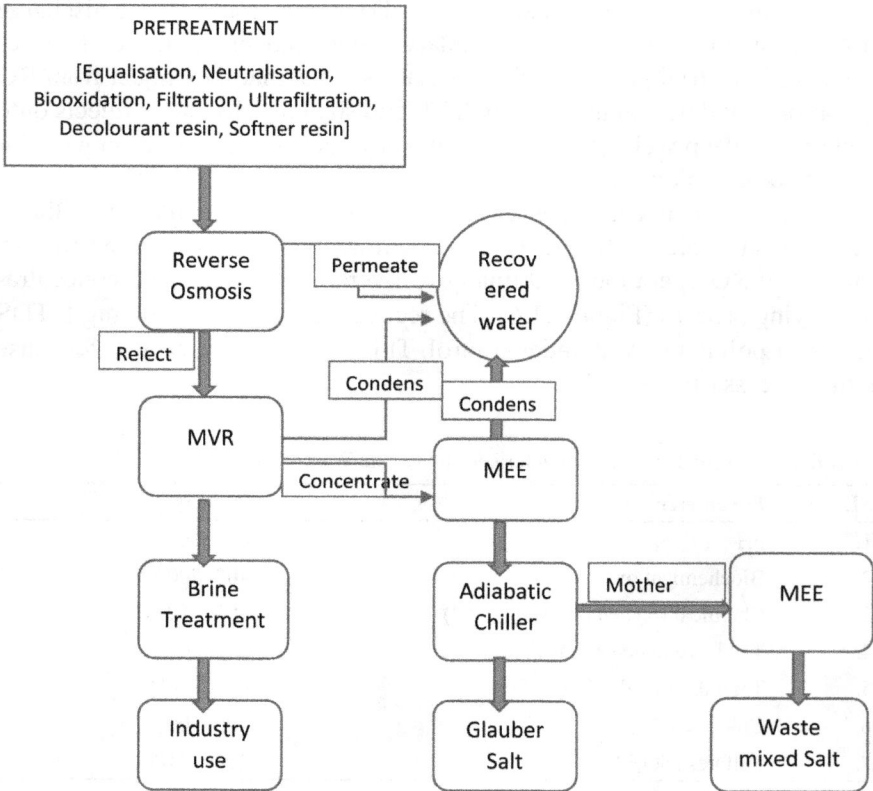

FIGURE 21.6   ZLD scheme adopted by textile units at Tirupur, Tamil Nadu.

21.4.4   ZLD IN TANNERIES

ZLD in the tannery industries is achieved through the implementation of advanced membrane and evaporation technology [35–38]. A schematic diagram (Figure 21.7) of ZLD based CETP for a tannery with MBR system and mechanical vapor recompression type evaporator (MVR-E) for brine concentration followed by MEE and crystallization of RO rejects. Ambur and Vaniyambadi Tannery cluster has installed 3 ZLD based CETPs of combined capacities of 7.2 MLD. In these CETPs feed for RO system is the effluent from MBR system. Evaporator is set up for RO reject management.

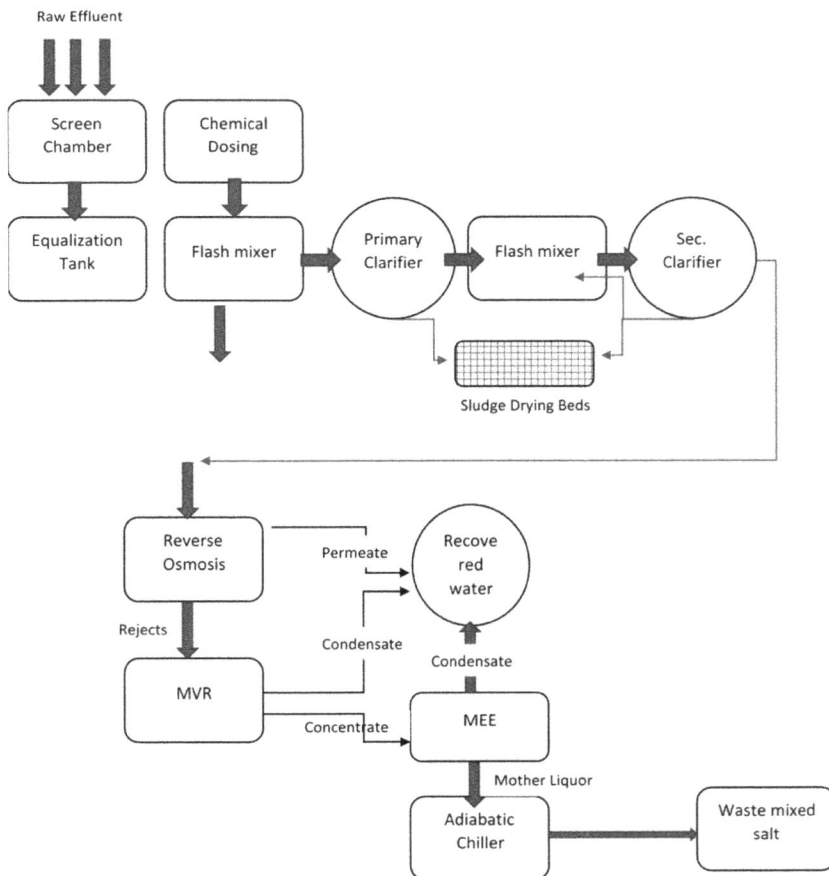

FIGURE 21.7   Flow diagram for tannery effluent management through ZLD technology.

21.4.5   REJECT MANAGEMENT

This is vital to the success of ZLD schemes at Industries [9]. Textile sector is by far the most evolved with respect to recovery of reusable salts from the ZLD system. More than 80% salts are recovered. There have been attempts to recover salts from tannery ZLD operations; however, the studies revealed the recovered salts (NaCl) is inferior to fresh salt in imparting leather its desirable quality. Thus, in the case of tanneries, the mixed salts are heaped or bagged and stored in cover sheds. This practice has space related limitations and hence effective alternative approaches need to be evolved urgently.

21.4.6   WATER RECYCLING THROUGH SEWAGE TREATMENT

Wastewater (sewage, gray water, canteen wastewater, and other lean effluents) from domestic and commercial establishments is necessary to be treated keeping in mind the use of treated wastewater. Often, it may not be required to apply expensive RO and thermal processes unless a particular use is intended which demands water of specific quality. Adequately treated wastewater can be reused for toilet flushing, gardening, floor washing and other applications (Figure 21.8). If treated sewage is planned for reuse by an industry, such as petrochemical unit or coal fired thermal power stations in their process operations, the sewage treated through primary and secondary processes requires further processing through RO and thermal processes (Figure 21.9). The state Water Boards are increasingly suggesting industries to reuse treated sewage in their operations. Thus, Chennai Petrochemicals Limited reuses treated sewage after RO-based tertiary treatment. Similarly, MAHAGENCO, Nagpur has been reusing a significant quantity of treated sewage in their cooling tower operations.

21.4.7   ENVIRONMENTAL IMPACTS OF ZLD

Apart from eliminating/reducing water pollution, ZLD is tasked to improve water reusability in industries. While rendering this service ZLD also has some negative impacts on the environment [21]. The rejects which constitute solids/salts generated by ZLD units may pose problems; (i) when stored in evaporation ponds, they can emit odor, may also affect wildlife if leakages occur and (ii) if disposed in landfills they may leach and contaminate groundwater [39]. It is necessary to use impervious liner materials in the landfill, and robust monitoring protocols are required to prevent contamination from stored solid wastes, if any. The various unit operations such as membrane technologies, TE methods, solid separation methods used in ZLD consume large amounts of grid energy. This implies ZLD indirectly causes greenhouse gases (GHG) emission [40, 41]. Though the amount of CO_2 produced by electricity generation varies depending on the fuel type [42, 43] assuming 939 g of CO_2 per kWh_e generated by bituminous coal, a mechanical vapor compression (MVC) brine concentrator which typically consumes 20–25 $kWhe/m^3$ will produce 19–23 kg of CO_2 per m^3 of feedwater.

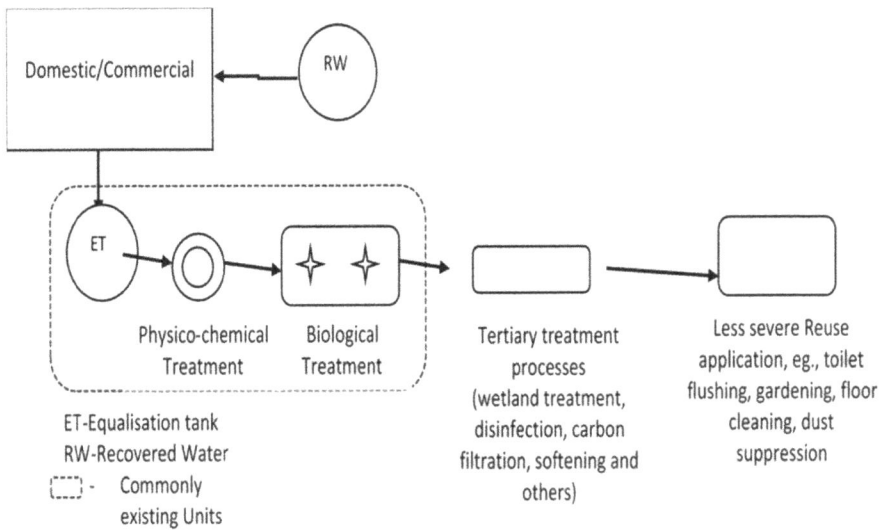

FIGURE 21.8 Schematics of physicochemical, biological, and water recovery-based sewage treatment plant (without salt recovery).

In order to make effluent treatment covering ZLD operations sustainable, it is important to substitute technologies with higher energy efficiency, which will reduce the GHG emissions. The GHG footprint of ZLD systems may be more reduced if technologies that utilize low-grade or renewable energy (e.g., waste heat, solar energy, geothermal energy [44–47]. ZLD and its impacts on the environment are still to be understood comprehensively. The feasibility studies for setting up ZLD based ETP or CETP should also consider a life-cycle assessment analysis of energy consumed versus GHG emissions [48]. It is possible that ZLD may become economically feasible and sustainable in the future when economic gains and environmental impacts are balanced.

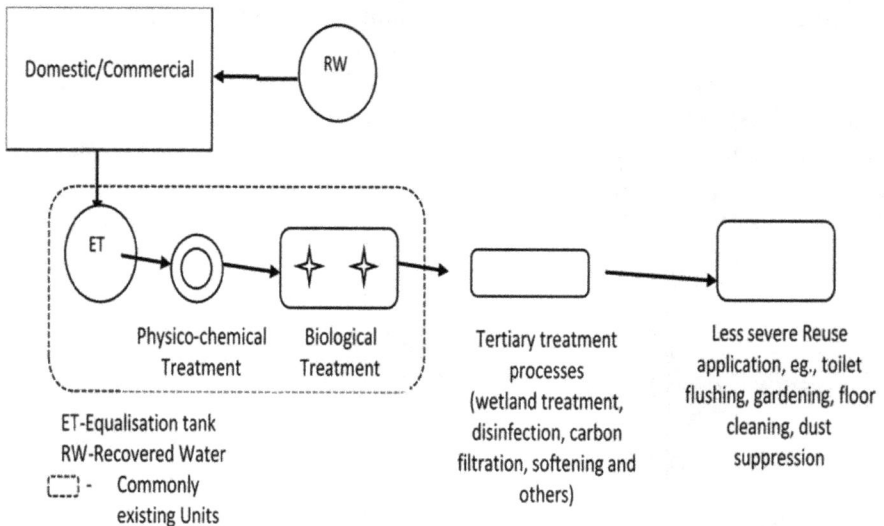

FIGURE 21.9 Schematics of physicochemical, biological, and water recovery through RO-based sewage treatment plant (with optional salt recovery unit).

21.4.8 CHALLENGES, BENEFITS, AND DRAWBACK OF ZLD

Table 21.5 provides challenges, benefits, and drawbacks of ZLD. Currently, ZLD plants, while recovering a good part of water for recycle also generate a lot of waste solids which must be either landfilled or stored in covered sheds. This is a serious limitation. Therefore, integrated approaches to make the best of use of recovered salts is necessary in consonance with the Zero Waste concept. While the benefits are obvious, the drawbacks must be addressed in order to economize the ZLD operations. Particularly, the clue for economization of ZLD is in the comparison of costs of energy consumption and water recovery between conventional and renewable-energy-powered processes [48].

21.4.9 ZLD MARKET

The predicted annual growth rate of ZLD market is 12% and is expected to reach $2.7–3.0 B market by 2030. The innovations in technology, rising freshwater pricing, and stricter environmental regulations due to increasing surface water contamination will further promote the ZLD market [34].

TABLE 21.5 Typical Challenges, Benefits, and Drawbacks of ZLD

ZLD Challenges	Benefits of ZLD	Drawback of ZLD
• ZWD and not ZLD • ZLD results in generation of hazardous solid wastes creating disposal challenges-need to think of zero waste disposal (ZWD) plants. Generate products/by-products out of the waste. • Economic viability-cost and availability of water, regulatory pressure is the real driving force. • High carbon foot print-is this environmentally sustainable? • High operating cost and financial impact on the industry and its regional/national/global competitiveness. • Technology shortcomings.	• Installing ZLD technology is beneficial for the plant's water management; encouraging close monitoring of water usage, avoiding wastage and promotes recycling by conventional and far less expensive solutions. • High operating costs can be justified by high recovery of water (>90–95%) and recovering of several by products from the salt. • A more sustainable growth of the industry while meeting most stringent regulatory norms. • Possibility of use of sewage for recovery of water, for Industrial and municipal use, using ZLD technologies. • Reduction in water demand from the industry frees up water for agriculture and domestic demands.	• Very high cost (both CAPEX and OPEX) • Custom-design on case-to-case basis • Complex waste streams need to be treated well before applying ZLD schemes • Industries perplexed as to which technologies fulfill ZLD requirements, which vendor satisfies all the requirements? • Lack of enough updating of suitable technologies.

21.5 CONCLUSIONS

The process of reclaiming water from wastewater constitutes ZLD technology. ZLD often proves to be beneficial to industry and municipal organizations as well as the environment because it helps recycle water, and the left-over solids can be safely disposed of into landfills. ZLD is often achievable by deploying reverse osmosis, evaporators, crystallizers, dryers, etc. This necessarily recovers solids which are otherwise present in a dissolved state in a waste stream. Both recovered water and solids may be reusable for further process applications in a suitable case, e.g., textile sector. While recovered water is always reusable, the solids sometimes pose problem for disposal due to their mixed nature. ZLD is a costly technological challenge, and the focus must be on the process modifications leading to waste minimization. Reduction in effluent volume is critical to control capital costs of ZLD unit. ZLD technology needs to be optimized for each case until a satisfactory techno-economical solution is found. Reject management is yet another intriguing problem with high costs. Focus on utilization of recovered salts to make up for some costs is necessary. In the future, an industry will be ranked based on the ratio between reuse water and freshwater consumption. Maximizing this ratio helps industry to move towards sustainability and contributes to cost savings also. Going by the judicial position on the status of environment in India, coupled with impending water scarcity, water economics, and regulatory requirements, most industries will have to adapt and implement ZLD for their future existence.

ACKNOWLEDGMENTS

N. N. Rao is happy that his tenure in CSIR-NEERI provided him much-needed industry exposure and helped develop perspectives on ZLD. Sumita N Rao thanks The Principal, PIET for encouragement.

KEYWORDS

- reject management
- reverse osmosis
- solids disposal
- thermal systems
- wastewater treatment
- water recovery

REFERENCES

1. *National Water Policy*, (2012). Ministry of Water Resources, Government of India.
2. Twenty Third Report Standing Committee on Water Resources, (2017–2018). Sixteenth Lok Sabha, Ministry of Water Resources, River Development & Ganga Rejuvenation. *Socio-Economic Impact of Commercial Exploitation of Water by Industries*. Loksabha Secretariat.
3. Report of the Sub Group of Chief Ministers on Swachh Bharat Abhiyaan, (2015).
4. General Standards for Discharge of Environmental Pollutants, (1986). *Part-A: Effluents*. The Environmental Rules.
5. *Ministry of Environment, Forest, and Climate Change (MOEFCC) Notification*, (2017). New Delhi, The Gazette of India, Extraordinary, Part II-Section 3-Sub-Section (I).

6. Bassia, N., Kumar, D. M., Sharma, A., & Pardha-Saradhi, P., (2014). Status of wetlands in India: A review of extent, ecosystem benefits, threats and management strategies. *J. Hydrol. Reg. Stud., 2*, 1–19.

7. Sanyal, P., Chakraborty, S. K., & Ghosh, P. B., (2015). Phytoremediation of sewage-fed wetlands of East-Kolkata, India - A case study. *Int. Res. J. Environ., 4*(1), 80–89.

8. EPA/625/1-88/022, (1988). *EPA Design Manual, Constructed Wetlands and Aquatic Plant Systems for Municipal Wastewater Treatment.*.

9. Central Pollution Control Board, (2015). *Guidelines on Techno-Economic Feasibility of Implementation of Zero Liquid Discharge (ZLD) for Water Polluting Industries*. CPCB. Pages 1–25.

10. Mutz, D., (2015). *Common Effluent Treatment Plants: Overview, Technologies and Case Examples.* Indo-German Environment Partnership (IGEP) Program.

11. GPCB-IADD, (2016). *Zero Liquid Discharge Technologies Guidance Manual.* GPCB and ILFS Academy of Applied Development (IAAD).

12. Association of Washington Business, (2013). *Treatment Technology Review and Assessment.* Association of Washington Cities Washington State Association of Counties, 213512.

13. WERF5T10, Water Environment Research Foundation (WERF), (2012). *Demonstration of Membrane Zero Liquid Discharge for Drinking Water Systems, A Literature Review*. Philip Brandhuber, November, 2014.

14. (2014). *Proceedings of 'Seminar and Training Workshop on Zero Liquid Discharge Policy Experience and Emergent Technologies.'* Mahatma Mandir, Gandhinagar.

15. Sustainability Outlook, (2016). *March to Sustainability: Zero Liquid Discharge Market Opportunities and Outlook for India by 2020.* http://sustainabilityoutlook.in/content/market/march-sustainability-zero-liquid-discharge-756544 (accessed on 30 June 2021).

16. Graham, J., Adam, Z., Winnie, S., Parameshwaran, R., Behrooz, M., & Michael, D. N., (2008). *Evaluation and Selection of Available Processes for a Zero-Liquid Discharge System for the Perris*. California, Ground Water Basin Desalination and Water Purification Research and Development Program Report No. 149.

17. Anthony, B., (2015). Advances in filtration systems for wastewater. *Filtration + Separation*, 28–33.

18. RPSEA Project 07122-12, (2009). *An Integrated Framework for Treatment and Management of produced Water, Technical Assessment of Produced Water Treatment Technologies* (1st edn.).

19. Pilutti, M. P. E., & Julia, E. N., (2003). *BAH03-029-Technical and Cost Review of Commercially Available, MF/UF Membrane Products.* International Desalination Association.

20. Ran, S., Wilbert, V. D. B., Sebastiaan, G., & Jozef, H., (2011). Wastewater reuse through RO: A case study of four RO plants producing industrial water. *Desalination and Water Treatment, 34*, 408–415.

21. Tiezheng, T., & Menachem, E., (2016). The global rise of zero liquid discharge for wastewater management: Drivers, technologies, and future directions. *Environ. Sci. Technol., 50*, 6846–6855.

22. Mehta, S., (2011). *Zero Liquid Discharge Technologies and Economics Enviro-Vision.* Volume 52, Issue 5, September–October 2015, pp. 28–33.

23. Craig, R. B., (2010). *Reverse Osmosis Membranes for Wastewater Reclamation.* Thesis.

24. Richard, L. S., (2014). A primer on reverse osmosis technology. *Chem. Eng.* Desalitech, Inc.

25. Sajid, H. I., (2012). *Case Study of a Zero Liquid Discharge Facility in Textile Dyeing Effluents at Tirupur.* A one-day National Workshop on CETPs-Hyderabad.

26. Water Reuse Foundation, (2008). *Survey of High Recovery and Zero Liquid Discharge Technologies for Water Utility.* Michael Mickley, Report 02-06.

27. Muhammad, Y., & Wontae, L., (2019). Zero-liquid discharge (ZLD) technology for resource recovery from wastewater: A review. *Sci. Total Environ., 681,* 551–563.

28. Vishnu, G., Palanisamy, S., & Joseph, K., (2008). Assessment of field-scale zero liquid discharge treatment systems for recovery of water and salt from textile effluents. *J. Clean. Prod., 16,* 1081–1089.

29. Dasaiah, S., Yakkala, K., & Battala, G., (2018). Zero liquid discharge (ZLD) system: An approach towards sustainable development. *Int. J. Recent. Sci. Res., 9*(1), 23145–23147.

30. Bluetech Insight Report, (2015). *Brine Management and ZLD – Innovation Landscape and Technology Review.* Paul O'Callaghan and Michael Mickley, pp. 5–6.

31. *Markets and Markets Report,* (2019). Report CODECH 5288, zero liquid discharge systems market by system (conventional, hybrid), process (pretreatment, filtration, evaporation & crystallization), end-use industry (energy & power, chemicals & petrochemicals, food & beverages), and region. Global Forecast to 2023.

32. William, A. S., (2011). PE, HPD LLC (power sector, coal). *Fundamentals of Zero Liquid Discharge System Design.* POWER Magazine, October 1, 2011.

33. Bostjancic, J., & Ludlum, R., (2013). *Getting to Zero Discharge: How to Recycle That Last Bit of Really Bad Wastewater.* Technical Paper, TO1041EN.

34. Water Online, News, (2017). *Eliminating Wastewater: Zero-Liquid Discharge Market to $2.7B in 2030.*

35. Roy, C., (2012). *A Study on Environmental Compliance of Indian Leather Industry & its Far-Reaching Impact on Leather Exports.* National Seminar, "Environmental Audit" organized by the Department of Commerce, Kaliyaganj College, West Bengal.

36. RANITEC, (2019). *Best Practice-Zero Liquid Discharge System in Tannery CETP in Tamil Nadu.* Ranipet Tannery Effluent Treatment Company Limited. https://www.tnpcb. gov.in (accessed on 30 June 2021).

37. Virapan, S., Saravanane, R., & Murugaiyan, V., (2016). Zero liquid discharge (ZLD) in industrial wastewaters in India-Need for sustainable technologies and a validated case study. *Int. J. Environ. Eng. Manage., 7,* 25–33.

38. TNPCB, (2017). *Tanneries-Common Effluent Treatment Plant Monitoring Data-Status.* https://www.tnpcb.gov.in (accessed on 30 June 2021).

39. Younos, T., (2005). Environmental issues of desalination. *J. Contemp. Water Res. Educ., 132*(1), 11–18.

40. Zhang, Y., Ghyselbrecht, K., Vanherpe, R., Meesschaert, B., Pinoy, L., & Van, D. B. B., (2012). RO concentrate minimization by electrodialysis: Techno-economic analysis and environmental concerns. *J. Environ. Manage., 107,* 28–36.

41. Stokes, J. R., & Horvath, A., (2009). Energy and air emission effects of water supply. *Environ. Sci. Technol., 43*(8), 2680–2687.

42. IPCC Guidelines for National Greenhouse Gas Inventories, (2006). *Energy* (Vol. 2). Intergovernmental Panel on Climate Change, Geneva, Switzerland, IPCC.

43. EPA, (2019). *AVERT, U.S. National Weighted Average CO_2 Marginal Emission Rate, Year 2018 Data.* U.S. Environmental Protection Agency, Washington, DC.

44. Xevgenos, D., Moustakas, K., Malamis, D., & Loizidou, M., (2016). An overview on desalination & sustainability: Renewable energy-driven desalination and brine management. *Desalin. Water Treat., 57*(5), 2304–2314.

45. Shaffer, D. L., Chavez, L. H. A., Ben-Sasson, M., Castrillon, S. R. V., Yip, N. Y., & Elimelech, M., (2013). Desalination and reuse of high-salinity shale gas produced water: Drivers, technologies, and future directions. *Environ. Sci. Technol., 47*(17), 9569–9583.
46. Gingerich, D. B., & Mauter, M. S., (2015). Quantity, quality, and availability of waste heat from United States thermal power generation. *Environ. Sci. Technol., 49*(14), 8297–8306.
47. Zhou, X. S., Gingerich, D. B., & Mauter, M. S., (2015). Water treatment capacity of forward-osmosis systems utilizing power-plant waste heat. *Ind. Eng. Chem. Res., 54*(24), 6378–6389.
48. Ali, A. K., & Lawrence, L. K., (2013). Energy consumption and water production cost of conventional and renewable-energy-powered desalination processes. *Renew. Sust. Energ. Rev., 24*, 343–356.

CHAPTER 22

Significance of Eco-Labeling in Textile and Clothing Product for Green Environment

PRERANA RATNAPARKHI[1] and SANDHYA DEOLE[2]

[1]Department of Home-Economics, C.P. and Berar College, Nagpur, Maharashtra, India, E-mail: preranaratnaparkhi@gmail.com

[2]Department of Home-Economics, Vasantrao Naik Government Institute of Arts and Social Science, Nagpur, Maharashtra, India

ABSTRACT

The textile industry is answerable for a lot of contamination, but it has begun to take a more holistic, environmentally friendly approach in recent years. Eco-label is important to convey environmental impact on textile and clothing products. It connects with people groups to segregate between products that are destructive to the environment. The main aim of the study is to evaluate the environmental significance regarding ecolabels and discuss the impact of availability of the ecolabels in the market. Advertisement, availability, and price these three factors are important to purchase in product. These factors are also responsible in the field of textile and clothing products with ecolabels. These factors enhance the consumer understanding regarding the information of ecolabels and encourage them for purchasing ecolabels on textile and clothing products.

22.1 INTRODUCTION

Natural and social issues apply emphatically influence the textile and clothing market. Now a day, the three primary concerns are ecology, human biology, and waste disposal. Some unrefined materials utilized in the clothing business are plants that are especially severe for nature, for their growth utilization of various fertilizers and pesticides. Other than this, the business' finishing strategies (shading, printing, and washing) use big proportions of chemical substances, so it poses a threat to the environment.

Environmental issues emerge at all phases of the textile and clothing supply chain. Most of the environmental impression of textiles happens during production, generally because of measure of chemical compounds required to produce finished fabrics. The finished piece of clothing may contain a limited amount of leftover processing synthetic substance [1]. The innovative development in worldwide textile industries has been rapid, but the textile industry in India has largely been driven by small units that training age-old strategies for bleaching and dyeing, which adversely affect the balance of the local ecology [2]. In spite of the fact that garments value continues falling, firms' benefits are consistently rising. This is because of the quickly developing number of nations and makers that are prepared products to the stores of firms. The ecolabel is one of the pointers that practical utilization and creation, and eventually, sustainable development. Eco-labeling is just one sort of environmental labeling, to arrangement of data to consumers about the environmental quality of a product. Eco-labeling is a moderately recent phenomenon having developed in popularity worldwide since Germany's Blue Angel ecolabel was presented in 1977. By 1992, eco-labeling was explicitly endorsed by the 156 nations who were signatories to Agenda 21 at the Earth Summit (United Nations Conference on Environment and Development, Rio de Janeiro, 1992) [3]. These eco-labeling programs are planned at conserving environmental resources through different channels, for example, promoting consumer awareness, making markets for green goods, and providing consumers with the capacity to make educated choices [4].

22.1.1   ECO LABELS IN INDIA

To improve awareness about the environmental effects of products, the Ministry of Environment and Forests, Government of India (GOI) has initiated a plan in 1991, which is fundamentally a plan of labeling eco-friendly product [5]. This plan targets recognizing through the 'Eco-Mark,' any product that is made, utilized, or discarded in a way that significantly decreases the unfriendly impact that it would otherwise have on the environment, with the Earthen Pot as the logo of this plan. The GOI has likewise developed intentional eco norms for the ecolabels of the textile things and the models for environmentally friendly textile in consultation with the Indian Textile Trade and Industry (notified in the Gazette on October 8, 1996 by Ministry of Environment and Forests) [6]. While eco-labeling in different nations is picking up popularity, the activities taken by the GOI are yet waiting for a breakthrough [7].

22.1.2 WHY ECO-LABELING IS IMPORTANT?

Biological and social issues apply positive influence on the textile and clothing market. Some raw materials utilized in the textile business are plants that are especially troublesome for nature, as their development regularly includes the utilization of different manures and pesticides. Other than this the business' completing procedures (coloring, printing, and washing) devour tremendous measures of compound substances. Eco-labeling is a compelling method of informing customers about the environmental impacts of chosen products and the decisions they can make. It empowers people to segregate between products that are destructive to the environment and those more compatible with environmental objectives. An ecolabel makes the customer more aware of the advantage of specific products. It also promotes energy efficiency, waste minimization and product stewardship [8]. "Eco-labeling" is a voluntary method of environmental performance certification and labeling that is practiced around the world. An ecolabel identifies products or services proven environmentally preferable overall, within a specific product or service category. Before the coronavirus emergency fashion and sustainability was the greatest popularity. Sustainability and single utilize plastic will be less critical to many consumers where cleanliness and neatness is even more a need to spread of the infection. Annick Ireland, founder of sustainable multi-brand e-commerce site Immaculate Vegan, works with excess of 100 little brands and notes that these labels brands are confronting huge difficulties to keep up tasks: "A lot of the conversation around sustainability is led by a large number of small brands, and small businesses face a real challenge in the short term. We are still selling, but there is less demand and so there is no cash flow for us or for the brands we sell" [9].

22.1.3 CLASSIFICATION OF ECO LABELS

Globally, the classification of ecolabels is either required or voluntary; and besides, accreditation should be possible freely or not [10]. Mandatory environmental labeling is executed by autonomous outsider and affirmed by legislation-this is commonly the situation for execution issues, for example, water or energy consumption [11]. Voluntary eco-labeling, in any case, is classified by the International Standards Organization (ISO) into three classifications and can include first and outsider labeling [12]. ISO 14024 Type I labels are outsider freely certified labels for certified items and are granted to governments or organizations endorsed by worldwide governments, For example, the EU Flower and the Forest Stewardship Committee labels [11]. ISO 14021 Type II labels incorporate outsider labels checked by self-declarations of financially dependent partner organizations, for example, the Better Cotton Initiative, Blue sign, German Öekotex [13]. ISO Type II labels generally pass on messages, for example, degradable, environment friendly, reused, ozone-friendly, and earth-friendly [14]. ISO 14025 Type III voluntary projects give evaluated natural information of a product, within a set of parameters that are dictated by a qualified outsider, based on the appraisal of products' lifecycles [13]. There are no ecolabels in textile and clothing executed by mandatory. Ecolabels are commonly given either by government maintained or private endeavors once it has been exhibited that the product of the applicant has met the guidelines [15].

22.1.3.1 GOVERNMENT ECO-LABELS

- Blue Angel (Germany), used in textile material: sacks, cleaning clothes, sleeping pads, napkins, material floor covers.
- Eco Mark (Japan), huge potential for development of the environment by utilizing the product.
- Environmental decision (Canada).
- White Swan (Nordic countries), used for eco-friendly baby diapers.
- EU-eco-labels-A voluntary scheme designed to encourage businesses to market products and services that are kinder to the environment and for European consumers-including public and private purchasers-to easily identify them. Textile material classification: mattress, shoes, textile.
- Eco-Mark (India), for textile things.
- Green Label (Singapore).

22.1.3.2 PRIVATE ECOLABELS

- Eco-Tex, Oeko-Tex (Germany) for textile raw materials, intermediate, and end products at all stages of production.

- Green Seal (US), used for building and construction products, food products, institutional products, household products, etc.

22.2   OBJECTIVES OF STUDY

- To know the awareness regarding ecolabels in textile and clothing product;
- To know the environmental significance regarding ecolabels;
- To know the availability of textile and clothing products with ecolabels in the market.

22.3   HYPOTHESIS

- Awareness regarding ecolabels on textile and clothing product is differ according to age, education, income, and profession.
- Environmental significance regarding ecolabels on textile and clothing product is differ according to age, education, income, and profession.

22.4   RESEARCH METHODOLOGY

The survey methods were used to collect the information regarding awareness, significance of ecolabel in textile and clothing product and also the availability of eco-labeled material in the local market. The survey was limited to the Nagpur city. Around 30 samples were selected on the basis of purposive sampling methods to collect the information. To collect the information regarding general information of the respondent questionnaire was framed and to collect the information regarding awareness, significance, and availability of the Eco labeled textile and clothing product in local market, scales were prepared in Google form the scales were directed as strongly agree, agree, undecided, disagree, strongly disagree. At that point inquiries regarding the awareness about eco-labeled product, information about the concept of eco-labeled product, and abstract information about a similar issue. The scales were validated through RM experts after a pilot study. The data was found reliable as per the Reliability Statistics by excel. The prepared questionnaire and scales in Google form were sending to the respondent via WhatsApp messenger. The responses of the respondent classified and analyzes. To draw the inferences student 't' test was used, and results are discussed in further section.

22.5   RESULT AND DISCUSSION

22.5.1   *GENERAL INFORMATION OF THE RESPONDENT*

1. Age Group: Many studies support those younger individuals are likely to be more sensitive to environmental issues (Figure 22.1). But in this study facing environmental challenges with required changes, 30–40 age group are more acceptable for reforming their minds than 20–30 age group.

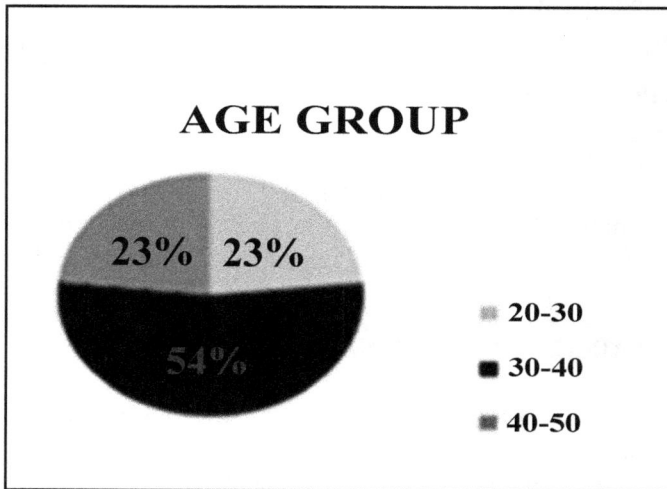

FIGURE 22.1   Age group pie chart.

2. Educational Qualification: The level of education of respondent effectively works on ecological awareness and significance about eco-labeled textile product (Figure 22.2). Around 63% postgraduate were much clearer and full perspective understanding on environmental issue than another educational category.
3. Monthly Income: The monthly income affects the respondent purchasing behavior towards ecolabel textile product (Figure 22.4). Around 36% of the respondent from 50,000–100,000 income group are ready to spend their expenses on eco-labeled textile products.
4. Employment Status: The awareness of eco-labeled product significance of environment and availability in markets these factors were highly effective on the respondent employment status (Figure 22.3). Around

60% respondent which are in service more conscious about their purchase decision of eco-labeled textile product than another respondent.

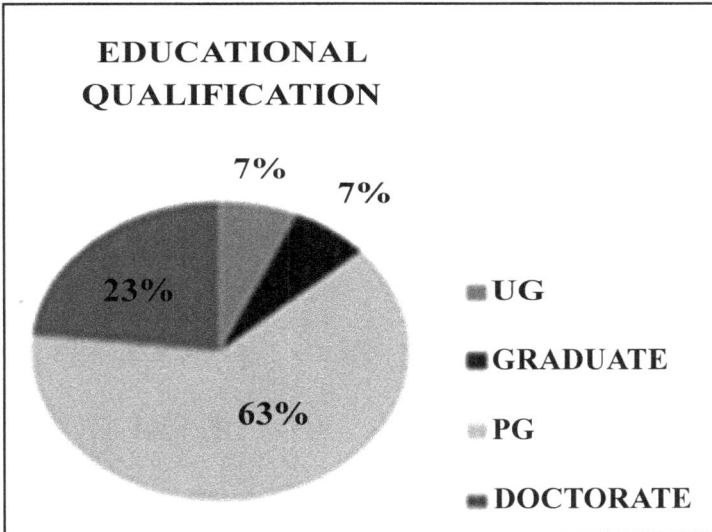

FIGURE 22.2   Educational qualification pie chart.

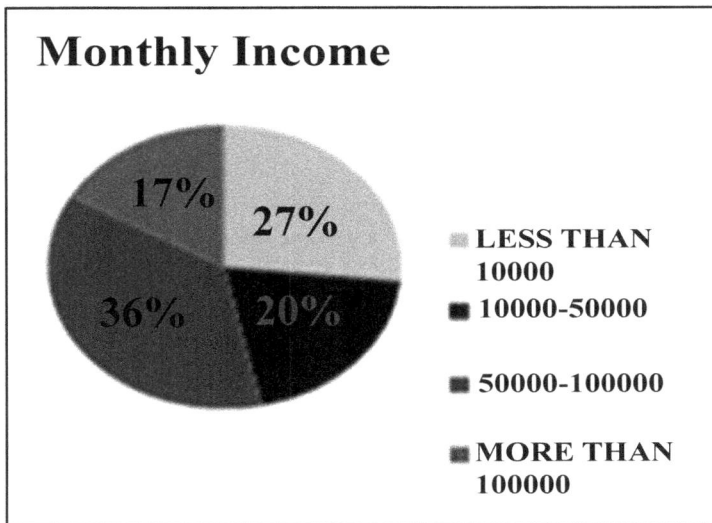

FIGURE 22.3   Monthly income pie chart.

EMPLOYMENT STATUS

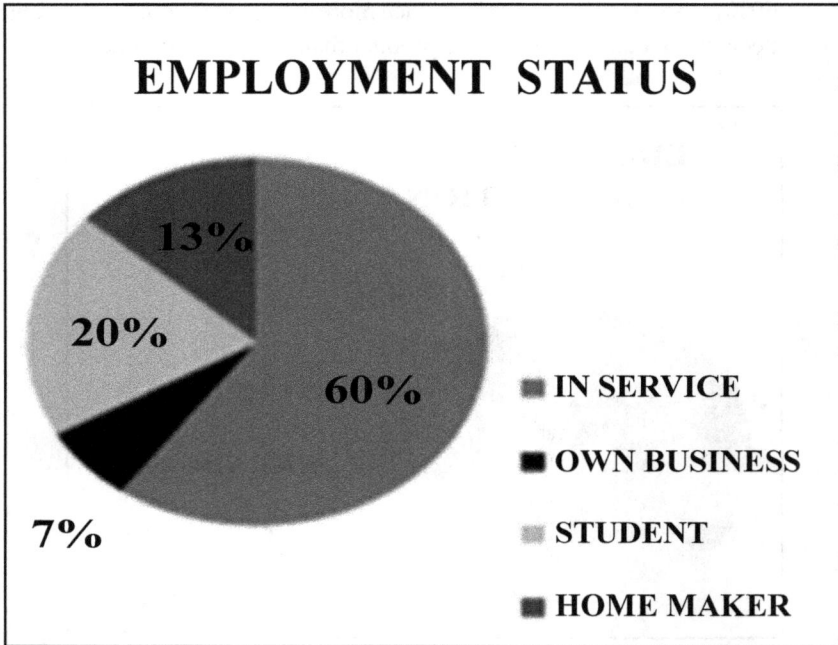

FIGURE 22.4   Employment status pie chart.

22.5.2   *AWARENESS REGARDING ECO-LABEL TEXTILE AND CLOTHING PRODUCT*

Awareness on Eco-label is likewise one of the elements of guiding consumer to buy environmental product. Figure 22.5 shows that most of the respondents know the information regarding ecolabels as well as aware about various symbols and certifications of ecolabels. Maximum respondents are strongly agreeing that ecolabels are one of the ways to create awareness about green products. All these facts show the awareness regarding ecolabels in textile and clothing products, but the uncertainty is seen regarding the clear concept of ecolabels. Respondent needs to comprehend about the ecological advantage of textile product through Eco-label, besides they have to comprehend what specific activities should be taken accordingly. If the state government should take some efforts to promote manufacturing textile and clothing products with ecolabels, then the respondent mindset will be more positive regarding green products.

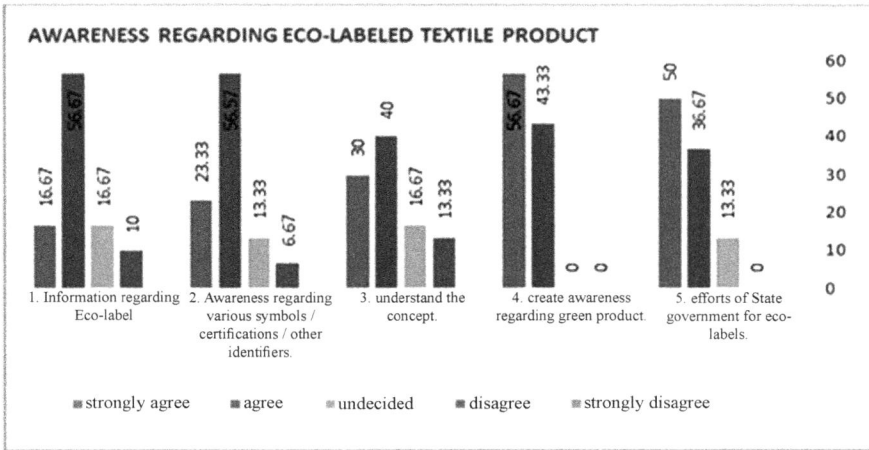

AWARENESS REGARDING ECO-LABELED TEXTILE PRODUCT

Legend: strongly agree | agree | undecided | disagree | strongly disagree

Categories:
1. Information regarding Eco-label
2. Awareness regarding various symbols / certifications / other identifiers.
3. understand the concept.
4. create awareness regarding green product.
5. efforts of State government for eco-labels.

FIGURE 22.5   Information regarding ecolabels.

22.5.3   AWARENESS REGARDING ECO-LABEL TEXTILE AND CLOTHING PRODUCT

In Figure 22.6, the factor environmental significance shows that maximum respondents are strongly accepted that eco-labeled textile products are beneficial for health and even strongly agreed that these products reduce the harmful effect on the environment. Respondents have the significance of advertisement regarding ecolabels, but still very few advertisements are seen. Maximum respondents have significance of the advertisement to spread the information regarding ecolabel textile and clothing product. Likewise, the study shows that respondents have significance of eco label products and even worried about health and environmental issues. So that the promotion of ecolabel textile product is very essential (Table 22.1).

Availability of ecolabel product in the market is an external factor, its 'role cannot be neglected as one of impact factors to the effectiveness of Eco-labels. It is seen that in the local market the Ecolabel textile and clothing products are available but in less quantity. Consumers are agreeing to purchase ecolabel textile product if easily available, labels give enough information to confirmed the greenness of the product and environmentally safeness. Consumer ready to pay more for eco-labeled Textile and Clothing products. It is clear that the availability of Eco-labeled textile product in the markets and information about eco labeled product both have a significant relationship. Consumer purchase behavior and purchase intention for eco-labeled product is high. It means it is the time for marketers and companies to take action to fulfill consumers' demand.

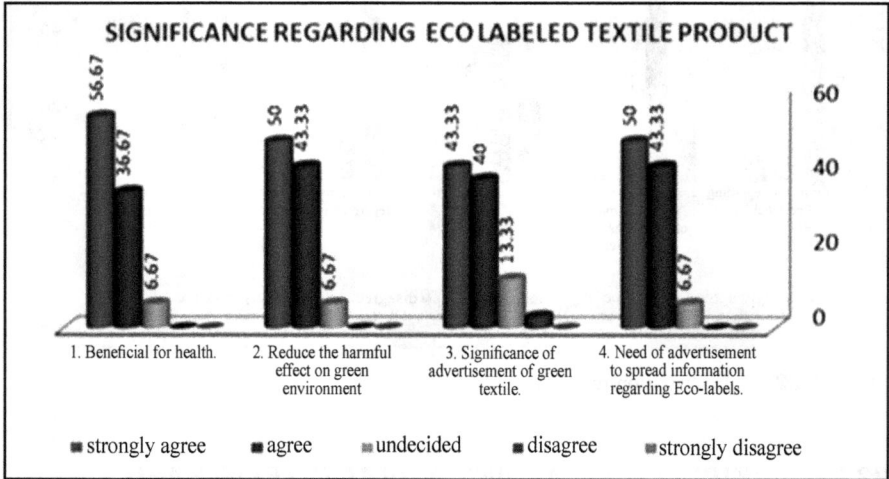

FIGURE 22.6 Responses for ecolabels.

TABLE 22.1 Impact of Availability of Eco-Label Textile and Clothing Product on Purchasing Behavior

Impact of Availability of Eco-Labeled Products on Purchasing Behavior					
SL. No.	Statements	SA	AG	UD	DA	SD
1.	In local market the ecolabel textile and clothing products are available.	13.33%	40%	30%	13.33%	3.33%
2.	I would prefer to buy eco-labeled textile and clothing products, if I have enough information which confirmed their greenness.	43.33%	43.33%	10%	3.33%	0%
3.	While purchasing textile and clothing products, I thoroughly read label to see if contents are environmentally safe.	30%	43.33%	6.67%	16.67%	3.33%
4.	I available, I would seek out eco-labeled clothes whenever possible, I buy products which I consider environmentally safe.	40%	50%	2.675%	3.33%	0%
5.	I would pay more for ecolabels on textile and clothing products.	33.33%	30%	33.33%	3.33%	0%

22.6   STATISTICAL ANALYSIS

22.6.1   AWARENESS REGARDING ECO-LABELED PRODUCT

According to all age groups (Table 22.2), the calculated value is less than the table value. Hence, the null hypothesis is accepted. It is clear that the awareness of the respondents regarding eco-labeled textile products for green environment does not differ according to age.

TABLE 22.2   Awareness of the Respondents Regarding Eco Labeled Textile Product

	Groups	No.	Mean	S.D.	Significance Value	Level of Significance 5%	Result
Age Wise	20–30	7	20.14	1.46	+0.29	2.04	<
	30–40	16	20.38	2.36			
	20–30	07	20.14	1.46	+1.57	2.04	<
	40–50	07	21.43	1.62			
	30–40	16	20.38	2.36	+ 1.23	2.04	<
	40–50	07	21.43	1.62			

The statistics in Table 22.3 show that the education is the important factor regarding awareness of the eco-labeled product. The result shows that, in all educational groups except the graduate and doctorate group, the calculated value is more than the table value. Hence, the null hypothesis is rejected. It is clear that the awareness of the respondents regarding eco-labeled textile product for green environment is differ according to education. It is concluded that highly educated respondents have high awareness regarding eco-labeled textile product.

It is seen that according to 50,000–100,000 and more than 100,000 income group (Table 22.4), the calculated value is more than the table value. It is clear that the awareness of the respondents regarding eco labeled textile product for green environment is more in higher-income group.

According to profession in all groups calculated value is less than the table value-wise (Table 22.5) hence, accept the null hypothesis. It is concluded that the professional criteria are not affected the awareness regarding eco-labeled textile product.

22.6.2   SIGNIFICANCE REGARDING ECO LABELED TEXTILE PRODUCT

According to all age groups (Table 22.6), the calculated value is less than the table value. Hence, the null hypothesis is accepted. It is clear that the

TABLE 23.3   Impact of Education on Awareness of the Eco Labeled Product

	Groups	No.	Mean	SD	Significance Value	Level of Significance 5%	Result
	UG	2	19	0	+ 4.98	2.04	>
	Graduate	2	21.5	0.71			
	UG	2	19	0	+ 2.42	2.04	>
	PG	19	20.11	2			
Education Wise	UG	2	19	0	+4.16	2.04	>
	Doctorate	7	22	1.91			
	Graduate	2	21.5	0.71	+ 2.04	2.04	=
	PG	19	20.11	2			
	Graduate	2	21.5	0.71	+0.57	2.04	<
	Doctorate	7	22	1.91			
	PG	19	20.11	2	+2.21	2.04	>
	Doctorate	7	22	1.91			

TABLE 23.4   Impact of Income on Awareness of the Eco Labeled Product

	Groups	No.	Mean	S.D.	Significance Value	Level of Significance 5%	Result
	Less than 10000	08	20.38	2.45	+0.21	2.04	<
	10000–50000	06	20.67	2.73			
	Less than 10000	08	20.38	2.45	+0.30	2.04	<
	50000–100000	11	20.09	1.45			
	Less than 10000.	08	20.38	2.45	+1.30	2.04	<
Income Wise	More than 100000	05	21.8	1.48			
	10000–50000	06	20.67	2.73	+0.48	2.04	<
	50000–100000	11	20.09	1.45			
	10000–50000	06	20.67	2.73	+0.87	2.04	<
	More than 100000	05	21.8	1.48			
	50000–100000	11	20.09	1.45	+2.16	2.04	>
	More than 100000	05	21.8	1.48			

TABLE 23.5   Impact of Profession on Awareness of the Eco Labeled Product

	Groups	No.	Mean	S.D.	Significance Value	Level of Significance 5%	Result
Profession Wise	In service	18	20.89	2.23	+ 0.05	2.04	<
	Own business	02	21	2.83			
	In service	18	20.89	2.23	+2.03	2.04	<
	Students	06	19.33	1.37			
	In service	18	20.89	2.23	+0.90	2.04	<
	Homemaker	04	20.25	1.96			
	Own business	02	21	2.83	+0.80	2.04	<
	Students	06	19.33	1.37			
	Own business	02	21	2.83	+0.36	2.04	<
	Homemaker	04	20.25	1.96			
	Students	06	19.33	1.37	+1.24	2.04	<
	Homemaker	04	20.25	1.96			

TABLE 23.6   Impact of Age on Significance of the Eco Labeled Product

	Groups	No.	Mean	S.D.	Significance Value	Level of Significance 5%	Result
AgeWise	20–30	7	16.57	2.15	+1.49	2.04	<
	30–40	16	18.06	2.35			
	20–30	07	16.57	2.15	+1.01	2.04	<
	40–50	07	17.57	1.51			
	30–40	16	18.06	2.35	+ 0.60	2.04	<
	40–50	07	17.57	1.51			

significance of the respondents regarding eco-labeled textile products for green environment is not differ according to age.

The statistics in Table 22.7 show that education is the important factor regarding the significance of the eco-labeled product. The result shows that, in the graduate, undergraduate and doctorate group, the calculated value is more than the table value. Hence, the null hypothesis is rejected. It is clear that the significance of the respondents regarding eco-labeled textile products for green environment is differ according to education. It is concluded that highly educated respondents have high significance regarding the eco-labeled textile product. The consumer has obtained information on environmental insurance through advanced education and higher goal to buy a green product.

TABLE 22.7 Impact of Education on Significance of the Eco Labeled Product

	Groups	No.	Mean	SD	Significance Value	Level of Significance 5%	Result
Education Wise	UG	2	14	0	+ 3.01	2.04	>
	Graduate	2	17	1.41			
	UG	2	14	0	+ 8.36	2.04	>
	PG	19	18.16	2.17			
	UG	2	14	0	+5.44	2.04	>
	Doctorate	7	17.29	1.60			
	Graduate	2	17	1.41	+ 1.04	2.04	<
	PG	19	18.16	2.17			
	Graduate	2	17	1.41	+0.25	2.04	<
	Doctorate	7	17.29	1.60			
	PG	19	18.16	2.17	+1.11	2.04	<
	Doctorate	7	17.29	1.60			

According to income in all groups (Table 22.8), the calculated value is less than the table value hence, accept the null hypothesis. It is concluded that the income criteria are not affected the significance regarding eco-labeled textile product.

TABLE 22.8 Impact of Income on Significance of the Eco Labeled Product

	Groups	No.	Mean	S.D.	Significance Value	Level of Significance 5%	Result
Income Wise	Less than 10000	08	18.13	2.42	+0.28	2.04	<
	10000–50000	06	17.83	1.60			
	Less than 10000	08	18.13	2.42	+1.35	2.04	<
	50000–100000	11	16.64	2.34			
	Less than 10000	08	18.13	2.42	+0.43	2.04	<
	More than 100000	05	18.60	1.52			
	10000–50000	06	17.83	1.60	+1.24	2.04	<
	50000–100000	11	16.64	2.34			
	10000–50000	06	17.83	1.60	+0.82	2.04	<
	More than 100000	05	18.60	1.52			
	50000–100000	11	16.64	2.34	+2.00	2.04	<
	More than 100000	05	18.60	1.52			

TABLE 22.9   Impact of Profession on Significance of the Eco Labeled Product

	Groups	No.	Mean	S.D.	Significance Value	Level of Significance 5%	Result
	In service	18	18.21	1.58	+ 0.40	2.04	<
	Own business	02	17	4.24			
	In service	18	18.21	1.58	+1.18	2.04	<
Profession Wise	Students	06	16.83	2.71			
	In service	18	18.21	1.58	+1.93	2.04	<
	Homemaker	04	16	2.16			
	Own business	02	17	4.24	+0.05	2.04	<
	Students	06	16.83	2.71			
	Own business	02	17	4.24	+0.31	2.04	<
	Homemaker	04	16	2.16			
	Students	06	16.83	2.71	+0.54	2.04	<
	Homemaker	04	16	2.16			

According to profession in all groups (Table 22.9), the calculated value is less than the table value-wise hence, accept the null hypothesis. It is concluded that the profession criteria are not affected the significance regarding eco-labeled textile product.

22.7   CONCLUSION

The ecolabel is one of the most important factors in the textile industry that evaluate usage and creation for earth improvement. Eco-labeling is only one sort of environmental labeling, and suggests specially information to consumer about the eco Textile and clothing product. This study worked on three-component awareness, significance, and availability of the market. A statistical analysis of respondent's awareness of criteria such as age, education, income, and profession reveal that, the higher education, more awareness seen in the respondent, which shows that education criteria has more aware about the ecolabel textile product than other criteria. With the same about income, if you have enough money, you can buy eco-labels textile product. So even respondent from higher income group who knew about eco-labels product they will definitely buy such products. While conducting a statistical analysis of the component of significance, it was noticed that apart from age, money, and job, education is the only criterion in which the respondent has the highest significance. Buying an eco-labeled product can help keep the environment safe.

This study has suggested that an individual having some worry about nature would have a more preference for buying an ecolabel textile product. The significance of a useful program must do for the people to clarifying the importance of an ecolabel in respondent purchase decisions. It should be done with the overall exertion to build awareness, significance, and accessibility of eco-labeled products to protect nature. It is surprising no uncertainty, after launching a national ecolabel since 1996, the level of customers' awareness with respect to the nearby ecolabel is practically insignificant.

Another finding, where the awareness for green products is still low is that this study had the option to reveal some restricted parts of respondent response to the utilization of ecolabel in their accessibility in market with environment friendly aspects. In general, this study has demonstrated the beneficial outcome of awareness of ecolabel textile products between the significance of ecolabel and availability in the market.

KEYWORDS

- **Eco-Labels**
- **Environmental Significance**
- **Textile & Clothing**

REFERENCES

1. Walters, A.; Santillo, D.; Johnston, P. An overview of textiles processing and related environmental concerns, Greenpeace Research Laboratories. *Technical Note. 2005*, 08, 28.
2. Koszewska, M. Social and Eco-labeling of Textile and Clothing Goods as Means of Communication and Product Differentiation. *FIBRES & TEXTILES in Eastern Europe. 2011, 19*(4), 20–26.
3. Appleton, A. E. (ed.), Labelling Strategies in Environmental Policy, 1997, Kluwer Law International, Boston.
4. Childs, C.; Whiting, S. Eco-labelling and the Green Consumer. Working papers from Sustainable Business Publications series. University of Bradford, West Yorkshire, 1998.
5. Challa, L. Impact of textiles and clothing industry on environment: approach towards eco-friendly textiles. March 2007.
6. Textile Committee, Ministry of Textile, Government of India, Textile Testing and Technical Services (3TS) What is ecolabeling of textiles? URL http://textilescommittee. nic.in/faq/textile-testing-technical-services-3ts (accessed on 28 February 2008).
7. Chaturvedi, S.; Nagpal, Environmental standards emerging issues and policy options. *Economic and Political Weekly. 2003, 4,* 66–74.

8. IISD (International Institute for Sustainable Development): A Global Guide, Benefits of Eco labeling, https://www.iisd.org/business/markets/eco_label_benefits.aspx.

9. Brown, H. Coronavirus and sustainability: how will fashion respond? Drapers: for all the fashion business. URL https://www.drapersonline.com/business-operations/coronavirus-and-sustainability- how-will-fashion-respond/7040147. Accessed 16 April 2020.

10. GEN (Global Eco-labelling Network). What is eco-labelling? URL: https://globalecolabelling.net/what-is-eco-labelling/Accessed 4 Jun. 2013.

11. Scheer, D.; Rubik, D.R.F.; Gold, S. Enabling developing countries to seize eco-label opportunities: project background paper. Heidelberg Germany: Institute for Ecological Economy research, 2008.

12. GEN (Global Eco-labelling Network). Annual report. URL: http://www.flipsnack.com/FB8F9BF569B/f7hfgl5j 2011, Accessed 4 Jun. 2013.

13. IISD (International Institute for sustainable development): A Global Guide, Eco-labeling-The ISO 14020 series URL:https://www.iisd.org/business/markets/eco_label_iso14020. aspx Accessed 5 Jun. 2013.

14. D'Souza, C.; Taghian, M.; Lamb, P. An empirical study on the influence of environmental labels on consumers. *Corporate Communications: An International Journal. 2006,* 11 (2), 162–173.

15. Hyvarinen, Eco-labeling and Environmentally Friendly Products and Production Methods Affecting the International Trade in Textiles and Clothing. 1999 [online]. Available from: http://www.intracen.org/btp/issues/atc/publications/ecolabatc.pdf. [Accessed 10th July 2010].

CHAPTER 23

Microwave Method for Regeneration of Spent-Activated Carbon from Pharmaceutical Industries

SUMITA N. RAO,[1] ADITI S. PANDEY,[1] MANGESH S. DHORE,[2] PRADEEP P. PIPALATKAR,[3] and BABUBHAI C. PATEL[4]

[1]Department of Applied Chemistry, Priyadarshini College of Engineering, Nagpur, Maharashtra, India,
E-mail: sumitarao2000@rediffmail.com (S. N. Rao)

[2]Department of Forensic Science, G.H. Raisoni University, Amravati, Maharashtra, India

[3]Pollution and Ecological Control Services, Nagpur, Maharashtra, India

[4]Nandesari Industries Association-CETP, Vadodara, Gujarat, India

ABSTRACT

Paracetamol is a pharmaceutical analgesic and antipyretic agent. The use of activated carbon is done for the removal of impurities and color by the process of adsorption in the treatment of wastewater generated from paracetamol industries. These industries generate a few 100 metric tons of spent activated carbon (SAC) annually, awaiting either regeneration or disposal. In the present study, the regeneration of SAC samples (Sample 1 and 2) from two pharmaceuticals industries was done by microwave irradiation at low (520 W) and medium (700 W) power. Each sample was exposed to microwave radiations for varied time, i.e., 5, 10, 15, 20, and 30 min. The adsorptive capacity of regenerated activated carbons (AC) was measured by loading them with methylene blue (MB) dye for calculating its methylene blue number (MBN). The MBN values and SEM images of treated samples were compared with that of fresh and SAC. The equilibrium adsorption data were used to calculate maximum monolayer coverage capacity and surface area (SA) of regenerated samples 1 and 2 by Langmuir isotherm. The results obtained show that the SAC can be regenerated successfully by more than 85% when it is subjected to microwave irradiation of 520 W and 700 W. The major advantage of this process is that a significantly high percentage of regeneration can be achieved within a short duration of time.

23.1   INTRODUCTION

During the last few decades, there is an increased demand for pharmaceuticals in the society leading to a rapid increase in its production. The pharmaceutical industry demands an exceptional level of purity in all processing aids. Paracetamol, or 4-hydroxyacetanilide, is a widely used drug due to its analgesic and antipyretic properties. In pharmaceutical industries, the use of activated carbon is done at various process steps with an aim to obtain high purity product [1]. It removes pollutants by adsorbing them during the drug manufacturing process. The wastewater generated from the industry is highly colored due to the presence of acetaminophen, intermediate aminophenol, and its oxidation products, etc. Removal of phenol and its derivatives present in paracetamol wastewater is also achieved by the use of powder activated carbons (AC) [2].

In paracetamol industries, removal of impurities and color is done by Activated carbon. The use of granular activated carbon for the removal of organic impurities is considered economical owing to its reuse during several adsorption-regeneration cycles. These industries generate a few 100 metric tons of spent activated carbon (SAC) annually, awaiting either regeneration or disposal. The 'spent' carbon, whose adsorptive capacity is diminished and it can no longer be used for the intended application without regeneration, is regarded as hazardous waste and sent either to landfill or incinerator. In most of the cases, the cost of replacing the saturated carbon would be prohibitive. Hence, it should be regenerated. The conventional methods of regeneration of SAC includes steam regeneration [3], thermal regeneration [4], chemical/ solvent regeneration and biological regeneration [5]. Although, thermal regeneration method regenerates the carbon very well but has several draw-backs. The alternative carbon reactivation/regeneration methods that can be used are Ex-situ/In-situ wet oxidations, Wet air oxidation [6, 7], Ultrasonic regeneration [8], Supercritical regeneration and Electrochemical regeneration. It has been reported that electrochemical regeneration process is more effective and economical as compared to thermal, ultrasonic, and base washing regeneration processes [9].

In the past few years, regeneration of activated carbon by microwave method has been found to be feasible owing to its ability of uniform and instantaneous heating. This method has many advantages over conventional treatment methods such as fast and precise temperature control, less space requirements, energy savings and greater efficiency [10, 11]. In microwave heating the materials receive energy through dipole rotations and ionic conduction [12]. SAC can be regenerated by microwave irradiation in less time because of excellent microwave absorbing properties by it [13]. In this study, two SAC samples obtained from pharmaceutical industries (Samples 1 and 2) were microwave irradiated at 520 W and 700 W power for different time intervals of 5, 10, 15, 20, and 30 min. Fresh, spent, and regenerated activated carbon samples were analyzed for their adsorption of methylene blue (MB). The adsorption capacity of activated carbon samples was expressed as their methylene blue number (MBN). SEM was also conducted to evaluate changes in surface morphology. The results of MBN show that SAC samples 1 and 2 can be regenerated by more than 85% within 10 min and 30 min, respectively, when exposed to microwave radiations. The equilibrium adsorption data were used to calculate maximum monolayer coverage capacity and total surface area (TSA) of regenerated samples 1 and 2 by Langmuir isotherm model [14, 15]. SEM analysis of the samples also clearly indicate that regeneration of SAC from pharmaceutical industries can be achieved using the microwave method.

23.2 MATERIALS AND METHODS

23.2.1 MATERIALS

M/s NIA-CETP, Nandesari, Vadodara, Gujarat, India supplied Fresh and SAC samples used in two different pharmaceutical industries. MB supplied by Merck was use as adsorbate. UV-Vis double beam spectrophotometer (Systronics AU-2701) was used for determining the concentration of MB. SEM analysis of the samples was done on VEGA3 TESCAN.

23.2.2 MICROWAVE IRRADIATION

The regeneration of SAC samples 1 and 2 was done under microwave heating at low (520 W) and medium (700 W) power. The samples were exposed to microwave radiations for varied time, i.e., 5, 10, 15, 20, and 30 min. During microwave heating, initially bright sparks described as micro-plasmas [16] were observed momentarily at the interface between silica crucible and carbon along with the release of odorous volatile compounds. Very soon, the carbon samples became red hot and remained so for the entire treatment period. Probably this resulted in the removal of the impurities from the carbon samples. The treated samples were then subjected to adsorption of MB dye for calculating its Methylene blue (MBN) number. The MBN of a treated sample was compared with that of fresh and SAC samples.

23.2.3   DETERMINATION OF METHYLENE BLUE NUMBER (MBN)

The MBN is the maximum amount of dye adsorbed on 1.0 g of adsorbent [17, 18]. In the present work, 250 mg of fresh, untreated, and treated AC of samples 1 and 2 of was stirred with 100 ml of 1000 ppm MB solution on a bottle shaker for 6, 15 and 24 h at room temperature. After the shaking period, the residual concentration of MB in each solution was determined using a UV-Vis double beam spectrophotometer at 663 nm. The MBN was calculated using Eqn. (1):

$$q_{eq} \text{ (mg g}^{-1}) = (C_0 - C_e) \times V/M \tag{1}$$

where; C_0 (ppm) is the initial concentration of the MB solution; C_e (ppm) is the residual concentration of the MB solution at equilibrium time; V(L) is the volume of the solution; and M (g) is the mass of the carbon sample taken.

23.2.4   LANGMUIR ADSORPTION ISOTHERM

According to the Langmuir isotherm, certain specific homogeneous sites within the adsorbent are occupied by the adsorbate. The equation of Langmuir adsorption is described as follows:

$$q_e = \frac{Q_o K C_e}{1 + K C_e} \tag{2}$$

where; q_e is the adsorption density at the equilibrium adsorbate concentration; C_e is the equilibrium concentration of adsorbate in solution; Q_0 is the maximum adsorption capacity corresponding to complete monolayer coverage; K is the Langmuir isotherm constant.

Langmuir adsorption parameters are calculated by converting the above equation into linear form as:

$$\frac{1}{q_e} = \frac{1}{Q_o} + \frac{1}{Q_o K C_e}$$
(3)

From the obtained experimental data, the values of Q_0 and K can be determined from the slope and intercept of the Langmuir plot of $1/q_e$ vs $1/C_e$. Using the value of approximate contact area of an adsorbate molecule, surface area (SA) of carbon can be calculated using following equation:

$$\text{Surface area of the adsorbent (m}^2\text{/g)} = \frac{Q_o\left(mg\,/\,g\right) \times N \times S\left(m^2\,/\,\text{molecule}\right)}{M}$$
(4)

where; Q_0 is the maximum amount of adsorbate adsorbed per gram of adsorbent for monolayer formation; M is the molecular weight of the adsorbate (mg/molecule); N is the Avogadro's number; and S is the contact SA by each molecule (m^2) [14, 15, 19, 20].

23.3 RESULTS AND DISCUSSIONS

The specifications of samples of fresh activated carbon (FAC) of both the pharmaceutical industries are given in Table 23.1.

TABLE 23.1 Specifications of Fresh Activated Carbon Samples

SL. No.	Test	Sample 1	Sample 2
1.	Description	Black color fine powder	Very fine free-flowing black powder
2.	pH (10% w/w suspension)	6.0 to 7.0	10.0 to 12.0
3.	Methylene blue number (Adsorption Value)	Not less than 220 mg/g	Not less than 340 mg/gm

23.3.1 METHYLENE BLUE NUMBER (MBN) ADSORPTION

MBN of fresh activated carbon (FAC-1) of sample 1 was found to be 285–310 mg/g, while that for spent activated carbon (SAC-1) was found to be 150–170 mg/g. The results of adsorptive capacity of MB dye of SAC by microwave radiation of 520 W and 700 W at 5, 10, 15, and 20 min are given in Table 23.2. It is evident from the values of MBN that the SAC is regen-erated successfully by more than 85% when it is subjected to microwave radiation of 520 W and 700 W for 5 and 10 min. On further increasing the time, a decrease in the value of MBN is observed.

TABLE 23.2   Methylene Blue Number Values of Regenerated Activated Carbon (Sample 1)*

SL. No.	Microwave Power (W)	Microwave Irradiation Time (min)	MBN (mg/g)
1.	520	5	224
2.		10	286
3.		15	157
4.		20	141
5.	700	5	271
6.		10	290
7.		15	227
8.		20	170

Note: MBN: FAC-1 = 285–310 mg/g; SAC-1 = 150–170 mg/g.

MBN of the fresh activated carbon (FAC-2) of sample 2 was 300–340 mg/g, while spent activated carbon (SAC-2) was found to be 20–40 mg/g. The results of MB uptake of SAC by microwave radiation of 520 W and 700 W at 5, 10, 15, 20, 25, and 30 min are given in Table 23.3. From the results obtained, it can be said that the SAC can be regenerated successfully >85% when it is subjected to microwave radiation of 520 and 700 W for 30 min.

SEM analysis was also performed on fresh, spent, and regenerated carbon (RAC) samples (1 and 2) shown in Figures 23.1(A–D) and 23.2(A–C), respectively.

23.3.2   LANGMUIR ADSORPTION ISOTHERM

The distribution between the activated carbon and MB dye solution, when the system was at equilibrium, is defined by the Langmuir adsorption isotherm and it helps to establish the adsorption capacity of activated carbon. As discussed in section 7.2.4 above, the values of Q_0 and K (Table 23.4) were calculated by fitting the obtained experimental data in to the linear form of the Langmuir adsorption Eqn. (3). Graphs were plotted between $1/C_e$ versus $1/q_e$ for FAC, spent carbon (SAC) and RAC, respectively of both the samples

1 and 2 for MB dye uptake, where, C_e is the concentration of dye in the equilibrium solution. q_e is the amount of dye adsorbed per gram of adsorbents. Figures 23.3–23.5 show the straight lines obtained by plotting $1/q_e$ vs $1/Ce$ for MB on FAC-1, SAC-1, and RAC-1 samples, respectively studied. Similarly, Figures 23.6–23.8 show the straight lines obtained by plotting $1/q_e$ vs $1/C_e$ for MB on FAC-2, SAC-2, and RAC-2 samples, respectively studied. As is evident from all the graphs, the value of R^2 in all the cases is > 0.9, indicating that Langmuir isotherm is favorable.

TABLE 23.3   Methylene Blue Number Values of Regenerated Activated Carbon (Sample 2)*

SL. No.	Microwave Power (W)	Microwave Irradiation Time (min)	MBN (mg/g)
1.	520	5	154
2.		10	243
3.		15	252
4.		20	271
5.		25	293
6.		30	321
7.	700	5	168
8.		10	200
9.		15	250
10.		20	300
11.		25	307
12.		30	328

*Note: MBN: FAC-2 = 300–340 mg/g, SAC-2 = 20–40 mg/g.

SA of each studied adsorbent was calculated using Eqn. (3) and reported in Table 23.4. The value of occupied SA of one molecule of MB (S, m^2/molecule) is taken as 197.2×10^{-20} m^2 [14, 15].

As is evident from Table 23.4, the values of Qo, the maximum adsorption capacity corresponding to complete monolayer coverage and SAs of regenerated activated carbons (RAC-1 and 2) of both the samples studied are almost comparable to fresh activated carbons (FAC-1 and 2), respectively. This indicates that the microwave technique for regeneration of SAC samples (SAC-1 and 2) is successful. The calculated value of SA of FAC-2 is found to be greater than FAC-1 (Table 23.4). But on comparing the values of SAs of SAC-1 and 2, it can be seen that SAC-2 has lesser SA than SAC-1. This indicates that probably a greater number of molecules of

FAC-1 (A)　　　　　　　　　SAC-1(B)

RAC-1-(C)　　　　　　　　　RAC-1-(D)

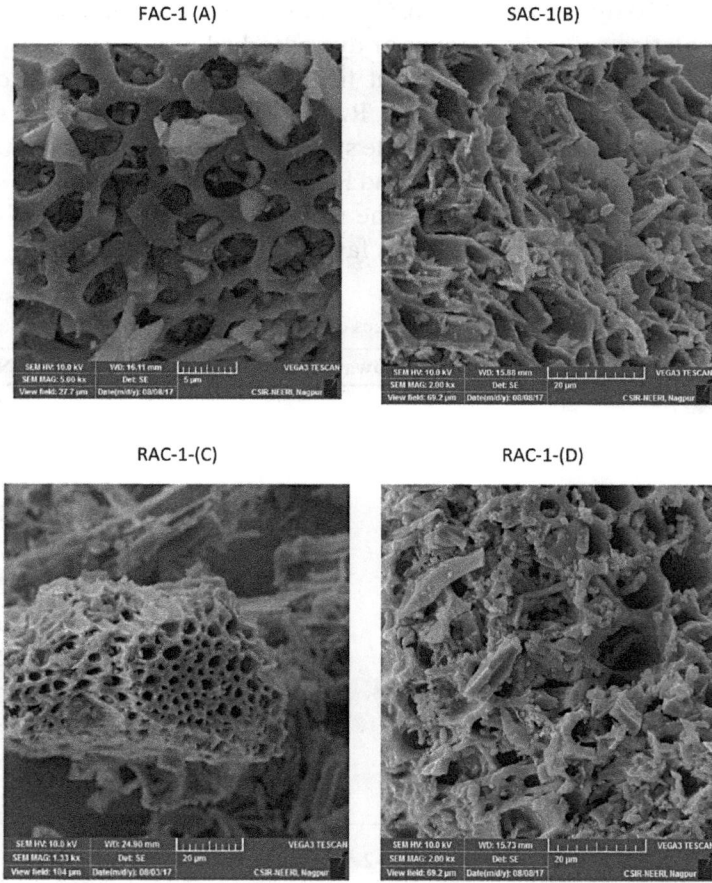

FIGURE 23.1　SEM images of fresh, spent, and regenerated activated carbon (sample 1).

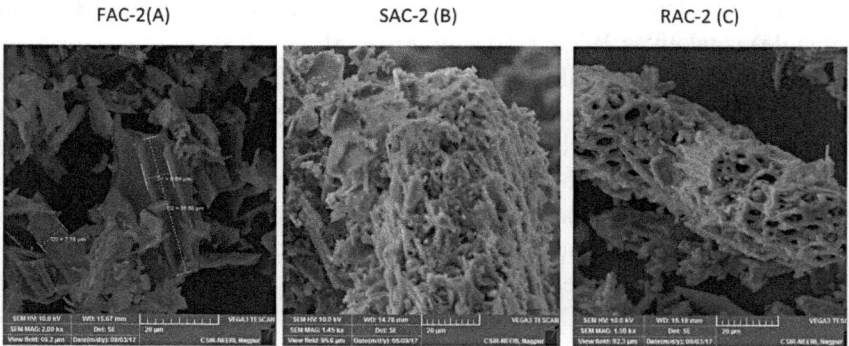

FAC-2(A)　　　　　SAC-2 (B)　　　　　RAC-2 (C)

FIGURE 23.2　SEM images of fresh, spent, and regenerated activated carbon (sample 2).

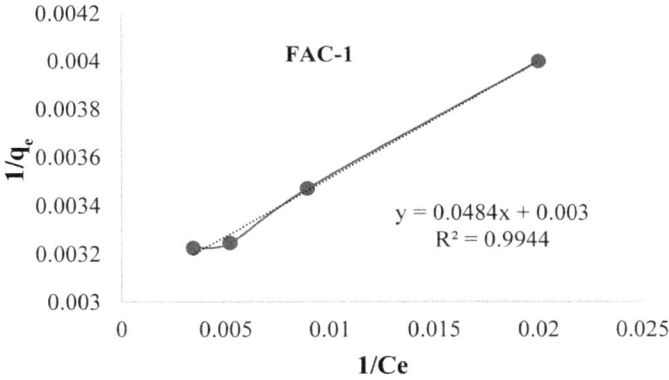

FIGURE 23.3   Langmuir adsorption isotherm of fresh activated carbon-1.

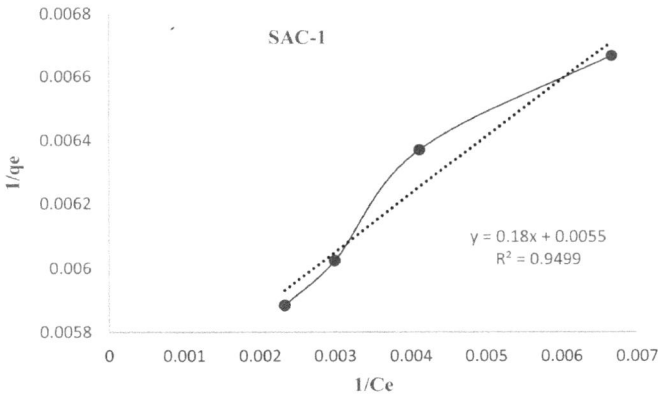

FIGURE 23.4   Langmuir adsorption isotherm of spent activated carbon-1.

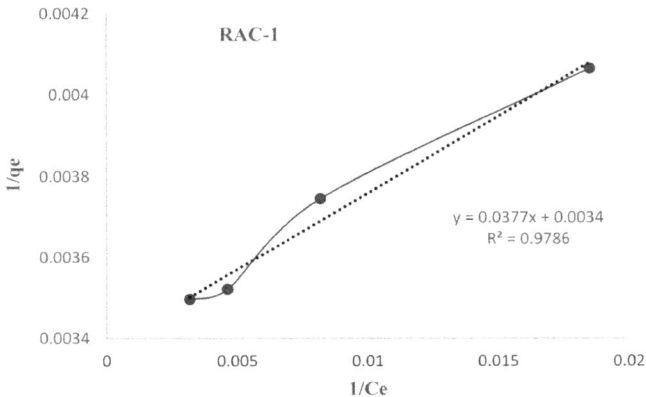

FIGURE 23.5   Langmuir adsorption isotherm of regenerated activated carbon-1.

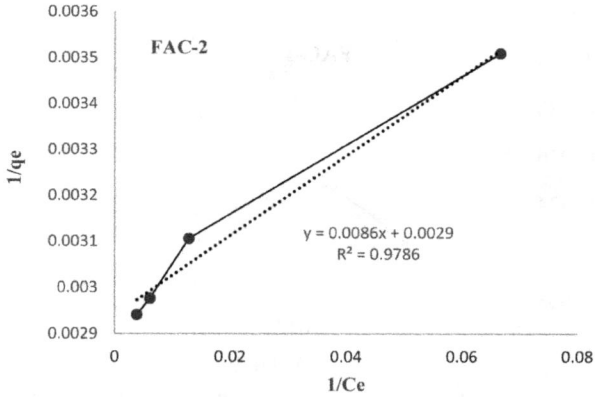

FIGURE 23.6 Langmuir adsorption isotherm of fresh activated carbon-2.

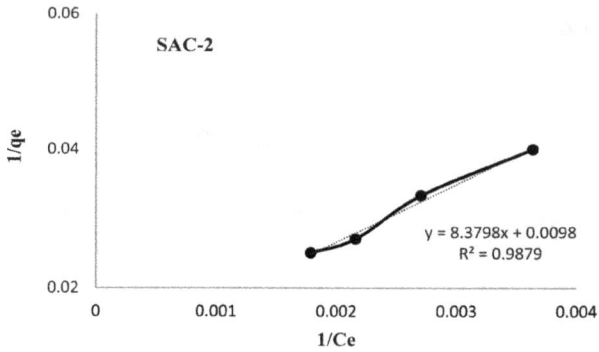

FIGURE 23.7 Langmuir adsorption isotherm of spent activated carbon-2.

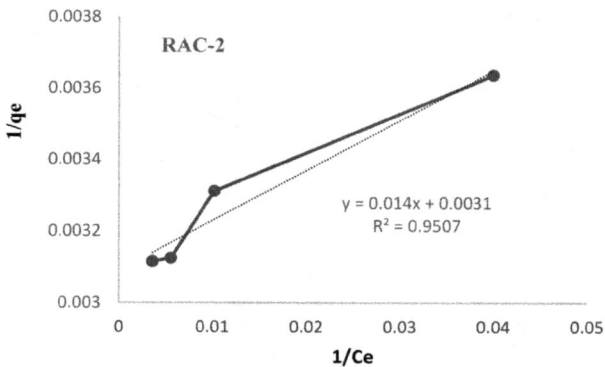

FIGURE 23.8 Langmuir adsorption isotherm of regenerated activated carbon-2.

adsorbate got adsorbed on the surface of FAC-2. Due to this reason, the time required for regeneration of SAC of samples 2 (SAC-2) was more time (30 min) as compared to SAC-1 (10 min). This is further supported by Langmuir constant (K) values also. The degree of interaction between adsorbate and adsorbent surface can be interpreted based on the K value. The higher value is indicative of the greater degree of interaction and vice versa. The K value obtained for FAC-2 is greater than FAC-1, suggesting that the adsorption capacity of the former sample is greater than the later one. Due to this, the K value of SAC-2 has decreased considerably, indicating that a greater number of molecules of adsorbate got adsorbed on the surface of the adsorbent.

TABLE 23.4 Calculated Values of Langmuir Isotherm Parameters of Each Studied Adsorbent

Parameter	FAC-1	SAC-1	RAC-1	FAC-2	SAC-2	RAC-2
Q_0 (mg/g)	312.50	181.82	294.12	344.83	102.04	322.52
K (L/mg)	0.225	0.031	0.095	0.337	0.0012	0.022
SA (m²/g)	1143.73	616.43	1050.87	1247.70	137.37	1151.16

Spent AC (Sample 1 and 2), respectively, were used for decolorization of paracetamol liquor. During this process activated carbon adsorbs different colored impurities present in liquor. The ease with which it can be regenerated depends upon the nature of adsorbate on activated carbon and number of molecules adsorbed on the surface of it. It is evident from the MBNs for Sample 1 that the SAC is regenerated successfully by more than 80% when it is subjected to microwave radiation of 520 W and 700 W for 5 and 10 min. On further increasing the time, a decrease in the value of MBN is observed indicating the probable commencement of structural disintegration. This observation is further supported by SEM image corresponding to relevant treatment time (Figure 23.1(D)). Further, from Table 23.2, it can be seen that regeneration of SAC (Sample 2) by more than 80% requires more time, i.e., 30 min. This difference in time required for the regeneration of SAC 1 and 2 may be due to the different nature and number of molecules of organic compounds adsorbed on them in both pharmaceutical industries and the adsorption capacity.

SEM analysis was performed on fresh, spent, and RAC samples 1 and 2 shown in Figures 23.1 and 23.2. The fresh carbon (FAC-1 and 2) displays porous structure with surface integrity [21, 22]. However, loss of surface structural uniformity is visible over the spent carbon (SAC 1 and 2). Microwave regeneration appears to open up the pores within a carbon particle regenerating them (RAC-1 and 2). Though the exact mechanism of regeneration through microwaves is still to be understood; it appears that microwaves generate 'dash' heat within a sample during which period organics get destroyed or evaporated or combusted. The steam or vapor pressure within a pore seems adequate to open it, a prime reason for observing a rapid increase in MBN value for RAC. The surface structure of the SAC is smooth and is covered by a lot of debris. On the other hand, regenerated activated carbon is comparatively free from surface debris, and internal pore structure is clearly visible.

23.4 CONCLUSION

On the basis of experiments conducted on the microwave regeneration of SAC samples supplied by M/s NIA, CETP, Baroda, and in the light of MB values, SEM images and values of Langmuir isotherm parameters, it can be concluded that microwave method of regeneration is practically feasible. The major advantage of this process is that a significantly high percentage of regeneration can be achieved within a short duration of time, when compared to thermal regeneration process.

ACKNOWLEDGMENT

The authors acknowledge the constant encouragement given by Dr. V. M. Nanoti, Principal, PIET, and Dr. N. N. Rao (Retired Chief Scientist, CSIR-NEERI, Nagpur) for valuable insights. We are grateful to Shri. Babubhai C. Patel, Chairman, Nandesari Industries Association-CETP, Vadodara, Gujarat for the financial support provided for the successful conduction of the experimental work.

KEYWORDS

- **Langmuir isotherm**
- **methylene blue number**
- **microwave**
- **regeneration**
- **spent activated carbon**
- **surface area**

REFERENCES

1. Ferreira, R. C., De Lima, H. H. C., Cândido, A. A., Couto, J. O. M., Arroyo, P. A., Carvalho, K. Q., De, G. F., & Gauze, M. A. S. D. B., (2015). Adsorption of paracetamol using activated carbon of dende and babassu coconut mesocarp. *International Journal of Biological, Biomolecular, Agricultural, Food and Biotechnological Engineering, 9*(7), 717–722.

2. Ruiz, B., Mestre, A. S., Fonseca, I. M., Carvalho, A. P., & Ania, C. O., (2010). Removal of an analgesic using activated carbons prepared from urban and industrial residues. *Chemical Engineering Journal, 163*(3), 249–255.

3. Shah, I. K., Pre, P., & Alappat, B. J., (2013). Steam regeneration of adsorbents: An experimental and technical review. *Chem. Sci. Trans., 2*(4), 1078–1088.

4. Guo, Y., & Erdeng, D., (2012). The effects of thermal regeneration conditions and inorganic compounds on the characteristics of activated carbon used in power plant. *Energy Procedia, 17*, 444–449.

5. Guo, D., Shi, Q., He, B., & Yuan, X., (2011). Different solvents for the regeneration of the exhausted activated carbon used in the treatment of coking wastewater. *Journal Of Hazardous Materials, 186*(2, 3), 1788–1793.

6. Shende, R. V., & Mahajani, V. V., (2002). Wet oxidative regeneration of activated carbon loaded with reactive dye. *Waste Management, 22*, 73–83.

7. Kiyokazu, O., Kazuyoshi, S., Toshihira, T., & Katsuyuki, N., (2007). Regeneration of spent activated carbon with adsorbed trichloroethylene using wet oxidation process. *Water Research, 4*, 1045–1051.

8. Guojun, Z., Sanfan, W., & Zhongzhou, L., (2003). Ultrasonic regeneration of granular activated carbon. *Environmental Engineering Science*, 57–65.

9. Chih-Huang, W., & Ming-Chien, H., (2008). Regeneration of granular activated carbon by an electrochemical process. *Separation and Purification Technology, 64*(2), 227–236.

10. Foo, K. Y., & Hameed, B. H., (2012). Microwave-assisted regeneration of activated carbon. *Bioresource Technology, 119*, 234–240.

11. Song, C., Jian, W., Hongying, X., Jinhui, P., ShiXing, W., & Libo, Z., (2016). Microwave-assisted regeneration of spent activated carbon from paracetamol wastewater plant using response surface methodology. *Desalination and Water Treatment., 57*, 18981–18991.

12. Hesas, R. H., Arami-Niya, A., Wan-Daud, W. M. A., & Sahu, J. N., (2013). Comparison of oil palm shell-based activated carbons produced by microwave and conventional heating methods using zinc chloride activation. *J. Anal. Appl. Pyrolysis, 104*, 176–184.

13. Yang, K. B., Peng, J. P., Srinivasakannan, C., Zhang, L. B., Xia, H. Y., & Duan, X. H., (2010). Preparation of high surface area activated carbon from coconut shells using microwave heating. *Bioresour. Technol., 101*, 6163–6169.

14. Hidayat, Y., Heraldy, E., & Patiha, M. F., (2016). The Langmuir isotherm adsorption equation: The monolayer approach. *IOP Conference Series Materials Science and Engineering, 107*(1), 12–67.

15. Itodo, A. U., Itodo, H. U., & Gafar, M. K., (2010). Estimation of specific surface area using Langmuir isotherm method. *J. Appl. Sci. Environ. Manage., 14*(4), 141–145.

16. Koshino, M., Tanaka, T., Solin, N., Suenaga, K., Isobe, & Nakamura, E., (2007). Imaging of single organic molecules in motion. *Science., 316*(5826) 853.

17. Kuang, Y., Zhang, X., & Zhou, S., (2020). Adsorption of methylene blue in water onto activated carbon by surfactant modification. *Water, 12*, 587.

18. Cleiton, A., Nunes, M., & Guerreiro, C., (2011). Estimation of surface area and pore volume of activated carbons by methylene blue and iodine numbers. *Quim. Nova., 34*(3), 472–476.

19. Mall, I. D., Srivastava, V. C., Agarwal, N. K., & Mishra, I. M., (2005). Removal of Congo red from aqueous solution by bagasse fly ash and activated carbon: Kinetic study and equilibrium isotherm analyses. *Chemosphere, 61*(4), 492–501.

20. Langmuir, I., (1916). The constitution and fundamental properties of solids and liquids. *Part-I. J. Am. Chem. Soc., 38*(11) 2221–2295.

21. Jujur, W. P., Nur, M., Asyari, M., & Nur, H., (2020). SEM, XRD and FTIR analyses of both Ultrasonic and heat generated activated carbon black microstructures. *Heliyon, 6*(3), 03546.

22. Wan, M. A., Wan, D., Wan, S., Wan, A., & Mohd, Z. S., (2003). Effect of activation temperature on pore development in activated carbon produced from palm shell. *Journal of Chemical Technology & Biotechnology, 78*(1), 1–5.

CHAPTER 24

Sustainable Waste Management: Smell Mitigation and Recovery of Fuel from Waste

VINAY DWIVEDI, P. SURYANARAYANA, and MEGHA H. TALWEKAR

JK Paper Mills, Unit: JKPM, Rayagada, Odisha, India,
E-mail: talwekarmegha@gmail.com (M. H. Talwekar)

ABSTRACT

In the light of rising energy costs, limited deposits of fossil fuel, and global warming due to CO_2 emissions, the efficient use of waste heat is continually gaining importance. In the Indian Paper industry, air emission is one of the major environmental concerns. Since the foul condensate treatment process in the evaporation plant produces stripped off-gases (SOG) that have high heating value but these gasses cannot be stored safely. Earlier, SOG were disposed of through incineration in the Lime Kiln. At JK Paper, we have adopted an approach where methanol is being separated from SOG and other contaminants through distillation and producing methanol of 80% purity that can be stored and fired in a controlled manner.

JK Paper is the first Indian Pulp and Paper Industry to install "SOG to Methanol" plant. With economic benefits, there are significant environmental impacts like mitigation of smell in nearby area and reduction in air pollution. This chapter helps to promote technological innovation and design production system that increases energy efficiency and reduce greenhouse gas emissions. This process has been developed to produce purified methanol that can be use as green renewable fuel, transforming an undesirable waste byproduct into potential energy for pulp mill.

24.1　INTRODUCTION

Methanol is produced during the cooking process in Kraft pulp mills, a shown in Figure 24.1. Methanol is formed as a byproduct of the kraft pulping process; in the digester, the hydroxyl ion reacts with a lignin Methoxyl group [3]:

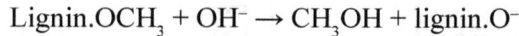

$$\text{Lignin.OCH}_3 + \text{OH}^- \rightarrow \text{CH}_3\text{OH} + \text{lignin.O}^-$$

It is released from the wood in the digester and is separated from pulp as weak black liquor. Evaporator increases dry solids content of the weak black liquor from about 16% to 75% for combustion in the Recovery Boiler. Modern evaporation plants are designed not only for concentrating black liquor for combustion in recovery boilers, but also for producing various clean condensate streams that can be reused 100% in the pulp mill to minimize overall water consumption. During evaporation, the methanol present in the weak black liquor is flashed and condensed into the evaporator condensate. Most evaporator train designs allow for the segregation of the condensates into three fractions (clean, intermediate, and foul) and concentration of the methanol into the smaller fraction of foul condensate. Foul condensate is typically 10% of the evaporator condensate for a modern evaporator train. Using a condensate treatment system, it is possible to reduce the methanol content of the foul portion of the condensate to the point where this treated condensate can also be reused in the mill [1].

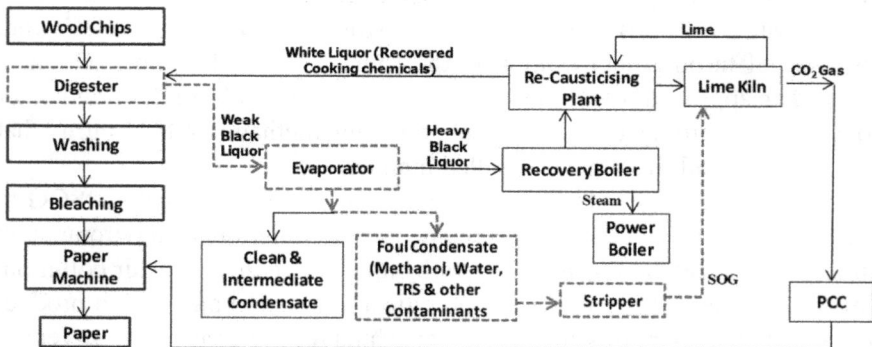

FIGURE 24.1　Manufacturing process of pulp and paper industry.

Methods for the removal or disposal of the methanol present in the mill condensates vary from mill to mill. In North America, these removal or disposal methods consist of sending the contaminated condensate to the secondary effluent treatment system where it will be handled as BOD, or to treat the contaminated condensate in a stripper system that will release the methanol as SOG [4]. Whereas, JK Paper were using a stripper column where the condensate flows counter current to either live steam or vapor from the concentrator. This releases the methanol into a gas form called stripped off-gas (SOG) and send to a condenser that also reduces moisture content of SOG (80% water, methanol, and remaining contaminants). Then these SOGs are incinerated in the lime kiln as shown in Figure 24.2.

Direct incineration of the SOG is not thermally efficient due to its high moisture content. The calorific value of methanol is higher than the water vapor rich SOG hence reduction in the fuel demand in the lime kiln. SOG is produced as gas that cannot be stored. Burning SOG in the line kiln can have other negative impacts as the SOG will not be produced a constant flow, which will cause variation in the kiln temperature profile. These changes in kiln temperature profiles have often been associated with the formation of rings and reduced capacity of lime kiln [1, 2].

JK Paper is the first Indian Pulp and Paper Industry to install "SOG to Methanol" plant. JK Paper condense the SOG to produce a lower moisture (around 15–20% water) liquid methanol stream that can be stored and burned as needed in a more controlled fashion. Some methanol is formed from the acid-catalyzed methanol elimination from hemicelluloses, while naturally occurring methanol is freed from the wood at the start of pulping. The amount of methanol produced is dependent on:

1. Wood Species: In general, lignin methoxyl groups are more numerous in hardwoods than softwoods; therefore more methanol is formed.
2. Hydroxyl Concentration: Higher alkali charge will produce more methanol.
3. Temperature: Higher cooking temperatures will produce more methanol.
4. Time: Longer cooking times will produce more methanol [3].

This chapter will focus on the benefits of converting the methanol to a liquid and new technology available to purify the methanol that makes it suitable as a green fuel, adding value to what is normally considered a waste product.

24.2   PROCEDURE

This process condenses the stripper off gas to lower the water content from 65% water content to about 20% water content and collects the condensed

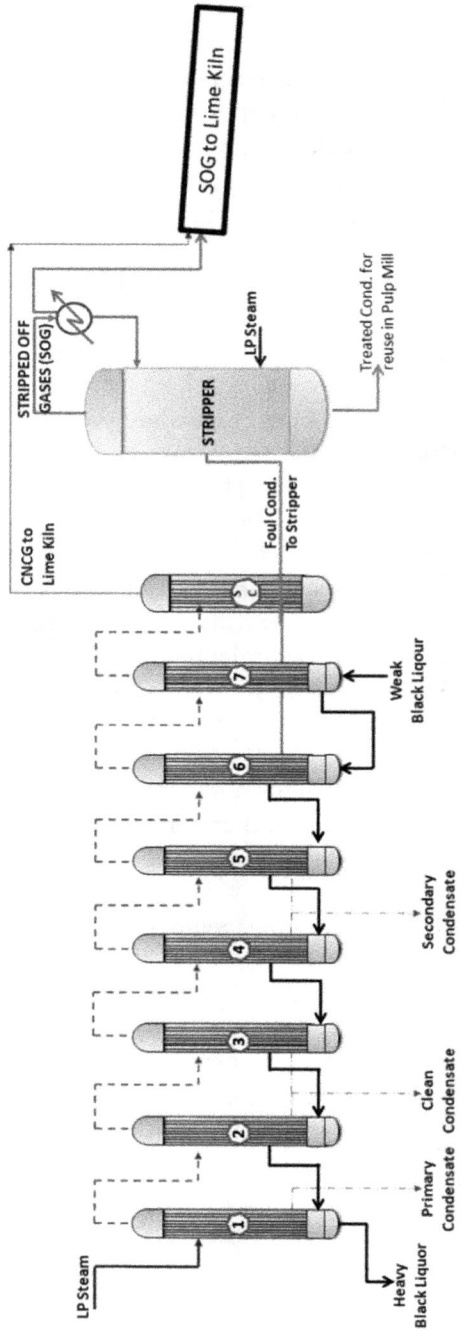

FIGURE 24.2   Process flow of stripped off-gases (SOG).

methanol in liquid form that can be easily stored. Liquid methanol is also considered to be a green fuel and can be used to replace natural gas or oil either as part of the main fuel or as a start-up fuel. Several of the recent new Pulp mills that can operate without any fossil fuel, use stored liquid methanol for start-up of combustion equipment [1].

Figure 24.3 depicts process flow of the methanol plant. JK Paper installed a methanol plant of a capacity of 330 kg/hr. The stripper off-gas, SOG, contains steam, methanol, and non-condensable sulfur compounds such as mercaptans, dimethyl sulfide, dimethyl disulfide, and hydrogen sulfide. From stripper outlet, the off-gas contains about 35% W/W methanol fraction and the rest is water vapor. SOG from stripper is being taken to the methanol column. The steam is injected at the bottom of the column to remove more methanol from the condensate. Methanol vapor is then directed to a water-cooled partial condenser that partially condense methanol to maintain reflux of the column. The rich methanol vapor is feed to the final condenser which produces liquid that is around 65% concentrated methanol. Finally, liquid methanol is being stored in tank for further use as per requirement. During this process, an uncondensed portion of the gas stream, which contains non-condensable sulfur compounds goes out as CNCG from the methanol plant condensers and is mixed with other CNCG (concentrated non-condensable gases) generated elsewhere in the existing process and incinerated in the lime kiln. Treated Foul condensate can be collected from the bottom of the column and typically reused as wash and/or dilution water in the fiber line and in the causticizing plant.

There is no need of a frequent quality control with sampling and analysis in the methanol Recovery System. The density indicator gives an indication of the methanol composition. The normal water content varies between 10–20%. The primary parameter to adjust the final methanol composition is by adjusting the Partial Condenser cooling temperature. The methanol concentration is determined by the gas temperature from the Partial Condenser. A decreased temperature results in a higher reflux to the Methanol Column. Too low temperature in the Partial Condenser results in larger steam requirements at the same time as the amount of methanol in the condensate is increased. If the methanol recovery system is operated in this way for an extended period, methanol is accumulated in the SOG. In the worst case the Methanol Column is overloaded. On the other hand, if the temperature in the Partial Condenser is set too high, the methanol produced will be too aqueous. JK Paper requires the following parameters during this process Table 24.1.

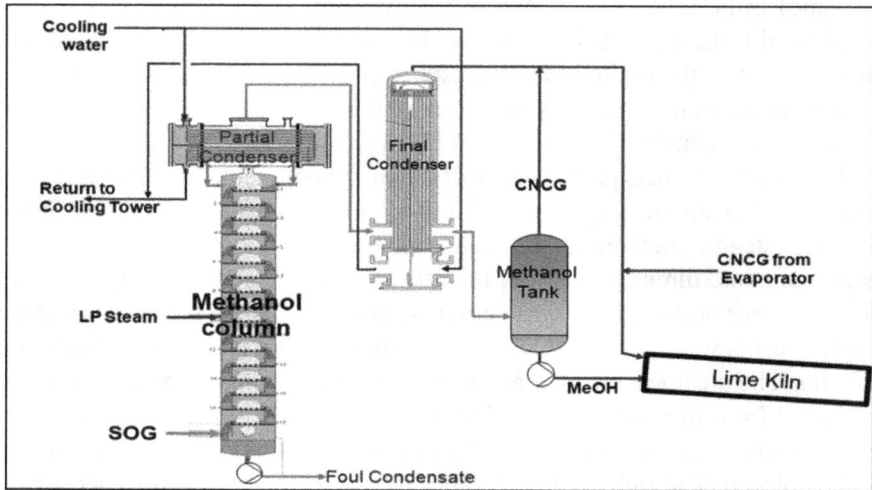

FIGURE 24.3   Process flow of SOG-methanol plant.

TABLE 24.1   Parameters Required for the Process

SOG Flow ultimately	1500–1600	m³/h
Cooling water requirement	53	t/h
Hot warm water returns to process	53	t/h
Net freshwater additional demand	0	kg/h
LP steam demand	245	kg/h
Foul condensate returns to stripper	1052	kg/h
Net fresh effluent generated	0	kg/h
Solid waste generated	0	kg/h
Gaseous effluent generated (CNCG)	75	m³/h
Installed power about	10	kWh

24.2.1 PROPERTIES OF METHANOL

- Explosive substance at 6.7 to 36.5% V/V in air;
- Boiling point 65°C;
- Density about 830 kg/m³;
- Toxic and highly flammable;
- Strong odor due to sulfurous compounds and ammonia;
- GCV 3500 kCal/kg.

24.3 RESULTS

In addition to economic benefits from purifying methanol from waste stream, there are significant environmental and process benefits:

1. Air: Bad smell in nearby area is being avoided as, SOG operation will not affect due to interruption in Lime Kiln Operation. Reduces GHG emission by 3,069,000 Kg of CO2/Annum.
2. Water: In this system, no extra water is used. Water for the methanol vapors condenser is being taken from the existing cooling tower and returned. Removing methanol from the mill condensate reduces effluent treatment costs and reduces freshwater usage. JK Paper is recovering 20 m3/day of condensate.
3. Solid: No solid waste generation. Around 80% of methanol instead of 20% of methanol in SOG. JK Paper is recovering 7–8 kg of Methanol/BDMT of pulp.

For a mill that has no good location for burning methanol as a gas (overloaded recovery boiler or operating problems if incinerated in kiln), producing liquid methanol could eliminate operating issues related to its disposal. A considerable portion of water vapor in the SOG is separated out and reuse in the pulp mill. Thereby unnecessary water vapor entry in to the lime kiln is reduced. SOG must be consumed online continuously, whereas by converting into methanol, it can be stored and used at mill requirements. The calorific value of methanol is higher than the water vapor rich SOG so, there can be some reduction in the fuel demand in the lime kiln as shown in Graph 10.1. JK Paper installed a methanol plant in March-18. Graph 10.1 depicts reduction in furnace oil consumption by 2.6 T/day after using methanol in limekiln instead of SOG (Figure 24.4).

24.4 CONCLUSIONS

Methanol produced in a Kraft pulp mill is more than a waste that requires disposal. If properly segregated, removed from the mill condensates, and made into a liquid fuel, it becomes a green fuel that can be efficiently combusted in the lime kiln or in other boilers. It can also be stored and become a startup fuel for mills that want to further reduce or eliminate their fossil fuel usage. Bio-Methanol produced at JK Paper reduces the

purchase of fossil fuel for lime kiln and eliminates the energy associated with transport. Adding an extra step to purify the methanol prior to
making it into a liquid also offers the potential to sell green methanol
that can be used as a transportation fuel or as a feedstock in the chemical
industry.

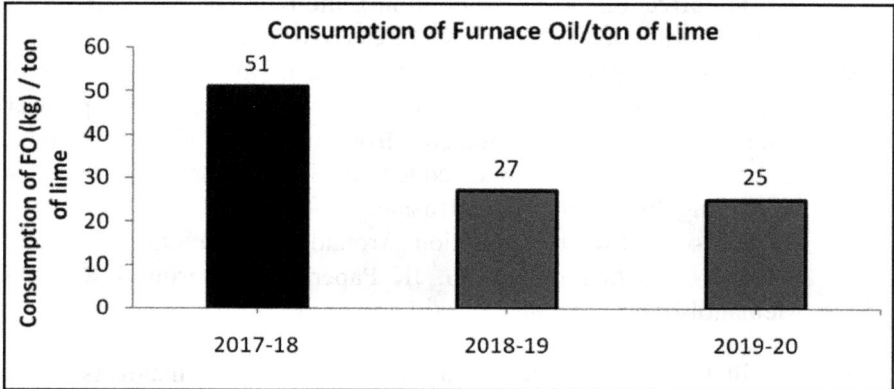

FIGURE 24.4   Consumption of furnace oil (kg) per ton of lime vs year.

KEYWORDS

- concentrated non-condensable gases
- evaporation
- methanol
- paper industry
- pulp
- stripped off-gases (SOG)

REFERENCES

1. Methanol-Waste Byproduct to Valuable Fuel. (2018). *Valmet.*
2. Honghi, T., (2008). Lime Kiln Chemistry and Effects on Kiln Operations. Pulp & Paper
   Centre and Department of Chemical Engineering and Applied Chemistry, University of
   Toronto, Canada, *TAPPI Kraft Recovery Course.*
3. Jensen, A., Trevor, I. P., & Percy, J., (2012). Methanol Purification System. *PEERS Conference.*

4. Zhu, J. Y., Yoon, S. H., Liu, P. H. X., & Chai, S., (2000). Methanol formation during alkaline wood pulping. *Tappi Journal.*

5. Bureau of Energy Efficiency, (2018). *Improving Energy Efficiency in Pulp & Paper Sector.* pp. 71–73, https://drive.google.com/open?id=1Ej2M1y51J4izzopA2ZRJbLnN_6-7_yXo (accessed on 31 July 2021).

6. CII Website, (2018). http://www.greenbusinesscentre.com/site/ciigbc/viewprest.jsp?event id=544472&event=544472&event=dd&dated=534429 (accessed on 30 June 2021).

CHAPTER 25

India Towards Cleaner Alternatives: Challenges and Solutions

ERENA SAYANKAR

Bhavans B.P. Vidya Mandir, Trimurti Nagar Nagpur – 440022, Maharashtra, India, E-mail: erenasayankar@gmail.com

ABSTRACT

Energy criterion decides the growth of the country; India is growing rapidly so does its waste. A population of 1.37 billion is generating a waste of 0.5 kg per person every day, which is a huge pile of waste. Since 1987, 15 WTE (waste to energy) plants have been set up across the country and half of these plants have since shutdown. To run a WTE plant successfully in India is still a challenge. In our country, around 75% of solid waste is often disposed of in unregulated dumps, landfills, or burned openly. Managing these wastes is extremely essential as these dumbs and landfills release harmful by-products which, if left untreated or uncheck, can leak into the soil resulting in contamination of soil and water bodies. Only WTE plants are not a solution in India, it should be a combination of technologies and treatment. The aim of this chapter is to discuss the challenges India is facing to manage waste for the aim of energy production as well as compare these methods with developed countries.

25.1 INTRODUCTION

There are many thermal and non-thermal technologies use to generate energy from waste some of the important technologies are discussed in subsections.

25.1.1 THERMAL

- Direct combustion (mass burn);
- RDF (refuse derived fuel);
- Pyrolysis;
- Convectional gasification;
- Plasma arc method.

Among these pyrolysis, gasification, and plasma arc methods are lesser used technologies, they provide lower emission than the mass-burn and the RDF.

25.1.2 NON-THERMAL

- Anaerobic digestions;
- Landfill gas;
- Hydrolysis and mechanical biological treatment [1].

25.2 GLOBAL LEADERS OF WTE

25.2.1 EUROPE

The largest and most enhanced market of WTE facilities in Europe. These plants produce around 69.5 million metric tons of MSW, generates 30 Twh of electricity and 55 Twh of heat. Especially Denmark and Sweden have remarkable developments in Waste management [2].

25.2.2 JAPAN

Japan dominates over 60% of the Asia-Pacific industries for WTE, they use many methods like stoker furnace, fluidized furnace, and gasification resource furnace [3]. There are 764 WTE plants in the country out of which 338 generates power, average processing capacity per plant is 158 TPD [4].

25.2.3 UNITED STATES

The country currently operating 75 WTE plants in its 23 states, managing around 90,000 TPD. This is the equivalent of a baseload electrical generation capacity of approximately 2,700 megawatts to meet the power needs of more than 2 million homes [5].

25.2.4   CHINA

After Europe, US, and Japan, China is the 4th largest user of WTE. China in recent years has increased the countries WTE capacity from 2 to 14 million tons of MSW [6]. Over the past few years, China leads the way with the largest installed WTE with more than 300 plants in operation [7].

25.3   INDIAN WASTE

In India, around 92 WTE plants with aggregated capacity around 250 MV have been set up for electricity generation from urban, agricultural, and industrial waste, but to make these plants successful we need tons of waste. India generates around 60 million tons of municipal solid waste (MSW) every year, but only 15% is useful for WTE. WTE plant's work is extremely challenging as the plants receives vast quantity of mixed waste, which is unsegregated, low in calorific value, high moisture content and have high inert content. Even such waste management is extremely difficult to handle because it leads to a lot of visual and odor pollution. Thus, plants are rejecting around 35 to 45% of the waste into land fillings. WTE plants cannot burn these wastes because it will cost them additional fuel which will make WTE plants more expensive to sustain. This is the main reason behind the closure of many WTE plants in India. According to the center for science and environment (CSE) (a Delhi-based non-profitable center), the calorific value of waste in India is 1,410–2,140 kcal/kg, but a waste should be at least 1800 kcal/kg for self-sustaining combustion. Garbage in successful WTE plants of developed countries range from 1900–3800 kcal/kg.

Around 1043 lakh tones per day of (TDP) MSW is generated across India, 1.11 lakh (TDP) is collected and only 35,602 (24.8%) processed. In addition, country generates around 25,940 (TPD) of plastic waste 60% remain uncollected [8].

25.3.1   SEGREGATION PROBLEMS

Presently in India, there is a lack of awareness among the citizens about proper disposal of wastes, systematic segregation of the waste is not being practiced, which creates a huge deflect in identifying a suitable technology to treat waste in our country. Hence, segregation of waste currently plays a vital role in deciding the success of any WTE project in India. Even if we import new technologies, still we need to customize many things in this system. At the same time, we have to develop a technology in India which is suitable for the Indian kind of waste and promote this technology in an integrated way for scientific disposal of waste. This initiative is essential at this juncture as such earlier efforts were not successful and found WTE technology is highly capital intensive. It is also expected that promoting a technology development in India will not only promote the 'Make in India' concept but also ensure sustainability in scientific disposal of waste.

25.4 WTE AND INDIA

WTE technologies have many ways to generate revenue for the country than any other power generation technology, but there are many factors that can have a negative impact on WTE plants like restrictive site size, accessibility of site, architectural plan to blend in with neighboring structures, noise consideration, possibility of contaminated soil, unavailability of skilled labors, among all these problems the most important problem is unavailability of segregated waste and plant burn these unsegregated waste which has direct influence on efficiency of electricity generation.

25.4.1 EXPENSIVE WTE

According to CSE, the cost of power generation from Indian WTE plants is the highest among all major sources of electricity, including many thermal power plants in the country. Indian WTE plants remain expensive despite of several subsidies, financial incentives, long-term lease on land and concession in customs duty on imported parts. Still, the price of electricity sell by the WTE plants are twice as compare to electricity from coal or solar plants, because excess of money is wasted in the segregation of waste and maintenance of the imported machineries [9].

25.4.2 ENERGY INDIA CAN GENERATE

Data given below indicates the State-wise calculation of MSW to energy generation in India. According to which India has the capacity to generate around 3000 MW energy through various methods like RDF, gasification, plasma arc method, pyrolysis, and biological treatments (Table 25.1).

TABLE 25.1 State Wise MSW Generation and Energy Generation from 2011 to 2020 [10]

State/Union Territory	Total MSW (T/day) (2011) p	Total MSW (T/day) (2015) p	Total MSW (T/day) (2020) p	Energy Potential 2011(p) (MW)	Energy Potential 2015(p) (MW)	Energy Potential 2020(p) (MW)
Arunachal Pradesh	13.56	14.2	14.94	0.27	0.28	0.31
Assam	341.73	358.51	378.49	6.8	7.13	7.91
Bihar	1956.78	2057.14	2170.08	38.94	40.94	45.35
Chhattisgarh	1077.02	1134.61	1201.69	21.43	22.58	25.12
Dadar and Nagar Haveli	25.75	30.11	35.35	0.51	0.6	0.74
Daman and Diu	25.63	30.09	37.12	0.51	0.6	0.78
Delhi	11873.06	13304.83	15326.68	236.27	264.77	320.33
Goa	221.92	245.28	272.54	4.42	4.88	5.7
Gujarat	7930.91	8342.24	8805.97	157.83	166.01	184.04
Haryana	2184.78	2325.63	2490.78	43.48	46.28	52.06
Himachal Pradesh	71.53	74.1	76.98	1.42	1.47	1.61
Jammu and Kashmir	746.24	782.6	820.75	14.85	15.57	17.15
Jharkhand	942.55	994.39	1056.53	18.76	19.79	22.08
Karnataka	8296.02	8628.03	8992.86	165.09	171.7	187.95
Kerala	1689.02	1733.49	1779.28	33.61	34.5	37.19
Lakshadweep	3.74	3.89	4.18	0.07	0.08	0.09
Madhya Pradesh	4633.63	4925.32	5271.18	92.21	98.01	110.17
Maharashtra	22434.35	23627.56	25033.24	446.44	470.19	523.19
Manipur	61.03	63.87	67.23	1.21	1.27	1.41
Meghalaya	54.25	56.78	59.76	1.08	1.13	1.25
Nagaland	14.52	15.38	16.18	0.29	0.31	0.34
Orissa	839.25	867.83	901.28	16.7	17.27	18.84
Pondicherry	185.66	217.02	259.47	3.69	4.32	5.42
Punjab	4645	4841.02	5051.64	92.44	96.34	105.58
Rajasthan	4671.89	4957.24	5286.89	92.97	98.65	110.5
Sikkim	14.71	15.38	16.17	0.29	0.31	0.34
Tamil Nadu	9501.77	9725.21	9948.8	189.09	193.53	207.93
Tripura	137.9	144.31	151.9	2.74	2.87	3.17
Uttar Pradesh	13651.39	14597.03	15736.25	271.66	290.48	328.89
Uttarakhand	424	447.71	474.58	8.44	8.91	9.92
West Bengal	12069.24	12504.27	13031.28	240.18	248.84	272.35
India	110,738.83	117065.07	124770.07	2203.69	2329.61	2607.71

25.5   CONCLUSION

India is on its way to become a developed country, the urbanization, industrialization, rapid economic growth and change in life style these all demands energy which brings environmental changes. The only solution in order to save our ecosystem is to reduce the quantity of waste and recycle our WTE, which is possible through proper disposal of these wastes and its systematic segregation. This systematic segregation is yet a challenge, but we can have a solution. We can make it possible by educating the citizens through social media, television ads, involving WTE topics in the latest curriculum for the students, practicing it in school, colleges, hospitals, and offices, which will make it a habit for all and they will practice the same in their homes. Government and researchers should work hand in hand to develop WTE technologies suitable for the waste produced in India, this will further minimize the expenses on WTE plants, and we will be able to generate affordable energy.

KEYWORDS

- **clean India**
- **municipal solid waste**
- **recycle**
- **refuse-derived fuel**
- **segregating waste**
- **thermal and non-thermal technologies**
- **tones per day**
- **waste to energy**

REFERENCES

1. Zafar, S. (2019). *Waste to Energy Conversion Routes.* Bioenergy Consult. www. bioenergyconsult.com/waste-to-energy-pathways (accessed on 30 June 2021).
2. Rogoff, M, J., (2019). *The Current Worldwide WTE Trend.* https://www.mswmanagement. com/collection/article/13036128/the-current-worldwide-wte-trend (accessed on 30 June 2021).
3. Park, H., (2018). *Municipal Solid Waste Incineration in Japan.* Submitted as course work for PH240 Stanford university. http://large.stanford.edu/courses/2018/ph240/park-h2/ (accessed on 30 June 2021).

4. Jackson, M., (2018). *Japan Energy from Waste Infrastructure Study Tour*. http://www.jacksonenvironment.com.au/2018-japan-energy-waste-infrastructure-study-tour/ (accessed on 30 June 2021).

5. American Own Energy Source, (2018). *American Society of Mechanical Engineering*. Integrated waste service association and solid waste association of North America. https://webcache.googleusercontent.com/search?q=cache:iwD4A6NNKwUJ:https://www.eia.gov/todayinenergy/detail.php%3Fid%3D25732+&cd=3&hl=en&ct=clnk&gl=in (accessed on 30 June 2021).

6. Themelis, N., & Zhang, Z., (2019). *WTE in China Hierarchy of Waste Management*. https://waste-management-world.com/a/wte-in-china (accessed on 30 June 2021).

7. Cortese, A. J., (2019). *China is Building the World's Largest WTE Plant in Shenzhen*. https://webcache.googleusercontent.com/search?q=cache:IIw6vceIK60J:https://pandaily.com/china-is-building-the-worlds-largest-waste-to-energy-plant-in-shenzhen/+&cd=1&hl=en&ct=clnk&gl=in (accessed on 30 June 2021).

8. Sambyal, S. S., Agrawal, R., & Shrivastava, R., (2019). *Trash-Fire Power Plants Wasted in India*. https://www.downtoearth.org.in/news/waste/trash-fired-power-plants-wasted-in-india-63984 (accessed on 30 June 2021).

9. Charles, R. K. J., Mary, A. B., Jenova, R., & Majid, M. A., (2019). *Sustainable Waste Management Through Waste to Energy Technologies in India-Opportunities and Environmental Impacts*. https://www.ijrer.org/ijrer/index.php/ijrer/article/view/9017 (accessed on 30 June 2021).

10. Saini, S., Rao, P., & Patil, Y., (2012). *City Based Analysis of MSW to Energy Generation in India, Calculation of State-wise Potential and Tariff Comparison with EU*. https://www.sciencedirect.com/science/article/pii/S1877042812007859 (accessed on 30 June 2021).

CHAPTER 26

Calculation of Ecological Footprint of a District Using An Emergy Analysis

LAXMI A. ZADGAONKAR, GAUTAMI KANT, and
SACHIN A. MANDAVGANE

*Department of Chemical Engineering, Visvesvaraya National Institute
of Technology, Nagpur, Maharashtra, India,
E-mail: laxmi.zadgaonkar@gmail.com (L. A. Zadgaonkar)*

ABSTRACT

Every anthropogenic activity involves interaction with the nature around. The activity demands goods or services of the ecosystem around for consumption or regulation of pollution it has generated. Population's well-being is affected by the resource availability and health. It becomes essential to measure the supply capacity of natural capital and human demand. The accounting helps in identifying challenges, setting targets, and making policy decisions for sustainability. Ecological footprint is one such method which helps in accounting the demand-supply process between human and environment. The present study calculates the ecological footprint and biocapacity of Nagpur district of Vidarbha, India using energy methodology. Consequently, the degree of sustainability is calculated in terms of ecological footprint index by estimating the gap between human consumption demand (in terms of ecological footprint) and nature's supply ability (in terms of biocapacity). The present chapter calculates ecological footprint in local hectares, thereby accounting for the impact on regional ecosystem rather than global ecosystem, making assessment more realistic. It was observed that the ecological footprint index of Nagpur is 65.66%. The positive value of the ecological footprint index shows that Nagpur is working within the sustainable limits.

26.1 INTRODUCTION

Urbanization is continuously growing at a faster rate as the population is shifting more towards cities. The share of the Metropolitan cities population has increased from 18.9% in 1951 to 42.3% in 2011 [1]. The increased rate of urbanization may have positive as well as negative impacts on the environment. With this growing rate of urbanization, it becomes essential to measure the supply capacity of natural capital and human demand. Just as financial accounting for an enterprise is important to keep track of its income and expenses, similarly environmental accounting is essential for a region to keep track of the resource consumption and its availability, waste generation, and natures assimilative capacity.

Ecological footprint is one such area-based indicator which quantifies the human consumption and waste discharge in relation to the carrying capacity of nature. Ecological footprint, invented by Wackernagel and Rees in the year 1996, is an important tool which evaluates quantitatively the pressure which human activity exerts on the environment [2]. Ecological footprint is a tool to assess the criteria of minimum sustainability. The area thus calculated is represented in terms of hectares of land. Various researchers have calculated the ecological footprint and biocapacities of nations [3, 4]. However, Huij-bregts et al., in their study has stated that ecological footprint in standalone is not enough for comprehensive sustainability study but needs integration with other tools [5]. Some researchers have integrated different methodologies to calculate ecological footprint. Mikulčić et al. calculated ecological footprint for different cement manufacturing units using energy analysis [6]. Siche et al. calculated ecological footprint using energy net primary productivity (NPP) [7]. Energy analysis incorporates material as well as energy flows and converts each flow into energy in the form of its solar equivalent, thus making it easy for fair comparison of its share in ecological footprint. Energy analysis helps in qualitative analysis in addition to quantitative analysis which ecological footprint or traditional energy analysis fails to do. Baral and Bakshi have stated the importance of energy quality and aggregation of metrics for fair comparison between alternatives for communicating findings in decision making [8].

The present study has used the modified ecological footprint method proposed by Zhao et al. to calculate ecological footprint and has used the method proposed by Siche et al. to calculate biocapacity. Another drawback of the conventional ecological footprint method is it calculates ecological footprint in global hectares. However, any alteration in the productivity of any nation will change the footprint values for local industry. An alternative is to use local productivity rather than global productivity in terms of regional energy density (RED).

The present study has calculated the ecological footprint of Nagpur district of Vidarbha, India. Every district has its own productivity and waste generation. Similarly, every district has its own biocapacity to support the given population. How well do we use earth's natural resources need measurements? How much is available and how much we are using will help in taking care of our natural capital. Moreover, the ecological footprint analysis will help in identifying the hotspots where necessary action needs to be taken during the decision-making process. The time-series data will help in predicting the sustainability of the regions. The ecological footprint will help in optimizing resource use by implementing the best possible alternative.

In the present study, it was observed that Nagpur district is operating within sustainable limits, i.e., the demand of resources out of human consumption and economic activity is less than the supply capacity of these resources from the ecosystem. However, with rapid urbanization, industrialization, and increase in population, it will be challenging to keep it within sustainable limits. The present study can be a benchmark for further study of optimized use of resources and economic activities. As a major part of Vidarbha comprises of agricultural production as its major occupation, application of integrated agricultural model replicating the circular economy will be helpful to maintain sustainability under positive limits.

26.2 BACKGROUND

26.2.1 ECOLOGICAL FOOTPRINT

Ecological footprint was first developed by Rees in the year 1992 [9] and then improvised by Wackernagel and Rees in the year 1996 [2]. Ecological footprint accounts for the provisioning of resources consumed in the process and assimilation of waste/emissions generated by the process in terms of biologically productive land. Ecological footprint is calculated based on two methodologies, namely compound and component methodology [4]. The basic difference between these two methodologies is that they follow different approaches and need different data sources for calculation. The compound method is used for calculation of ecological footprint of nations, regions, and utilizes top-down approach whereas the component method is used for the calculation of ecological footprint of product, process, or industries with the help of life cycle inventory and follows bottom-up approach.

26.2.2   *ENERGY ANALYSIS*

Energy evaluation is the concept first developed by Odum [10]. According to Odum, energy is the available energy of one kind used up directly or indirectly to produce a product or service. Energy converts all flows, energy as well as matter, into its solar energy equivalents. Solar energy is the principal energy input to earth. Hence energy is expressed in terms of solar equivalent joules (sej). The other two energy inputs to earth are tidal energy and crustal heat. These three forms the total energy input to earth. For converting any flow into its solar equivalent, the intensity factor called transformity or unit energy value (UEV) is used. The transformity or UEV is the amount of solar energy used up directly or indirectly to produce a unit of product or service. Its unit is sej/unit. Energy incorporates renewable inputs from the ecosystem in addition to economic inputs. Energy analysis converts all flows into the same units thus rendering fair comparison, ease in aggregation, and evaluation. Energy assigns quality to each flow by using transformity. Higher the transformity, higher the solar energy required for making a product, or service. Each product undergoes various transformations during its formation. With each transformation, the energy memory associated with it increases. In this hierarchy of transformation, the transformity increases with its position.

The various steps involved in the energy analysis are:

- Selection of system boundary and expansion of LCI by incorporating renewable energy flows from an ecosystem and other indirect economic flows like labor, expenditure, machinery, etc.
- Categorizing inputs into renewable, non-renewable, and purchased inputs (those taken from outside the system).
- Energy diagram depicting flows.
- Collection of relevant UEV data for different flows from literature or calculation of UEV if not available in the literature (corresponding to reference global energy baseline).
- Calculating energy for all inputs using UEV.
- Calculation of energy indices.

The global energy baseline is the total annual energy input to the geo biosphere which is equal to the sum of three primary driving forces to the geo biosphere namely solar radiation, tidal energy, and deep earth heat. These three energies are combined in terms of their solar equivalences. The various baselines that are calculated over the years are 9.44E+24 sej y^{-1}, 15.83E+24 sej y^{-1}, 12.0E+24 sej y^{-1}. The global energy baseline used in this chapter is 15.83E+24 sej y^{-1}.

26.3   METHODOLOGY

The method of assessment used here is the energy ecological footprint (EEF) proposed by Zhao et al. [11] which integrated the conventional ecological footprint and energy analysis. The ecological footprint is the effect of human activities and resources extracted in the given area and not human consumption in the given area [12]. Consequently, the present study calculates EEF based on production of the region and not human consumption. Note here that each district has been considered as a closed system for calculations, which implies that an exchange of products and resources between systems has been neglected. The methodology used for the calculation of biocapacity is the one proposed by Siche [7]. The steps involved in the calculation of ecological footprint index is as follows (see Figure 26.1):

- Quantifying the production of various resources using appropriate data sources.
- Conversion of each production unit into its solar energy unit (solar emjoules) using energy analysis.
- Calculating the EEF using RED.
- The total biocapacity of the region includes the incident renewable energy and natural capital of the region. Natural capital is the carbon sequestrated in the region over the span of one year. In addition to natural capital, every region has its own incident renewable energy.

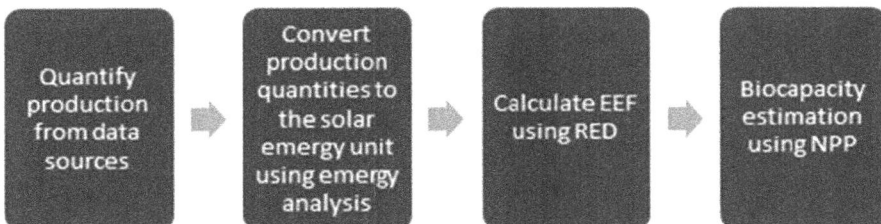

FIGURE 26.1   Methodology of calculating ecological footprint.

26.3.1 ENERGY ECOLOGICAL FOOTPRINT (EEF)

Energy analysis is a method of environmental accounting introduced by Odum [10]. Energy helps in quantifying the difference of quality between different types of energy and resources by converting them to a common unit. Here, all values have been expressed in terms of the solar energy equivalents or solar energy joules that is used to produce them directly or indirectly. This is done using the UEV in sej/J for each resource or energy type which can be obtained from literature or can be calculated. Energy in solar emjoules (sej) using the UEV is calculated for each flow by the relation.

$$Energy(sej) = Energy \ (J) \times UEV \ (sej/J)$$

The first step in EEF calculation involves calculation of RED which is defined as the amount of renewable energy per unit area per annum. RED for each district has been evaluated by calculating five types of renewable energy that is solar radiation energy, rain geopotential energy, rain chemical energy, wind kinetic energy and earth cycle energy incident on a region. Each energy flow is converted to its energy equivalent using UEVs. In order to avoid double-counting, only the highest of the five incident energy is considered as the total energy input due to renewable resources to a particular region. The RED is calculated using the following formula.

$$RED = Renewable \ energy \ input/Regional \ Area$$

The second step in calculation of EEF involves:

- Collection of data for agricultural, fisheries, poultry, dairy production in addition to electricity and waste generation. Each production unit corresponds to the annual production value.
- Conversion of each input to its energy equivalent using UEVs.
- EEF is calculated in hectares by using the following formula:

$$EEF = Sum \ of \ energy \ values \ of \ all \ resources/RED$$

Another useful indicator is the EEF per capita which is obtained by:

$$eef = EEF/population$$

26.3.2 BIOCAPACITY

Biocapacity is calculated as a function of natural renewable sources mentioned in Table 26.1 and the internal natural capital accumulated by each region. The latter is quantified using the NPP of different land types. So, the energy available from the natural capital storage is the biomass accumulated per annum. The following steps is used to calculate biocapacity of a district:

- Collection of land use land cover data.
- If the biomass contains 50% of carbon, after collecting NPP values of each type of land use, the natural capital is calculated as:

    Natural Capital (sej) = (Σ Area × NPP) × UEV of Biomass

- Total capacity is calculated as the sum of natural capital and the maximum of the five renewable energy inputs to the region.

    Biocapacity (m^2) = Total capacity/RED

26.3.3   ECOLOGICAL FOOTPRINT INDEX

Biocapacity and EEF are used to calculate the Ecological Footprint Index (EFI). EFI is a parameter that can be used to measure the sustainability of a region. Just as the financial accounting is done for any organization by calculating income and expenses and the n profitability or loss, similarly environmental accounting in terms of EFI is performed using Biocapacity and EEF. Biocapacity and EEF are calculated in the same unit (m^2) or in hectares, whichever is convenient, which is used in evaluating the ecological footprint index (EFI) as follows:

    EFI = ((Biocapacity – EEF)/Biocapacity) × 100

26.4   CASE STUDY

26.4.1   STUDY AREA DESCRIPTION

The present area of study is the Nagpur district of Vidarbha, India (Figure 26.2). The climate of this region is characterized by a hot summer and general dryness throughout the year except during the south-west monsoon season, i.e., June to September. Nagpur (21° 09' 23.58" N, 79° 05' 16.99" E) has a total area of 9860 km^2 and population of 46.5 lakhs in 2011. Average elevation is 339 m. Climate is characterized by a hot summer and general dryness throughout the year except from June to September. The mean minimum and maximum temperatures are

12°C and more than 45°C. Average annual rainfall lies between 1000 mm–1200 mm. There are large reserves of Manganese in this district.

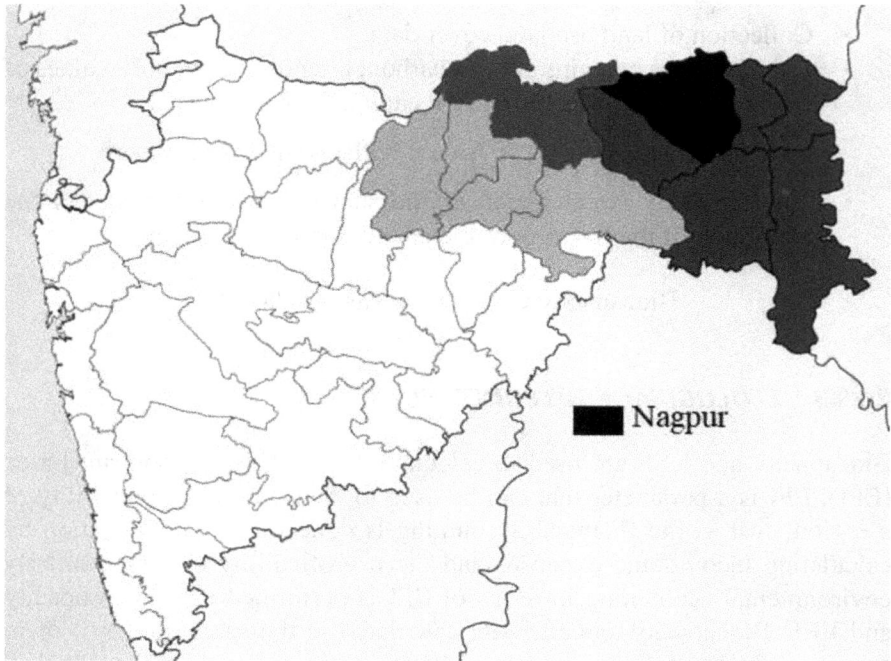

FIGURE 26.2   Nagpur district in Vidarbha, India.

26.4.2 DATA SOURCES

All data used for calculations has been obtained from various published reports of the government of Maharashtra. Annual production data for crops, poultry, and fisheries has been obtained from reports of the Central Research Institute for Dryland Agriculture [13] for 2008–2009; milk production from National Dairy Development Board [14] for 2008–2009; electricity from Maharashtra State Data Bank for 2011–2012 and wastewater, municipal solid waste (MSW) from the Maharashtra Pollution Control Board [15] for 2003–2004. The references from where the transformity values are used are mentioned in Table 26.1. While calculating biocapacity, wherever region specific data was not available, country specific data was assumed to be the same for the respective region.

TABLE 26.1   Source of UEV's

SL. No.	Resource	ECC	Unit	Transformity	Unit	UEV	Unit	Refs.
1.	Grain	1.62E+10	J/kg	8.30E+04	sej/J	1.34E+15	sej/t	[12]
2.	Wheat	1.57E+10	J/kg	6.80E+04	sej/J	1.07E+15	sej/t	
3.	Beans	1.85E+10	J/kg	6.90E+05	sej/J	1.28E+16	sej/t	
4.	Oil plants	2.55E+10	J/kg	6.90E+05	sej/J	1.76E+16	sej/t	
5.	Vegetables	2.50E+09	J/kg	2.70E+04	sej/J	6.75E+13	sej/t	
6.	Potato	4.20E+09	J/kg	8.20E+04	sej/J	3.44E+14	sej/t	
7.	Peanuts	2.55E+10	J/kg	6.90E+05	sej/J	1.76E+16	sej/t	
8.	Cotton	4.60E+09	J/kg	4.40E+06	sej/J	2.02E+16	sej/t	
9.	Fruits	3.30E+09	J/kg	5.30E+05	sej/J	1.75E+15	sej/t	
10.	Rice	1.55E+10	J/kg	6.03E+04	sej/J	9.35E+14	sej/t	[16]
11.	Oranges	1.76E+10	J/kg	1.09E+05	sej/J	1.92E+15	sej/t	
12.	Sugarcane	1.81E+10	J/kg	2.10E+04	sej/J	3.80E+14	sej/t	
13.	Milk	2.90E+09	J/kg	2.00E+06	sej/J	5.80E+15	sej/t	[12]
14.	Fishery	5.50E+09	J/kg	2.00E+06	sej/J	1.10E+16	sej/t	
15.	Meat	4.60E+09	J/kg	4.00E+06	sej/J	1.84E+16	sej/t	
16.	Waste solid	6.90E+08	J/kg	1.80E+06	sej/J	1.24E+15	sej/t	
17.	Waste water	5.00E+06	J/kg	8.60E+05	sej/J	4.30E+12	sej/t	
18.	Electricity	3.60E+06	J/kWh	1.60E+05	sej/J	5.76E+11	sej/kWh	

26.5   RESULTS AND DISCUSSION

RED calculation for Nagpur district is shown in Table 26.2.

TABLE 26.2   Renewable Energy Properties for Nagpur District (2008–2009)

Property	Formula	Energy (J)	UEV (sej/J)	Energy (sej)
Solar radiation energy	$S \times D \times (1 - A)$	6.26E+19	1[17]	6.26E+19
Rain geopotential energy	$S \times H \times P \times M \times g$	3.30E+16	8888[11]	2.92E+20
Rain chemical energy	$S \times P \times G \times M$	5.36E+16	15,444[11]	8.27E+20
Wind kinetic energy	$S \times M_{air} \times E \times r^3$	4.02E+16	623[11]	2.50E+19
Earth cycle energy	$S \times F$	2.34E+14	29,000[11]	6.79E+18

Since rain chemical energy has the highest energy value, it is used in further calculations to find the RED of Nagpur as 8.39E+10 sej/m^2. The EEF for Nagpur district is 1.71E+12 m^2 which is calculated in Table 26.3. Table 26.4 shows the biocapacity calculations for Nagpur district.

TABLE 26.3 EEF Calculation of Nagpur District

	Commodity	Production	Unit	EC (J/Unit)	Transformity (sej/J)	Energy (sej)	EEF (m²)	EEF (m²/cap)
Agricultural Crops	Soybean	242800	t	1.85E+10	6.90E+05	3.0993E+21	3.69E+10	7.94E+03
	Gram	57600	t	1.85E+10	6.90E+05	7.3526E+20	8.76E+09	1.88E+03
	Wheat	47500	t	1.57E+10	6.80E+04	5.0711E+19	6.04E+08	1.30E+02
	Green gram	14800	t	1.85E+10	6.90E+05	1.8892E+20	2.25E+09	4.84E+02
	Black gram	1900	t	1.85E+10	6.90E+05	2.4254E+19	2.89E+08	6.21E+01
	Pigeon pea	76900	t	1.85E+10	6.90E+05	9.8163E+20	1.17E+10	2.51E+03
	Jowar	92800	t	1.62E+10	8.30E+04	1.2478E+20	1.49E+09	3.20E+02
	Banana	6000	t	3.30E+09	5.30E+05	1.0494E+19	1.25E+08	2.69E+01
	Orange	6300	t	1.76E+10	1.09E+05	1.2086E+19	1.44E+08	3.09E+01
	Onion	13700	t	2.50E+09	2.70E+04	9.2475E+17	1.10E+07	2.37E+00
	Cotton	283300	t	4.60E+09	4.40E+06	5.734E+21	6.83E+10	1.47E+04
	Safflower	2400	t	2.55E+10	6.90E+05	4.2228E+19	5.03E+08	1.08E+02
Dairy	Milk	120000	t	2.90E+09	2.00E+06	6.96E+20	8.29E+09	1.78E+03
Meat	Fishery	10900000	t	5.50E+09	2.00E+06	1.199E+23	1.43E+12	3.07E+05
	Poultry	471300	t	4.60E+09	4.00E+06	8.6719E+21	1.03E+11	2.22E+04
Waste Generation	Solid waste	338793	t	6.90E+08	1.80E+06	4.2078E+20	5.01E+09	1.08E+03
	Waste water	114031475	t	5.00E+06	8.60E+05	4.9034E+20	5.84E+09	1.26E+03
Electricity	Electricity	3.847E+09	kWh	3.60E+06	1.60E+05	2.2159E+21	2.64E+10	5.67E+03
Total							**1.71E+12**	**3.67E+05**

TABLE 26.4   Biocapacity of Nagpur District

Land-use	Cropland	Forest	Grassland
Area (m²)	6.80E+09	1.59E+09	4.10E+08
NPP (gC/year)	450	658	191
UEV of biomass (sej/gC)	99600000000		
Natural capital (sej)	4.17E+23		
Energy from renewable source (sej)	8.27E+20		
Total capacity (sej)	4.18E+23		
RED (sej/m²)	8.39E+10		
Biocapacity (m²)	4.98E+12		
EEF (m²)	1.71E+12		
Ecological footprint index (EFI)	65.66%		

The biocapacity of Nagpur district comes to be around 4.98E+12 m². The EFI of Nagpur district is 65.66%. Thus, Nagpur district has its sustainability of 65.66% in the scale of 1 to 100%. Figure 26.3 shows the difference between biocapacity and ecological footprint of Nagpur district.

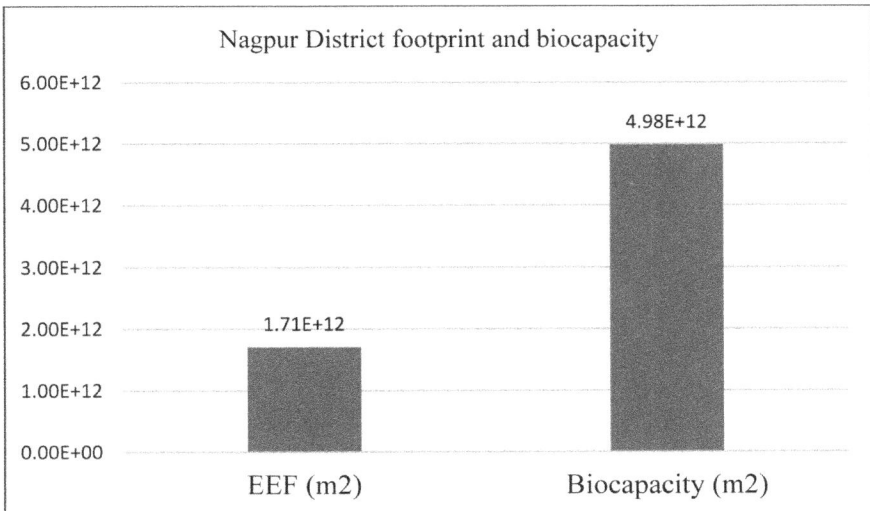

FIGURE 26.3   Energy ecological footprint and biocapacity of Nagpur district in Vidarbha, India.

26.6   CONCLUSION

In the present study, the modified ecological footprint method was used to calculate ecological footprint using RED and the biocapacity was calculated using the natural capital and incident renewable energy over the respective district. This method not only captured the resources extracted and supply capacity of the ecosystem but also compared the ecological footprint and biocapacity (carrying capacity). The comparison was possible as both the values possessed the same units (sej). Instead of relying only on the incident renewable energy as carrying capacity, the present study also included natural capital, which is the accumulated wealth that a region possesses over the years. Thus, making the study more realistic in nature. The present study also uses production data rather than consumption data for the calculation of ecological footprint as the sustainability of any region depends on the resources extracted and economic activities can reflect the actual sustainability.

It can be observed that the Nagpur district of Vidarbha is operating within sustainable limits, i.e., the demand of resources out of human consumption and economic activity is less than the supply capacity of these resources from the ecosystem. However, with rapid urbanization and an increase in population, it will be challenging to keep the districts within sustainable limits. It can be seen from Figure 26.4 that fisheries are a major contribution to the total EEF of Nagpur. Fisheries forms 83.63% of total EEF. The fisheries in Nagpur district are mostly inland fisheries. Inland fisheries are the most neglected sector in national or regional governance priority. As a result, it is not managed properly and very resource intensive sector. To decrease ecological footprint due to fisheries, there should be investment in evaluation and assessment methods, methods to improve resource efficiency, and management of water bodies across region. Moreover, the fisheries in Nagpur should be incorporated into regional development plans and policies keeping into mind its potential as food security, livelihoods, and human well-being.

The major occupation of the people in Vidarbha is agriculture. The application of integrated agricultural model can be helpful in optimizing resources and minimizing wastes. The present study can be a reference for maintaining or operating city under sustainable limits.

The present study may miss some resources values due to non-availability of data in statistics yearbook published by the Government of Maharashtra. Hence there may be some minor deviation in the results. Moreover, in reality, each system is a complex system. The present study has simplified the system for study. The present results can increase the reliability; however, the study needs further improvement.

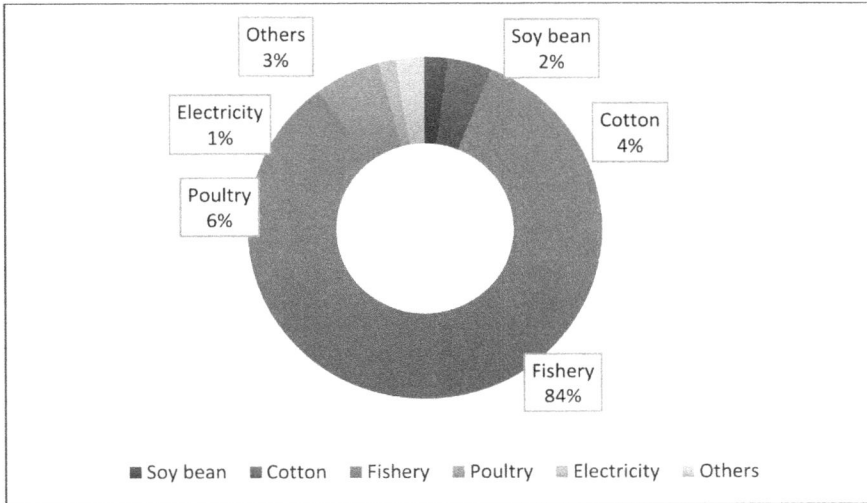

FIGURE 26.4   Footprint shares by resources.

KEYWORDS

- **biocapacity**
- **ecological footprint index**
- **energy**
- **energy ecological footprint**
- **net primary productivity**
- **regional energy density**
- **sustainability**

REFERENCES

1. Kumar, A., & Rai, A. K., (2014). Urbanization process, trend, pattern and its consequences in India. *Neo Geogr., 3*(4), 54–77.
2. Wackernagel, M., & Rees, W., (1996). *Our Ecological Footprint: Reducing Human Impact on Earth.* New Society Publishers.

3. Galli, A., Kitzes, J., Niccolucci, V., Wackernagel, M., Wada, Y., & Marchettini, N., (2012). Assessing the global environmental consequences of economic growth through the ecological footprint: A focus on China and India. *Ecol. Indic., 17,* 99–107. https://doi.org/10.1016/j.ecolind.2011.04.022.

4. Monfreda, C., Wackernagel, M., & Deumling, D., (2004). Establishing national natural capital accounts based on detailed ecological footprint and biological capacity assessments. *Land Use Policy, 21*(3), 231–246. https://doi.org/10.1016/j.landusepol.2003.10.009.

5. Huijbregts, M. A. J., Hellweg, S., Frischknecht, R., Hungerbühler, K., & Hendriks, A. J., (2008). Ecological footprint accounting in the life cycle assessment of products. *Ecol. Econ., 64*(4), 798–807. https://doi.org/10.1016/j.ecolecon.2007.04.017.

6. Mikulčić, H., Cabezas, H., Vujanović, M., & Duić, N., (2016). Environmental assessment of different cement manufacturing processes based on energy and ecological footprint analysis. *J. Clean. Prod.* https://doi.org/10.1016/j.jclepro.2016.01.087.

7. Siche, R., Agostinho, F., & Ortega, E., (2010). Energy net primary production (ENPP) as basis for calculation of ecological footprint. *Ecol. Indic., 10*(2), 475–483. https://doi.org/10.1016/j.ecolind.2009.07.018.

8. Baral, A., & Bakshi, B. R., (2009). Thermodynamic metrics for aggregation of natural resources in life cycle analysis: Insight via application to some transportation fuels. *Environ. Sci. Technol., 44*(2), 800–807. https://doi.org/10.1021/es902571b.

9. Rees, W., (1992). *Ecological Footprints and Appropriated Carrying Capacity: What Urban Economics Leaves Out* (pp. 121–130). Environ. Urban.

10. Odum, H. T., (1996). *Environmental Accounting Energy and Environmental Decision Making.* John Wiley & Sons Inc.: New York.

11. Zhao, S., Li, Z., & Li, W., (2005). A modified method of ecological footprint calculation and its application. *Ecol. Modell., 185*(1), 65–75. https://doi.org/10.1016/j.ecolmodel.2004.11.016.

12. Wei, W., Li, W., Song, Y., Xu, J., Wang, W., & Liu, C., (2019). The dynamic analysis and comparison of energy ecological footprint for the Qinghai-Tibet plateau: A case study of Qinghai province and Tibet. *Sustain., 11*(20). https://doi.org/10.3390/su11205587.

13. CRIDA. (2018–19). *Central Research Institute for Dryland Agriculture.* http://www.crida.in/CP-2012/statewiseplans/Maharastra(Pdf)/PDKV,Akola/Akola.pdf%0A (accessed on 30 June 2021).

14. NDDB. (2018–19). *National Dairy Development Board.* https://www.nddb.coop/sites/default/files/statistics/Mp%20States-ENG-2019.pdf (accessed on 31 July 2021).

15. MPCB. (2018–19). *Maharashtra Pollution Control Board.* https://www.nddb.coop/sites/default/files/statistics/Mp%20States-ENG-2019.pdf (accessed on 31 July 2021).

16. Wenjun, J., Qingwen, M. I. N., Shengkui, C., Dan, Z., & Yehong, S. U. N., (2011). *The Energy-Based Ecological Footprint (EEF) of Traditional Agricultural Areas in China: A Case Study of Congjiang County, Guizhou Province, 2*(1), pp. 56–65. https://doi.org/10.3969/j.issn.1674-764x.2011.01.009.

17. Odum, H. T., (2002). *Handbook of Energy Evaluation a Compendium of Data for Energy Computation Agriculture,* pp. 1–40.

CHAPTER 27

Green Audit: A Step Towards Sustainable Environment

SWAROOP LAXMI MUDLIAR

Department of Chemistry, Shri Ramdeobaba College of Engineering and Management, Nagpur-13, Maharashtra, India,
E-mail: swarooplaxmim71@gmail.com

ABSTRACT

The role of the Green audit is important in the aftermath of the current situation of the world, and it safeguards the globe from the environmental and ecological crisis. Thus, it becomes essential to focus on important parameters of the environment viz conservation of water, waste management, and follow a paperless pattern of working and use alternative sources of energy, like solar and wind energy generation and map biodiversity as a step towards sustainable development which ultimately leads to a sustainable environment. Thus, by allowing changes to the existing system and modifying the pattern of energy consumptions, a system could be modified to a greener system, thereby achieving a sustainable environment with a reduction in pollution levels as well. The aim of the green audit is to ensure that the green practices followed in the area are in accordance with the Environmental Policy adopted by the institution and stipulated standards by CPCB and MOEF. The methodology includes preparation and filling up of documentation, an actual visit to the site by the audit team and perform a physical check of the green initiatives claimed by the authorities, analysis, and discussion of the environment condition, interaction with persons who are responsible towards the green initiatives and suggest an environmental management plan to mitigate the outliers if any. A green audit helps in assessing the damage caused to the environment through harmful activities of a particular company. Green audits should provide the baseline status of the environment in terms of Air, Water, Noise pollution and if need be, made a comprehensive process.

27.1   INTRODUCTION

The history of green audit dates back to the 70's. The need of this study is to trace out the exercise that can cause detrimental effects on human health and the environment and thus leading to pollution. In India any upcoming project must get clearance from the Ministry of forest and environment before it is set up. Hence meticulous testing of land, water, air, and noise environment of the project must be conducted before granting clearance to the project. The Statutory pollution Control bodies in India, set the permissible limits for the various pollution causing parameters. These limits judge the vulnerability of the projects to cause harmful effects on the environment [1].

Further as a check-in India, we have the National Assessment and Accreditation Council (NAAC) which is a government organization that testifies the institution, companies, Industries ranking according to the institutes initiatives taken to reduce the harmful effect on the environment by the projects NAAC also grades these institutions as Grade A, Grade B, Grade C according to their performance in abiding by the green policy of their institute. Thus, the institutes stay motivated to improve their environmental condition and take measures such as solid waste management, Rainwater harvesting, E-waste recycling that will lead to substantial utilization of the resources.

27.1.1   OBJECTIVES OF GREEN AUDIT

The main objective of the Green audit is to promote the Environment Management and Conservation in the Institute, Company, or Industry campus. The purpose of green audit is to identify, quantify, describe, and prioritize the implementation of standards and permissible limits laid down by the Pollution Control Boards (PCB). This is essential to safeguard the environment and reduce the danger caused by pollutants to human health by analyzing the data collected and the extent of use of resources within the study area. The green audit allows the institutions, organizations to prepare a data of the existing condition of the environment and predict the extent of pollution that would be caused due to the activities of the organizations under study.

27.2   THE AREAS OF GREEN AUDITING

1.   The different segments of the audit are:

- Material management;
- Savings in energy by using alternative sources of energy;
- Air quality analysis;
- The initiatives taken to create awareness among staff and imparting training to them in management of the green areas.

2. Water sources and water effluent management is assessed by the following sources of water:
   - Raw water;
   - Drinking water;
   - Process water sewage water;
   - Rainstorm drain water.

3. Energy consumption and conservation is assessed by the following sources:
   - Vehicular load in the study area;
   - LPG usage;
   - Electricity consumption;
   - Batteries used;
   - Solar Energy generation and usage.

4. Waste management is assessed by the following sources:
   - Hazardous waste;
   - Nonbiodegradable solid waste;
   - Biomedical waste;
   - Kitchen waste;
   - Biodegradable waste.

5. E-waste management is assessed accounting for the used:
   - Computers;
   - Mobiles;
   - Pen drives.

6. Green belt development is assessed by the:
   - Plantation in the study area.

27.3   STUDY AREA

- Water management;
- Air pollution management;
- Noise pollution management;
- Energy use and conservation green belt area and biodiversity.

27.3.1 WATER MANAGEMENT

Water management is generally assessed by accounting for the various sources of water present in the audit region. This includes assessing the use of water to help in conserving it. and perform a water balance with regard to inlet water and outlet water. Water conservation is a key activity as water availability effects on the development of the campus as well as on all areas of development such as farming, industries, etc. Keeping this in view, water conservation activity is carried out.

27.3.2 AIR POLLUTION MANAGEMENT

This study is carried out to identify key air polluting devices, processes, equipment responsible for causing pollution to the environment, and their air quality parameters are determined and validated.

27.3.3 NOISE POLLUTION MANAGEMENT

This indicator will identify the noise quality of the environment and measures that could be taken to reduce this kind of pollution.

27.3.4 ENERGY USE AND CONSERVATION GREEN BELT AREA AND BIODIVERSITY

In order to address the energy consumption pattern study of the sources of energy, energy monitoring is usually done.

Thus, the aim of the audit is to:

- To save conventionally produced electric energy;
- Use of non-conventional source of energy;
- Use of carbon-neutral electricity;
- Minimization of electric expense.

27.4 BENEFITS OF GREEN AUDITING

- It will help improve the quality of the environment;

- Identify cost-effective methods through waste reduction and management;
- Highlight pre-existing and predictive complications;
- Conformity with applicable laws;
- Helps organizations create a better environment.

27.5 CERTIFICATION

National Environmental Policy [2] states that after successful completion of audit, an organization is awarded with the certificate mentioned earlier the green audit is a continuous process. The Environment maintenance department of an organization should adhere to this continuous process of green analysis and adhere to the permissible limits of air, water, noise prescribed by the statutory authorities and continue with this practice on a regular basis [2].

27.5.1 INITIATIVES BY ACADEMIC INSTITUTES FOR GREEN CAMPUS

As green auditing aims at improving the environmental status, the following measures can be taken by academic institutes to contribute toward a sustainable environment.

27.5.2 CREATING CARBON FOOTPRINT DATA

Carbon footprint is the total amount of greenhouse gases (GHGs) emitted by way of vehicles, generator sets in terms by a person, institute, company, state, or country. For calculation of carbon footprint, the basic data regarding direct and indirect sources of emission of GHGs is required. The GHG are calculated using the values of emission factors, like use of electricity, use of diesel generators for electricity generation or any other purpose, LPG consumption, food wasted and vehicular emission [3].

Tree plantation helps in achieving reduction in GHGs to a large extent. It is a widely accepted solution for the reduction of carbon footprint in the study area. Indigenous plants which help in building soil fertility and are suitable for the academic campus. Tree census should be conducted periodically [4].

The measures for biodiversity conservation also highlight the efforts towards the environment. Even though plantation is a regular activity in the academic institutions, following 'no vehicle' day, use of bicycles, public transport, rainwater harvesting are the many options. In the present times, energy consumed and heat produced by sophisticated instruments such as servers, CPU's mobiles are so high that it creates more harmful problems in the environment. Industries and organizations are concentrating more on Green Computing to reduce the harmful impacts of pollution. In this chapter, an outline to use green initiatives to reduce pollution and thereby taking a step towards a sustainable future is stressed upon [5].

27.5.3   NEED OF ENVIRONMENTAL AUDIT

In the present times, it is essential for each and every organization to perform green audit so as to reap the benefits like cost-cutting, enabling environmental problems and likewise plan responses to anticipated environmental risks, more efficient utilization of resource use, and financial savings. This practice thus leads to improvement in environmental practices and help the organization to save money too [6].

27.6   CONCLUSIONS

The environmental audit tends to:

- To improve the existing technologies with better input-output ratios.
- To prepare a well-documented environmental status report.
- To determine the efficiency of the environmental management information systems and the performance of operating equipment in the study area.
- To check and compare the compliance of the various environmental parameters with the existing national, local, or other laws and regulations existing that country.
- To reduce the human exposure to risks from environmental, health, and safety problems.
- To assess the small benefits being sacrificed for long term environmental benefits [7, 8].

Thus, frequent Green auditing can help the organization, industry, etc., to achieve business success enhancing environmental protection. And lead to minimization of health and safety risks, improved productivity, resource conservation and contribute their bit to environmental protection.

KEYWORDS

- **green audit**
- **green-computing**
- **greenhouse gases**
- **National Assessment and Accreditation Council**
- **post-audit**
- **pre-audit**
- **rainwater harvesting**
- **site-audit**

REFERENCES

1. Arora, P., (2017). Environmental audit: need of the hour. *International Journal of Advanced Research in Engineering & Management (IJAREM)*, *3*(4), 25–31.
2. *National Environmental Policy*, (2006) Government of India, Ministry of Environment and Forest.
3. Gogoi, A., (1995). Environmental audits: a mean to going green. *Development Alternatives*, *5*(4), 7.
4. Gowri, S., & Harikrishnan, V., (2014). Green computing: Analyzing power consumption using local cooling. *International Journal of Engineering Trends and Technology (IJETT)*, *15*(3), 105.
5. Silvennoinen, K., Heikkilä, L., Katajajuuri, J. M., & Reinikainen, A., (2015). Food waste volume and origin: Case studies in the Finnish food service sector. *Waste Management, 46,* 140.
6. Kulkarni, P., (2009). Combined Heat and Power Potential at the California Wastewater Treatment Plants. *California Energy Commission.* https://efiling.energy.ca.gov/GetDocument.aspx?tn=52418&DocumentContentId=9488 (accessed on 31 July 2021).
7. Kuo, J., & Dow, J., (2015). Biogas production from anaerobic digestion of food waste and relevant air quality implications. *Journal of the Air & Waste Management Association, 67*(9), (1000).
8. Kitila, A. W., (2015). Electronic waste management in educational institutions of Ambo Town, Ethiopia, East Africa. *International Journal of Sciences: Basic and Applied Research, 24*(4), 319.

CHAPTER 28

Evolution of Waste to Wealth: An Overture to Reduce Insidious Effects

KANCHAN D. GANVIR,[1] NIKHIL D. PACHKAWADE,[1] and JAYASHRI CHOPADE[2]

[1]Department of Mechanical Engineering, Priyadarshini Bhagwati College of Engineering, Nagpur, Maharashtra, India,
E-mail: kanchan.ganvir100@gmail.com (K. D. Ganvir)

[2]Department of Mechanical Engineering, Pimpri Chinchwad College of Engineering and Research Ravet, Pune, Maharashtra, India

ABSTRACT

Presently, disposing of the waste is the key agenda in conversations of rectifiable turn of events. Squanders and biomass fills are normally considered as reasonable vitality sources. Solid waste commonly called debris or rejected entities from everyday routine that is thrown away by every individual. Our standard exercises offer ascent to an assortment of strong squanders of various physio-synthetic qualities, which hurt the environmental factors gravely. But if appropriately oversaw and handled under sharp thought can be one of the ways to overcome. As technology is taking its ascent and innovation is propelling, squanders are likewise commutable to practicable vitality structures like hydrogen (biohydrogen), biogas, bio-liquor, and so on through waste-to-vitality advances. Tidying up of waste defilement is considerably more costly contrasted with its counteraction at each phase of conceivable pollution. The living creatures are confronting a prodigious test to appropriately deal with the squanders. Confronted with colossal volumes and weighty consumption for the executives, endeavors ought to be made to lessen squander volumes and produce income from treatment thereof. So revolutionary attempt has been made for converting of waste to wealth to decrease its adverse effects which are directly or indirectly harming living beings as well as the environment. The focus of the chapter is to do the survey of the literature majorly focusing on environmental management converting wastes for research streams that relate the indigenous habitat to tasks the board in an undeniably engaged way. The intention behind the chapter is to find out the methods for converting the waste into some usable form. Another most important aspect of the chapter is to find out an alternative energy source from the waste which should have environmental compatibility which would ultimately benefit socio-economic, health, and well-being to humankind.

28.1 INTRODUCTION

Solid waste and its contamination are the main reason for the existence of most harmful diseases, additionally improper waste management is another cause to stimulate environmental pollution. The World Health Organization has witnessed 22 sorts of ailments related with imprudent administration of metropolitan substantial trash. Likewise, there are social complexity of inappropriate waste administration excessively influencing helpless networks residing in ghettos and zones closer to the landfills and dump-locales [1]. A huge number of waste gatherers are presented to perilous substances while gathering debris in the landfill destinations truly affecting their physical health, pertaining to harmful disease. The inappropriate waste administration generally adds to air, land, and water defilement. Strong waste involves perishable, reclaimable, and inactive trash [2]. The biodegradable type of waste is the main part of urban solid waste and generally tainting comprising disposal of leftover food, grocery, fruit market organic products, fish, and meat market its dampness, framing intense ground for germs to increase. Succeeding to previous next is the recyclable waste which is generally large in volume and abundant involving newspapers, plastics, metal, glass, fiber, cardboard, wood board and so on. Such things whenever kept independently hold a bundle of significant worth, yet whenever blended in biodegradable and inactive scrap, make immense part of ravage contamination having bacteria and if haphazardly stuffed all together leads to an ecological calamity. Potential squanders generated from cleared residue, sand, cinders, and materials of building development and fixes [8]. If maintained separately are reusable and innocuous, yet their unpredictable involvement tends to fill the pockets for germs and irritations to hold up and multiply in nature, which would additionally be difficult to control.

28.2   SOLID WASTE: AN OVERVIEW OF ONGOING SCENARIO

The waste has consistently been the shadow side of the economy. Underway and utilization, it is what is dismissed as futile and desolate. Whatever the word trash, refuse, garbage flotsam and jetsam, wastes, and whichever might be the language, the importance is comparable. The social assignment of waste administration has been to dispose of it. The present waste is diverted through sewers and dustbins, dispatched noticeable all around through consuming, dumped in neglected quarries or the seas, onto midden or fly-tipped in canals or behind supports over [13]. As indicated study on, waste to generating energy, a normal 60 million tons of MSW waste is delivered each year by around 375 million people in India's citified area, of which more than 75% is disposed of capriciously at dump yards in a careless and casual way by metropolitan masters provoking issues of prosperity and biological defilement [12], covering 3,50,000 cubic meters of landfill space each day. Considering the all-inclusive wastage of around 160 million tons by 2030, consumed area via landfills for a long time can be calculated as 45,400 hectares of significant area of ground, which the nation cannot manage the cost of [19]. The team report thinks of it as basic to limit the waste going to landfills by at any rate 75% through management of Municipal strong waste utilizing suitable advancements.

28.3   NEED FOR WASTE MANAGEMENT

The most extreme requirement for squander the board is the current circumstance the world is confronting. To deal with the current possibilities and difficulties of urban waste administration, an incorporated way to deal with squander the executives including arranging, financing, development, and activity of offices for the isolation, hodgepodge, transportation, reusing, treatment, and last removal of the waste ought to be thought of. Despite striking improvement in social, financial, and preservation segments, SWM settings in India have suffered commonly whole. The impromptu part has a key capacity in safeguarding an allurement from spend, with generally 90% of suffering waste at present forsaken rather than fittingly landfilled there is an exceptional need to move to more reasonable solid waste organization, and this requires new organization plans and waste organization workplaces [7]. Current SWM structures are luxurious, with abuse adversely appalling general wellbeing, the earth inciting monetary annihilation (Table 28.1).

The challenge has many prospects and elements, following are the most required one:

- Source isolation should be energized and commanded;
- Productivity must be purchased to assortment of waste;
- Logical treatment/preparing of waste commanded alongside definite removal.

TABLE 28.1   Tabular Form Showing Source, Generator, and Contents of Waste

Sources	Generator of Waste	Contents of Solid Waste
Urban	Individual and multiple households/residence.	Leftover food, paper scrap, plastic trash, textile wastes, leather squanders, yard debris, wood, glass, metals, ashes, electronic wastes, gadgets, scrap batteries, other harmful wastes, daily household trash.
Foundation	Institutions, schools/colleges, clinics, and hospitals, government, and private office.	Paper waste, cardboard, plastics, wood, food wastes, glass, metals, special wastes, unsafe scraps.
Commercial	Shops, goods, inns, cafes, markets, offices, malls, etc.	Paper, cardboard, plastics, wood, food wastes, glass, metals, special wastes, hazardous wastes.
Construction and demolition waste	Construction sites, road repairs, renovation sites, demolition of buildings.	Wood, steel, concrete debris, glass, sand, tiles, bituminous concrete.
Municipal services	Highway sweeping, camouflaging, cleaning of parks, beaches, other leisure areas.	Sweeping of street; drain silt; landscape and leaves tree bits and pieces; debris from sea beaches, trash from park and other amusement areas.

28.4   GENERATION OF WASTE IN INDIA

India being a populated is encountering a fast ascent in expansion while staying a nation with carnal, climatic, topographical, biological, public, social, and semantic decent variety. The number of inhabitants in India was 1252 million out of 2013, contrasted, and 1028 million out of 2001. Populace development is a significant supporter of expanding Municipal strong waste in India.

Table 28.2 from sources indicates statistics of generation of trash from major cities, Squander age rate relies upon components, for example, populace thickness, monetary status, level of business movement, culture, and city/locale. Gives information on Municipal strong waste age in various urban communities, demonstrating high waste age in Mumbai, Maharashtra. At that point, Delhi indicating second most noteworthy, third-most elevated Kolkata West Bengal, trailed by Chennai, Bangalore, Hyderabad, Ahmadabad.

Solid Waste Management removal is a basic phase of improvement in India. It is a necessary to build offices to extravagance and discard intensifying measures of Municipal strong waste. Over 90% of excess is recognized in India to derelict in an inappropriate way. It is assessed that roughly 1400 km^2 was involved to dissipate dumps in 1997 and this is obligatory increment in future.

TABLE 28.2 Statistics of Generation of Wastes in Major Cities in India

Place	Population (2011) × 100	Total Waste Generated in Tons Per Day	Waste Generation (Kg Per Capita Per Day)
Ahmadabad	6.3	2300	0.36
Hyderabad	7.7	4200	0.54
Bangalore	8.4	3700	0.45
Chennai	8.6	4500	0.52
Kolkata	14.1	3670	0.26
Delhi	16.3	5800	0.41
Mumbai	18.4	6500	0.35

28.5 WASTE DUMPING HEALTH HAZARDS AND ENVIRONMENTAL IMPACTS

Squander dumps impacts affect the earth and general wellbeing. Open landfills liberation methane from the crumbling of recyclable excess under absence of oxygen circumstances. Methane causes blazes, flares, and is a significant supporter of an unnatural weather change [11]. There are likewise issues related with scent and movement of leachates to accepting waters. Smell is a hard dispute, especially throughout the mid-year when degrees of hotness in India can surpass 45°C. Disposed of drains at landfills gather water, permitting mosquitoes to rise, expanding the danger illnesses, for example, intestinal illness, respiratory disorder, and severe fever. Abandoned consumption of waste at dump locales delivers fine particles which are significant reason of respiratory illness and cause consume cloud Open consuming of MSW and tires transmits 22000 huge amounts of poisons into the environment around Mumbai consistently The effects of helpless waste administration on general wellbeing are very much reported, with expanded rates of nose and throat diseases, breathing challenges, aggravation, bacterial contaminations, paleness, diminished insusceptibility, sensitivities, asthma, and different diseases (Figures 28.1 and 28.2).

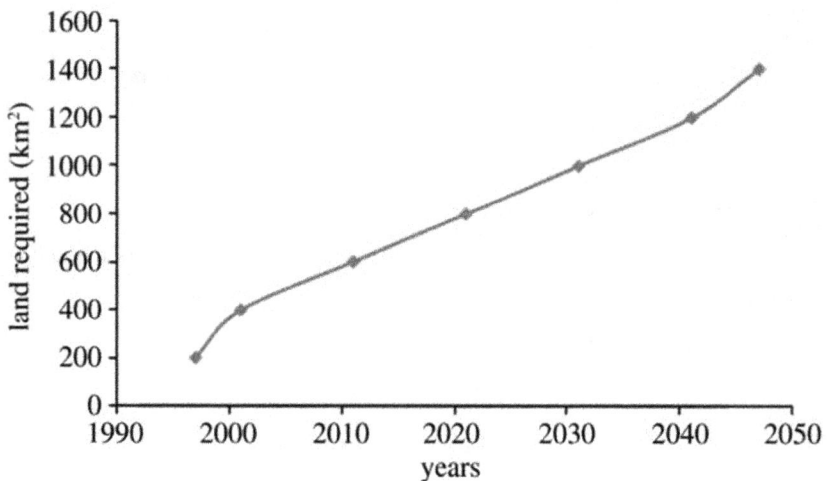

FIGURE 28.1   Graphical representation of years vs. land required in km.

28.6   WASTE MANAGEMENT TECHNIQUES

The advantages of achieving eco-modernization, the base requirement is to reduce the squander so as to bring it with more prominent productivity. Proper reusing and waste decrease will change the entire framework, It depends on appropriate insight as opposed to incorporated information and on advancement that is broadly scattered across assortment, preparing waste management techniques. Therefore, the change can appear to be far harder to accomplish than the direct modernization of existing frameworks. Managing waste incorporate the activities from inception to final disposition Another practice which can be inculcated is proper collection, proper transportation, and treatment of waste with regular monitoring. The advisable exercise of MSW is the 3R's-Reduce, Reuse, and Recycle.

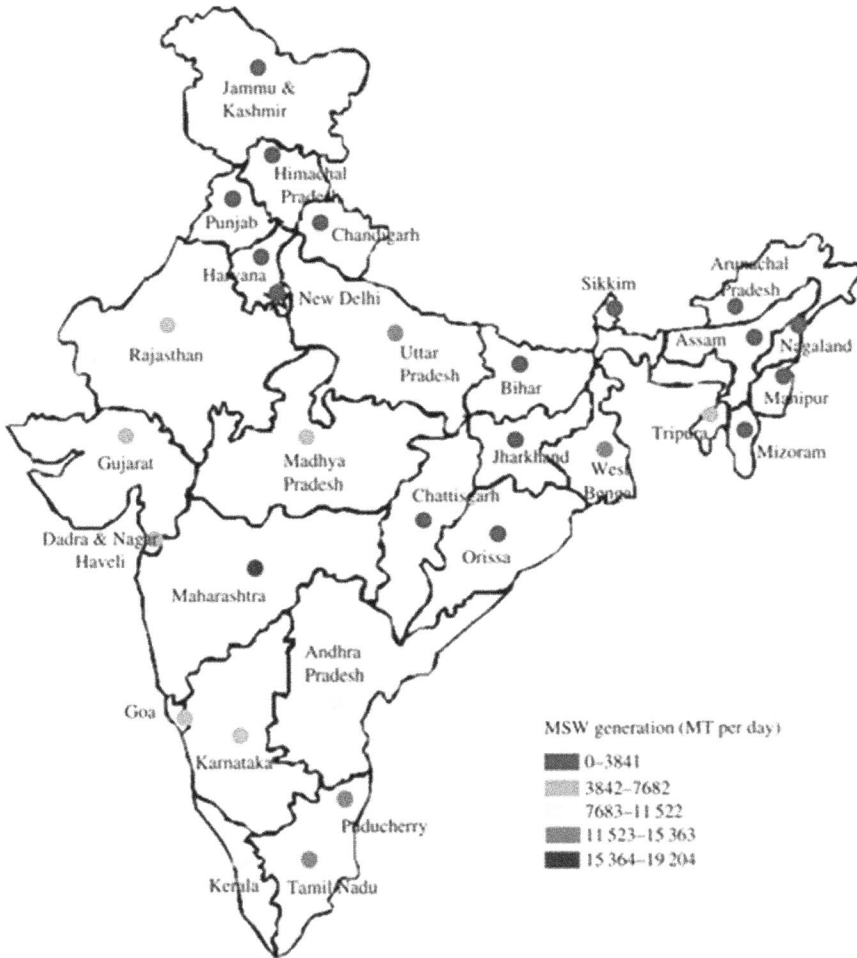

FIGURE 28.2   Region representing municipal solid waste in India.

28.6.1   *REDUCE*

- Decrease waste generation at your place;
- Make maximum use of resources will tend to waste less;
- Minimize plastic use and removals;
- Decrease plastic shopping bags to maximum extent;
- Decline packaging henceforth production of less scrap;
- Packaging and dead product can be returned to manufacturers.

28.6.2 REUSE

- Make use of old items such as jars from grocery to store foods, vegetables, fruits, etc.;
- Reuse of disposed things with safety at personal level;
- Reuse packaging and wrapping;
- Reuse maximum tems as much as possible and Natural resources as well.

28.6.3 REPROCESS

Reprocess the scrap materials back to consumption cycle, and for source recovery, segregation is a key for reprocessing and recognizing "waste to wealth" (Figure 28.3).

FIGURE 28.3 Reduce, reuse, and recycle.

28.7 WASTE MANAGEMENT AND ENGINEERED LANDFILLS IN INDIA

The guidelines given by municipal solid waste (MSW) directions 2000 to guarantee appropriate surplus administration in India and original refreshed current rules have as of late been distributed. Public specialists are liable for actualizing these principles and creating a foundation for assortment, stockpiling, isolation, transportation, handling, and removal of MSW. Chandigarh is a primary city to create Solid Waste Management in an arranged manner and has improved waste administration contrasted and other Indian urban communities [17]. As the MSW management guidelines before some years and instructions as a task to force on WTE focuses on the idea of specialized manuals, Housing ministry and Urban Affairs has arranged this Ready Reckoner for "Waste to Wealth" agenda as a Swachha Bharat Mission activity for the direction of leaders and all different partners in straightforward and effectively justifiable organization for squander handling and asset recuperation thereof.

The UN Environmental Program portrays landfill as the controlled expulsion of MSW aground so contact among waste and nature is by and large lessened, with trash evacuation pressed in an especially described district. Structured landfill allows the protected evacuation of extra MSW shoreward, yet guards ground and surface water from pollution and escapes air transmission, wind-blown litter, smell, fire hazards, issues with animals, fowls, and various aggravations/rodents, and decreases ozone exhausting substance outpourings and grade dubiousness issues. Fittingly directed structured landfills should override dumps in India. This would basically moderate the regular impact of waste (Table 28.3).

TABLE 28.3   Landfills in India

Place/City	Landfill Numbers	Landfill Area (hectares)
Chennai	1	467.5
Coimbatore	3	295
Surat	2	205
Mumbai	2	147
Hyderabad	1	127.5
Ahmadabad	2	86

28.8   AREAS ON WHICH WORK SHOULD BE DONE

Local Bodies of urban sector, Urban Areas outgrowth, Census Towns and Census Commission of India, Notified Areas, Notified Industrial Townships, Areas heavily influenced by Indian Railways, Airports/Airbuses, Ports, and Harbors, school universities, Defense Establishments, Special Economic Zones, State, and Central Government Organizations, Places of travelers and Pilgrimage [18]. Spots of Religious and recorded significance as might be informed by the Government now and again.

28.9   OBJECTIVE OF THE RESEARCH

The principle goal of the exploration is to comprehend the difficulties presented by waste and its unfavorable effect on condition and wellbeing. Another goal is the way exercises produce squander order of various sorts of waste and components for their removal; by improvising idea of 3R which is presently altered as idea of 5R: Refuse, Reduce, Reuse, Recycle, Recover, dangers presented to wellbeing and condition and security measures to be embraced in dealing with squander imaginative cycles that produce items from squander making riches, the extent of occupation age through enterprise.

28.10   INITIATIVE BY PEERS FOR ENCOURAGING WASTE TO WEALTH

28.10.1   WASTE TO COMPOST

Fertilizing soil includes the cracking of natural waste by microorganisms within the sight of air, warmth, and dampness. This should be possible by bringing things down a peg in nuclear families or for an immense extension depending on the measure of waste to be taken care of and space open. Microorganisms, parasites follow up on the misfortune to change over it into sugars, starch, and regular acids followed by high-temperature venerating minute creatures, achieving a consistent thing called City Compost.

28.10.2   WASTE TO FUEL

Refuse derived fuel (RDF) ordinarily comprises of the leftover dry ignitable division of the MSW like paper, material, clothes, calfskin, elastic, non-recyclable plastic, jute, multi-layered bundling and other complex bundling, cellophane, thermocol, melamine, coconut shells, and other high calorific parts of MSW. The reasonableness of RDF use as a fuel or asset is subject to specific boundaries like:

- Ash content;
- Sulfur content;
- Calorific value;

- Chlorine content;
- Water content.

The necessary explicit creation and qualities of derived fuel or co-preparing is dictated by sorting kettle/heater, heats accomplished in the heater, and the related vent gas the executives' frameworks.

28.10.3   WASTE TO ENERGY (WTE)

Structured the cycle to change over waste plastic into significant things like radiator fuel and liquefied petroleum gas. The cycle depends on the rule of irregular de-polymerization and includes specific breaking of Carbon-Carbon bonds. The two plastics and oil inferred fills are hydrocarbons. Notwithstanding, the plastic particles have longer carbon chains than those in LPG, petroleum, and diesel fills [20]. Thus, it is possible to change over waste plastic into stimulates. The cycle is a warm specific parting response of the enormous nuclear mass polymer carbon manacles under an oxygen-free condition and creates little sub-atomic weight specks.

28.10.4   WASTE TO CONSTRUCTION

A significant reform has passed by the government that the plastic waste must be used in Road Construction for all street engineers in the nation. Certain measures of waste plastic, with bituminous blends for street development. Street designers will presently need to utilize squander plastic alongside hot blends for developing bitumen streets inside 30 km of fringe of any city that has a populace of more than six lakh. In as of late delivered rules for engineers, the administration has made it compulsory that if there should arise an occurrence of non-accessibility of waste plastic, the designer needs to look for the street transport and parkways service's endorsement power developing just bitumen streets [1]. The squander is lonely into huge solid pieces, merged C&D squander rendering to measure and unrecyclable resources like plastic and wood, which are sent to a waste-to-vitality plant in Okhla. The plant consumes physical segregation for grander malleable pieces just as an attractive separator for metallic items. The waste is squashed, washed, and used to prepare blend solid, kerbstones, concrete blocks, tarmac squares, empty blocks and made sand.

28.11   ALTERNATIVE TECHNOLOGIES FOR PROCESSING OF WASTE

28.11.1   BIOCHEMICAL TECHNOLOGY

Bio-synthetic transformation of biodegradable MSW can be ordered into fertilizing the soil and biomethanation. Fertilizing the soil is a high-impact measure in which naturally degradable squanders are changed over through organic change to yield stable granular material usually called City Compost which could be utilized as soil conditioner and supplement supplier. Biomethanation is an anaerobic slurry-stage measure that can be utilized to recuperate supplements and vitality contained in biodegradable squanders [3]. Biogas can be utilized as a wellspring of warm vitality or to create power by utilizing gas motors or gas igniting.

28.11.2   BIODEGRADABLE WASTE TECHNOLOGIES

A brief glance at any environmental framework shows boundless cooperative energy between creatures. Squander from one living element regularly makes a sound situation for another creature, giving sustenance and conditions helpful for its endurance and development. Rotted natural material (fertilizer) is ordinarily utilized as compost or fertilizer for developing plants. It improves soil structure and gives supplements [5]. The cycle of composting requires making a stack of wet natural issue known as green squanders and trusting that the materials will separate into humus following a time of weeks or months. The decay can be quickened by other living beings, for example, microscopic organism's growths, creepy crawlies, worms, and so forth and abiotic components like temperature, dampness, and oxygen, bright light, and so on.

28.11.3   ORGANIC AGRICULTURE

With more noteworthy affectability to issues identified with reasonable turn of events, agro-environmental cultivating techniques are picking up in fame. These depend on environmental cycles to support the strength of soil just as regarding cultivating as an incorporated, all-encompassing, interconnected cycle of food creation by streamlining the ranch in plan and intently weave supplement and asset reusing. Rather than manufactured manures and pesticides, fertilizer, green compost, and bone feast are key fixing in natural cultivating and furthermore on non-synthetic methods of irritation and infection control. Awareness towards a pleasing way of life has seen natural homestead creation and exchange rising as a significant segment in India as in different pieces of the creating scene [7].

28.11.4   USE OF BIOFUELS

Other significant heading in offer for maintainability was utilization of bio-fills from biomass. Biofuel creation is a spotless low carbon innovation for the transformation of natural squander into clean sustainable biogas and a wellspring of natural manure. Biogas acquired by anaerobic assimilation of steers manure and other free and verdant natural issue/squanders can be utilized as vitality hotspot for cooking, lighting, refrigeration, and power age and transport applications.

28.11.5   COMPOSTING TECHNOLOGIES

Cycle Biology Different living beings effectively available in metropolitan hard squanders are known to make light of a dominating job in breaking ecological elements of MSW. A progression of contagious development and movement among the microorganisms, growths, actinomycetes, leaven, and so on happens during the cycle, whereby the earth made by one network of microorganisms supports the action of a replacement gathering. Various sorts of thusly dynamic at various occasions and areas inside the windrow relying on the accessibility of substrate, oxygen flexibly, and dampness substance of the natural issue.

The creation cycle of fertilizer includes the accompanying cycles:

- Windrow formation under the monsoon sheds;
- Rough separation;
- Straining;
- Final product;
- Wrapping.

28.11.5.1   VARIOUS TYPES OF COMPOSTING

1. **Windrow Composting:** Windrow fertilizing the soil is the creation of fertilizer by heaping eco-friendly squander, in long lines. This

strategy is suitable to carrying mammoth mass of fertilizer. These stay consistently gone over to improve absorbency and oxygen content, blend or eliminate dankness, and more sweltering chunks of the heap [13]. Windrow treating the soil is a generally utilized fertilizing the soil strategy.

2. **Vermi Composting:** Vermi fertilizer (or vermin-manure) is the outcome of the soil cycle treatment which utilizes the different types of insects, typically red wigglers, white worms, and different nightcrawlers, which feed in a blend of disintegrating vegetable food squander, and delivery droppings known as vermicast is the finished result of the breakdown of natural issue by a nightcrawler.

3. **In-Vessel Composting:** In-vessel preparing the dirt usually is a technique that restricts the getting the dirt tackles a structure, holder, or container. In-vessel treating the dirt structures can involve metal or plastic tanks or strong safe houses in which air current and heat can be controlled, using the principles of a "bioreactor." Generally, the air course is metered in by methods for secured chambers that grant common air to be imbued under pressure, with the vapor being removed through a bio-channel, with heat and suddenness conditions checked using tests in the bulk to permit upkeep of ideal oxygen devouring weakening conditions.

4. **Aerated Static Pile Composting:** Circulated air through Static Pile treating the soil, alludes to the framework used to biodegrade natural material without physical control during treating the soil. The mixed squander is typically positioned on punctured channeling, giving air course to controlled air circulation. It might be in windrows, open or secured, or in shut holders [14]. Concerning multifaceted nature and cost, circulated air through frameworks is most generally utilized by bigger, expertly oversaw fertilizing the soil offices, despite the fact that the strategy may run from minuscule, basic frameworks to huge, capital serious, mechanical establishments.

5. **Trench Composting:** It is the way toward covering natural waste legitimately into soil. Digging is a great technique to use in mix with developing yearly plants, particularly plants like cabbage, corn, and so forth. It likewise supports the improvement of profound, water moderating root frames. Digging uses anaerobic disintegration to make a concealed crowd of supplement rich humus for plants. This is a slower fertilizing the soil cycle than that in an all-around oversaw windrow, however the dug materials will hold more nitrogen during the cycle.

28.11.5.2 WASTE TO ENERGY (WTE) TECHNOLOGY

Squander to-energy innovation uses trash or garbage to deliver electric and warmth vitality through composite transformation techniques [16]. Waste to vitality innovation gives an elective wellspring of sustainable power sources in a world with restricted or tested fossil stores (Figures 28.4 and 28.5).

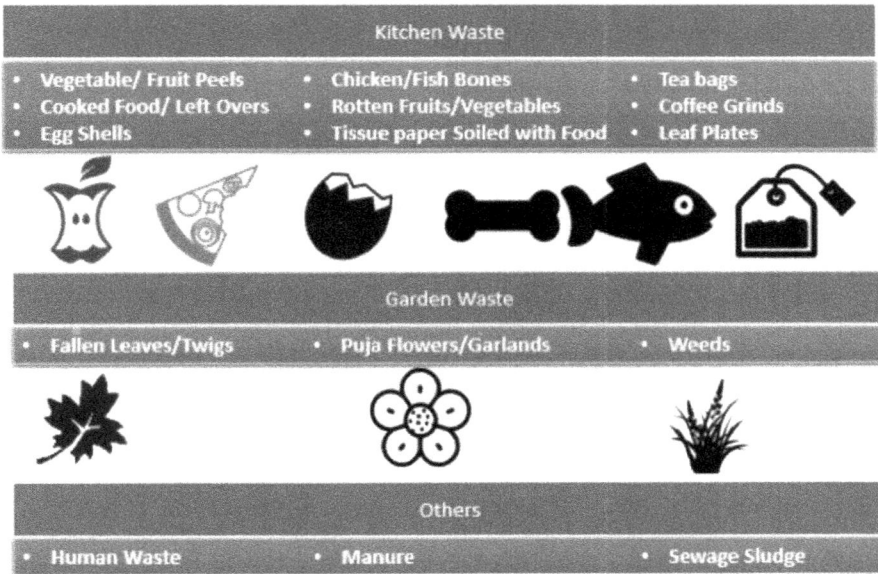

FIGURE 28.4 Common sources of waste in daily routine.

28.11.5.3 LANDFILL

Following are the methods to select a landfill—logical investigation about territory of the plot, geology, below groundwater level, region of the varieties of the water, for instance, stream, and the thickness of another squander. Clearing the dumpsite of all ground spread.

- Exhuming of the earth (the volume of the uncovering relies upon the aftereffects of initial step).
- Improvement of the bank all around the landfill site. Advancement of the liner and leachate the board structure.

- Development of high-density polyethylene (HDPE) which dodges leachate to flee into nature from the dumpsite.
- Foundation of the leachate and methane extraction pipe.
- Foundation of force plant generator and emitting structure or blower station.

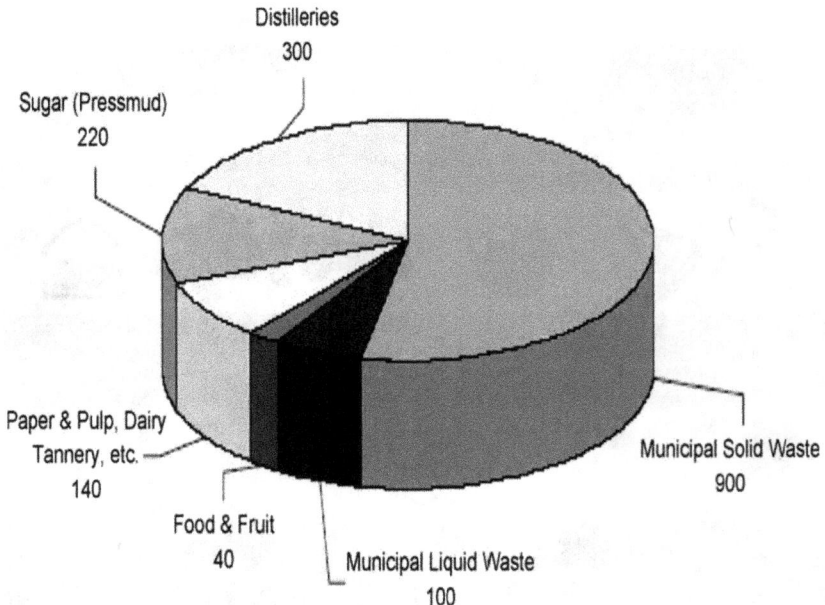

FIGURE 28.5   Percentage representation of different sources of waste.

28.11.5.4   INCINERATION

A pyre office measures squander in the accompanying advances:

- Squander is dumped into a waste pit by dump trucks.
- An overhead crane gets the debris from the waste pit into a holder which deals with a moving cross-section that experiences the incinerator warmer. The waste is resolved to fire by using 08 to 11 GJ of fuel 73% of waste garbage is scattered.
- An extra 2 GJ of gasoline is utilized to diminish the creation of perilous mixes, for example, organo-chlorines. Squander is changed into strong debris and reused for additional application.

- Imperativeness is eliminated from the toxic fumes conveyed by making steam in a radiator. Steam is used to turn a turbine to deliver power.
- Poisonous fumes go through an electrostatic precipitator to permit the residue to resolve. The harmful products at that point experience pipe gas and other cleaning frameworks before being delivered into the climate through chimneys.

28.11.5.5   PLASMA ARC GASIFICATION AND VITRIFICATION

Vitrification is the change of a material into a goblet, in other means a non-translucent undefined tough plasma bend office measures squander in the accompanying advances:
- Squander is gotten and weighed inside a regulation structure.
- Huge bits of salvaged material and risky squanders are isolated from the squander stream.
- Squander was destroyed into tiny segments and benefited from a transport into a vaporizing container; Process squander heat from the last plasma circular segment transformation stage is utilized to gasify squander.
- Gas is taken care of into a plasma circular segment transformation container, where a plasma light separates the gas stream at a natural level, changing over it to a vitality rich combination gas. Fly trash is softened in an alternate plasma container (vitrification measure) and is cooled into a glass-like solid known as slag.
- Blend gas is cleansed to isolate harmful constituents Purified amalgamation gas is utilized to fuel a responding motor-generator set.
- Abundance measure from the heat which responds motor and plasma curve to produce power which drives a steam turbine.

28.11.5.6   PLASTIC WASTE MANAGEMENT TECHNOLOGY

The suggestion for technology are:

- Fuel from animal excreta, for example, bovine waste, bio fills.
- Treatment of e-waste.
- Distinguishing proof, documentation, and grouping of the measure of electronic and electrical items left unused in one's zone and post for infers for evacuation and moreover re-use.

- Strong squander preparing and removal.
- Planning briquettes and verifying effective use for ground less estate.
- Restricting wastage of water framework through improving water upkeep in the soil using distinctive squanders, e.g., Wood shavings, briquettes, paddy husk, coconut ensemble, and so forth.
- Bioremediation of wastewater.
- Aquaculture utilizing dim or potentially dark water.
- Treatment of kitchen squander water for evacuation of fat, oil grease.
- Planning characteristic colors from leaves, blossoms, products of weeds.
- Change of discarded blooms into supportive things.
- Separating oil from the seeds of weeds, to be utilized for brightening.
- Maintenance of ground dampness utilizing chippings, briquettes, pod, coconut husk, Hyacinth in crop fields.
- Using ensilage as substance for mushroom advancement.
- Utilization of graphite from utilized batteries (dry cells).
- Improving fruitfulness of land utilizing different squanders.
- Plastic rot by pathogen.
- Inspecting properties of wipe for water support and creating plants.
- Use of disposed of wrinkled cardboard for production of shingle, placard, etc.
- Planning tiles and board sheets from tetra pack/plastic containers.
- Getting ready manure and other helpful items from the corpses.
- Creative procedures for changing over agribusiness squanders into valuable items.

28.12   BENEFITS OF RESEARCH

Around 100% waste is changed over into esteem included items. Answer to the waste plastic issue can change overall budgetary circumstance by saving countless money in import of grungy oil, no initial cleaning or disengagement, easy shipment of unwieldy waste, sulfur content in the fuel delivered is under 0.002 ppm. There are no noticeable unwanted rays seen, neither sewer water noticed, in fact purified liquid is obtained (Figure 28.6) [19].

28.13   CONCLUSION

People advancement and predominantly the improvement of metropolis is creating SWM in India a noteworthy issue. We need better frameworks for estimating various types of result. Just as the monetary impacts of reusing, it can make social and natural profits that add to personal satisfaction and sound networks. However, these characteristics, regardless of being strategy objectives for governments over the world, are infrequently considered along with monetary arranging or natural guideline [6]. Accordingly, the business motivating forces for venture make unreasonable results and have driven firms, neighborhood specialists and local government to move towards burning and infer that reusing is a costly and intrinsically restricted type of action. This will possibly change when we can account appropriately for the fluctuated impacts of our various decisions. Over the entire of society, business is increasing another job in creating social and ecological results. Until we can gauge this commitment appropriately, we will not locate a maintainable method of overseeing or directing the economy [3]. Fourth, making riches from squander includes shoppers and householders as dynamic members in the framework and focuses on another type of citizenship accordingly. The possibility that creation and utilization are being re-joined through people's experience is presently spreading generally, having been first anticipated in the last part of the 1960s and the 1970s. In regions where it produces social, open, and ecological products, we can perceive how such action makes another connection between the individual and the state: pragmatic, dynamic, and restricted, a type of direct association that offers individuals the chance to have any kind of effect and Ensures that the entire of what they do is more noteworthy than the aggregate of its parts. Once more, during a period of far-reaching voter alienation, this has colossal criticalness.

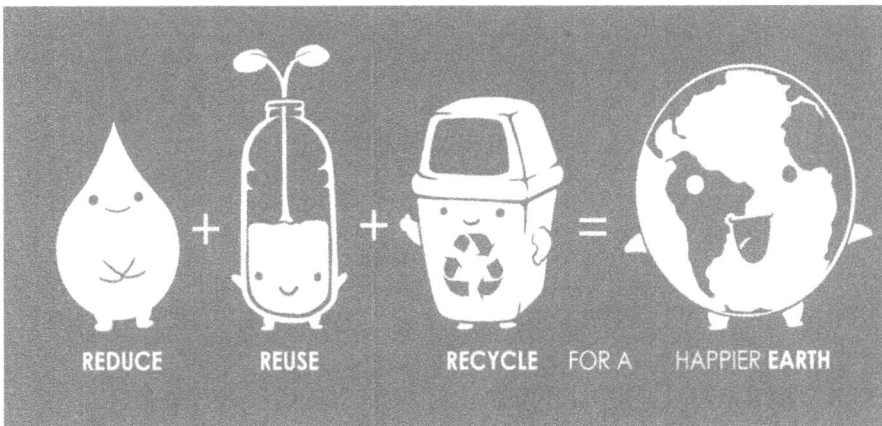

FIGURE 28.6 Combining 3R for better and happier earth.

KEYWORDS

- high-density polyethylene
- hygiene benefits
- refuse-derived fuel
- solid waste
- sustainability
- waste management environment
- waste to wealth technologies

REFERENCES

1. Chen, S., Mulgrew, B., & Grant, P. M., (1993). A clustering technique for digital communications channel equalization using radial basis function networks. *IEEE Trans. on Neural Networks, 4,* 570–578.

2. Duncombe, J. U., (1959). Infrared navigation Part I: An assessment of feasibility. *IEEE Trans. Electron Devices.* ED-1134-39.

3. Lin, C. Y., Wu, M., Bloom, J. A., Cox, I. J., & Miller, M., (2001). Rotation, scale, and translation resilient public watermarking for images. *IEEE Trans. Image Process, 10*(5), 767–782.

4. Adewole, A. T., (2009). Waste management towards sustainable development in Nigeria: A case study of Lagos state. *International NGO Journal, 4*(4), 173–179.

5. Enete, I. C., (2010). *Potential Impacts of Climate Change on Solid Waste Management in Nigeria.* Earthzine.

6. Gowan, T., (1997). American untouchables: Homeless scavengers in San Francisco's underground economy. *International Journal of Sociology and Social Policy, 17*(3, 4), 159.

7. Nabegu, A. B., (2010). An analysis of municipal solid waste in Kano metropolis. *Nigeria. J Hum Ecol., 31*(2), 111–119.

8. Ogwueleka, T. C., (2009). Municipal solid waste characteristics and management in Nigeria. *Iran. J. Environ. Health. Sci. Eng., 6*(3), 173–180.

9. Rotich, H., Zhao, Y., & Dung, J., (2006). Municipal solid waste management challenges in developing countries - Kenya case study. *Waste Management., 26,* 92–100.

10. Scheinberg, A., & Anschütz, J., (2007). Slim pickin's: Supporting waste pickers in the ecological modernization of urban waste management systems. *International Journal of Technology Management and Sustainable Development, 5,* 257–27.

11. Samson, M., (2008). *Refusing to be Cast Aside: Waste Pickers Organizing Around the World.* Cambridge, MA: WEIGO.

12. Srinivas, H., (2011). *Solid Waste Management: Glossary.* The Global Development Research Center.

13. Wilson, D. C., Velis, C., & Cheeseman, C. R., (2006). Role of informal sector in recycling in waste management in developing countries. *Habitat International, 30,* 797–808.

14. Zurbrugg, C., (1999). *The Challenge of Solid Waste Disposal in Developing Countries* (p. 4). SANDEC News, EAWAG.
15. Asokan, P., Saxena, M., Asolekar, S. R., & Sharma, A., (2009). Cross-sector waste recycling opportunities. In: Shailesh, K. A., & Amit, R., (eds.), *Proc. International Seminar on Waste to Wealth Green Building Materials and Construction Technologies using agricultural and Industrial Wastes*. New Delhi, BMTPC. 278292.
16. Asokan, P., (2007). *Application of Some Inorganic Residues in Management of Hazardous Jarosite Waste*. PhD. Thesis (ENV-PHD-07-701). Department of Environmental Science and Engineering, Indian Institute of Technology Bombay, Powai, Mumbai,
17. Central Pollution Control Board (CPCB), (2000). *Report on Management of Municipal Solid Waste*. Delhi. India,
18. Cheeseman, C. R., Butcher, E. J., Sollars, C. J., & Perry, R., (1993). Heavy metal leaching from hydroxide, sulphide and silicate stabilized / solidified wastes. *Waste Management, 13*, 545–552.
19. Cheeseman, C. R., Makinde, A., & Bethanis, S., (2005). Properties of lightweight aggregate produced by rapid sintering of incinerator bottom ash. *Resources Conservation and Recycling, 43*, 147–162.
20. Ding, M., Schuiling, R. D., & Van, D. S. H. A., (2002). Self-sealing isolation and immobilization: Agrochemical approach to solve the environmental problem of wastes acidic jarosite. *Applied Geochemistry, 17*, 93–104.

CHAPTER 29

Utilization of Alkyd Resin in Water-Thinnable Paints for Low Volatile Organic Compounds Emission

BHARATI BURANDE

Department of Applied Chemistry, Priyadarshini College of Engineering, Nagpur, Maharashtra, India
E-mail: bharatiburande@gmail.com

ABSTRACT

Petroleum solvent-based coatings contain higher levels of organic compounds than water-based coatings, when these compounds evaporate; they release Volatile Organic Compounds (VOCs) into the atmosphere resulting in a toxic impact on the environment and human health. Stricter environmental regulations have required paint manufacturers to dramatically reduce the levels of VOCs in their paints. The best remedies to address these problems are the waterborne paints. In this chapter, we have attempted to formulate waterborne paints which contain low VOCs. Alkyd resin is one of the most commonplace sorts of binder inside the paint and coating industry, prepared from the reagents including petroleum products, phthalic anhydrides, and organic solvents. However, the excessive use of those reagents harms the environment. In this event, the involved industries are seeking to replace these hazardous chemicals with alternative sources which are of vegetative nature. In the existing chapter, we have tried vegetable oils, herbs, plant extracts as alternative substitutes for petroleum products. The fundamental instinct of the use of those substitutes in the synthesis of products is as they are generally nonhazardous and degradable in nature. We have prepared alkyd resin with much less than 15% of petroleum ingredients and 80–90% renewable plant products like castor oil, rosin, etc. Then it has been thoroughly analyzed for its physicochemical and spectroscopic properties. Alkyd resin with perfect properties has further been used for the preparation of water thinnable paints. In the composition of paints, we have been replaced approximately 50% organic solvents with water without sacrificing technical performance. All paint samples have been analyzed for their physicochemical and film properties. Further, the properties of these samples had been in comparison with the commercial sample, which shows good to excellent results.

29.1 INTRODUCTION

Paints, surfactants, inks, germicides, etc., industries depend mainly on the petroleum-based solvents, for example, acrylic emulsions in paints. The excessive use of the solvents creates a global pressure on extraction and prices of petroleum products. It also causes several environmental problems. Recent study shows that these solvents can be partially or fully replaced by water, for example, in water thinnable paints. The water replacement capability in alkyd-based paints enhances if latter is synthesized from natural products and plant extracts [1–4].

In the present work, first we have synthesis alkyd resin from castor oil and rosin as the raw materials. Resin is prepared using less than 15% of petroleum ingredients and 80–90% renewable plant products like castor oil, rosin, etc. Further, suitable resin has been selected for the preparation of paint. Extensively use of natural raw materials for the preparation of useful products is more feasible in India as these raw materials abundantly available and they are inexpensive. It is found these materials are suitable substitute for vinyl and other petroleum products [2]. In the following subsection, we will give a brief account of natural raw materials, castor oil, and rosin.

29.1.1 CASTOR OIL

Castor oil [5, 6] is a triglyceride of fatty acids, contains 87–90% ricinoleic acid, cis-12 hydroxyoctadec-9-enoic acid ($C_{18}H_{34}O_3$), derived from the beans of the castor plant, Ricinus Communis L. of the family Euphorbiaceae. The chemical structure of typical castor oil is shown in Figure 29.1.

It is found widely in tropical and semi-tropical countries, either wild or cultivated. It is non-edible oil and has enormous applications in industries. It is well documented that the reaction of castor oil with maleic anhydride and monobasic alcohol gives a configuration suggesting a new type of alkyd with alcohol coupled to the acid chain through maleic acid. This resin composition acts as excellent softeners in lacquer varnishes [7].

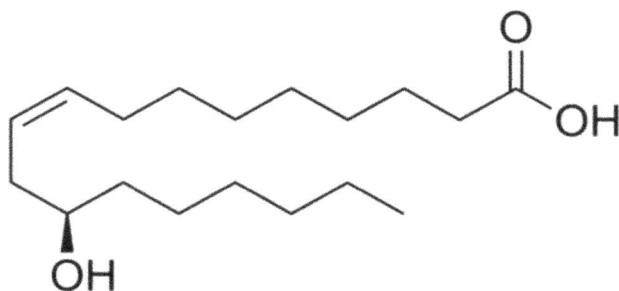

.**FIGURE 29.1** Chemical structure of castor oil.

29.1.2 ROSIN

Rosin [6] is the residue obtained from the distillation of pine exudate. Rosin acids are monocarboxylic having molecular formula $C_{20}H_{30}O_2$. It contains 90% rosin acids and the remaining are esters, aldehyde, and alcohol. It is monobasic acid it controls the polymerization reaction as a chain stopper. It has a natural tendency of making excellent emulsion for development of water-based compositions. It is also abundantly available and widely used in a large number of industrial products like paints, detergents, cosmetics, and pharmaceuticals. Alkyds obtained from rosin have good water and environmental resistance. Rosin and its derivatives [9] have some inherent properties such as excellent film-forming property, ability to form moisture-protective films, resistance to acids, excellent binding ability, non-conductor of electricity, etc. The chemical structures of typical rosin are shown in Figure 29.2.

An anti-tacking agent is a necessary component in a coating system to prevent tackiness of the dosage forms during the manufacturing process. Adherence to the substrate ensures the coating will be durable, long-lasting, and effective. Anti-tack coatings are applied externally to reduce the adhesive condition (stickiness) of a given material, fast drying coatings. [10]. The paints and primers based on these special alkyds have excellent resistance to water, xylene, and mineral turpentine.

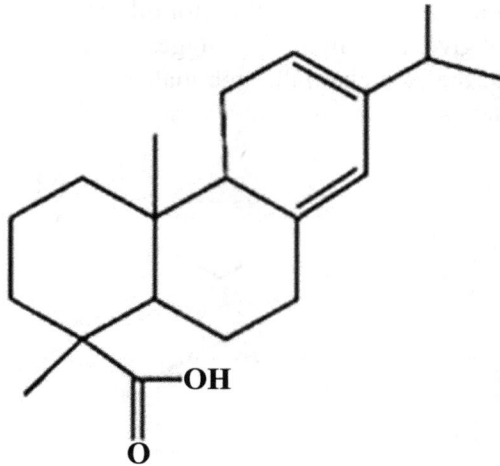

FIGURE 29.2   Chemical structure of rosin.

29.1.3   ALKYD RESIN

Alkyd resin [11] is polyester of polyhydroxyl alcohols and polycarboxyl acids chemically combined with the acids of various drying, semi-drying, non-drying oils in different proportions. All this alkyd possesses most of the desirable properties which are required for protective coatings. They are ideal riders for pigmented coatings because they have good wetting and dispersing properties. The coatings prepared from alkyd resins are comparatively low in cost, have excellent properties such as durability, flexibility, toughness, gloss retention and color retention. They also act as good solvents and exhibits remarkable heat resistance property.

29.1.4   WATER THINNABLE COATINGS

Although, the coating thinned with organic solvents gives desirable results but latter posed some serious problems, namely:

- Solvents are volatile, may easily catches fire;
- They are toxic, lead atmospheric pollution;
- They are expensive;
- They do not form part of the final film.

These problems are overcome when paints are thinned by water [12, 13].

In the context of the growing emphasis on preservation of petrochemical substances, water-soluble binders play an important role in the surface coating industry [14]. Current research is mainly focused on the synthesis and characterization of alkyd resins (polymer) using various types of polybasic acids, polyhydric alcohols, and fatty acids and their application in water thinnable paints.

29.2 EXPERIMENTAL STUDIES

29.2.1 SYNTHESIS OF ALKYD RESINS

Synthesis of alkyd resin based on compositions like castor oil, rosin, glycerol, pentaerythritol, sorbitol, maleic anhydride, phthalic anhydride, and benzoic acid were carried out in a glass reactor in this research work. Four different compositions (see Table 29.1) were prepared by batch process.

TABLE 29.1   Composition of Alkyd Resins based on Castor Oil (Composition % by Weight)

Ingredients	A1	A2	A3	A4
Castor oil	39.44	39.04	35.45	35.89
Rosin	39.36	38.98	40.55	37.62
Glycerol	6.60	6.53	1.0	6.0
Pentaerythritol	4.67	4.62	–	4.25
Sorbitol	–	–	16.01	–
Maleic anhydride	3.94	3.90	3.26	5.40
phthalic anhydride	5.99	5.93	3.73	9.84
Benzoic acid	–	1.00	1.00	1.00
Sodium bisulfate	0.5	0.5	0.5	1.5
Sodium bisulfite	0.1	0.1	0.1	0.5
% Yield	95.5	96.3	93.6	95.0

Various steps involved in the synthesis of alkyd resin are discussed in subsections.

29.2.1.1 FIRST STEP

The castor oil along with rosin, maleic anhydride, glycerol, and catalyst were heated nearly about 2 hours at 160°C to 275°C. In this step dehydration of castor oil (DCO) takes place. Herein we have used some portion of maleic anhydride (approximately 50% of actual requirement) and the remaining was used in later the stage.

29.2.1.2   SECOND STEP

Pentaerythritol was added at slightly higher temperature 200°C to 260°C and then the mixture was heated for 4 to 5 hours. This stage leads OH groups of pentaerythritol to reacts with acidic groups of rosin and maleic anhydride. This process permits bodying and polymerization of oil.

Note: If sorbitol is used as an ingredient, then this is added at a lower temperature of 100–110°C and then the heating is continued for 4 to 5 hours.

29.2.1.3   THIRD STEP

The remaining maleic anhydride, phthalic anhydride, benzoic acid, and 5% solvent was added, and the reaction was continued for 4 to 6 hours so that acids can react with the OH group of pentaerythritol or glycerol. This step was carried out to get polymerization reaction in a controlled manner. The presence of maleic adducts and rosin esters are expected to make the reaction product more viscous and give exceptional durability characteristics. The main chemical reaction involved in the formation of alkyd resin is the polymerization of monoglyceride of castor oil with phthalic and maleic anhydrides, as shown in Figure 29.3.

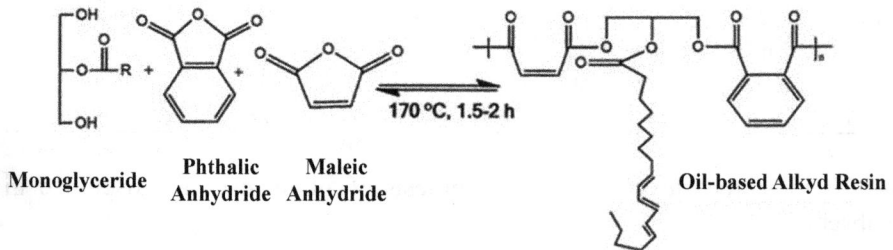

| Monoglyceride | Phthalic Anhydride | Maleic Anhydride | | Oil-based Alkyd Resin |

FIGURE 29.3   Chemical reaction of alkyd resin formation.

The extent of the reaction was determined by checking the acid value and viscosity periodically. After attaining the desired acid value, the sample was cooled to 80°C. At this temperature, the sample was thinned with solvent, filtered, and stored in airtight container. The result of the analysis was tabulated in Table 29.2. The requisite parameters of alkyd resin such as acid value [15], iodine value [16], peroxide value [17], and hydroxide value [18] were obtained by physicochemical analysis.

29.2.2   PREPARATION OF WATER THINNABLE PAINTS FROM ALKYD RESIN

The water thinnable paint was prepared in a laboratory pot mill of 10 dm^3 capacity. At first the pot mill was charged with pigment (TiO_2, ZnO), extender (polyvinyl alcohol), a part of binder (alkyd resin), solvent (xylene, turpentine, and butanol) and water. A thick paste of mixture (i.e., more viscous) was maintained in the ball mill. The 50% volume of the ball mill was occupied by steel balls of 1.27 cm (1/2") diameter. The ball mill was run for 8 hours then the paint was tested for fineness of grind. The expected thickness should be 1–2 micrometers for final application. The grinding was continued to get desirable thickness. The sample was removed from the mill. Additional alkyd resin was added, and the viscosity was adjusted by solvent and water. This sample was filtered through wire mesh and stored in airtight containers.

The prepared paints were studied for physicochemical properties like viscosity [19], density [20], volatile content [21], and solid content [22]. The film properties, hiding power [23], drying time [24], finish (gloss) [25], scratch hardness [26], adhesion [27] and rub resistance test [28] were identified. Further, resistive tendencies of the paint films to water, seawater, chemicals, detergent [29, 30], and various solvents [8] have been tested.

29.3   RESULTS AND DISCUSSION

29.3.1   ALKYD RESINS

- The prepared alkyd resin contains 35–39% of castor oil and 37–40% of rosin.
- In conventional short oil alkyd, a high percentage of anhydrides are used (40–50%). These anhydrides are mainly based on petroleum-based products. In our product only 6–15% of phthalic anhydride has been used. Hence, we have successfully reduced the petroleum based phthalic anhydride by 75% and substituted it with renewable rosin without affecting the performance characteristics.

- Dehydration of castor oil (DCO) with a very high quantity of anhydride is risky and sometimes goes out of control. Here we have easily controlled the reaction by using rosin, a smaller quantity of anhydride and chain stopper benzoic acids. It was observed that benzoic acid regulates the polymerization reaction in an excellent manner.
- In the final best formulation A4, the quantity of polyol was 10% (including pentaerythritol was only 4% and glycerol was 6%), which is less than conventional resins (15–20%). Moreover, pentaerythritol is very expensive. Therefore, the use of the smaller quantity of pentaerythritol is a cost-saving.
- The catalysts, sodium bisulfate (1.5%) and sodium bisulfate (0.5%) gave better performance characteristics. The proportion of catalysts has been fixed by repeated experimentation and analysis of products. With these proportions of catalysts, the yield of resin is 94 to 95%.
- In commercial practices, 16–20 hours are required for the synthesis of a short oil alkyd. In our case, the reaction time was 12–13 hours.
- IR spectra of alkyd resin indicate the presence of carboxylic group, OH group, aliphatic group, and aromatic ring in the structure of polymer. UV spectrum of alkyd resin shows the presence of conjugated double bond, aromatic or heteroaromatic system. NMR predicts the presence of aromatic protons, aliphatic protons, and hydroxyl proton.

29.3.2 ANALYSIS OF ALKYD RESINS

The analysis of alkyd resin is shown in Table 29.2. It is observed that the prepared resin with a lower acid number is not very suitable for converting them into water thinnable coatings. For getting water thinnable compositions, the acid value should be slightly higher than 30. Therefore, we have designed our formulation in such a way that we get the acid value around 30 in the final product (A4). The higher acid value products are soluble in water by reaction with ammonia.

Thus, in our formulation, the acid values are on the higher side 10 to 38. The iodine value of the samples indicates DCO. The peroxide value is very low, indicating that all available peroxide groups have been utilized in oxidative polymerization reaction. The higher hydroxide value (77 to 120) ensures excellent adhesive nature of the samples.

29.3.3   WATER THINNABLE PAINTS

Several formulations (compositions) of water thinnable paints based on alkyd resin (A4) have been studied (Table 29.3). Few remarkable properties of the water thinnable paints are listed below:

- Paints based on a combination of alkyd resin and polyvinyl alcohol has been formulated successfully. The resin content in paints is in the range of 12–20%.
- The useful pigment binder ratio for paints is 1.25 to 3.12. Thus, prepared alkyd resin can be used over a range of pigment binder ratios.
- The incorporation of PVA about 3% improves water intake. We prepared successfully paints with water intake up to 50% and binder 8–10%.
- Addition of water up to 10 to 35% reduces the cost of paint without affecting the performance characteristics.
- The solvent content in paint varies from 25–50% while water content 10 to 50%. Thus, the organic solvent reduced approximately 50%.
- The fire hazard and the toxicity of the paint has also been reduced due to the water intake.
- The water thinnable paints have excellent hiding power, hardness, adhesion, stability, and resistance to water seawater and chemical (see Tables 29.4 and 29.5).
- The main features of these formulations are their excellent hardness and film properties which are obtained even at lower resin content of 10 to 25%. Normally, 35 to 50% binder required for getting useful technical properties. Thus, in our composition we saved 10 to 15% binder.
- The stability of paints is excellent. Generally, water thinnable coatings unstable. However, we have obtained stable products with stability more than six month.
- We have developed a generation of paint which is lower in cost, excellent in technical performance compared to commercial sample and thinnable with water as solvent.
- The following E2, E3, and E4 formulations were compared with commercial samples in all physicochemical and film properties and can be adjudged as best compositions from technical performance viewpoint.
- All prepared paint formulations mentioned in tables are showing fast surface dry and hard dry without addition of driers. In commercial preparation mostly driers are added for achieving this purpose/property.

TABLE 29.3 Composition of Water Thinnable Paints based on Alkyd Resin (A4) (Composition % by Weight)

Component	E1	E2	E3	E4	E5	E6
TiO_2	28.17	21.98	20.0	18.39	11.41	7.33
ZnO	2.58	2.02	1.83	1.67	1.05	0.66
Alkyd (A4)	19.55	15.25	13.87	12.76	8.0	5.08
Polyvinyl alcohol (PVA)	–	1.21	1.65	2.02	3.5	2.2
Ammonia	–	0.55	0.50	0.46	0.3	0.2
Xylene	35.28	30.77	30.32	30.06	28.05	24.57
Mineral turpentine	14.11	12.31	12.13	12.02	11.22	9.83
Butanol	1.01	0.87	0.86	0.85	0.80	0.7
Water	–	15.67	18.18	21.73	35.67	50.00
Pigment binder ratio	1.573	1.574	1.574	1.572	1.557	1.573

TABLE 29.4 Physicochemical and Film Properties of Water Thinnable Paints based on Alkyd Resin (A4)

Analysis	E1	E2	E3	E4	E5	E6	Asian Paint	Nerolac Paint
Viscosity	60	60	60	60	60	60	130	160
Density (G/cc)	1.18	1.15	1.13	1.11	1.01	1.006	1.112	0.928
Hiding Power (m^2/L)	18.80	17.04	16.26	14.8	11.23	10.01	14.43	14.28
Percentage of Solids	50%	45.08%	39.17%	37.40%	26.13%	18.29%	59.02%	48.03%
Surface Dry (in min)	5	5	5	5	10	10	180	120
Hard Dry (in min)	10	15	18	20	35	55	480	360
Scratch Hardness (gm)	>1000	>1000	>1000	>1000	>1000	900	>1000	>1000
Finish	Matt	Matt	Matt	Matt	Matt	Matt	Gloss	Gloss
Adhesion Test	Excellent	Excellent	Excellent	Excellent	Excellent	Excellent	Excellent	Excellent
Stability Test (90 Days)	Excellent	Excellent	Excellent	Excellent	Excellent	Excellent	Excellent	Excellent

TABLE 29.5 Film Resistance Properties of Water Thinnable Paints based on Alkyd Resin (A4)

Analysis		E1	E2	E3	E4	E5	E6	Asian Paint	Nerolac Paint
Water	15 Days	Excellent	Excellent	Excellent	Excellent	Excellent	Excellent	Excellent	Excellent
	1 Month	Excellent	Excellent	Excellent	Good	Good	Good	Excellent	Good
Sea Water	15 Days	Excellent	Excellent	Good	Good	Good	Good	Excellent	Poor
	1 Month	Good	Good	Good	Good	Good	Poor	Good	Poor
3% NaOH (30 min)		Excellent	Excellent	Excellent	Excellent	Excellent	Excellent	Excellent	Excellent
3% H_2SO_4 (48 hrs)		Excellent	Excellent	Excellent	Good	Good	Good	Excellent	Excellent
3% Detergent (30 min)		Excellent	Excellent	Excellent	Excellent	Good	Good	Excellent	Excellent
Xylene (15 min)		Good	Good	Good	Good	Good	Good	Excellent	Excellent

29.4 CONCLUSION

In the present chapter, we have demonstrated the experiment of how the petroleum ingredients can be replaced by natural resources such as castor oil and rosin in synthesis of resins (polymers) and maximum amount of organic solvent by water in the preparation of water thinnable paints. In the present investigation, we establish that the products developed by our techniques exhibit satisfactory performance. In order to obtain the products at par with commercial ones with zero environmental harm, further exploration has been needed.

ACKNOWLEDGMENTS

The author is also highly indebted to Dr. B. B. Gogte, Ex-Professor, LIT Nagpur, for his valuable guidance and suggestion. The author is also gratefully acknowledged the financial support by M S Jotun A/S Norway Government for this research work.

KEYWORDS

- **alkyd resin**
- **castor oil**
- **global pressure**
- **rosin**
- **volatile organic compounds**
- **water thinnable paints**

REFERENCES

1. Burande, B., (2017). Antifouling paints based on plant extracts. *IJRAET, 5*(9, 10), 12–16.
2. Burande, B., Dhakite, P. A., & Gogte, B. B., (2017). Development of green ecofriendly products based on natural vegetation. *IJAIR, 4*(1), 132–134.
3. Phate, B., & Gogte, B. B., (2005). *Wall Finishes Based on Novel Short Oil Rosinated Alkyd* (pp. 71–76). Paint India.
4. Dhakite, P. A., Burande, B. C., & Gogte, B. B., (2014). Ecofriendly malenized oils for liquid detergent. *IJRBAT, 1*(2), 239–249.
5. Joshi, A., (1991). *Castor Oil: King of Indian Exports, 41*(3), 41–46. Paint India.
6. Kirk-Othmer, (1979). *Encyclopedia of Chemical Technology* (3rd edn., pp. 1–12). John Wiley and Sons: Inc. New York.
7. Solomon, D. H. (1967). *The Chemistry of Organic Film Formers*, John Wiley & Sons, New York.
8. *Indian Standard Method*, (1980). 6.01 D 2792-69, for solvent and fuel resistance of paints.
9. Sudhra, S., Fould, I. S., Gray, C. N., Koh, D., & Gardiner, K., (1994). *Ann. Occup. Hyg., 38*, 385.
10. Gogte, B. B., Agrawal, A. Paint India, 44(10), 51.
11. Payne, H. F., (1954). *Organic Coatings Technology* (Vol. 1, pp. 279–280). John Wiley and Sons: Inc. New York.
12. Taylor, C. J., & Marks, S., (1972). *Paint Technology Manual-Part III Convertible Coatings* (pp. 92–94). Chapman and Hall Publications: London.
13. http.//webumr.edu/jstoffer/CHEM 381/Chap.33.html (accessed on 30 June 2021).
14. Gooch, J. W. (2002). *Emulsification and Polymerization of Alkyd Resins*; Kluwer Academic: New York.
15. *ASTM Standard Method*, (1981). 6.01, D 1639-70, for acid value of organic coating materials. American Society for Testing Materials, Philadelphia.
16. *ASTM Standard Method*, (1979). 6.03, D 1959-69, for iodine value of drying oils, varnishes, resins.
17. AOCS, (1946). *Official and Tentative Methods of the American Oil Chemist's Society* (2nd edn.). Cd 8–53, for peroxide value of fatty oils and acids.
18. *ASTM Standard Method*, (1979). 6.03, D. 1957-63, for hydroxyl value of fatty oils and acids.

19. *ASTM Standard Method*, (1982). 6.01, D 1200-82, for viscosity of paint, varnishes and lacquers by ford cup method.
20. *ASTM Standard Method*, (1982). 6.03, D 1963-74, for specific gravity of drying oils, varnishes, resins.
21. *ASTM Standard Method*, (1981). 6.01, D 2369-81, for volatile content of paints.
22. *ASTM Standard Method*, (1980). 6.02, D 1259-61, for non-volatile content of resin.
23. *ASTM Standard Method*, (1980). 6.01, D 280580, for hiding power of paint.
24. *ASTM Standard Method*, (1974). 6.01, D, 1640-69, for drying or film formation of organic coating at room temperature.
25. *ASTM Standard Method*, (1980). 6.01, D. 523-80, (for determination of gloss.
26. Morgan, W. M., (1996). *Outlines of Paint Technology* (3rd edn., p. 438). CBSW Publishers and Distributors: New Delhi.
27. Taylor and Marks, (1965). *Paint Technology Manuals: The Testing of Paints* (p. 80). B.S. 391, Performance test for protective Coatings.
28. *Indian Standard Specification,* (1969). I.S. -428, for dry rub resistance test.
29. *Indian Standard Specifications,* (1964). I.S. 101, for water resistance.
30. *ASTM Standard Method*, (1981). 6.01, D1647-70, (for resistance of water and chemicals of dried films of varnish and paint.

CHAPTER 30

Biosorption Studies of Indigo Carmine Dye from Its Aqueous Solution Using Agro-Waste

PRANAY RAUT,[1,2] DHRUTI. S. PATTANAYAK,[1] PRAVIN D. PATIL,[3] DHARM PAL,[1] and V. K. SINGH[1]

[1]Department of Chemical Engineering, NIT Raipur, Raipur–492010, Chhattisgarh, India

[2]Department of Biotechnology, Priyadarshini College of Engineering, Nagpur–19, Maharashtra, India, E-mail: pranay.raut@gmail.com

[3]Department of Basic Science and Humanities, NMIMS Mukesh Patel School of Technology Management and Engineering, Mumbai–400056, Maharashtra, India

ABSTRACT

The pigeon pea (*Cajanus Cajan*) husk CCH fine particles were utilized as a biosorbent to eradicate Indigo Carmine (IC) dye out of its solution in aqueous media. FT Infrared Spectroscopy was tested to depict the features of biosorbent CCH. The effect of diverse factors like original dye concentration, a dose of biosorbent, time of contact, pH, and temperature on eradication of dye was examined. Reaction kinetic data and equilibrium data of biosorption were represented well by kinetic model for Pseudo second-order reaction and isotherm model of Langmuir for biosorption, respectively. Thermodynamic criterion like variation in Gibb's free energy (ΔG), variation in enthalpy (ΔH), and variation in entropy (ΔS) were also studied, and their values were calculated as 25.976 KJ/mol., 24293.5 J/mol.K as well as 85.7506 KJ/mol, consequently. The above findings showed that pigeon pea (*cajanus Cajan*) husk powder is a viable and economical biosorbent that may be utilized to get rid of dye IC from its aqueous solution.

30.1　INTRODUCTION

Industries like textile, printing, paint use dyes, which are then eventually released into waste. When released in water, dyes from the waste affect human health and Environment even at a low level [1]. Indigo Carmine (IC) is derived from Indigo Disulfonate blue dye by sulfonation, which gives water-soluble compound [2]. It is used as an analytical test indicator, coloring agents in the food industry like confectionery, beverages, and cosmetics additive in Pharmaceutical tablet manufacturing [10]. IC is an anionic dye (acid dye) having a molar mass of 466.36 gm per mole and has a topmost frequency of 609 nm [1, 18].

A lot of Chemical methods like Electrochemical oxidation, advanced oxidation process [3], and ozonation [4], some biological methods like using microorganisms [5] have also been used to remove dye from wastewater. High investment and operating costs are among the major disadvantages that the above methods offer that may differ as per their effectiveness [2]. Simplicity, low cost, and flexible operational parameters are key factors that make adsorption one of the most frequently used processes. Several non-conventional low-cost adsorbents like citrus limetta peel, Zea Mays Cob, Grivellia Robusta leaves, Terminalia Catappa shell, Rice husk, Brazil nutshell, Calcium hydroxide turned out to be increasingly investigated for their IC dye removal capacity from water [1, 6–11]. This work attempt has been made to clear away IC dye from its aqueous solution using Pigeon Pea (*Cajanus cajan*) husk as an economical biosorbent by conducting the batch study. The response of numerous factors like the point of zero charge, time of contact, temperature, dose of biosorbent, and original concentration of dye [19] has also been explored.

30.2　METHODS AND MATERIALS

30.2.1　SAMPLE PREPARATION

Indigo Carmine dye (CI No. 73015, Molecular Weight-466.36 gm/mol) was procured from HiMedia Limited to prepare an aqueous solution using deionized water. Pigeon pea (*Cajanus cajan*) husk (CCH) was acquired from neighborhood farmers and then entirely washed using tap water to eliminate dust and soil. It was subsequently dehydrated in an oven heated by circulating hot air at 90°C for 24 hours and crumbled in a high-speed blender to prepare powder [19]. The Sample was then stored in a closed bottle without any pre-treatment.

30.2.2   CHARACTERIZATION OF ADSORBENT

30.2.2.1   FOURIER TRANSFORM INFRARED SPECTROSCOPY

FTIR spectra of biosorbent CCH afore and later biosorption was obtained by Perkin Elmer Spectrum-BX USA spectrometer. It was used to study functional groups on biosorbent.

30.2.2.2   ZERO CHARGE POINT ON SURFACE

Zero charge on the surface (pH_{zpc}) of biosorbent CCH was measured using standard method of solid addition [9, 12]. Precisely 0.05 liter of 0.1 M KNO_3 solution then shifted in a succession of 250 ml conical Erlenmeyer flasks. 0.1 N NaOH or HCL solution was used to alter the pH of the solution roughly from value 2 to 12. The volume was adjusted to 100 ml using 0.1 M KNO_3. Initial pH was measured by Systronics India 361 pH system model pH meter. An absolute 1 gram of biosorbent CCH was mixed in the solution kept in the individual flask. Flasks were kept shaking for 3 hrs at 150 rpm. After equilibrium reached, the ultimate pH value of the supernatant was tested. A curve representing a change in pH vs. initial pH value helped to calculate the point of zero charge [20].

30.2.2.3   BATCH STUDY

The outcome of parameters like time of contact, original concentration value, amount of biosorbent, and temperature on the separation efficiency of dye IC was reviewed trying batch study. The study was performed in a 250 mL Erlenmeyer conical flask containing 100 ml IC solution at the given initial concentration. A predefined quantity of biosorbent CCH was taken in each flask and agitated at 30°C for different contact times at 150 rpm. Dye from Solution in aqueous media was separated using centrifuge at 4000 rpm for five minutes, the final concentration of clear solution obtained by centrifugation was measured by monitoring its absorbance spectrophotometrically at 610 nm. The outcome of using different amounts of CCH powder for the eradication of IC dye was investigated by taking different amounts of CCH powder (0.5 gm to 2.5 gm) in 100 ml of 20 ppm dye solution. The influence of contact time was also examined by adding 1 gm of CCH powder in 100 ml of IC dye solution kept in a flask at 4.74 pH. The initial IC dye concentration effect was measured by adding 1 gm of CCH powder in 100 ml of IC dye solution of different concentrations (20–100 ppm) at pH 4.72. Similarly, the impact of different temperatures (30°C, 50°C, and 70°C) on the ability of dye expulsion of CCH powder was also investigated.

The amount of adsorption at equilibrium can be measured by Eqn. (1).

$$q_e = \frac{C_0 - C_e)V}{W} \tag{1}$$

where; C_0 and C_e are the original and steady-state concentrations (milligram/liter), respectively; V is the cubic measure of solution in aqueous media in liter; and W is the amount of biosorbent (gram).

Percent elimination of dye color using adsorbent is given from Eqn. (2).

$$\text{Removal } \% = \frac{C_0 - C_e}{C_0} \tag{2}$$

where; C_e is the steady-state concentration (milligram/liter) of dye solution.

30.2.2.4 KINETICS OF BIOSORPTION

The study of kinetics on biosorption of IC dye was performed employing Lagergren's kinetic model of pseudo-first-order and pseudo-second-order to acquiesce the data obtained from experiments. The kinetic model of Pseudo-first-order assumed that the accessibility of free spots on the exterior part of biosorbent has a direct correlation with the biosorption rate; whereas the kinetic model of pseudo-second-order speculate that the whole biosorption process is based on second-order chemisorption [30].

Equation for kinetic model of Pseudo-first-order is endowed as:

$$\log (q_e - q_t) = \log q_e - \left(\frac{K_1}{2.303}\right)t \tag{3}$$

Equation for kinetic model of Pseudo-second-order is endowed as:

$$\frac{t}{qt} = \frac{1}{k_2 q^2_e} + \left(\frac{1}{q_e}\right)t \tag{4}$$

where; K_1 is taken as the constant value of rate of pseudo-first-order adsorption (minute inverse); and K_2 is considered as the constant value of rate of pseudo-second-order adsorption (gm/mg. min). The values of q_e and q_t are the uptake of adsorbate on biosorbent (mg/gm) at steady-state and at any time t (minute), consequently [12, 22, 30].

30.2.2.5   ADSORPTION ISOTHERM

Various types of adsorption isotherms were developed using theoretical and empirical models. This study investigates the interrelation of the concentration with uptake of dye using Langmuir [23], Freundlich [24], and Temkin [25] isotherms given by Eqns. (5), (6), and (7), respectively.

$$\frac{C_e}{q_e} = \frac{1}{K_L q_m} + \frac{1}{q_m} C_e \tag{5}$$

$$\log q_e = \log K_F + \frac{1}{n} \log C_e \tag{6}$$

$$q_e = B \ln A + B \ln C_e \tag{7}$$

where in Eqn. (5); q_e is the quantity of IC dye biosorbed at steady state (mg/gm); C_e is the steady-state concentration of the IC dye solution (milligram/Litre); q_m is the utmost magnitude of biosorption (milligram/gram); and K_L is the Langmuir sorption constant allied with sorption energy.

In Eqn. (6); k_f is the Freundlich isotherm constant showing magnitude of sorption, and $1/n$ is the multiformity parameter indicating favorability of adsorption.

In Eqn. (7); A is the equilibrium binding energy constant (L/gm); and B is the Temkin constant, i.e., Heat of sorption (KJ/mol).

30.2.2.6   THERMODYNAMIC STUDY

Thermodynamic variables like variations in Gibb's Energy (ΔG°), variation in Enthalpy (ΔH°), and variation in Entropy (ΔS°) were computed using the equations mentioned here:

$$\Delta G^\circ = -2.303 RT \ln K_D \tag{8}$$

$$K_D = \frac{C_s}{C_e} \tag{9}$$

$$\Delta G^\circ = \Delta H^\circ - T\Delta S^\circ \tag{10}$$

$$\ln K_D = \frac{AS^\circ}{R} - \frac{\Delta H^\circ}{RT} \tag{11}$$

where; K_D is the distribution coefficient; C_s is the amount of dye in solution at steady-state adsorbed onto biosorbent; and C_e is the concentration of solution of dye at steady-state into its aqueous solution (milligram/liter); R is the ideal gas constant (8.314 Joule/mol. kelvin).

ΔH^0 and ΔS^0 were measured using the graph of the natural logarithm of distribution coefficient versus temperature inverse. The value of ΔH^0 and ΔS^0 was the gradient and interpose of the plot, consequently [18, 26].

30.3 RESULT AND DISCUSSION

30.3.1 FT-IR

Figures 30.1 and 30.2 show FTIR spectrum of CCH powder afore and later biosorption. A prominent and intense peak at 3301.2 cm⁻¹ as well as 3347 cm⁻¹. In Figures 30.1 and 30.2 show the stretching of the –OH bond. While peak at 1614 cm⁻¹ and 1621 cm⁻¹. In Figures 30.1 and 30.2 show the stretching of the double bond between carbon and oxygen (C=O bond). The value of apex at 1028 cm⁻¹ and 1032 cm⁻¹ shows the C-OH widening of alcoholic and carboxylic acid groups [13].

30.3.2 ZERO CHARGE POINT ON THE SURFACE

The zero charge point on the surface is a crucial parameter to assess linear pH sensitivity and adsorptive capabilities [14, 21]. Figure 30.3 reveals that pH_{zpc} of CCH powder was found to be 7.35. It means that CCH powder bears a net positive surface charge at a pH value lower than 7.35, which is ideal for removing IC dye, an anionic dye, from its solution in aqueous media [1].

30.3.3 EFFECT OF ADSORBENT DOSE

Figure 30.4 shows that IC dye removal has increased to a specific limit of the amount, i.e., 1 gm. Thereafter, no effective removal takes place due to CCH powder's conglomeration. It means that there is no remarkable increment in functional surface area (SA) and adsorption sites in adsorbent [12].

30.3.4 IMPACT OF ORIGINAL CONCENTRATION

Figure 30.5 depicts that the expulsion efficiency of CCH decreases from 95.4% to 16.08% with a rise in concentration from 20 mg/lit to 100 mg/lit, which may arise as a result of the deposition of dye particles at large concentrations values.

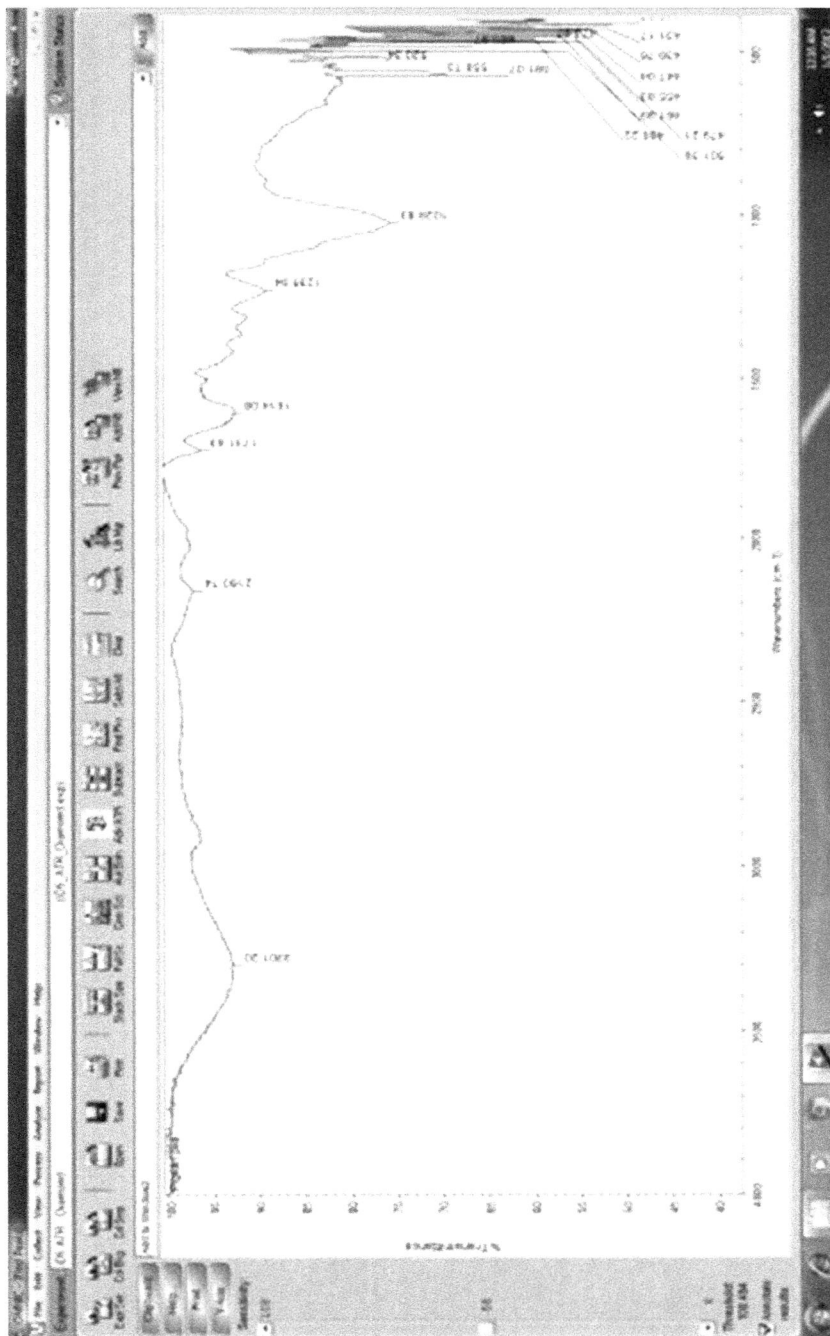

FIGURE 30.1   FTIR spectrum before adsorption.

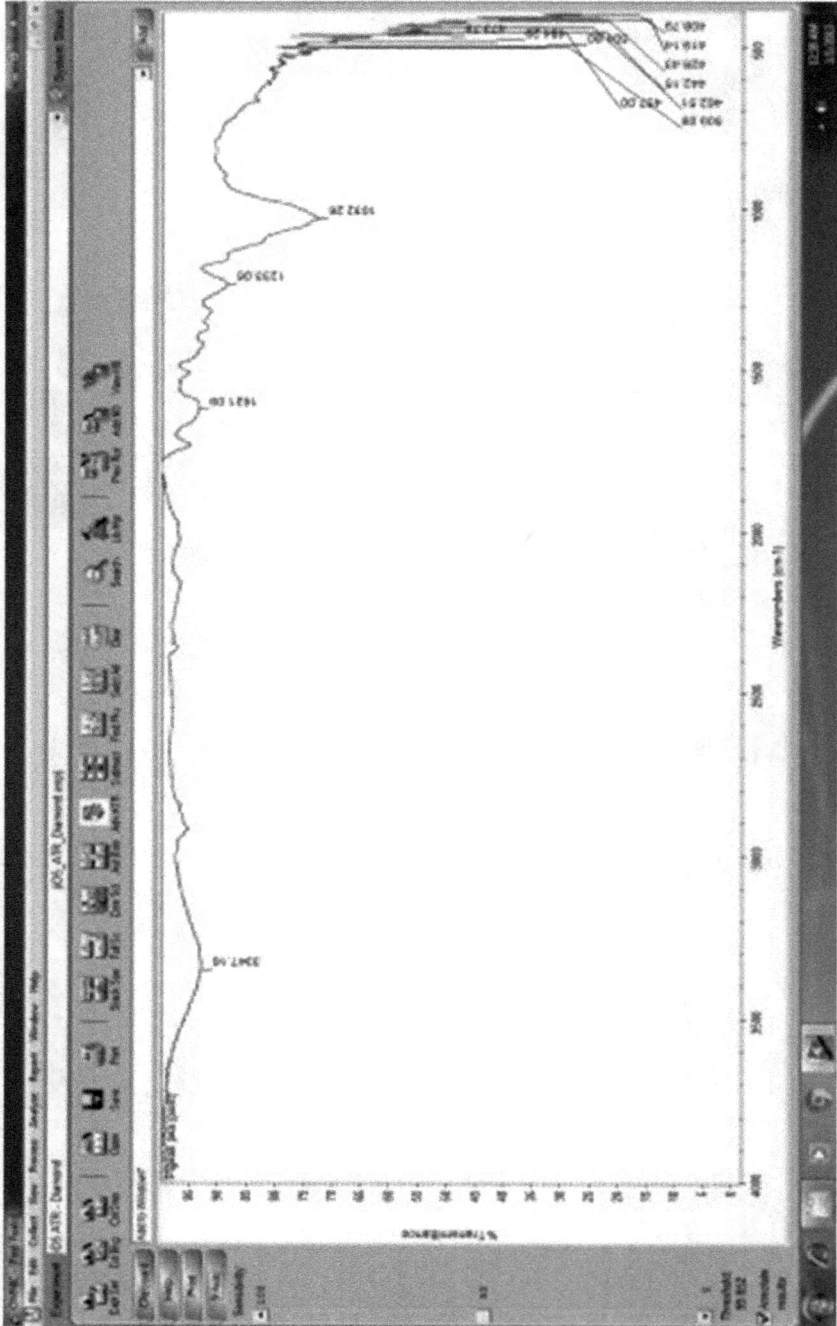

FIGURE 30.2   FTIR spectrum of CCH after adsorption.

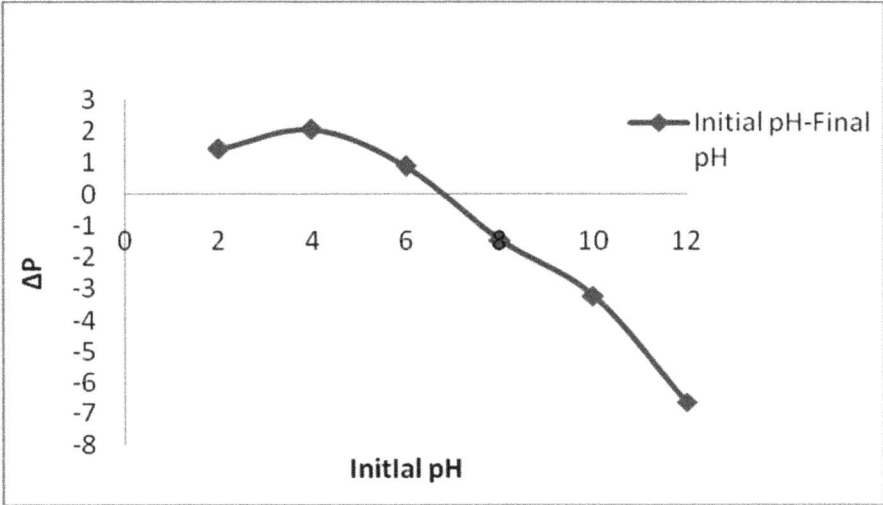

FIGURE 30.3   Point of zero charges of pigeon pea husk.

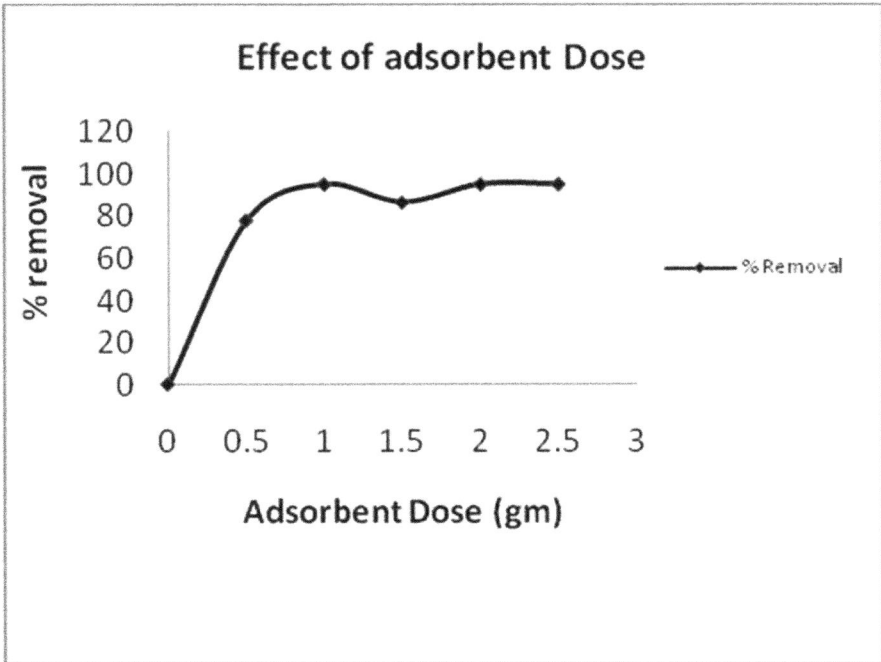

FIGURE 30.4   Impact of dose of adsorbent.

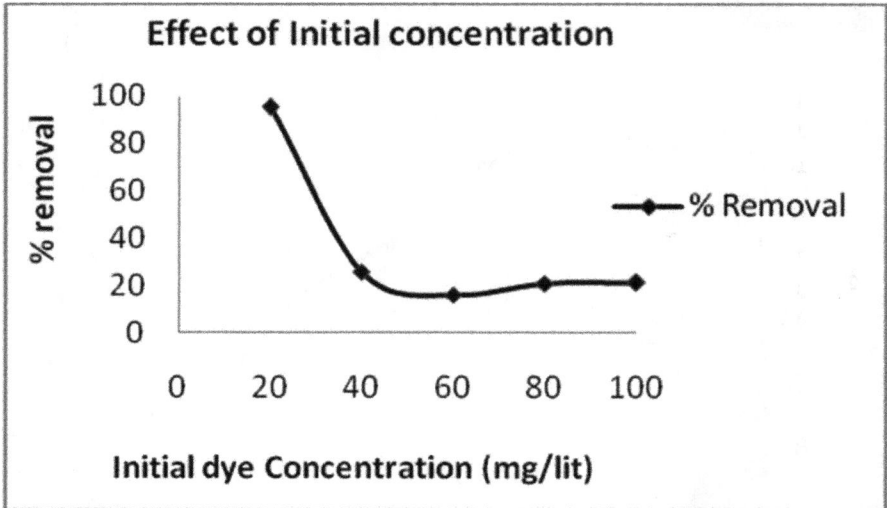

FIGURE 30.5 Impact of original concentration.

30.3.5 IMPACT OF CONTACT TIME

Figure 30.6 shows that nearly 95% of Adsorption efficiency as the time of contact altered from 20 min to 120 min. It becomes steady after 120 minutes for 20 ppm dye concentration for 1 gm of pigeon pea husk.

30.3.6 EFFECT OF TEMPERATURE

Figure 30.7 shows that an increase in temperature values are 30°C, 50°C and 70°C has no significant effect on dye removal. It slightly increases from 53% to 60%, which indicates an endothermic process.

30.3.7 ADSORPTION ISOTHERM

30.3.7.1 FREUNDLICH ISOTHERM

From Freundlich isotherms, correlation coefficient value, i.e., R^2 was calculated as 0.941 (Figure 30.8).

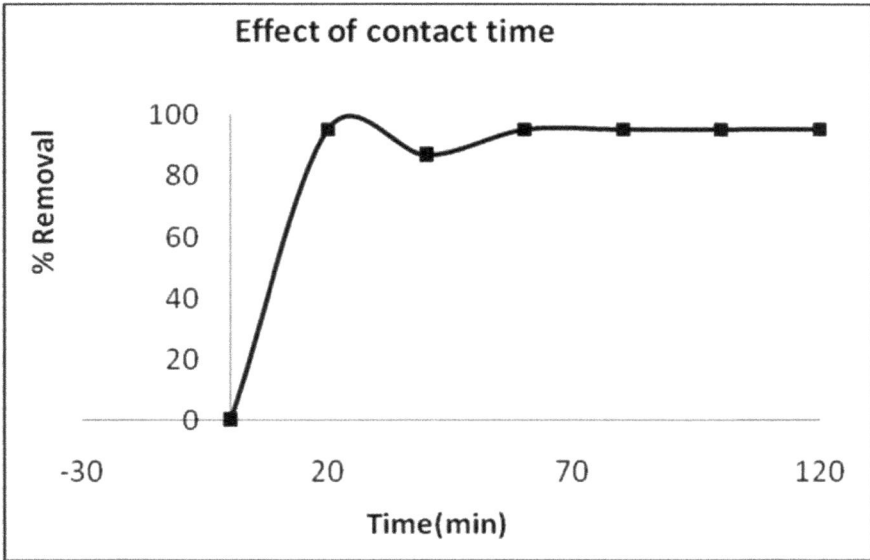

FIGURE 30.6   Impact of contact time.

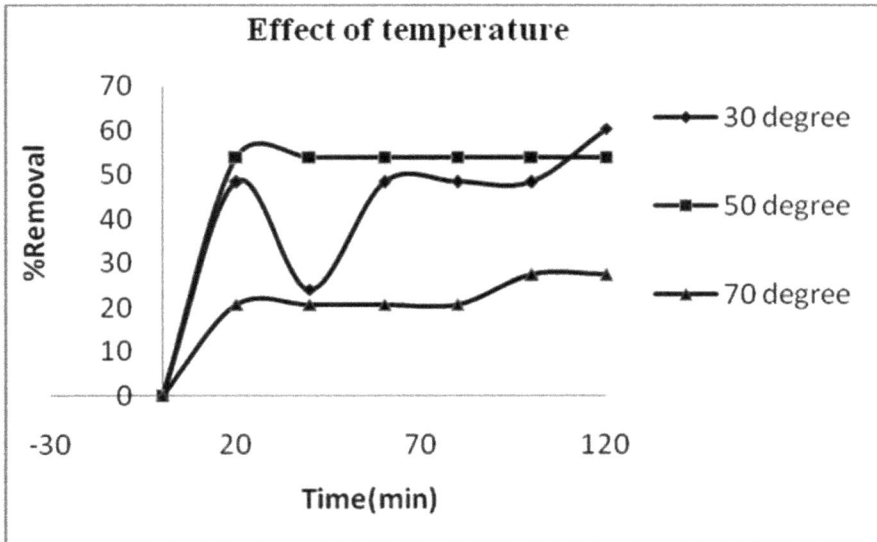

FIGURE 30.7   Effect of temperature.

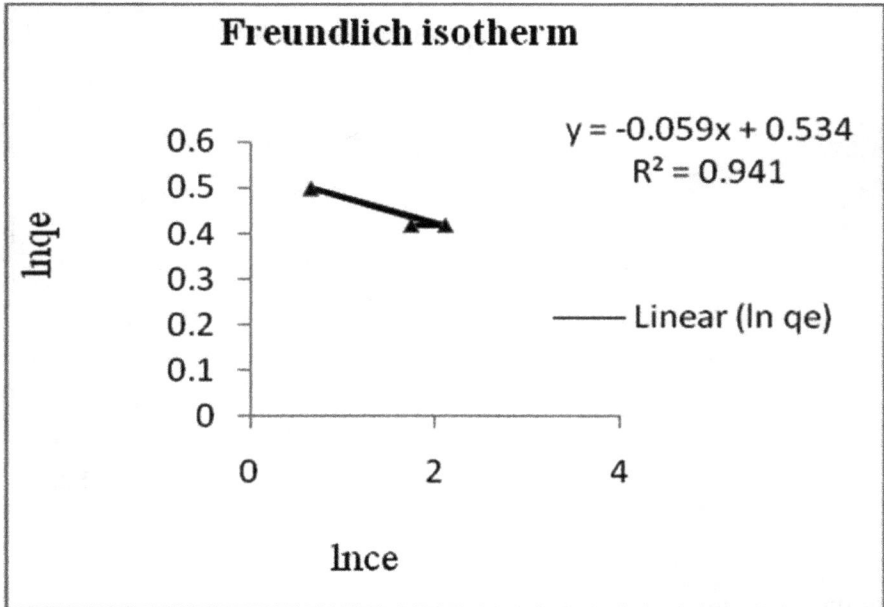

FIGURE 30.8   Freundlich isotherm.

30.3.7.2   *LANGMUIR ISOTHERM*

Figure 30.9 shows that for Langmuir isotherm correlation coefficient value of R^2 was calculated as 0.999, which indicates it best-suited isotherm.

30.3.7.3   *TEMKIN ISOTHERM*

Temkin isotherm indicates a decline in all molecules' heat of adsorption occurs linearly, showing homogenous binding energy over adsorption sites [22]. The correlation coefficient was found to be 0.941 (Figure 30.10).

30.3.8   *KINETIC STUDY*

30.3.8.1   *MODEL FOR KINETICS OF PSEUDO FIRST-ORDER REACTION*

Model for kinetics of Pseudo-first-order biosorption reaction analysis showed very low observed correlation coefficient (R^2) values (0.375–0.612).

It highlights that the biosorption of IC dye onto CCH powder does not obey the kinetic model for pseudo-first-order (Figure 30.11).

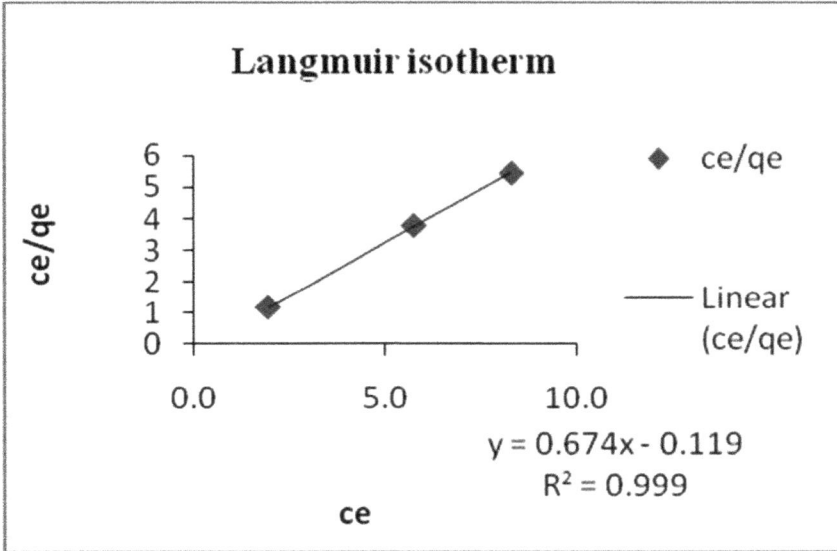

FIGURE 30.9 Langmuir isotherm.

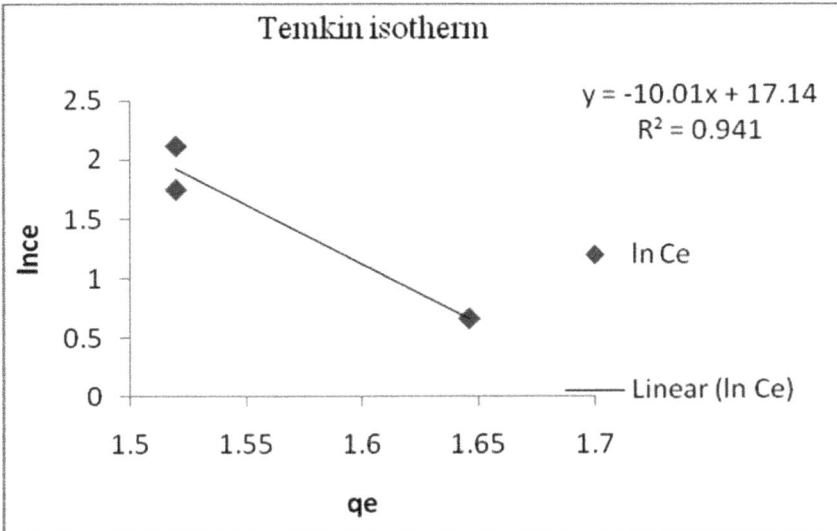

FIGURE 30.10 Temkin isotherm.

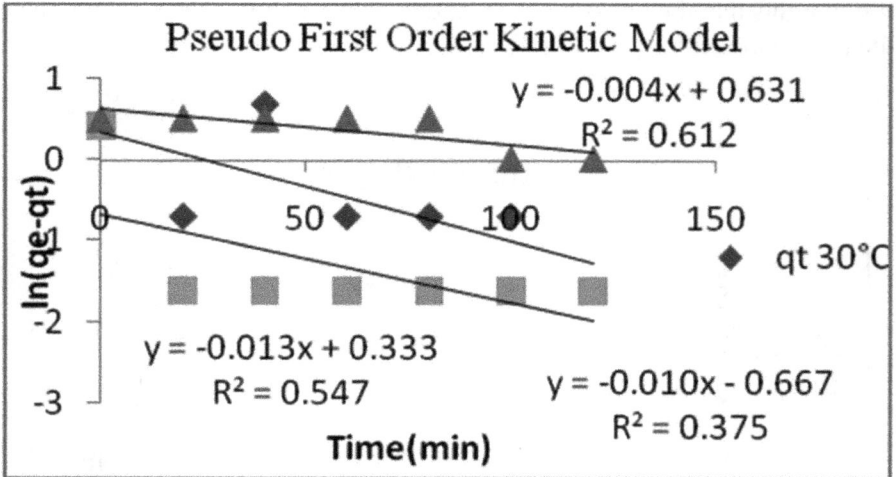

FIGURE 30.11   Pseudo-first-order kinetic model.

30.3.8.2   KINETIC MODEL FOR PSEUDO SECOND-ORDER REACTION

The higher equivalence coefficient (R^2) values, i.e., 0.809, 0.942, and 1 at a temperature of 30°C, 50°C, 70°C imply that the kinetic model for pseudo-second-order is very well be fitted for the biosorption of IC dye upon CCH powder (Figure 30.12).

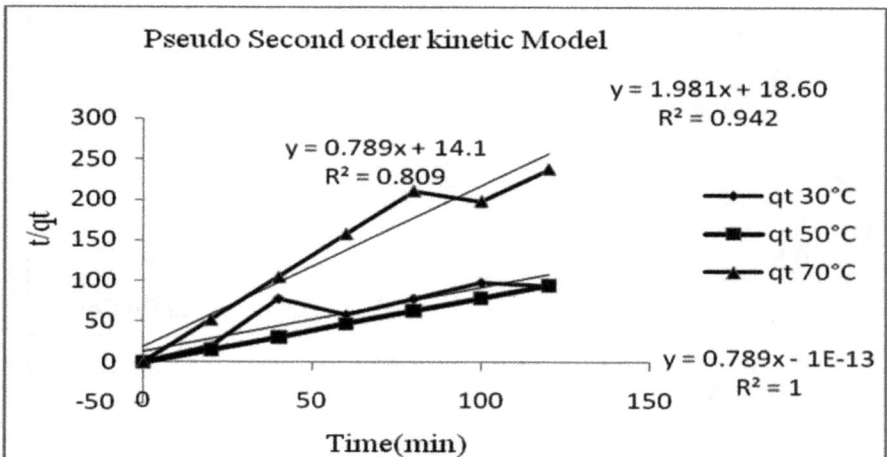

FIGURE 30.12   Pseudo-second-order kinetic model.

30.3.9 STUDY OF THERMODYNAMICS

The temperature dependency of the equilibrium constant has been evaluated using an analysis of thermodynamic parameters. The positive value of Entropy change (ΔS) revealed that the elevated impermanence of IC dye molecules at the particle-solution assemblage. It signifies an improvement in the adsorbed species' degree of freedom favoring IC dye adsorption [15–17, 28, 29]. The calculated values of variation in Gibb's free Energy (ΔG), variation in Entropy (ΔS), and variation in Enthalpy (ΔH) were 25.976 KJ/mol, 85.7506 KJ/mol, and 24293.5 J/mol.K consequently. A positive value of enthalpy variation showed that the biosorption requires heat to adsorb dye molecules onto the surface of CCH particles (Figure 30.13).

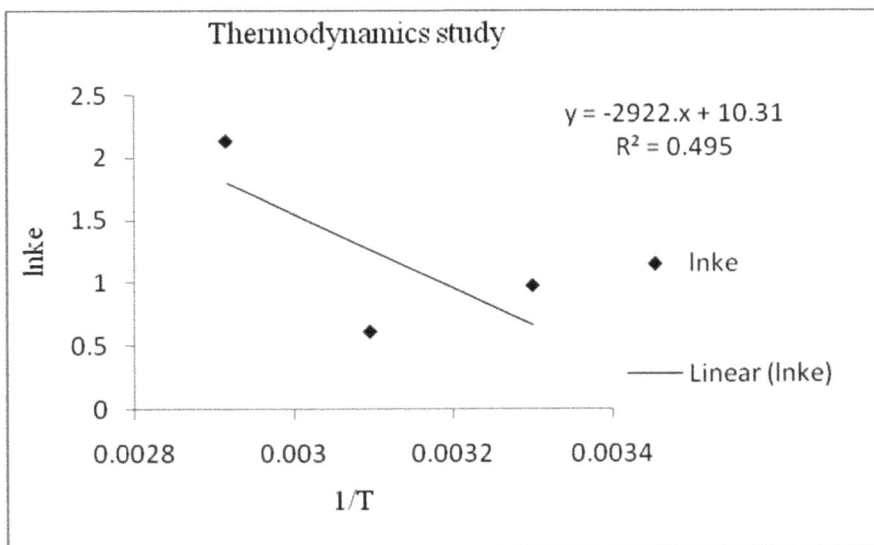

FIGURE 30.13   Thermodynamic study.

30.4   CONCLUSION

This study revealed that *Cajanus cajan* husk (CCH) powder was a potential adsorbent for removing the dye named IC out of its solution in aqueous media. The removal of IC dye was significant at a high amount of biosorbent and consequently lowered with an elevated original concentration of dye. Isotherm data of IC dye on CCH powder showed sorption process followed

Langmuir isotherm (R^2 = 0.999) forming a monolayer on CCH powder's surface. The biosorption process was very well delineated by model of kinetics for pseudo-second-order reaction as per the calculated values of correlation coefficient. The thermodynamic analysis indicated that IC dye's adsorption on CCH powder was feasible, spontaneous, and endothermic.

KEYWORDS

- agro-waste
- biosorbent
- electrochemical oxidation
- isotherm
- kinetic model
- thermodynamic study

REFERENCES

1. Hevira, L., Zilfa, Rahmayeni, Ighalo, J. O., & Zein, R., (2020). Biosorption of indigo carmine from aqueous solution by *Terminalia catappa* shell. *Journal of Environmental Chemical Engineering.* doi-https://doi.org/10.1016/j.jece.2020.104290.
2. Harrache, Z., (2018). Thermodynamic and kinetics studies on adsorption of indigo carmine from aqueous solution by activated carbon. *Microchemical Journal.* doi-doi: 10.1016/j.microc. 2018.09.004.
3. Palma-Goyes, R. E., (2014). Comparative degradation of Indigo Carmine by Electrochemical oxidation and advanced oxidation process. *Electrochimica Acta., 140*, 427–433.
4. Bernal, (2013). Ozonation of indigo carmine catalyzed with Fe-pillared clay. *International Journal of Photoenergy.* doi-http://dx.doi.org/10.1155/2013/918025.
5. Park, S., (2012). Isolation and characterization of alkaliphilic and thermotolerant bacteria that reduce insoluble indigo to soluble leuco-indigo from indigo dye vat. *Journal of Korean Society for Applied Biological Chemistry, 55*(1), 83–88.
6. Singh, H., Chauhan, G., Jain, A. K., & Sharma, S. K., (2016). Adsorptive potential of agricultural wastes for removal of dyes from aqueous solutions. *Journal of Environmental Chemical Engineering.* doi: http://dx.doi.org/10.1016/j.jece.201611.030.
7. Venkata, R. P., Pydiraju, P., Madhuri, V., Vineeth, S., Rahimuddin, S., & Vangalapat, I. M., (2020). Removal of indigo carmine dye from aqueous solution by adsorption on biomass of *Grevillea Robusta* leaves. *Materials Today: Proceedings., 26*(2), 3020–3023.
8. Lakshmi, U. R., Srivastava, V. C., Mall, I. D., & Lataye, D. H., (2009). Rice husk ash as an effective adsorbent: Evaluation of adsorptive characteristics for indigo carmine dye. *Journal of Environmental Management., 90,* 710–720.

9. Britoa, S. M., Andrade, H. M. C., Soares, L. F., & Rafael, P. D. A., (2010). Brazil nut shells as a new biosorbent to remove methylene blue and indigo carmine from aqueous solutions. *Journal of Hazardous Materials, 174*, 84–92.
10. Thimmasandra, N. R., Devarahosahally, V. K., Ashwathaiah, A., & Manasa, T. R., (2015). Calcium hydroxide as low cost adsorbent for the effective removal of indigo carmine dye in water. *Journal of Saudi Chemical Society, 58*(2). doi: 10.1016/j.jscs.2015.03.001.
11. Ho, Y. S., & Mckay, G., (1998). Kinetic models for the sorption of dye from aqueous solution by wood. *Process Safety and Environmental Protection, 76*(2), 183–191. doi-https://doi.org/10.1205/095758298529326.
12. Hameed, B. H., (2009). Removal of cationic dye from aqueous solution using jackfruit peel as non-conventional low-cost adsorbent. *Journal of Hazardous Material, 162*, 344–350.
13. Venkata, R. D. K., Harikishore, K. R. D., & Jae, S. Y. K., (2012). Pigeon peas hulls waste as potential adsorbent for removal of Pb(II) and Ni(II) from water. *Chemical Engineering Journal, 197*, 24–33.
14. Terzyk, A. P., Rychlicki, G., Biniak, S., & Łukaszewicz, J. P., (2003). New correlations between the composition of the surface layer of carbon and its physicochemical properties exposed while paracetamol is adsorbed at different temperatures and pH. *Journal of Colloid and Interface Science, 257*(1), 13–30.
15. Prado, A. G. S., Torres, J. D., Faria, E. A., & Dias, S. C. L., (2004). Comparative adsorption studies of indigo carmine dye on chitin and chitosan. *Journal of Colloid Interface Science, 277*(1), 43–47.
16. Banerjee, S., Sharma, G. C., Gautama, R. K., Chattopadhyaya, M. C., Upadhyay, S. N., & Sharma, Y. C., (2016). Removal of malachite green, a hazardous dye from aqueous solutions using *Avena sativa (*oat) hull as a potential adsorbent. *Journal of Molecular Liquid, 213*, 162–172.
17. Ahmad, M. A., Ahmad, N., & Bello, O. S., (2015). Adsorption kinetic studies for the removal of synthetic dye using durian seed activated carbon. *Journal of Dispersion Science and Technology, 36*, 670–684.
18. Chowdhury, S., & Das, P., (2012). Utilization of a domestic waste-Eggshells for removal of hazardous Malachite Green from aqueous solutions. *Environmental Progress & Sustainable Energy, 31*(3), 415–425.
19. Gong, R., (2008). Kinetics and thermodynamics of basic dye sorption on phosphoric acid esterifying soybean hull with solid-phase preparation technique. *Bioresource Technology, 99*, 4510–4514.
20. Udoetok, I. A., . Dimmick, R. M., Wilson, L. D., & Headley, J. V., (2015). Adsorption properties of cross-linked cellulose-epichlorohydrin polymers in aqueous solution. *Carbohydrate Polymers, 136*, 329–340.
21. Lina, J., Zhanb, Y., Zhub, Z., & Xinga, Y., (2011). Adsorption of tannic acid from aqueous solution onto surfactant-modified zeolite. *Journal of Hazardous Materials, 193*, 102–111.
22. Hameed, B. H., & El-Khaiary, M. I., (2008). Malachite green adsorption by rattan sawdust: Isotherm, kinetic and mechanism modeling. *Journal of Hazardous Materials, 159*, 574–579.
23. Dabrowski, A., (2001). Adsorption—from theory to practice. *Advances in Colloid and Interface Science, 93*(1–3), 135–224.
24. Boparai, H. K., Joseph, M., & O'Carroll, D. M., (2011). Kinetics and thermodynamics of cadmium ion removal by adsorption onto nano zerovalent iron particles. *Journal of Hazardous Materials, 186*(1), 458–465.

25. Okoli, C. A., Onukwuli, O. D., Okey-Onyesolu, C. F., & Okoye, C. C., (2015). Adsorptive removal of dyes from synthetic wastewater using activated carbon from tamarind seed. *European Scientific Journal, 11*, 190–221.
26. Srilakshmi, C., & Saraf, R., (2016). Ag-doped hydroxyapatite as efficient adsorbent for removal of Congo red dye from aqueous solution: Synthesis, kinetic and equilibrium adsorption isotherm analysis. *Microporous and Mesoporous Materials, 219*, 134–144.
27. Kara, A., & Demirbel, E., (2012). Kinetic, isotherm and thermodynamic analysis on adsorption of Cr (VI) ions from aqueous solutions by synthesis and characterization of magnetic-poly (divinylbenzene-vinyl imidazole) microbeads. *Water Air Soil Pollution, 223*(5), 2387–2403.
28. Bhatnagar, A., Kumar, E., Minocha, A. K., Byong-Hun, J., Song, H., & Yong-Chan, S., (2009). Removal of anionic dyes from water using *Citrus Limonum* (Lemon) peel: Equilibrium studies and kinetic modeling. *Separation Science and Technology, 44*(2), 316–334.
29. Anirudhan, T. S., & Radhakrishnan, P. G., (2007). Chromium(III) removal from water and wastewater using a carboxylate-functionalized cation exchanger prepared from a lignocellulosic residue. *Journal of Colloid and Interface Science, 316*, 268–276.
30. Chen, S., Yue, Q., Gao, B., Li, Q., & Xu, X., (2011). Removal of Cr (VI) from aqueous solution using modified corn stalks: Characteristic, equilibrium, kinetic and thermodynamic study. *Chem. Eng. J., 168*, 909–917.

CHAPTER 31

Valuable Products from Thermal Degradation of Industrial Waste Lignin in the Presence of Chemical Additives

RUPALI A. NANDANWAR,[1] A. R. CHAUDHARI,[1] and J. D. EKHE[2]

[1]Department of Applied Chemistry, Priyadarshini Bhagwati College of Engineering, Nagpur, Maharashtra, India,
E-mail: rupalianandanwar@gmail.com (R. A. Nandanwar)

[2]Department of Chemistry, Visveswaraya National Institute of Technology, Nagpur, Maharashtra, India

ABSTRACT

Pulp and paper industries produce lignin as waste material in very large quantities, and it causes a great disposal problem which poses an environmental burden. The structure of lignin shows a three-dimensional branched polymer with aromatic phenolic units with various functional groups; it degrades slowly and has the potential to produce valuable chemicals. In this study, the waste lignin obtained from Simplex Paper Mills, Gondia, Maharashtra, India, was purified and characterized by CHN analysis, FTIR spectroscopy, and thermal studies. The degradation of this waste lignin was carried out at 400–500°C without and with the chemical additives such as $ZnCl_2$ and K_2CO_3 separately in N_2 atmosphere. The degradation of lignin yielded gaseous products, distillate, tarry material, and highly porous carbonaceous material. The distillate was analyzed through GC-MS studies, and it showed the presence of various phenolic compounds having immense industrial applications. The influence of the chemical additives used on the formation of phenolic compounds was studied. Thus, this study indicated that industrial waste lignin (IWL) has the potential to produce valuable chemicals when subjected to thermal degradation.

31.1 INTRODUCTION

In the components of wood, together with cellulose and hemicellulose, lignin is also the main component. It is a valuable resource for energy and chemicals, also it is considered as the important source of organic raw material [1, 2]. Lignin consisting about 4–35 wt.% of most biomass. In the pulp and paper industry, during the Kraft process of delignification, lignin is separated from cellulose and hemicellulose. Around 70 million tons of lignin is produced per year. Out of this large amount of lignin produced only 3–5% of lignin was utilized commercially, and the rest amount of lignin is only burnt out. Thus, for the pulp and paper industries, the disposal of this produced lignin is big problem [3, 4]. However, due to the generation of very large quantities of lignin, it has the potential to be used as an alternative source of chemicals and hydrocarbon fuels [5–8]. Due to the chemistry and structure of lignin, it is considered as the important raw material for a number of applications. Many researchers worked a lot for the utilization of industrial waste lignin (IWL), e.g., hydrogenation [9, 10], alkali fusion [11], pyrolysis [12] and low molecular weight fuel additives can be prepared from lignin [13, 14].

Lignin is a highly branched three-dimensional phenolic polymer consisting of main three phenyl propane units such as trans-coniferyl, trans-sinapyl, and p-coumaryl alcohol. The lignin contains a high amount of aromatic units, thus it has the potential to be used as a raw material for the production of low molecular weight phenolic compounds. The conversion of abundant lignocellulosic biomass into useful component is achieved by the process of thermal degradation, which is one of the promising thermochemical technologies [15, 16]. Due to the structure of lignin and its complex composition, its degradation is mainly depending [17] on its nature, reaction temperature, rate of heating and degradation conditions. Thus, it also affects the temperature domain of degradation, the conversion ratio and the amount of products obtained. The study of degradation kinetics for lignocellulosic biomass says that the decomposition of lignin components starts at low temperature as compared to the carbohydrates, but it covers the whole temperature range up to 900°C. At lower temperatures, only volatile products are released due to certain chemical reactions (dehydration, dehydrogenation, deoxygenation, and decarboxylation) resulting from the breaking of weaker bonds and condensation reaction takes place [18]. With the increase in temperature, the rearrangement takes place and produces volatiles (Syngas; CO and H_2). At temperature between 250–400°C, phenolic components are the main volatile products along with syngas. From the TGA experiments of lignin, it is clear that the amount of C-C bonds in its structure enhances the formation of carbonaceous residue. In the present work, the lignin was isolated from paper mill waste and then purified and subjected to various techniques of characterization. The amount of Carbon, Hydrogen, Nitrogen, Sulfur, and Oxygen was determined by the elemental analysis of lignin. The presence of functional groups in alkali lignin was determined using FTIR technique. TGA-DTA studies were carried out to know the temperature range for active thermal degradation of lignin. The lignin was subjected to degradation in absence and in the presence of $ZnCl_2$ and K_2CO_3 catalysts separately in an inert atmosphere. The distillate obtained after degradation of lignin was extracted in methanol and the low molecular weight compounds obtained were analyzed using GC-MS.

31.2 MATERIALS AND METHODS

31.2.1 ISOLATION AND PURIFICATION OF LIGNIN FROM INDUSTRIAL WASTE BLACK LIQUOR

The black liquor was obtained from Simplex Paper Mills, Gondia, Maharashtra. It was filtered and then acidified with dilute HCl (up to pH = 5), and then allowed to precipitate for 30 mins. It was then filtered and washed with plenty of distilled water, and then it was dried in oven for overnight at 80°C and weighed. From 100 ml of black liquor, about 5 g of crude lignin was obtained. The crude alkali lignin was dissolved in a minimum quantity of 1,4-dioxane and the dissolved lignin was re-precipitated from water. The residue obtained was in the form of shiny brown crystals. It was dried and weighed (3.5 g) and 1,4-dioxane was removed by vacuum distillation. Now the residue is referred as pure lignin [19, 20].

31.2.2 ELEMENTAL ANALYSIS OF PURE LIGNIN

The pure lignin was subjected to its elemental analysis for the determination of Carbon, Hydrogen, Nitrogen, Sulfur, and Oxygen.

31.2.3 FTIR CHARACTERIZATION OF PURE LIGNIN

An infrared spectrum of pure lignin was recorded for the characterization of various functional groups present.

31.2.4   THERMO-GRAVIMETRIC ANALYSIS OF PURE LIGNIN

TGA was recorded to select the most useful range of temperature where maximum thermal cleavage would occur to produce greater yields of low molecular weight compounds. The samples were scanned using simultaneous TGA-DTA. It was set to have heating rate 15°C/min.

31.2.5   THERMAL DEGRADATION OF LIGNIN IN N2 ATMOSPHERE

About 80 g of purified lignin was taken in a two-necked round bottom flask fitted with distilling head, condenser, collector for distillate and a thermometer. It was heated to about 500°C for 5 hours in N_2 atmosphere. The distillate and solid carbonaceous material formed was collected, and distillate was extracted in methanol. It was further analyzed using GC-MS. The pure lignin was subjected to similar degradation by using $ZnCl_2$ and K_2CO_3 catalysts separately and the temperature was maintained at 500°C. The impregnation ratio is 1:1 for pure lignin with the catalysts separately (impregnation for 1 hour).

31.2.6   ANALYSIS OF DEGRADED PRODUCTS BY GC-MS (GAS CHROMATOGRAM MASS SPECTROMETER)

The organic compounds formed during the thermal degradation of pure lignin and degradation of lignin in the presence of $ZnCl_2$ and K_2CO_3 were analyzed by GC-MS.

31.3   RESULTS AND DISCUSSION

31.3.1   LIGNIN ISOLATION AND ITS PURIFICATION

The crude lignin was isolated from industrial black liquor which on purification yielded pure solid lignin (about 70%).

31.3.2   ELEMENTAL ANALYSIS OF PURE LIGNIN

The elemental analysis of pure lignin gave the following composition mentioned in Table 31.1.

TABLE 31.1 Elemental Analysis of Pure Lignin

Elements	C	H	O	N	S
Percentage of Composition (%)	62.32	5.91	31.0	0.1	0.1

31.3.3 FTIR ANALYSIS

The FTIR spectrum of pure lignin as depicted in Figure 31.1 indicates some characteristics features which are helpful for structural investigations. The spectrum clearly indicated the presence of peaks due to hydrogen-bonded O-H, C-H stretching in methoxyl group, methyl, and methylene group, aromatic skeletal vibrations due to guiacyl group, carbonyl stretching in unconjugated ketone and carboxyl group, aromatic C-H in-plane deformation and aromatic C-H out of plane, etc. All these bands along with their assignments are tabulated in Table 31.2.

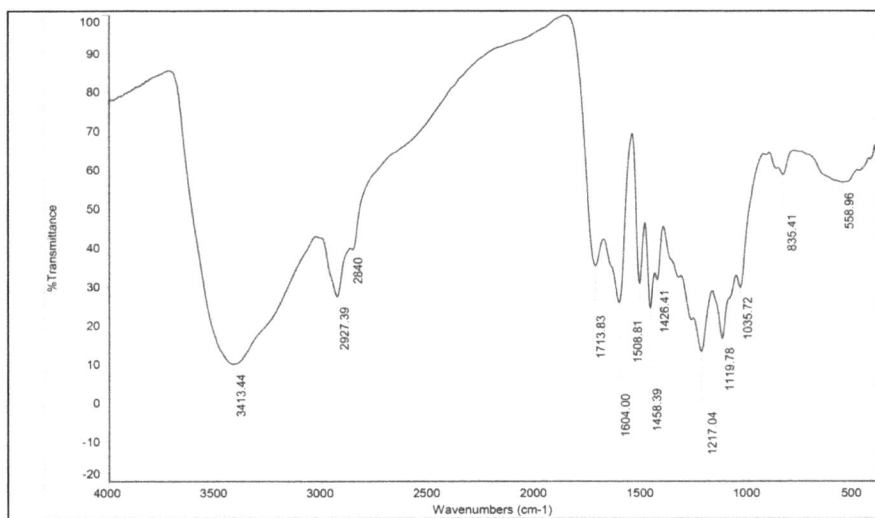

FIGURE 31.1 FTIR of pure lignin.

31.3.4 THERMOGRAVIMETRIC ANALYSIS

TGA of the purified lignin showed that maximum thermal cleavage occurs around 350°C. It was also observed from the DTA, that the degradation at this temperature was found to be exothermic process along with the evolution of gaseous products [21, 22] during the thermal degradation. Mostly CO, CO_2, methane, ethane, and other such gases are expected to escape.

TABLE 31.2 IR Characterization of Pure Lignin

SL. No.	Peaks Obtained	Nature of Band Intensity	Assignments
1.	3413 cm^{-1}	S	-OH Hydrogen bonded
2.	2927 cm^{-1}	S	-CH stretching in methyl and methylene groups
3.	2840 cm^{-1}	S	Symmetric stretch for-CH_3 of methoxyl group
4.	1713 cm^{-1}	M	Carbonyl stretching-unconjugated ketone and carboxyl groups
5.	1604 cm^{-1}	S	Aromatic skeletal vibrations
6.	1508 cm^{-1}	S	Aromatic skeletal vibrations
7.	1458 cm^{-1}	S	Aliphatic CH bending, aromatic skeletal vibrations
8.	1426 cm^{-1}	M	Aromatic skeletal vibrations
9.	1325 cm^{-1}	W	Syringyl ring breathing with CO stretching
10.	1217 cm^{-1}	S	Guaiacyl ring breathing with CO stretching
11.	1119 cm^{-1}	S	C-O stretching for secondary alcohols and aliphatic ether, aromatic CH in-plane deformation, syringyl type
12.	1035 cm^{-1}	M	Aromatic CH in-plane deformation, guaiacyl type and C-O stretching for primary alcohol
13.	835 cm^{-1}	M	Aromatic C-H out of plane bending

Note: S: strong; M: medium; W: weak.

The thermogram of pure lignin in air has clearly indicated about 7.48% loss up to 101.23°C. It seems mainly due to the loss of moisture and some minor other gases due to rearrangements within the lignin molecule. The thermogram showed further gradual loss up to 40.6% at 400.94°C, 50.67% at 475.13°C. In this temperature range, the elimination of phenolic aldehydes are likely to take place. The maximum weight loss occurred at 750.63°C (Figure 31.2).

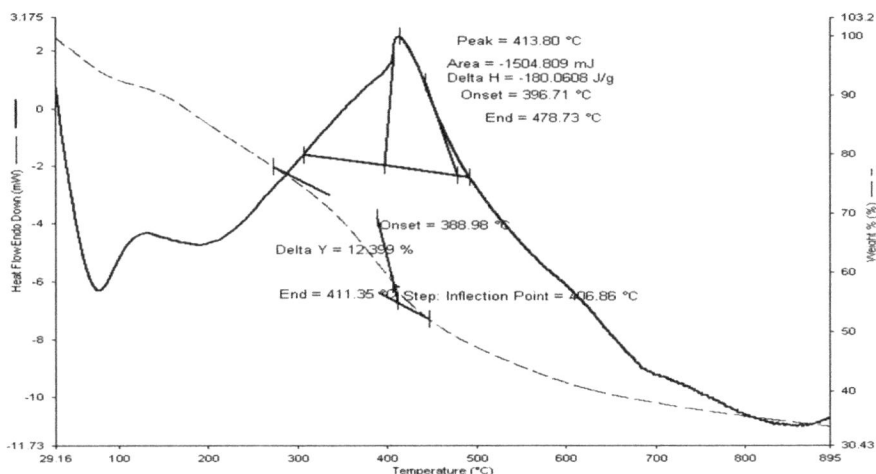

FIGURE 31.2   TGA and DTA of pure lignin.

31.3.5   *THERMAL DEGRADATION OF LIGNIN*

Degradation, initially only aqueous distillate comes out up to 150°C and then the thick fumes were observed. With continuous degradation, a yellow oily liquid emerged out. When the temperature rises to 400°C, a dark tarry material was observed. After the degradation gets completed, a carbonaceous material was collected. When the thermal degradation was carried without catalyst, 31.25% of aqueous distillate, 52.5% of carbonaceous material and 15.25% of gaseous products were obtained. There was an observed increase in the rate of the reaction when the thermal degradation was carried out in the presence of $ZnCl_2$ and K_2CO_3. It has been attributed to the greater surface area (SA) and uniform heating due to the presence of catalyst. Lignin on degradation in the presence of $ZnCl_2$ gives 32.5% of liquids, 58% of carbonaceous material, and 9.5% of gaseous products. In the degradation of lignin with K_2CO_3 catalyst, the yields were 28% liquids, 51.9% of carbonaceous material and 20.1% of gaseous products.

The effect of catalysts showed the variation in the yield of liquid and gaseous products. The comparative yield of coke and distillate obtained by thermal degradation of lignin is given in Table 31.3.

The comparative yield reveals that the lignin degradation process is influenced by the use of catalysts and it also improves the destructive distillation process. The lignin degradation during thermolysis gives the formation of a great variety of substituted phenols, such as alcohol, carboxylic acid, aldehydes, ketones, and different re-condensation products. In this study, the maximum yield of distillate is obtained in the presence of $ZnCl_2$, which matches with the earlier reports [23]. The organic components in the distillate were extracted in methanol and analyzed through GC-MS.

TABLE 31.3 Effect of Catalysts on Degradation Products

SL. No.	Coke Type	Coke	Aqueous Distillate	Gaseous Products
1.	Only lignin	52.5%	31.25%	15.25%
2.	Lignin + $ZnCl_2$	58%	32.5%	9.5%
3.	Lignin + K_2CO_3	51.9%	28%	20.1%

31.3.6 ANALYSIS OF LIGNIN DERIVED LIQUID PRODUCTS BY GC-MS (WITHOUT CATALYST)

The total ion chromatogram of liquid products obtained from pure lignin (without catalyst) up to retention time 39 minutes. The chromatogram showed total 37 peaks and out of which 21 are substitute phenol, rest are mono and dicarboxylic acid, cyclic esters, etc. The details of the identified compounds, their peaks, their relative peak area (%) and the retention time for all identified peaks and are listed in Table 31.4.

TABLE 31.4 Mass Spectral Data of Lignin Derived Products (Without Catalyst)

Peak No.	Retention Time	Compound	Percentage of Area (%)
1.	5.256	Carbolic acid	11.42
2.	6.444	Phenol, 3-methyl	3.78
3.	6.785	Phenol, 3-methyl	7.76
4.	7.052	Phenol, 3-methyl	12.14
5.	7.298	Phenol, 3-methyl	0.36
6.	7.773	Phenol, 2-methyl	0.58
7.	7.936	Phenol, 2,3-dimethyl	1.58
8.	8.231	Phenol, 4-ethyl	9.45
9.	8.441	2-methoxy-6-methyl phenol	0.88
10.	8.559	Azulene	6.50
11.	8.660	Phenol, 2-methoxy-4-methyl	5.13

TABLE 31.4 *(Continued)*

Peak No.	Retention Time	Compound	Percentage of Area (%)
12.	8.719	1,2-benezenediol	1.86
13.	9.000	Benzofuran	0.24
14.	9.134	Phenol,4-(1-methylethyl)	1.32
15.	9.302	Phenol,2-ethyl,5-methyl	1.37
16.	9.680	1,2-benzenediol	2.18
17.	9.935	Phenol,4-ethyl-2-methoxy	9.11
18.	10.053	2-ethoxy-4-methyl phenol	1.10
19.	10.358	2,5-Diethyl phenol	0.40
20.	10.480	3-Methoxy-5-methyl phenol	0.29
21.	10.642	Benzene,4-ethyl-1,2-dimethoxy	0.57
22.	10.782	2-Methoxy-6-methyl phenol	0.30
23.	10.946	Phenol,2,6-dimethoxy	6.95
24.	11.066	Phenol, 3-methoxy-2,5,6-trimethyl	0.77
25.	11.153	Phenol, 2-methoxy-4-propyl	0.65
26.	11.349	4-Ethyl Catechol	1.28
27.	11.607	Methyl isodehydracetate	0.27
28.	11.689	2,3-dimethoxy toluene	0.45
29.	12.203	1,2,4-trimethoxy benzene	3.67
30.	13.202	Benzene,1,2,3-trimethoxy-5-methyl	3.04
31.	13.517	Ethanone,1-(4-hydroxy-3-methoxy phenyl)	0.24
32.	13.701	1-(2-Hydroxy-4,5-dimethoxy-phenyl)	0.47
33.	13.822	2,3-Dimethoxy benzoic acid	0.56
34.	14.199	2,4-Hexadienedioic acid 3,4-diethyl-dimethyl	0.55
35.	15.025	Ethanone,1-(4-hydroxy-3-methoxy phenyl)	0.72
36.	15.633	1-(2-Hydroxy-4,5-dimethoxy-phenyl)	0.92
37.	18.196	9,10-Anthraquinone	0.45

The lignin derived products also contain Azulene, Benzofuran, 1,2-Benzendiol, 2–4, dimethoxytoulene, methyl isodehydracetate, 4-ethyl catechol, 2,3-dimethoxy benzoic acid, 9,10-anthroquinone. Due to degradation of ether linkages of lignin, the lignin derived products comprise of condensed type linkages (Figure 31.3).

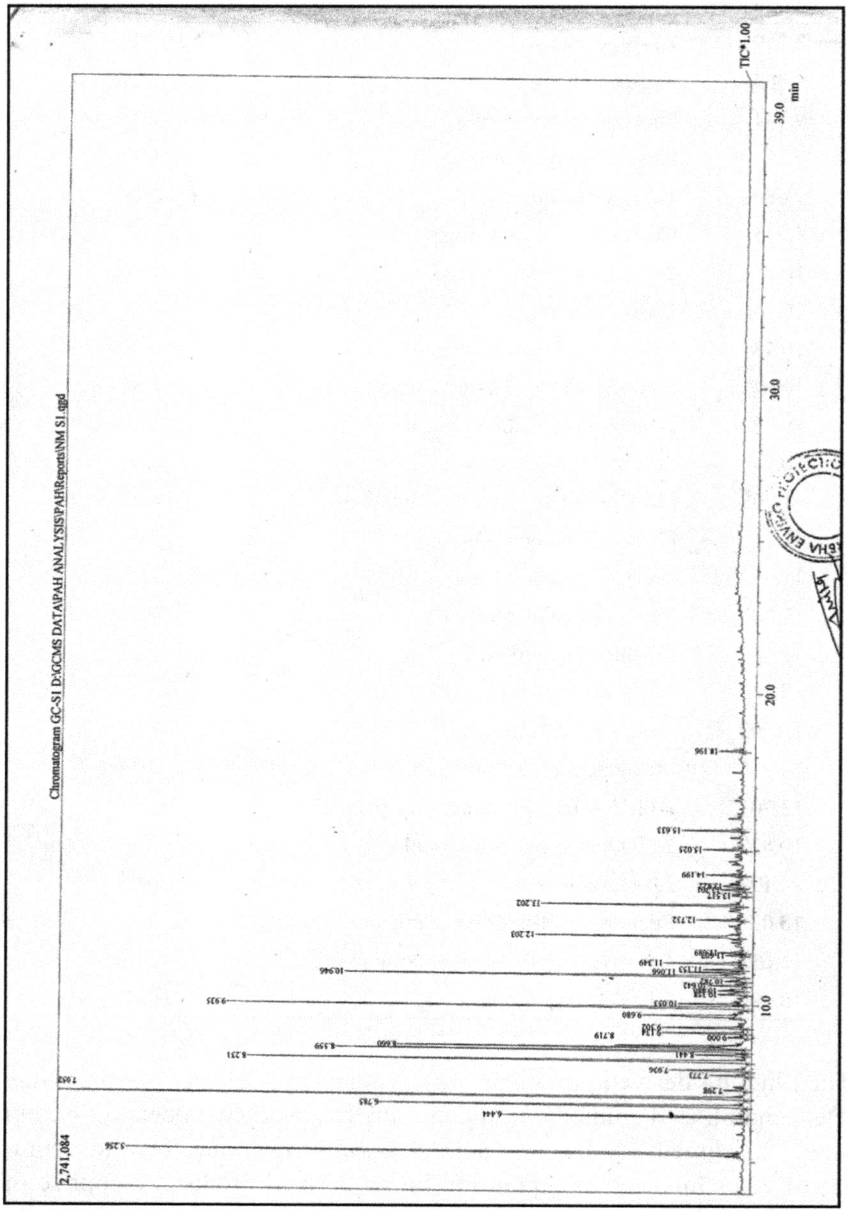

FIGURE 31.3　Gas chromatogram of lignin.

31.3.7 ANALYSIS OF LIGNIN DERIVED LIQUID PRODUCTS BY GC-MS (WITH ZNCL2 AS CATALYST)

Figure 31.4 shows the total ion chromatogram of aqueous distillate obtained from lignin with $ZnCl_2$ catalyst up to retention time 39 minutes. There are 34 peaks obtained from which 20 are substitute phenol, remaining are mono and dicarboxylic acid, cyclic esters, etc. The details of the identified compounds, their peaks, their relative peak area (%) and the retention time for all identified peaks and are listed in Table 31.5.

TABLE 31.5 Mass Spectral Data of Lignin Derived Products with $ZnCl_2$ Catalyst

Peak No.	Retention Time	Compound	Percentage of Area (%)
1.	5.234	Phosphonic acid	2.45
2.	6.201	2-cyclopentene1-one,2,3-dimethyl	0.87
3.	6.286	2-pyrrolidinone, 1-methyl	1.33
4.	6.456	Phenol,3-methyl	4.75
5.	6.784	Phenol,3-methyl	6.41
6.	7.041	2-cyclopentene-1-one,2,3,4-trimethyl	1.46
7.	7.301	Phenol, 2,3-dimethyl	1.60
8.	7.784	Phenol, 2-ethyl	3.76
9.	7.947	Phenol, 2,3-dimethyl	7.12
10.	8.244	Phenol, 3-ethyl	12.43
11.	8.415	Phenol, 2,3-dimethyl	1.02
12.	8.562	Azulene	6.76
13.	8.639	Phenol,2,3-dimethyl	1.36
14.	8.809	Phenol,2,3,5-trimethyl	1.91
15.	9.151	Phenol, 4(-1-methylethyl)	3.31
16.	9.313	Phenol, 2-ethyl,5-methyl	6.74
17.	9.374	Phenol, 2-ethyl,6-methyl	0.82
18.	9.614	Phenol,3 ethyl,5-methyl	2.71
19.	9.736	Phenol,3,4,5-trimethyl	1.38
20.	9.805	Phenol,2,3,6-trimethyl	1.27
21.	9.899	2,5-diethyl phenol	1.97

TABLE 31.5 *(Continued)*

Peak No.	Retention Time	Compound	Percentage of Area (%)
22.	10.119	Phenol,2-methyl-5-(1-methylethyl)	2.23
23.	10.372	2,5-diethyl phenol	5.98
24.	10.842	phenol 3,5-diethyl	0.80
25.	11.065	Phenol, 2-(1,1-dimethylethyl),5methyl	3.37
26.	11.177	1H-Indole,1,3-dimethyl	1.32
27.	11.450	1H-Indole,7-methyl	2.71
28.	11.795	2,3-dimethyl,1,4-methoxyphenyl acetonitrile	0.70
29.	12.044	Benzene, 3-ethyl,1,2,4,5,-tetramethyl	0.74
30.	12.687	2-(5-Isopropyl,2-methylphenyl)ethanol	0.83
31.	13.033	Butylated Hydroxytoluene	5.53
32.	16.342	Dibenzo[b,e]7,8-diazabicyclo	2.15
33.	16.702	pentadecanone,6,10,14-trimethyl	0.96
34.	32.844	Cholestanol	1.33

In the presence of $ZnCl_2$ catalyst, the degradation of lignin is in a somewhat different pattern as the volume of aqueous distillate as well as the weight of carbonaceous residue increases than the degradation products obtained from lignin without catalyst. Figure 31.4 represents the total ion chromatogram for lignin degradation in the presence of $ZnCl_2$.

31.3.8   ANALYSIS OF LIGNIN DERIVED LIQUID PRODUCTS BY GC-MS (WITH K2CO3 AS CATALYST)

When lignin was subjected to the degradation in presence of K_2CO_3 as a catalyst, the liquid products obtained are shown in Figure 31.5. The liquid products are obtained up to the retention time of 39 minutes. Total 43 peaks are obtained from which 28 peaks are substitute phenols whereas remaining are mono and dicarboxylic acids, cyclic esters, etc. The details of the identified compounds, their peaks, their relative peak area (%) and the retention time for all identified peaks and are listed in Table 31.6.

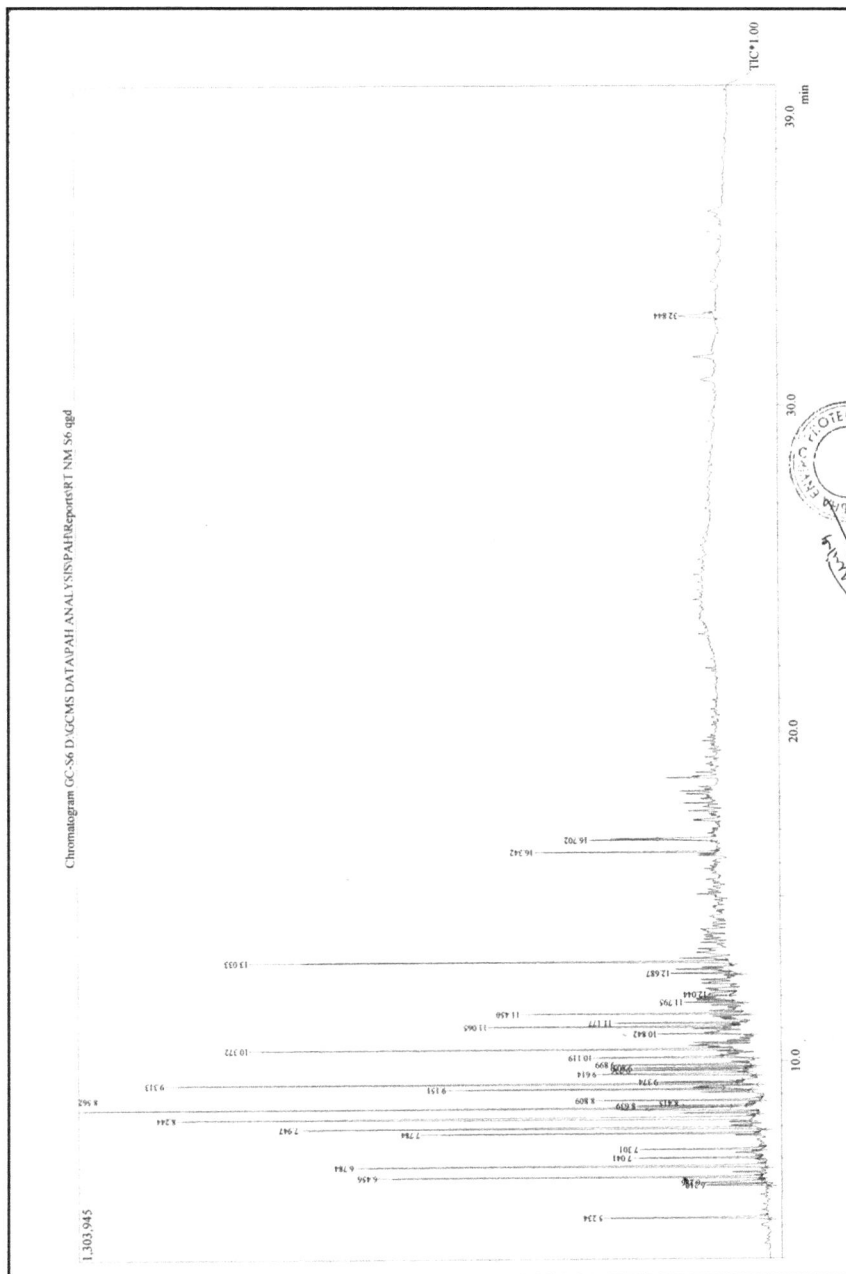

FIGURE 31.4 Gas chromatogram of lignin in the presence of ZnCl2.

TABLE 31.6   Mass Spectral Data of Lignin Derived Products with K_2CO_3 Catalyst

Peak No.	Retention Time	Compound	Percentage of Area (%)
1.	5.199	Carbolic acid	14.24
2.	6.252	2-cyclopentene-1-one,2,3-dimethyl	0.45
3.	6.422	Phenol,3-methyl	13.39
4.	6.649	2-cyclopentene-1-one,2,3,4-trimethyl	0.17
5.	6.766	Phenol,3-methyl	12.61
6.	7.009	Phenol,2-methoxy	3.07
7.	7.269	Phenol,2,3-dimethyl	1.86
8.	7.760	Phenol, 2-ethyl	3.85
9.	7.936	Phenol,2,3-dimethyl	10.42
10.	8.223	Phenol, 3-ethyl	10.96
11.	8.401	Phenol,2,3-dimethyl	1.17
12.	8.542	Azulene	4.96
13.	8.627	Phenol,2,3-dimethyl	1.79
14.	8.793	Phenol,2,3,5-trimethyl	0.96
15.	9.042	Phenol,2-propyl	0.62
16.	9.136	Phenol,4-(1-methylethyl)	1.95
17.	9.191	Phenol,4-(1-methylethyl)	0.40
18.	9.301	Phenol,2-ethyl,5-methyl	3.43
19.	9.360	Phenol, 3-ethyl,5-methyl	0.61
20.	9.608	Phenol, 3-ethyl,5-methyl	2.32
21.	9.727	Phenol, 3-ethyl,5-methyl	0.85
22.	9.796	Phenol,2,3,6-trimethyl	1.07
23.	9.915	Phenol, 4-ethyl,2-methoxy	1.89
24.	10.125	Phenol, 2-methyl-5-(1-methylethyl)	1.11
25.	10.234	Phenol,2,3,6-trimethyl	0.21
26.	10.365	2,5-diethyl phenol	1.25
27.	10.437	Phenol,2,4,5-trimethyl	0.16
28.	10.514	Benzene,4-ethyl-1,2-dimethoxy	0.33
29.	10.572	Phenol,4-(1-methyl propyl)	0.38
30.	10.638	Phenol,3-methyl,6-propyl	0.18
31.	10.830	Phenol,3-methyl,6-propyl	0.46
32.	10.916	1H-inden-1-one,2,4,5,6,7,7a-hexahydro	0.18
33.	10.994	Benzene,1-ethoxy-4-ethyl	0.23
34.	11.065	2,5-diethyl phenol	0.71

TABLE 31.6 *(Continued)*

Peak No.	Retention Time	Compound	Percentage of Area (%)
35.	11.438	1-(3-Methoxymethyl-2-methylphenyl)ethanol	0.35
36.	11.536	Benzene,1-methoxy-4-(1-methylpropyl)	0.14
37.	11.921	2-propanone,1-(4-methylphenyl)	0.14
38.	12.227	Benzaldehyde,4-(1-methylethyl)	0.16
39.	12.682	Benzene,1-(methoxymethyl)-4-(1-methylethyl)	0.15
40.	13.027	Butylated hydroxytoluene	0.34
41.	13.683	Pentainethyl phenyl acetonitrile	0.15
42.	15.767	1-Naphthol,5,7-dimethyl	0.16
43.	18.195	Anthroquinone	0.17

From all the above chromatograms, it is clear that the yield of liquid products (aqueous distillate) was similar in absence and in the presence of catalysts. The results obtained were matched accordingly with the results already reported by some researchers [24]. The products, phenol, 2 methoxy-4 propyl are composed of phenyl propane unit (C_6–C_3). In the structure of lignin, the β-O-4 linkage can be cleaved easily; thus through the cleavage of ether linkages, these products would be obtained after the degradation of lignin. Formation of the products like phenol, 2-ethyl, 5-methyl, Benzene,1,2,3,-trimethoxy 5 methyl, ethanone 1-(4-hydroxy,3-5-dimethoxy phenyl) showed that the cleavage of $C_β$/C γ and $C_α$/$C_β$ linkages of lignin takes place. The remaining products having carbon chain of C_4 to C_{10} comprise of the simple and branched acids and acid esters. Due to the preferential degradation of ether linkages, the lignin derived products mainly consist of condensed type of linkages. The other lignin derived monomeric products detected are catechol, phenol, m, p-cresol, and o-cresol, etc.

It is observed that, in the presence of $ZnCl_2$, the yield of liquid products obtained is somewhat more than the yield obtained by degradation of pure lignin. It may happen as $ZnCl_2$ helps to reduce gaseous products by converting those to little bigger molecules that could produce liquid fraction in more proportion.

31.4 CONCLUSION

The above studies indicated that the industrial waste kraft lignin have a potential as a raw material for the chemical industry. Various organic compounds obtained after thermal degradation have a great significance. The

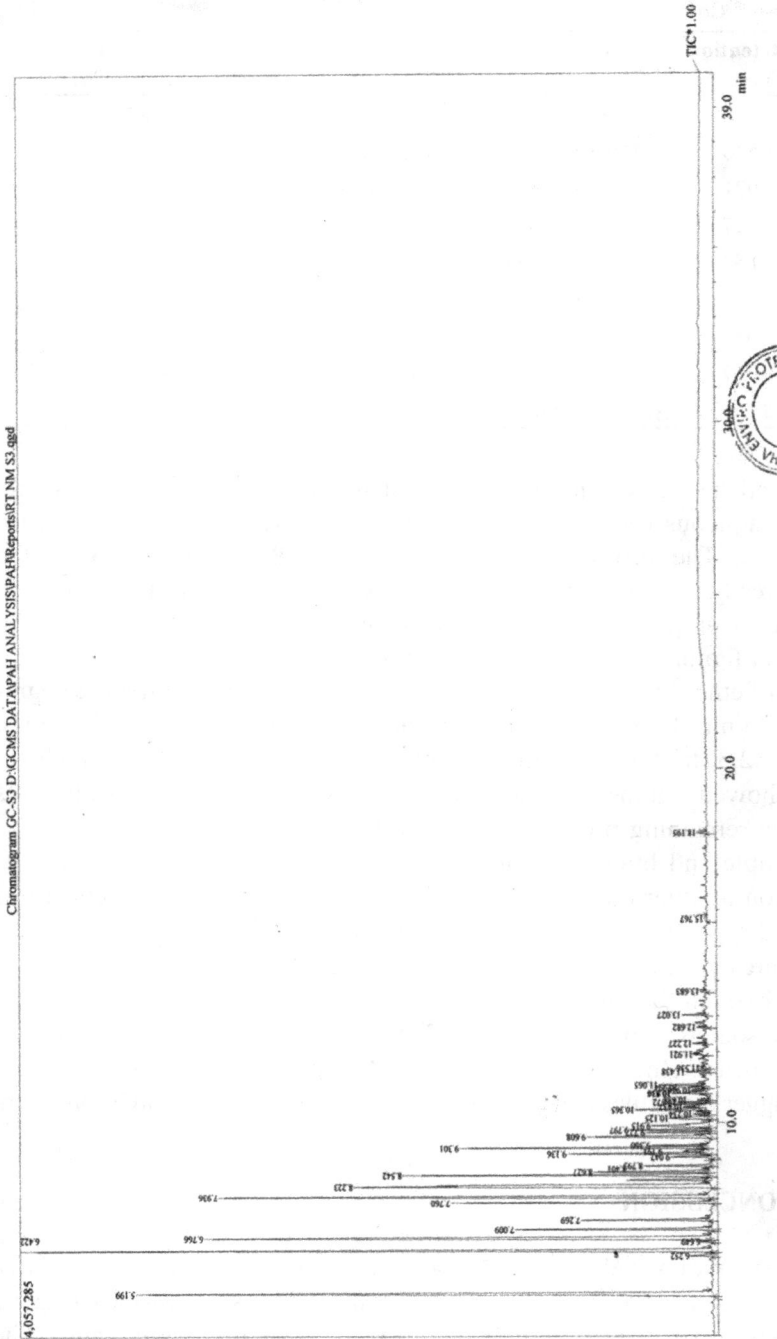

FIGURE 31.5   Gas chromatogram of lignin in the presence of K2CO3.

degradation of lignin produces relatively high amounts of phenol derivatives and a small amount of low molecular weight alcohols, acids, and esters. Phenols obtained can be used in the production of Bakelite (Phenol form-aldehyde resin) and alkylbenzene compounds can be used as high-octane additives in gasoline. In the presence of catalyst, the conversion of solid lignin into degradation products is high as compared to that of pure lignin. The use of $ZnCl_2$ catalyst produces more liquid products as compared to the use of K_2CO_3. The maximum cleavage of lignin molecules occurs in the temperature range 300–400°C. Thus pyrolysis-GC-MS are powerful tools to rapidly chemically characterize various types of lignin degradation products.

KEYWORDS

- **chemical additives**
- **gas chromatogram mass spectrometer**
- **hydrogenation**
- **lignin**
- **phenolic compounds**
- **thermal degradation**

REFERENCES

1. Gosselink, R. J. A., De Jong, E., Guran, B., & Abächerli, A., (2004). Analytical protocols for characterization of sulphur-free lignin. *Ind. Crop. Prod., 20*, 121.
2. Lora, J. H., & Glasser, W. G., (2002). Recent industrial applications of lignin: A sustainable alternative to nonrenewable materials. *Journal of Polym. Environ., 10*, 39.
3. Xiao, B., Sun, X. F., & Runcang, (2001). The chemical modification of lignins with succinic anhydride in aqueous systems. *Polymer degradation and Stability, 71*, 233–231.
4. Pranda, J., Brenzy, R., & Micko, M. M., (1990). *1st European Workshop*, pp. 91–96.
5. Karagoz, S., Bhaskar, T., Muto, A., & Sakata, Y., (2005). *Fuel, 84*, 875.
6. Piskorz, J., Majerski, P., Radlein, D., & Scott, D. S., (1989). Conversion of lignins to hydrocarbon fuels. *Energ. Fuel., 3*, 723.
7. Oasmaa, A., & Johansson, A., (1993). Catalytic hydrotreating of lignin with water-soluble molybdenum catalyst. *Energ. Fuel., 7*, 426.
8. Huber, G. W., Iborra, S., & Corma, A., (2006). Synthesis of transportation fuels from biomass: chemistry, catalysts, and Engineering. *Chem. Rev., 106*, 4044.
9. Brewer, C. P., Cooke, L. M., & Hibbert, H., (1948). *J. Am. Chem. Soc., 70*, 57.
10. Shabtai, J., Zmierczak, W., Kadangode, S., Chornet, E., & Johnson, D. K., (1999). *Proceedings Fourth Biomass Conference of the Americas* (p. 811). Oxford, U.K.

11. Phillips, M., & Goss, M. J., (1938). *Journal of Biological Chemistry, 125*, 241–246.
12. Naae, D., Whittington, L., & Davis, C., (1992). *Presented at the Symposium on Alteration and Utilization of Lignin.* San Francisco, CA.
13. Meier, D., Ante, R., & Faix, O., (1992). Catalytic hydropyrolysis of lignin: Influence of reactions conditions on the formation and composition of liquid products. *Bioresource Technology, 40*(2), 171–177.
14. Ronal, W., Breau, T., & Breau, J., (1996). Hydrocracking of solvolysis lignin in batch reactor. *Fuel, 75*(7), 795–800.
15. Mohan, D., Pittman, C. U. Jr., & Steele, P. H., (2006). Pyrolysis of wood/biomass for bio-oil: A critical review. *Energy Fuels, 20*, 848–889.
16. Lu, Q., Zhang Z. F., Dong, C. Q., & Zhu, X. F., (2010). Catalytic upgrading of biomass fast pyrolysis vapors with nano metal oxides: An analytical Py-GC/MS Study. *Energies, 3*, 1805–1820.
17. Li, J., Li, B., & Zhang, X., (2002). Comparative studies of thermal degradation between larch lignin and Manchurian ash lignin. *Polym. Degrad. Stabil., 78*, 279.
18. Orfao, J. J. M., Antunes, F. J. A., & Figueiredo, J. L., (1999). Pyrolysis kinetics of lignocellulosic materials—three independent reactions model. *Fuel, 78*, 349–358.
19. Argyropoulos, D. S., Sun, Y., & Palns, E., (2002). *Journal of Pulp and Paper Science, 28*, 50.
20. Sarkanen, K. V., & Ludwig, C. H., (1971). *Lignin, Occurrence, Formation, Structure* (p. 698). Wiley Interscience, NY, USA.
21. Nagpurkar, L., Chaudhari, A., & Ekhe, J., (2002). Formation of industrially important chemicals from thermal and microwave assisted oxidative degradation of industrial waste lignin. *Asian J. of Chemistry, 14*, 1387.
22. Gillet, A., & Urlings, J., (1952*). Chim. And Ind. (Paris), 68*, 55.
23. Gonzalez, E., Cordero, T., Rodrigues, J., & Rodrigues, J. J., (1997). Development of porosity upon chemical activation of kraft lignin with $ZnCl_2$. *Ind. Eng. Chem. Res., 36*, 4832–4838.
24. Maldure, A. V., & Ekhe, J. D., (2013). Pyrolysis of purified kraft lignin in the presence of $AlCl_3$ and $ZnCl_2$. *J. Environ. Chem Eng., 1*, 844–849.

CHAPTER 32

The Study of Indigenous Knowledge Adopted for Fencing by Farmers and Tribes of Chhattisgarh, India

LEENA GANVIR

Priyadarshini Institute of Architecture and Design Studies, Nagpur, Maharashtra, India, E-mail: leenaganvir2000@yahoo.co.in

ABSTRACT

Indigenous knowledge is considered as the oldest practices by villagers. This skill is spent in fight for their existence, to make their bread, shed, and to attain power of their own lives. Indigenous communities have been in charge for the development of many technologies, have significantly contributed and shared their knowledge to the communities. Indigenous knowledge is a key to sustainable approach. Indigenous knowledge is conveyed as the distinctive knowledge limited to a particular tradition or community. This knowledge is developed and spread by people in a period of time. It is an attempt to manage their agriculture, biological, frugal, and community environments. Farmers use this skill in live fencing and the species used to safeguard their day-to-day lives. In agriculture, fences are used for varied applications like to keep animals in or out of an area. Fences can be constructed from a wide variety of plant species, materials, depending on topography, location, and animals to be confined, mainly helpful to limit animals and walker access. Fencing is an old agriculture green practice, is less understood and rarely recorded in India. These techniques hold a deep understanding of locally available materials and the art of using them with the help of their indigenous knowledge. This chapter talks about sustainable practices adopted through various indigenous techniques and materials used for fencing in rural India.

32.1 INTRODUCTION

Indigenous knowledge is an asset for any country. In India, various techniques of Fences are practiced by various tribes and farmers. These techniques and practices are passed generation from to generations and from mouth to mouth, giving basis to strong routes of knowledge to society. Indigenous knowledge has a wide outlook of the environment contributing to sustainability by using available natural resources [1]. Live fences not only act as protective structures to safeguard garden produce, intrusion of wandering animals and trespassing. These also add contributions to the environment and mankind due to high biodiversity. Live fencing involves the arrangement of medicinal plants, wood, food, fruit, and vegetables. It is also useful to regulate entry of livestock such as shelter, wind force and increasing soil fertility level [2]. These indigenous knowledge and practices are endangered due to use of foreign technologies. The other drawback of technologies can be overlooking skills, problem solving strategies, and expertise in certain knowledge. Indigenous knowledge is the result of applied dedication in everyday life. It is all the time strengthened by experience and trial and error. Indigenous knowledge systems have sustainable ways of using natural resources and have a broad perspective of the environment and ecosystem [3]. Crop damaged by stray animals is increasing day by day and to overcome these issue farmers adopt techniques which result in death of wildlife (Table 32.1).

TABLE 32.1   State Wise Percentage of Crops Damaged by Intrusion of Animals

SL. No.	States in India	Percentage of Crops Damaged by Intrusion of Animals
1.	Nanda Devi Biosphere reserve	50% to 60% by wild animals
2.	Himachal Pradesh	60% to 70%
3.	Kerala	Up to 30% every year
4.	Chhattisgarh	Hike by 13% in 2012 up to 60%
5.	Karnataka	Declined by 75% using preventive measures
6.	Haryana	58% of the total crops

Crops damaged by wild animals or domestic animals depend on the season, crop pattern, density of population. Looking at crops damaged by the intrusion of animals, the use of different fence techniques has evolved from time to time to secure their crops [18–22].

Fencing, although is one of the old agroforestry practices, focus is always in live fence practices. As per the different climatic zones in India,

Chhattisgarh fall under hot and humid climatic zone. Here live fencing is considered as a focal part of agroforestry practices. A study considering other fence techniques in hot and humid climates needs to study.

This study is based mainly on Indigenous techniques of fencing used by farmers of India. This study explores their knowledge of live and dry fence species and how they incorporate in their fencing. The chapter attempts to study and compile various breeds used in fencing (Live and Dry) in Chhattisgarh state.

32.2   STUDY AREA

Chhattisgarh is a state with more than 40 tribes (Figure 32.1) as a native population. These tribal communities are struggling to survive with minimum resources available. Their Indigenous knowledge is capital for them to get food and shelter. Vegetation in the Chhattisgarh forest is mostly thorny trees and tall deciduous trees. Plants and trees have a special place and high cultural value in the lives of farmers and tribes [6]. Chhattisgarh traditional farmers have broad knowledge about their live fencing techniques by using these trees and plants. These practices have deep roots of knowledge based on their natural resources and nearby forest. This includes knowledge-based upon practices, trees, and shrubs used, and construction techniques inheriting from their society.

It has a major variety of strong cultural heritage and biodiversity. Total geographical area covered with forests in the Chhattisgarh district is 44%. This area has excessive rice cultivation [4]. It has hot and humid weather and receives rainfall from the southwest monsoon. There are 129 tree species, 48 shrub species, 50 herb species available in Chhattisgarh state. The species of the forest are majorly Lagerstroemia parviflora, *Shorea robusta* (Sal local name). The Green State of India has 44% of its area under dense forests. The other important tree species are *Madhuca longifolia* (Mahua local name), *Abies pindrow* (Achar local name), *Bambusa vulgaris* (bamboo), -*Cassia fistula* (Amaltas), *Syzygium cumini* (Jamun), *Phyllanthus emblica* (Amla), and *Diospyros melanoxylon* (Tendu). Deep jungles of Chhattisgarh are home to a variety of plants [5, 24].

32.3   LITERATURE STUDY

The fencing enactment was usually single breed and more than one breed. Live fence method depends on the breeds used in live fencing around agriculture plants, around back gardens and throughout farms. This depends on the landscapes being protected. The fencing practices were divided into two types single breed and breeds used in varieties, depending on the species used in fencing [7].

FIGURE 32.1　Map of India.

Usually, two types of fence techniques are used by Tribes and farmers of Chhattisgarh, mainly live fence and dry fence. Live fence is done with the help of planting small plants and bushes. Dry fence is usually done by using wood trunk, stems, straws, and ropes.

32.3.1   LIVE FENCE

Live fencing is employed by live wooden breeds for fences. This fencing is practiced by using fencing material like individual fence poles fastened with wire. It may also be a heavy plantation of hedgerows with no interlinked wire. This fence technique may also involve the use of natural shrubs and trees used as fences. The motive of live fence tree breeds includes timber craftwork, fence pole, and remedial use, and dirt protection, diversion of winds, firewood, and food. Live fences are planted to prevent their farm-lands from wild animals, livestock, and homes from animals. Live fencing is recognized as a knowledge that gives a variety of facilities and products. Live fences are diversified depending on cultural uses and composition of species used. This includes knowledge about species rooting ability, wind protection, quality of wood and its durability. Various methods employed in the management of live fences include local knowledge, pruning, and various uses of pruned material [8].

Farmers use a variety of plants and shrubs which are locally available and easy to grow for no cost. Due to lack of technology advancement and revenue, these tribes have managed to evolve the fence techniques with the available resources. Their indigenous knowledge of materials and techniques give the architecture a different sense to build form. Thus, they contribute a lot towards the sustainability and ecosystem. Common species used for live fence are *Acacia eatechu, Ipomea carnea, Lawsonia inermis* (henna), *Calotropis gigantea* (ruii shrub), Pigeon pea (red gram), *Bambusa vulgaris* (bamboo tree). These plants are used as live fence to protect their fields and livestock. Split red gram shrubs are usually planted in farms to mark barriers and divide farms. These shrubs not only prevent farms to keep away animals but also widely used to separate adjacent fields for distinct purposes. Split Red gram is a famous food crop in Chhattisgarh. This Live Fence crop is popularly used by farmers and tribes to protect their farmland from entry by-passers and animals.

Several plant species are used for live fencing. *Lawsonia inermis*/Henna plants are also used as a farmland plot boundary. This type of live fence has medicinal value too. Dried leaves of this plant are used for medicinal purposes. This plant spreads densely and thus prevents the animals from entering. This plant not only retains water but also used as a visual barrier as it is usually planted densely [9]. Farmlands are usually protected by a small ditch in the periphery where *Ipomoea carnea* (morning glory) is planted. *Ipomoea carnea* plant which grows up to 2-meter height. This plant grows very fast and is a good barrier for farmlands. It is planted as a weed in cultivated fields, such as paddy fields. This plant strongly competes for resources such as nutrients and water. Due to its foul smell of flowers, animals do not enter the farmlands [10]. Tribes of Chhattisgarh have good knowledge of plants with good medicinal value. Local tribes of Chhattisgarh plant Calotropis Gigantea as live fences in their farmlands. The flowers of this plant are offered to Lord Hanuman (Popular God of India) and serve religious purposes also. The leaves of this plant have pain healing properties and dry stems are used to cover fences. This shrub can be seen widely all over this region. The practice of planting trees with dual-purpose shows their awareness and knowledge based on their learning [11].

32.3.2   DRY FENCE

Dry fence technique may be used as permanent cover to boundaries. It may be temporary grassland dividers or to enclose a house, fences require careful organization and fabrication for its planned use. This also ensures long life and low preservation cost. This type of fence involves use of various plants and shrubs in dry state. Fence can be used to make their livestock safe, so most of the time, farmers give importance to expense, to facilitate construction and durability of the fence when considering fencing. The indigenous knowledge involves the use of either one species or group of species woven together along with the use of various construction techniques to make fence durable. A fence in a farm or home requires knowledge of good materials, the right method of construction and proper decision making. Different fencing work faces distinct issues [12]. Two major species of trees and plants of Chhattisgarh Forest are Tectona grandis (Teak) and *Shorea robusta* (Sal). Various other species are *Madhuca indica* (Mahua), *Diospyros melanoxylon, Pterocarpus marsupium, Terminalia tomentosa, Anogeissus latifolia, Emblica officinalis*, and *Dendrocalamus strictus, Vachollia nilotica* commonly known as babul, *Anogeissus pendula* constitute a significant chunk of middle canopy of the Chhattisgarh Forest.

The tribes and farmers have well established knowledge of trees and species available in the forest, and they use it extensively to serve their livelihood and prevent their houses and farms as well. They use thorns of various trees as fences to keep away wild animals. Huge forest land and availability of various species of trees helps the tribes to use them in construction of Dry fence. Trunk of tree is used for post or support to fence. Dry stem of a tree is used to bind posts together. Popularly bamboo (*Dendrocalamus strictus*), Sal (*Shorea robusta*), *Acacia nilotica* commonly known as babul, Neem (*Azadirachta indica*) are used for the Post material. A ditch of 300 to 450 mm deep is dug and trunks are fixed together in regular intervals. The post and dry stems are bonded together with ropes of coconut tree leaves. The soil is rammed to fill the hole and keep the post in position [13].

32.3.3   CONSTRUCTION TECHNIQUE OF DRY FENCE

The construction of dry fence involves the indigenous knowledge of post spacing, depth of ditch and types of species used. The choice of dry fence depends on tradition from years, budget, and knowledge-based on durability and life of dry fence. The holes are dug along the periphery of the area to be covered. The holes can be dug by hand, and the soil is rammed back around the post using a crowbar. Sometimes locally available stones are used to support the fence. Then the trunks are paced in holes at required intervals. The trunks are placed closely and at regular intervals as well. The coconut rope is used to bind these trunks together [14]. Construction of dry fence is done with tree trunk and dry stems of tree. Here local grass and coconut rope is used to tie them together. Figure 32.2 shows compound gates are also made out of bamboo (locally called Kamachi) and ballies (Tree trunk) to add security to their huts. Dry stems are usually collected from fallen trees or from forest nearby.

The use of dry stems, thorns of trees, bamboo mats are common practice in fence construction. Although due to lack of resources, material used for the fence construction is usually taken from nature. The understanding of material properties, strength, behavior, elasticity limit, load, and durability shows that local tribes of Chhattisgarh have excelled in using natural resources to the fullest.

Bamboo is a sustainable, renewable, environment friendly and widely available. Bamboo turns up to a very suitable building material due to its adaptability to climatic conditions, rapid growth. Most of the time, bamboo is used on fences. Bamboo is a versatile material with its high strength and availability [15]. The elasticity and strength of bamboo is well understood and used in dry fences. Bamboo is used in green as well as in dry state. Green bamboo is stretched and bent without breaking it into pieces, without the use of Sutali (coconut rope is used to tie the members). Nails are used minimum and the members are bonded together with cane rope. The perception and use of bamboo are renewable in nature. The community knowledge of using bamboo in houses will be greatly encouraged with the matter of long term and green building practices. This also helps in reducing global and climate change issues. Bamboo mat is used as a common material in fences by tribes. Mat is made from bamboo sticks cut in long slices and woven to mat. These mats not only cover the boundary but also give privacy to the space. Here built the environment is greatly created by the use of bamboo mats [16, 25].

FIGURE 32.2   Dry fence.

32.3.4   *TREE LEAVES USED TO COVER FENCE*

The use of tree leaves to cover fence is another popular practice adopted by tribes of Chhattisgarh. Large palms with high straight single trunks, full canopies can be good choices planted near a tall fence, either in row or single. The most commonly used are thorns of *Vachellia nilotica* (babul), Arecaceae (Palm tree), Cyatheales (Fern) to cover fence. To keep wild animals away from Livestock and their farms, farmers, and tribes prefer dry thorn stems as a fence material [17]. Palm leaves have multiple uses like covering thatch roof, dry fence as well as in catchment areas. It is observed that most species are used in live fence is safeguard their crops and houses from wild animals. Out of which few species are used both in live fence and dry fence.

Table 32.2 shows the various plant species used in Live and Dry fences [7]. Out of 229 totals of trees, shrubs, and herbs species available in Chhattisgarh state, it is observed that 29 species of live fencing, tree trunks, and dried tree stems are used in fences as they are easily available. Few trees have thorns which are used as fence cover to stay away from animals from farms and houses. Sometimes green leaves of trees are used to cover fences. This shows knowledge of these people about uses of various species available in their surroundings and the best way of utilizing it. It is observed that timber species dominate in both areas of live and dry fence as well.

32.4   ANALYSIS

After this study, it is clear that locals of Chhattisgarh use their expanded knowledge of tree species and plants in their possible means. India experiences different climates and hence is divided into six major climatic zones. As Chhattisgarh falls in Hot and Humid climate, the use of different species in fences varies as climate changes. Use of these species in fences differs from those used in an extremely opposite climate. Himachal Pradesh falls under the cold and cloudy climate of India and Chhattisgarh extreme opposite of Himachal Pradesh. Both are extremely different climatic zones and species variation is clearly seen as plant species and their growth has a strong contribution to climate [24]. After Comparing these two extreme climates fence techniques, it is observed that traditional knowledge of using similar plant species remains the same.

Table 32.3 shows different species which are most commonly used in both Chhattisgarh and Himachal Pradesh. All these common species are used both in live and dry fences. This shows that indigenous knowledge is similar even in extreme climates and use of thorns, dry leaves, stems are popularly used all over. Plants listed above are common in both climates, and the knowledge of using these plant species is also known to indigenous people.

TABLE 32.2   List of Plant Species as Fences

SL. No.	Species Name (Local Name)	Live Fence	Dry Fence
1.	*Acacia caesia*	Yes	No
2.	*Acacia eatechu* (Khair)	Yes	No
3.	*Adhatoda vasica*	Yes	No
4.	*Anogeissus latifolia* (Dhawra)	No	Yes
5.	*Anogeissus pendula* (Hingla)	No	Yes
6.	*Azadirachta indica* (Neem)	No	Yes
7.	*Bambusa vulgaris* (Bamboo)	Yes	Yes
8.	*Calotropis gigantea* (Ruii)	Yes	No
9.	*Diospyros melanoxylon* (Tendu)	No	Yes
10.	*Emblica officinalis* (Amla)	No	Yes
11.	*Erythrina variegate*	Yes	No
12.	*Euphorbia tithymaloides*	Yes	No
13.	*Ficus benghalensis* (Bargad)	No	Yes
14.	*Gliricidia sepium*	Yes	No
15.	*Hygrophila auriculata*	Yes	No
16.	*Ipomea carnea* (Beshram)	Yes	No
17.	*Jatropha curcas*	Yes	No
18.	*Lawsonia inermis* (Henna)	Yes	No
19.	*Madhuca indica* (Mahua)	No	Yes
20.	*Morus alba*	Yes	No
21.	Pigeon pea, red gram	Yes	Yes
22.	Pteridophytes (Fern)	Yes	Yes
23.	*Pterocarpus marsupium* (Bija)	No	Yes
24.	*Shorea robusta* (Sal)	No	Yes
25.	*Tectona grandis* (Teak)	No	Yes
26.	*Terminalia tomentosa* (Saja)	No	Yes
27.	*Thevetia peruviana*	Yes	No
28.	*Vachollia nilotica* (Babool)	No	Yes
29.	*Zizyphus mauritiana* (Ber)	No	Yes

TABLE 32.3 Comparison of Plant Species Commonly Used as Fences in Chhattisgarh and Himachal Pradesh

SL. No.	Species Name	Live Fence	Dry Fence
1.	*Acacia eatechu* (Khair)	Yes	No
2.	*Adhatoda vasica*	Yes	No
3.	*Bambusa vulgaris* (Bamboo)	Yes	Yes
4.	*Ipomea carnea* (Beshram)	Yes	No
5.	*Jatropha curcas*	Yes	No
6.	*Lawsonia inermis* (Henna)	Yes	No
7.	*Vachollia nilotica* (Babool)	No	Yes
8.	*Zizyphus mauritiana* (Ber)	No	Yes

32.5 DISCUSSION

Chhattisgarh state experiences hot and humid climate with a large variety of species used in fencing practices. Few plants are good for medicinal use; few are good in keeping animals away either by their thorns or by smell. After comparing species in Chhattisgarh and Himachal Pradesh, it is observed that few species grow in both climates and are used in similar fence methods. Common species in both the climate are *Acacia eatechu, Adhatoda vasica, Bambusa vulgaris, Ipomea carnea, Jatropha curcas, Lawsonia inermis, Vachollia nilotica, and Zizyphus mauritiana.* Out of 61 species available for fence in the form of tree, shrub, thorns, and climbers in the Himachal Pradesh region, 17 species are most commonly used in all forms of fence. In Chhattisgarh, 29 species are widely used as they are safe to plant near the neighborhood. These after comparing different plant species that grow in both the regions, it is seen that eight species are commonly used. Even in varied climates, this traditional knowledge helps in preserving nature irrespective of different plant species used.

32.6 CONCLUSION

It is observed that species which are available in deep forest zones are rarely used in fences. Plants which can be easily grown, less dangerous to mankind and their availability are popular among communities. Some species, although are very dangerous to be used as fences, still locals prefer them planting near their homes to avoid wild animal's intrusion. These practices can be further modified to match the urban scenario so as to utilize the natural resources. Extreme differences in climate also differ the use of species and so as their use in fences. Although it is clearly seen that few species are commonly used as they grow in both the regions. This custom is irrespective of climatic change as it has been practiced as old indigenous knowledge of fences. In modern days, village settings have been modified to take advantage of modern technologies. Fence wires with electrical supply are used which is not at all good for wildlife. Wild animals when come in contact with these wires, current flowing through these wires causes death to the wildlife. New technology also involves the use of solar powered dry fences, leads to wild life losses, and increase in numbers. Hence it is proved that fencing techniques adopted as Indigenous techniques remain environment friendly and support sustainable approach. Live fence, dry fence, thorns, dried leaves are different ways of using fences, but the purpose of using them remains the same all over India. These indigenous practices do not harm wildlife and promote sustainable practices in tribal settings. Thus, these indigenous practices are more like green building practices, which are greatly helpful to develop a sustainable environment over modern-day practices.

KEYWORDS

- dry fence
- green practices
- indigenous knowledge
- *Shorea robusta*
- sustainable materials
- traditional techniques

REFERENCES

1. Singh, R., & Sureja, A., (2008). Indigenous knowledge and sustainable agricultural resources management under rainfed agro ecosystem. *Indian Journal of Traditional Knowledge, 7*(4), 642–654.
2. Hitinayake, H. M. G. S. B., Priyadarshana, G. V. U., & Waidyarathna, D. M. K., (2018). Living fences, a widespread agroforestry practice in Sri Lanka: Two cases from dry and

intermediate zones. *International Journal of Environment, Agriculture and Biotechnology (IJEAB), 3*(2), 701–709.

3. Senanayake, S. G. J. N., (2006). Indigenous knowledge as a key to sustainable development. *The Journal of Agricultural Sciences, 2*(1), 87–94.

4. *Chhattisgarh Action Plan for Climate Change*, (2013). Government of Chhattisgarh. Final Draft.

5. Singh, B., (2008). *Studies on Prospects and Potentials of Ecotourism in Chhattisgarh with Special Reference to Biodiversity Conservation of Protected Areas*. An Institutional Repository of Indian National Agricultural Research System. Krishikosh.

6. Singh, B. P., (2017). Biodiversity, tribal knowledge, and life in India. *Environment and Social Psychology, 2*(1), 1–10.

7. Choudhury, P. R., Rai, P., Patnaik, U. S., & Sitaram, R., (2005). Live fencing practices in the tribal-dominated eastern ghats of India. *Agroforest System, 63*, 111–123.

8. Jayavanan, K., Pushpakumara, D. K. N. G., & Sivachandran, S., (2014). Role of live fence agroforestry in Jaffna peninsula. *Tropical Agriculturist, 162*, 25–44.

9. AESA-based IPM Package, (2014). *AESA based IPM-Redgram*. Department of Agriculture and Corporation, Ministry of Agriculture, Government of India.

10. Saxena, P. K., Nanda, D., Gupta, R., Nitin, K., & Tyagi, N., (2017). A review on ipomoea carnea: An exploration. *International Research Journal of Pharmacy, 8*(6).

11. Sarkar, S., Chakraverty, R., & Ghosh, A., (2014). *Calotropis gigantea* Linn. A complete busket of Indian traditional medicine. *International Journal of Pharmacy Research and Science, 2*(1), 7–17.

12. Buschermohle, M. J., Wills, J. B., Gill, W. W., & Lane, C. D., (2001). *Planning and Building Fences on the Farm*. The University of Tennessee Agricultural Extension Service. 2001-2.5M-5/01(Rep) E12-4315-00-02-01.

13. Worley, J., (2015). *Fences for the Farm*. The University of Georgia College of Agricultural and Environmental Sciences., Circular 774 by University of Georgia Extension.

14. Harvey, C. A., Tucker, N. J., Estrada, A., (2004). Live fences, isolated trees, and windbreaks: tools for conserving biodiversity in fragmented tropical landscapes. *Agroforestry and Biodiversity Conservation in tropical Landscapes*. 261–289.

15. Raj, D., & Agarwal, B., (2014). Bamboo as a building material. *Journal of Civil Engineering and Environmental Technology, 3*, 256–61.

16. Das, P., Korde, C., Sudhakar, P., & Satya, S., (2012). Traditional bamboo houses of north-eastern region: A field study of Assam & Mizoram. *Key Engineering Material, 517*, 197–202.

17. Forline, L. C., (2000). Framework for enhancing the use of indigenous knowledge. *Indigenous Knowledge & Development Monitor, 8*(3).

18. Rao, K. S., Maikhuri, R. K., Nautiyal, S., & Saxena, K. G., (2002). Crop damage and livestock depredation by wildlife: A case study from Nanda Devi biosphere reserve, India. *Journal of Environmental Management, 66*, 317–327.

19. Mehta, P., Negi, A., Chaudhary, R., Janjhua, Y., & Thakur, P., (2018). A study on managing crop damage by wild animals in Himachal Pradesh. *International Journal of Agriculture Sciences, 10*(12), 6438–6442.

20. Mittal, S., (2016). *Crop Damage by Wild Animals Declined 33%: Government*. Indiaspend.

21. Jayson, E. A., (1999). *Studies on Crop Damage by Wild Animals in Kerala and Evaluation of Control Measures* (pp. 12–34). Kerala forest research institute peechi, Thrissur.

22. Chauhan, N. P. S., & Singh, R., (1990). Crop damage by overabundant populations of nilgai and blackbuck in Haryana (India) and its management. *Proceedings of the Fourteenth Vertebrate Pest Conference, 13.*

23. *India State of Forest Report,* (2019). 11.5 Chhattisgarh.

24. Sharma, P., & Devi, U., (2013). Ethnobotanical uses of biofencing plants in Himachal Pradesh, Northwest Himalaya. *Pakistan Journal of Biological Sciences, 16*(24), 1957–1963.

25. Panda, T., Mishra, N., Pradhan, B. K., & Mohanty, R., (2018). Live fencing: An eco-friendly boundary wall in Bhadrak district of Odisha, India. *International Journal of Conservation Science, 9*(2), 301–310.

CHAPTER 33

Catalytic Potential of Phyto-Synthesized Silver Nanoparticles for the Degradation of Pollutants

DHRUTI SUNDAR PATTANAYAK,[1] DHARM PAL,[1]
CHANDRAKANT THAKUR,[1] PRANAY RAUT,[1] and K. L. WASEWAR[2]

*[1]Department of Chemical Engineering, National Institute of Technology Raipur, Raipur – 492010, Chhattisgarh, India,
E-mail: dpsingh.che@nitrr.ac.in (D. Pal)*

[2]Department of Chemical Engineering, VNIT, Nagpur–440010, Maharashtra, India

ABSTRACT

Green nanotechnology provides a wide area for research in the field of science and engineering. The development of an inexpensive and environmentally responsive preparation method for the production of nano-size materials, particularly metal nanoparticles is a promising area of research. In the recent past, the green synthesis of metal nanoparticles has attracted the researchers as it is a hazard free, less toxic, quick, and environmentally benign synthesis techniques. Among the metal nanoparticles silver nanoparticle (AgNPs) has especially fascinated the researchers owing to its dynamic properties. Researches have been carried out on the synthesis of AgNPs through biological, chemical, and physical methods. Among these, the biological method is one of the promising methods because it is benign to the natural environment. Bacteria, plant, fungus, etc., were used for biosynthesis of AgNPs. However, plant-based synthesis of AgNPs is the best biogenic procedure, due to some drawbacks associated with biosynthesis using bacteria. It is difficult to maintain the bacterial culture under specific environment and to avoid contamination before utilization.

In the past decade, researchers have attempted the production of AgNPs by using different parts of several plants. Different biomolecules such as protein, amino acid, enzymes, cyclic peptides, flavonoids, polyphenols, and other substances available in plant as the natural reductants are mainly responsible for the synthesis of AgNPs. Phytosynthesized AgNPs has shown enhanced activity and favorable performance in biomedical, antimicrobial, diagnostics, and environmental applications. The effluent from various industries poses a serious threat to the receiving environment. The photocatalytic system had shown outstanding efficiency for the degradation of different kinds of pollutants from wastewater with economic feasibility. AgNPs can play a specific role in different catalytic reactions. This review provides a systematic and brief account on the eco-friendly, Phyto-assisted synthesis of AgNPs and its catalytic potential for degradation of various environmental pollutants.

33.1 INTRODUCTION

In the last few years nanotechnology have created a base for research in diverse area of science and engineering. The potential difference of nano scale materials (1–100 nm) from bulk material is the ratio between large surface area (SA) and volume. Metal nanoparticles, basically silver nanoparticles (AgNPs) are the outstanding candidates in nanotechnology owing to its impressive and dynamic properties shown in the application in the fields such as drug delivery, water treatment, catalysis, biosensing, as antibiotic agents, etc., [1]. Several methods were developed for the production of noble AgNPs, such as physical, chemical, and biological methods. The chemical reduction method is frequently used method for the preparation of AgNPs, but toxic and costly chemicals are always needed during complete synthesis process. So, there is a need of green and environmentally benign method for minimization of hazardous chemicals uses and to overcome the adverse situation arises in conventional method. The bio-reduction methods are eco-friendly, less expensive, and environmentally benign, cost-effective, and simple [2]. Therefore, in the recent past, the researchers are influenced to synthesize AgNPs by using biogenic material.

The biological source like plants, algae, fungi, and bacteria were used by different researchers for the synthesis of AgNPs. Among biological source, plant mediated synthesis is more effective than others as plants are less responsive to metal toxicity. Generally, plant extracts are used for the synthesis of AgNPs and seems to be a most popular green method. The phyto-synthesized route is low cost, environmentally friendly and more advantageous over chemical reduction. Different types of plants have been effectively introduced for reducing silver ion (Ag+) to silver metal nanoparticle (Ag^0) [3]. The phytochemicals such as polyphenolic, anthocyanin, proteins, saccharides, terpenes, oils, vitamins, enzymes, etc., acts as reducing agent, capping agents, and stabilizing agents. These phytoconstituents is responsible for metal ion reduction and prevent the nuclei from aggregation and more accountable for the shape and size of nanoparticles [4, 5]. Now, all over the world is facing a serious concern like water pollution, especially in developing countries. Both Organic and inorganic pollutants are the important sources of water pollution due to their toxic nature, non-biodegradability, and higher chemical stability. The AgNPs have kept more attention from various metal NPs as it is relatively cheap and its market availability is more. So, the use of AgNPs as a catalyst for the catalytic reduction is the ideal path for removing the pollutants from wastewater [6]. In present time, AgNPs are extensively used to enhance catalytic potential in redox reactions (reduction and oxidation) process [7]. The AgNPs are the renowned catalysts and encourage the electron transfer through alternative pathway by requiring less activation energy. It is primarily important for catalytic application in industrial purposes [8].

33.1.1 PLANT MEDIATED SYNTHESIS OF SILVER NANOPARTICLES (AgNPs)

Biosynthesis could be green alternatives to chemical and physical methods for the synthesis of AgNPs. Bacteria, plants, fungi are used as biomaterial for the synthesis of AgNPs. Among them plant fascinated synthesis method is the better strategy as compared with bacteria and fungi. Huge availability and abundance of plants has drawn the attention of researchers for biogenic synthesis of AgNPs. Plant parts such as leaves, flowers, fruits, seed, bark, stem, etc., were used in the synthesis process. Basically, some phytochemicals or plant metabolites available in plant parts are accountable for the synthesis of NPs. The phytochemicals such as protein, amino acid, enzymes, cyclic peptides, flavonoids, polyphenols, and other substances contained in plant extracts helps at the time of nanoparticle synthesis. These phytochemicals play a key role in reducing Ag^+ ion to Ag^0 NPs. The plant metabolites play a significant role in all the three stages of AgNps formation such as reduction, capping, and stabilization [9]. The formation of AgNPs passes through three stages, namely, reduction phase, growth phase and stabilization phase.

Figure 33.1 shows the hypothetical mechanism of formation of AgNPs through various stages.

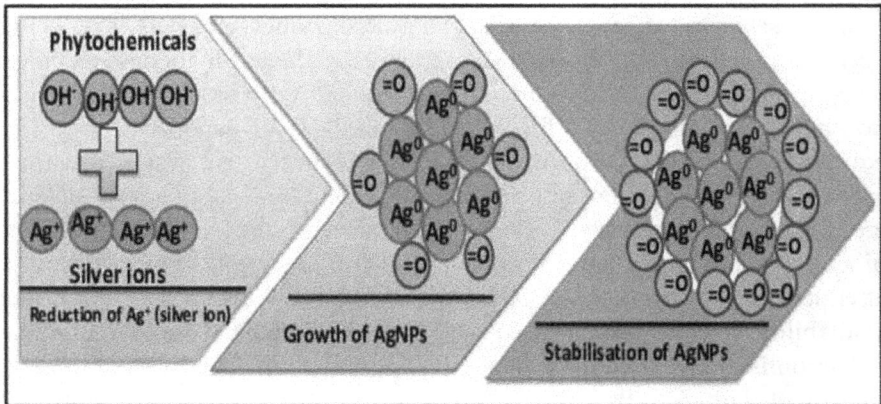

FIGURE 33.1 Stages of phyto-synthesis (green synthesis) of silver nanoparticles.

The formation of AgNPs is established through different characterization techniques using sophisticated instruments. The size, shape, surface morphology, stability, etc., could be revealed after performing the various characterizations. Several researchers have reported the synthesis of AgNPs using plant extract. The phyto-synthesized AgNPs mostly show spherical structure with size in the nano range. Different plants re-reported which have been used for AgNPs synthesis. A brief accounts of different plant used for AgNPs synthesis, plant part, shape, and size of AgNPs are summarized in Table 33.1.

A brief account of different works reported on AgNPs synthesis by using various plants is presented in Table 33.1. The Ag⁺ were reduced to AgNPs by using fruit extract of Mortiño (Vaccinium floribundum Kunth) [1]; Alpinia nigra [18]. The reduction of silver ion to AgNPs were carried out by using the bark extract of *Dillenia indica* [2]; bulb extract of *Crocus Haussknechtii* Bois [10]; stem bark extract of *Soymida febrifuga* [22]; ribbed stem extract of *Angelica gigas* [27]; and nut extract of *Areca catechu* [36]. Similarly, several researches have been published where the Ag⁺ reduces to AgNPs by using the seed extract of Grape [3]; *Cuminum cyminum* [11]; *Dimocarpus longan* [14]; *Trachyspermum Ammi* (Ajwain) [29]. Also, the silver ion reduces to AgNPs by using the flower extract of *Verbascum thapsus* [26]; Cassia auriculata [32];

TABLE 33.1 Phyto-Synthesis of Silver Nanoparticles (AgNPs) by Using Different Plant Extracts

SL. No.	Plant Name	Plant Part	Shape	Size	References
1.	Mortiño (*Vaccinium floribundum* Kunth)	Fruits	Sph.*	20.5 + 1.5 nm	[1]
2.	*Dillenia indica*	Bark	Sph.*	15–35 nm	[2]
3.	Grape	Seed	–	–	[3]
4.	Piper betle	Leaves	–	–	[5]
5.	*Ginkgo biloba*	Leaves	Sph.*	20–40 nm	[6]
6.	Tulsi	Leaves	–	5–10 nm	[7]
7.	*Phyllanthus amarus*	Leaves	–	–	[8]
8.	*Crocus haussknechtii* Bois	Bulb	Sph.*	10–25 nm	[10]
9.	*Cuminum cyminum*	Seed	Sph.*	16 + 2 nm	[11]
10.	*Momordica charantia*	Leaves	Sph.*	16 nm	[12]
11.	*Erigeron bonariensis*	Leaves	Sph.*	13 nm	[13]
12.	*Dimocarpus longan*	Seed	–	40 nm	[14]
13.	*Convolvulus arvensis*	Leaves	Sph.*	28 nm	[15]
14.	*Convolvulus arvensis*	Leaves	Sph.*	10–30 nm	[16]
15.	*Prosopis juliflora*	Leaves	–	10–20 nm	[17]
16.	*Alpinia nigra*	Fruits	Sph.*	6 nm	[18]
17.	*Ficus panda*	Leaves	Sph.*	12–36 nm	[19]
18.	*Polygonum hydropiper*	Leaves	Sph.*	60 nm	[20]
19.	*Parkia speciosa*	Leaves	Sph.*	35 nm	[21]
20.	*Soymida febrifuga*	Stem bark	Sph.*	10–30 nm	[22]
21.	*Ehretia laevis*	Leaves	–	5–20 nm	[23]
22.	*Passiflora edulis* F. Flavicarpa	Leaves	Sph.*	7 nm	[24]

TABLE 33.1 *(Continued)*

SL. No.	Plant Name	Plant Part	Shape	Size	References
23.	*Verbascum thapsus*	Flower	Sph.*	23.2 nm	[26]
24.	*Angelica gigas*	Ribbed stem	Sph.*	20–80 nm	[27]
25.	*Dolichos lablab*	Leaves	Sph.*	9 nm	[28]
26.	*Trachyspermum ammi* (Ajwain)	Seed	–	–	[29]
27.	*Solanum nigurum*	Leaves	Sph.*	3.46 nm	[30]
28.	*Lippia citriodora*	Leaves	–	23.8	[31]
29.	*Cassia auriculata*	Flower	Sph.* and triangle	10–35 nm	[32]
30.	*Paederia foetida* Linn.	Leaves	Sph.*	5–25 nm	[33]
31.	*Lawsonia inermis*	Leaves	Sph.*	18 nm	[34]
32.	*Allium ampeloprasum* L.	Leaves	Quasi-Sph.*, Sph.*, ellipsoidal, hexagonal, and irregular	2–43 nm	[35]
33.	*Areca catechu*	Nuts	Sph.* and irregular	12 nm	[36]
34.	*Thunbergia grandiflora*	Flower	Sph.*	–	[37]
35.	*Biophytum sensitivum*	Leaves	Sph.*	19.06 nm	[38]

Note: Sph.*: spherical.

Thunbergia grandiflora [37]. In some reported works the reduction of silver ion to AgNPs were achieved by using the leaf extract of *Piper betle* [5]; *Ginkgo biloba* [6]; Tulsi [7]; *Phyllanthus amarus* [8]; *Momordica charantia* [12]; *Erigeron bonariensis* [13]; *Convolvulus arvensis* [15, 16]; *Prosopis juliflora* [17]; *Ficus panda* [19]; *Polygonum hydropiper* [20]; *Parkia speciosa* [21]; *Ehretia laevis* [23]; *Passiflora edulis F. Flavicarpa* [24]; *Dolichos lablab* [28]; *Solanum nigurum* [30]; *Lippia citriodora* [31]; *Paederia foetida* Linn. [33]; *Lawsonia inermis* [34]; *Allium ampeloprasum* L. [35]; *Biophytum sensitivum* [38].

33.1.2 APPLICATION OF PHYTOSYNTHESIZED SILVER NANOPARTICLES (AgNPs)

There are multifacet applications of phyto-synthesized AgNPs with the advantages of less toxic and benign to the environment. Plant mediated synthesis of AgNPs is the promising method over traditional method on account of environment safety. The modular and dynamic properties enhance the uses of AgNPs in different fields. The AgNPs shows potential efficiency on various applications like antimicrobial, bio-medical, sensing, catalysis, etc. A brief discussion about the catalytic degradation potential of environmental pollutants by Phyto-synthesized AgNPs is appended below.

33.1.3 CATALYTIC ACTIVITIES OF PHYTOSYNTHESIZED SILVER NANOPARTICLES (AgNPs) AND IT IS MECHANISM OF ACTION

The AgNPs are utilized in different water treatment process owing to its catalytic potential. Mostly, AgNPs is used at industrial scale for the treatment of environmental pollutants. Different researchers have proposed mechanism of action for degrading of the pollutants. Degradation of pollutants occurs in two ways, i.e., reduction, and oxidation. The basic mechanism of degradation of pollutant via reduction in presence of phyto-synthesized AgNPs is represented in Figure 33.2.

The catalytic mechanism is based on redox reaction. When the light irradiates on the synthesized AgNPs, the electron gets excited by absorbing photon energy and the excited electron moves from valance band to empty conduction band. Due to which electron hole-pair started at valence band. The hole-pair and excited electrons produce free radicals such as $OH^•$, $O_2^{-•}$ and $HOO^•$ by reacting with water and dissolved oxygen which is available in reaction solution. Ultimately, the free radicals catalyze the degradation of environmental pollutants and convert them from toxic to non-toxic or less toxic [18]. Different researchers have studied the catalytic degradation of various environmental pollutants by making use of phyto-synthesized AgNPs.

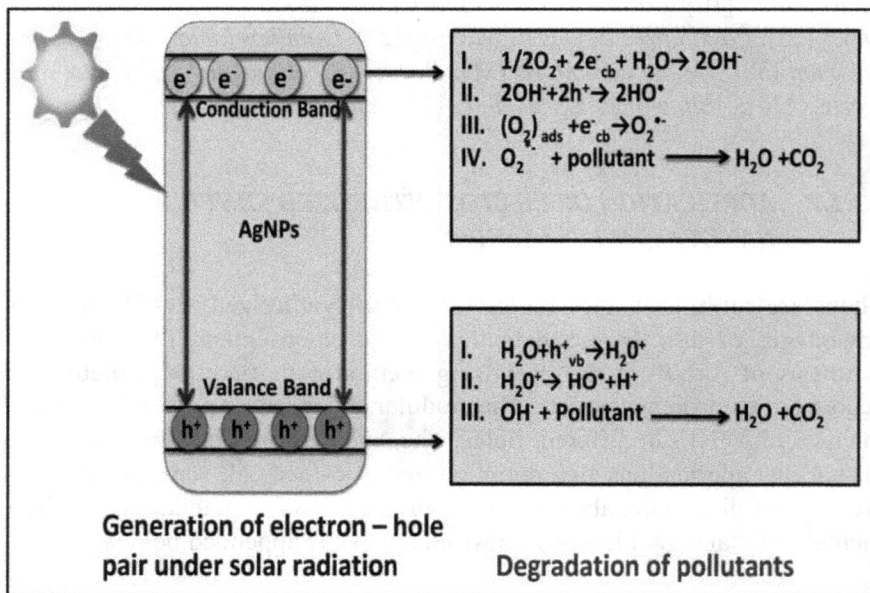

e⁻ e⁻ e⁻ e-
Conduction Band

AgNPs

Valance Band

h⁺ h⁺ h⁺ h⁺

I. $1/2O_2 + 2e^-_{cb} + H_2O \rightarrow 2OH^-$
II. $2OH^- + 2h^+ \rightarrow 2HO^\bullet$
III. $(O_2)_{ads} + e^-_{cb} \rightarrow O_2^{\bullet -}$
IV. $O_2^{\bullet -} + pollutant \longrightarrow H_2O + CO_2$

I. $H_2O + h^+_{vb} \rightarrow H_2O^+$
II. $H_2O^+ \rightarrow HO^\bullet + H^+$
III. $OH^- + Pollutant \longrightarrow H_2O + CO_2$

Generation of electron – hole pair under solar radiation   **Degradation of pollutants**

FIGURE 33.2   The photocatalytic degradation mechanism of pollutants.

A brief account of tested environmental pollutant, name of biomaterial used for AgNPs synthesis, method of degradation is listed in Table 33.2 followed by the detailed discussion on catalytic activity with plausible mechanism.

Kumar et al., have reported Phyto-synthesized AgNPs using *Andean Mortiño* berry as bio-reductant and the photocatalytic property was studied against methylene blue (MB) under sunlight. The proposed mechanism is that the surface hydroxyl groups enhance the acceptance of photo-induced electron and holes and form the superoxide ion. ($O_2^{-\bullet}$) reacting and hydroxyl ion (OH•). So, the photocatalytic process gets influenced [1]. Mohanty et al., discussed about synthesis of AgNPs using extraction of *Dillenia indica* which act as both in the process of reduction and capping. Besides these, green synthesized AgNPs nanoparticles forms an excited free radical scavenger and act as a reducing agent to convert 4-Nitrophenol to 4-aminophenol and MB at 25°C because of faster electron transfer in the presence of catalyst fastens the reaction rate [2]. Ping et al., suggested that using grape seeds extract AgNPs could be synthesized having moderate temperature and shown catalytic activity against Orange 26 by using strong reducing agent $NaBH_4$. The moderate temperature helps to maintain the size of nanoparticles and breaking the double bond of Azo dye occurring in the transition state, which leads to easy degradation. After that the amino group in the transition state can be further decayed. Uniform shape of AgNPs having enhanced SA which increases the rate of dye degradation along with two by-products aniline and formaldehyde [3]. Ankamwar et al., accounted for the formation of AgNPs using plant extracts of Piper betle plant and photocatalytic study carried out for conversion of 4-nitrophenol to 4-aminophenol using $NaBH_4$ at 0–4°C and measured its catalytic activity and a comparative study was made with the AuNPs synthesized by same process. AgNPs shows better catalytic property over AuNPs in the translation of 4-NP to 4-AP [5].

Wang et al., discussed about using the extract of Ginkgo Biloba leaves AgNPs for biosynthesis and produced spherical shape (20–40 nm). The formed NPs show catalytic activity in favor of four typical pollutants, i.e., 4-nitrophenol, Congo red, methyl orange and Rhodamine-B by $NaBH_4$ and within 100. min the reaction finished completely and follows pseudo first-order reaction kinetics as well as having potential of five times reusability with more than 70% activity [6]. Singh et al., formed about the production of biogenic AgNPs using Tulsi leaf extract utilizing sun radiation at mild temperature. They have also studied the catalytic property of biogenic AgNPs used for 100% conversion of 4-NP to 4-AP under alkaline condition using $NaBH_4$ as a reductant and the E_a and TOF value were calculated to be 40.25 kJ mol^{-1} and 6.3×10^{17} molecules g^{-1}s^{-1}, respectively with additional advantages of reusability of catalyst [7]. Ajitha et al. synthesized the biogenic AgNPs using *P. amarus* leaf extracts and studied its photocatalytic behavior against Rhodamine B dye and suggested that the mechanism involves that the formed photocatalyst prevent the further catalysis by coating them and act as a capping agent forming electron-rich hydroxyl group [8]. Mosaviniya et al., have synthesized spherical shaped AgNPs with dimeter 10–25 nm using fresh bulb of Crocus Haussknechtii Boisplant and shown the catalytic property against Congo Red Dye in the presence of $NaBH_4$ [10].

Choudhary et al. the AgNPs using the Cuminum cyminum seed's extract and the synthesized nanoparticles was used as nanocatalyst in presence of $NaBH_4$ as a reductant for reduction of four dye pollutant namely, MB, methyl red (MR), Rhodamine-B, and 4-nitrophenol [11]. Ajitha et al., studied the green synthesis of Momordica charantia leaf extract 's in which both reduction and stabilization takes place and catalytic activity was evaluated via negatively charged COO$^-$ group of proteins [12]. Kumar et al. have reported,

TABLE 33.2 Catalytic Activity of Silver Nanoparticles Synthesizing by Using Plant Extracts

SL. No.	Plant Name	Test Pollutants	Method	References
1.	Mortiño (*Vaccinium floribundum* Kunth)	Methylene blue	Redn.*	[1]
2.	*Dillenia indica*	4-nitrophenol and methylene blue	Redn.*	[2]
3.	Grape	Direct Orange 26	Redn.*	[3]
4.	Piper betle	4-nitrophenol and 4-aminophenol	Redn.*	[5]
5.	*Ginkgo biloba*	4-nitrophenol, Congo red, methyl orange and rhodamine B	Redn.*	[6]
6.	Tulsi	4-Nitrophenol	Redn.*	[7]
7.	*Phyllanthus amarus*	Rhodamine B.	Redn.*	[8]
8.	*Crocus Haussknechtii* Bois	Congo red	Redn.*	[10]
9.	*Cuminum cyminum*	Methylene blue (MB), methyl red (MR), rhodamine-B (Rh-B) and 4-nitrophenol (4-NP).	Redn.*	[11]
10.	*Momordica charantia*	Methylene blue	Redn.*	[12]
11.	*Erigeron bonariensis*	Acridine orange	Redn.*	[13]
12.	*Dimocarpus longan*	Methylene blue, 4-nitrophenol (4-NP)	Redn.*	[14]
13.	*Convolvulus arvensis*	Methylene blue (MB)	Redn.*	[15]
14.	*Convolvulus arvensis*	Azo dyes	Redn.*	[16]
15.	*Prosopis juliflora*	Methylene blue (MB) and Congo red (CR)	Redn.*	[17]
16.	*Alpinia nigra*	Methyl orange, Rhodamine B and Orange G	Redn.*	[18]
17.	*Ficus panda*	Methylene blue	Redn.*	[19]
18.	*Polygonum hydropiper*	Methylene blue	Redn.*	[20]
19.	*Parkia speciosa*	Methylene blue	Redn.*	[21]
20.	*Soymida febrifuga*	Methylene blue, rhodamine B and eosin Y	Redn.*	[22]

TABLE 33.2 (Continued)

SL. No.	Plant Name	Test Pollutants	Method	References
21.	Ehretia laevis	Alcohol	oxidation	[23]
22.	Passiflora edulis F. Flavicarpa	Methyl orange and methylene blue	Redn.*	[24]
23.	Verbascum thapsus	Nitrobenzene	Redn.*	[26]
24.	Angelica gigas	Eosin Y (EY) and malachite green (MG)	Redn.*	[27]
25.	Dolichos lablab	4-Nitrophenol	Redn.*	[28]
26.	Trachyspermum ammi (Ajwain)	p-nitrophenol	Redn.*	[29]
27.	Solanum nigurum	4-Nitrophenol	Redn.*	[30]
28.	Lippia citriodora	Methylene blue	Redn.*	[31]
29.	Cassia auriculata	4-nitrophenol and methyl orange	Redn.*	[32]
30.	Paederia foetida Linn.	Rhodamine B (RhB).	Redn.*	[33]
31.	Lawsonia inermis	4-nitrophenol	Redn.*	[34]
32.	Allium ampeloprasum L.	4-nitrophenol	Redn.*	[35]
33.	Areca catechu	Methylene blue	Redn.*	[36]
34.	Thunbergia grandiflora	Congo red	Redn.*	[37]
35.	Biophytum sensitivum	Methyl orange and methylene blue	Redn.*	[38]

Source: Redn.*: reduction.

synthesized aqueous extract of Erigeron bonariensis AgNPs which is stable and spherical in natures which have provided both reduction and stabilizing property and gave catalytic action towards degradation of Acridine Orange (AO) without using any external reducing agent, and followed pseudo-first-order kinetics [13]. Khan et al., have synthesized AgNPs using Dimocarpus longan plant's seed extract and found out that the seed extract contains high levels of polyphenolic compound, which show both reducing and stabilizing properties. The nanoparticles show the catalytic property for nitro-phenol, which reduces to amino-phenol and photocatalytic degradation of MB due to its small size and capping phytochemicals [14]. Hamedi et al., reported that photo-synthesized AgNPs was synthesized by using *Convolvulus arvensis* leaf extract and the catalytic performance was measured by reducing the MB dye with the rate constant (k) 0.108 min^{-1} [15].

Rasheed et al. reported about biosynthesized AgNPs using the Convolvulus Arvensis extract and investigated catalytic performance employing synthesizing AgNPs that act as reducing agents for three synthetic azo dyes using [16]. Arya et al., synthesized AgNPs using leaf extract of Prosopis juliflora and showed catalytic potential degradation of Congo Red, MB, and conversion of 4-Nitophenol to nontoxic 4-Aminophenol [17]. Baruah et al., studied AgNPs prepared using the fruit extract of Alpinia nigra, which is rich in saponins, glycosides, alkaloids, and steroids. The catalytic activity was measured against Methyl orange, Rhodamine B and Orange G, respectively. The catalysis mechanism depends on radiation of light, due to which electron get excited from valence band of Ag to the CB of the Ag which leads to Ag act as good oxidizing agent and formed e$^-$-h$^+$ pair. The excited electron formed superoxide ion ($O_2^{-•}$) reacting with oxygen, hydroxyl ion (OH$^•$) and HOO$^•$ at the surface of the catalyst, which is known as active sites. Ultimately which are catalyze the pollutant later [18]. Bonnia et al., focused on the biogenic synthesized AgNPs using the extract of Polygonum hydropiper and the catalytic efficiency was measured against MB dye with the strong reducing agent NaBH$_4$. Catalytic activity carried out through electron transfer process [20].

Ravichandran et al., reported the bio-inspired AgNPs synthesized from P. Speciosa leaves extract and act as catalyst in the presence of visible light for degradation of MB dye up to 40% [21]. Sowmyya et al., have studied the biosynthesized stem bark extract of *Soymida febrifuga* AgNPs forming a dark brown color. The catalytic activity was studied against MB, Rhodamine B and Eosin Y [22]. Warghane et al., have investigated synthesized green AgNPs utilizing the extract of *Ehretia laevis* plant and studied catalytic performance towards p-cresol and catalytic oxidation of alcohols containing aromatic ring [23]. Thamas et al., reported that phyto-synthesized AgNPs was synthesized using the leaf extract of *Passiflora edulis. F. flavicarpa (P. edulis)* and the catalytic performance was measured against organic dye. The mechanism behind the catalytic activity was suggested as the fall of solar irradiation on the NPs (suspended in the reaction mixture), causes electrons to get excited and reduced the dissolved oxygen present in the solution. After reduction they became superoxide radical and lead to degrading of organic dye. They were able to convert toxic organic compounds to less toxic byproducts. Therefore, phyto-synthesized AgNPs act as an efficient catalyst for degradation of the Methyl Orange and MB [24]. Elemike et al., have produced spherical AgNPs by flower extract of Verbascum Thapsus plant in which both reduction and stabilization takes place. The produced AgNPs show tremendous catalytic potential against nitrobenzene [26].

Chokkalingam, et al. have reported that AgNPs was synthesized using stem aqueous extract of Angelica Gigas and studied the catalytic reduction against Eosin Y (EY) and malachite green (MG) dye [27]. Kahsay et al. produced an aqueous leaf extract of Dolichos lablab of AgNPs having spherical shape with particle size 4–16 nm and showed the catalytic property for reduction of nitrophenol [28]. Chouhan et al., studied the biogenic AgNPs which was synthesized by using Trachyspermum Ammi (Ajwain)'s seed extract. The catalytic performance was observed by converting p-nitrophenol to p-aminophenol using reductant; $NaBH_4$ [29]. Vijilvani et al., have studied AgNPs which was formed by using Solanum nigrum leaves extract and show effective catalytic performance enabling complete reduction of 4-NP and the production of 4-Aminophenol. The phytosynthesized AgNPs requires very less time for complete conversion compared to others metal nanoparticles such as AuNPs and PdNPs. The AgNPs exhibit smaller size which increases the speed for reduction and due to the smaller particle, size time taken for reduction is comparatively less [30]. Muthu et al., carried out the research on ecofriendly AgNPs synthesized using C. auriculata flower extract and measured the catalytic performance against 4-nitrophenol and methyl orange [32].

Bhuyan et al., have studied biogenic AgNPs which was synthesized using aqueous extracts of *Paederia foetida* Linn. that act as a reductant, as well as a stabilizing agent. The catalytic activity was tested against Rhodamine B (RhB) which was potentially reduced by AgNPs in the presence of $NaBH_4$. The reported mechanism state that phyto-synthesized AgNPs leads to the formation of BH_4^- ion which get adsorbed on the surface of 4-nitrophenolate ions [33]. Ajitha et al., biosynthesized AgNPs by using the leaf extract of Lawsonia inermis plant and the catalytic performance was measured against 4-nitrophenol with $NaBH_4$ as a catalytic reagent. The mechanism behind the catalytic activity assumed that the BH_4^- ion and 4-nitrophenolate ions were adsorbed on the surface of phyto-synthesized AgNPs. The electron transfers in between two charged ions on the surface of catalyst cause the reduction of 4-NP to 4-AP [34]. Khoshnamvand et al., have discussed the synthesized biogenic AgNPs using leaf extract of A. ampeloprasum plant and the catalytic activity was observed against hazardous organic pollutants like 4-nitrophenol. As a result, after reduction by catalyst the toxic 4-nitrophenol converted to less toxic 4-aminophenol [35]. Vinay et al., produced the green synthesized AgNPs using the Areca catechu. The catalytic performance was tested against MB dye, and they have found potential result under visible light [36]. Varadavenkatesan et al., studied the phyto-synthesized AgNPs formed using flower of Thunbergia grandiflora plant and their catalytic activity was also measured against Congo red (CR) dye with $NaBH_4$. Here, AgNPs play the role of electron transferor in between $NaBH_4$ and Congo Red [37].

From the above literatures, it is a common consensus among most of the researchers that Phyto-synthesized AgNPs acts as a best reductant for environmental pollutant degradation, particularly dye degradation. In most of the study $NaBH_4$ acts as a strong reducing agent in the presence of the AgNPs. Catalytic mechanism is based on light irradiation on the synthesized AgNPs. Consequently, the electron gets excited and moves from valance band to empty conduction band. The formed electron hole-pair which used as both oxidizing and reducing agent is due to the reaction with oxygen in the reaction mixture and generation of free radicals. Those free radicals are mainly dependable for catalyzing the decomposition of the environmental pollutants and converting them from toxic to non-toxic or less toxic substances. Besides this, in some cases, AgNPs act as electron transporter in between $NaBH_4$ and pollutants.

33.2 CONCLUSION

Green technology is the ultimate path for today's; as it is keeping all sides of nature balanced. This review highlights the recent research's activities related to plant extract-based AgNPs and its photocatalytic property applied in various fields. The plants part such as leaves, stem, flowers, roots, fruits, and barks are being used extensively for AgNPs synthesis which is lower in cost, easily available and eco-friendly in nature. The photocatalytic properties of plant extract synthesize AgNPs has shown a new path for different application which in environmental friendly manner. The phyto-synthesized AgNPs act as a promising reductant for environmental pollutant degradation basically for dye degradation. The photocatalytic activity shown by the AgNPs by acting as capping agent and reductants is providing excellent result for degradation of organic pollutant, which can be employed at industrial scale in coming days and can show us a new era to our future life.

CONFLICT OF INTEREST

The author declared no conflict of interest.

KEYWORDS

- **catalytic properties**
- **eco-friendly**
- **methylene blue**
- **phytochemicals**
- **photosynthesis**
- **reductant**
- **silver nanoparticles (AgNPs)**

REFERENCES

1. Kumar, B., Vizuete, K. S., Sharma, V., Debut, A., & Cumbal, L., (2019). Eco-friendly synthesis of monodispersed silver nanoparticles using Andean Mortiño berry as reductant and its photocatalytic activity. *Vacuum, 160*, 272–278.
2. Mohanty, A. S., & Jena, B. S., (2017). Innate catalytic and free radical scavenging activities of silver nanoparticles synthesized using *Dillenia indica* bark extract. *J. Colloid Interface Sci., 496*, 513–521.
3. Ping, Y., (2018). Green synthesis of silver nanoparticles using grape seed extract and their application for reductive catalysis of direct orange 26. *J. Ind. Eng. Chem., 58*, 74–79.
4. Sable, S. V., Kawade, S., Ranade, S., & Joshi, S., (2020). Bioreduction mechanism of silver nanoparticles. *Mater. Sci. Eng., C107*, 110299.
5. Ankamwar, B., Kamble, V., Sur, U. K., & Santra, C., (2016). Spectrophotometric evaluation of surface morphology dependent catalytic activity of biosynthesized silver and gold nanoparticles using UV-vis spectra: A comparative kinetic study. *Appl. Surf. Sci., 366*, 275–283.
6. Wang, F., (2019). Extract of Ginkgo biloba leaves mediated biosynthesis of catalytically active and recyclable silver nanoparticles. *Colloids Surfaces A Physicochem. Eng. Asp., 563*, 31–36.

480 Sustainable Engineering, Energy, and the Environment

7. Singh, J., Mehta, A., Rawat, M., & Basu, S., (2018). Green synthesis of silver nanoparticles using sun-dried tulsi leaves and its catalytic application for 4-nitrophenol reduction. *J. Environ. Chem. Eng., 6*, 1468–1474.

8. Ajitha, B., Kumar, Y. A., Jeon, H., & Won, C., (2018). Synthesis of silver nanoparticles in an eco-friendly way using *Phyllanthus* amarus leaf extract: Antimicrobial and catalytic activity. *Adv. Powder Technol., 29*, 86–93.

9. Shankar, P. D., (2016). A review on the biosynthesis of metallic nanoparticles (gold and silver) using bio-components of microalgae: Formation mechanism and applications. *Enzyme Microb. Technol., 95*, 28–44.

10. Mosaviniya, M., Kikhavani, T., Tanzifi, M., & Tavakkoli, M., (2019). Facile green synthesis of silver nanoparticles using *Crocus haussknechtii* Bois bulb extract: Catalytic activity and antibacterial properties. *Colloid Interface Sci. Commun., 33*, 100211.

11. Choudhary, M. K., Kataria, J., & Sharma, S., (2018). Evaluation of the kinetic and catalytic properties of biogenically synthesized silver nanoparticles. *J. Clean. Prod., 198*, 882–890.

12. Ajitha, B., Kumar, Y. A., & Reddy, P. S., (2015). Biosynthesis of silver nanoparticles using *Momordica charantia* leaf broth: Evaluation of their innate antimicrobial and catalytic activities. *J. Photochem. Photobiol. B Biol., 146*, 1–9.

13. Kumar, V., Singh, D. K., Mohan, S., & Hasan, S. H., (2016). Photo-induced biosynthesis of silver nanoparticles using aqueous extract of *Erigeron bonariensis* and its catalytic activity against acridine orange. *J. Photochem. Photobiol. B Biol., 155*, 39–50.

14. Khan, F. U., (2016). Antioxidant and catalytic applications of silver nanoparticles using *Dimocarpus longan* seed extract as a reducing and stabilizing agent. *J. Photochem. Photobiol. B Biol., 164*, 344–351.

15. Hamedi, S., Shojaosadati, S. A., & Mohammadi, A., (2017). Evaluation of the catalytic, antibacterial and anti-biofilm activities of the *Convolvulus arvensis* extract functionalized silver nanoparticles. *J. Photochem. Photobiol. B Biol., 167*, 36–44.

16. Rasheed, T., (2018). Catalytic potential of bio-synthesized silver nanoparticles using *Convolvulus arvensis* extract for the degradation of environmental pollutants. *J. Photochem. Photobiol. B Biol., 181*, 44–52.

17. Arya, G., Kumari, R. M., Sharma, N., Gupta, N., & Kumar, A., (2019). Catalytic, antibacterial and antibiofilm efficacy of biosynthesized silver nanoparticles using *Prosopis juliflora* leaf extract along with their wound healing potential. *J. Photochem. Photobiol. B Biol., 190*, 50–58.

18. Baruah, D., Narayan, R., Yadav, S., Yadav, A., & Moni, A., (2019). Alpinia nigra fruits mediated synthesis of silver nanoparticles and their antimicrobial and photocatalytic activities. *J. Photochem. Photobiol. B Biol., 201*, 111649.

19. Tripathi, R. M., Kumar, N., Shrivastav, A., Singh, P., & Shrivastav, B. R., (2013). Catalytic activity of biogenic silver nanoparticles synthesized by Ficus panda leaf extract. *J. Mol. Catal. B Enzym., 96*, 75–80.

20. Bonnia, N. N., (2016). Green biosynthesis of silver nanoparticles using 'polygonum hydropiper' and study its catalytic degradation of methylene blue. *Procedia Chem., 19*, 594–602.

21. Ravichandran, V., (2019). Results in Physics photocatalytic activity of Parkia speciosa leaves extract mediated silver nanoparticles. *Results Phys., 15*, 1065.

22. Sowmyya, T., & Vijaya, L. G., (2018). Spectroscopic investigation on catalytic and bactericidal properties of biogenic silver nanoparticles synthesized using *Soymida febrifuga* aqueous stem bark extract. *J. Environ. Chem. Eng., 6*, 3590–3601.

23. Warghane, U. K., & Dhankar, R. P., (2019). Novel biosynthesis of silver nanoparticles for catalytic oxidation of Alcohols containing aromatic ring. *Mater. Today Proc., 15*, 526–535.
24. Thomas, B., Vithiya, S. M. B., & Prasad, T. A. A., (2019). Antioxidant and photocatalytic activity of aqueous leaf extract mediated green synthesis of silver nanoparticles using *Passiflora* edulis f. flavicarpa. *Mater. Today Proc., 14*, 239–247.
25. Chand, K., (2020). Green synthesis, characterization, and photocatalytic application of silver nanoparticles synthesized by various plant extracts. *Arab. J. Chem.*
26. Elemike, E. E., Onwudiwe, D. C., & Mkhize, Z., (2016). Eco-friendly synthesis of AgNPs using *Verbascum thapsus* extract and its photocatalytic activity. *Mater. Lett., 185*, 452–455.
27. Chokkalingam, M., (2019). Optik photocatalytic degradation of industrial dyes using Ag and Au nanoparticles synthesized from *Angelica gigas* ribbed stem extracts. *Opt. - Int. J. Light Electron Opt., 185*, 1213–1219.
28. Hagos, M., Ramadevi, D., & Kumar, Y. P., (2018). Open Nano Synthesis of silver nanoparticles using aqueous extract of Dolichos lablab for reduction of 4-Nitrophenol, antimicrobial and anticancer activities. *OpenNano, 3*, 28–37.
29. Chouhan, N., Ameta, R., & Meena, R. K., (2017). Biogenic silver nanoparticles from *Trachyspermum ammi* (Ajwain) seeds extract for catalytic reduction of p-nitrophenol to p-aminophenol in excess of NaBH4. *J. Mol. Liq., 230*, 74–84.
30. Vijilvani, C., (2020). Antimicrobial and catalytic activities of biosynthesized gold, silver and palladium nanoparticles from *Solanum nigurum* leaves. *J. Photochem. Photobiol. B Biol., 202*, 111713.
31. Elemike, E. E., Onwudiwe, D. C., Ekennia, A. C., Ehiri, R. C., & Nnaji, N. J., (2017). Photosynthesis of silver nanoparticles using aqueous leaf extracts of *Lippia citriodora*: Antimicrobial, larvicidal and photocatalytic evaluations. *Mater. Sci. Eng., C75*, 980–989.
32. Muthu, K., & Priya, S., (2017). Green synthesis, characterization and catalytic activity of silver nanoparticles using *Cassia auriculata* flower extract separated fraction. *Spectrochim. Acta - Part A Mol. Biomol. Spectrosc., 179*, 66–72.
33. Bhuyan, B., Paul, A., Paul, B., Sankar, S., & Dutta, P., (2017). *Paederia foetida* Linn. promoted biogenic gold and silver nanoparticles: Synthesis, characterization, photocatalytic and *in vitro* efficacy against clinically isolated pathogens. *J. Photochem. Photobiol. B Biol., 173*, 210–215.
34. Ajitha, B., (2016). Instant biosynthesis of silver nanoparticles using *Lawsonia inermis* leaf extract: Innate catalytic, antimicrobial and antioxidant activities. *J. Mol. Liq., 219*, 474–481.
35. Khoshnamvand, M., Huo, C., & Liu, J., (2019). Silver nanoparticles synthesized using *Allium ampeloprasum* L. leaf extract: Characterization and performance in catalytic reduction of 4-nitrophenol and antioxidant activity. *J. Mol. Struct., 1175*, 90–96.
36. Vinay, S. P., & Chandrasekhar, N., (2019). Science direct facile green chemistry synthesis of ag nanoparticles using areca catechu extracts for the antimicrobial activity and photocatalytic degradation of methylene blue dye. *Mater. Today Proc., 9*, 499–505.
37. Varadavenkatesan, T., Selvaraj, R., & Vinayagam, R., (2019). Proceedings green synthesis of silver nanoparticles using *Thunbergia grandiflora* flower extract and its catalytic action in reduction of Congo red dye. *Mater. Today Proc.*, 10–13.
38. Joseph, S., & Mathew, B., (2015). Microwave-assisted green synthesis of silver nanoparticles and the study on catalytic activity in the degradation of dyes. *J. Mol. Liq., 204*, 184–191.

CHAPTER 34

Adsorption Studies of Activated Carbon Prepared from Industrial Waste Lignin for the Removal of Malachite Green

ARCHANA R. CHAUDHARI,[1] MONITA A. BEDMOHATA,[2] and
SHRIPAL P. SINGH[3]

[1]*Department of Applied Chemistry, Priyadarshini Bhagwati College of
Engineering, Nagpur, Maharashtra, India, E-mail: arcbce@gmail.com*

[2]*Department of Applied Chemistry, G H Raisoni Institute of Engineering
and Technology, Nagpur, Maharashtra, India*

[3]*CSIR-Central Institute of Mining and Fuel Research Regional Center,
Bilaspur, Chhattisgarh, India*

ABSTRACT

Activated carbon has been prepared from industrial waste lignin (IWL)
using $ZnCl_2$ as an activating agent. This prepared activated carbon was
utilized for adsorption of the malachite green (MG) dye. Here an endeavor
has been made to study the applicability of IWL as potential dye adsorbent
for the removal of MG from aqueous solution. The systematic optimization involved the study of factors such as effect of adsorbent doses, pH, and
contact time. Langmuir, Freundlich, and B.E.T. isotherms were applied to
the data obtained at equilibrium. A maximum adsorption capacity of 111.11
mg/g of MG was achieved by activated carbon obtained from IWL using
$ZnCl_2$. The adsorption isotherm data was found to fit well into Langmuir
and B.E.T. isotherm models with R^2 values 0.99. The kinetic data obtained at
various concentrations was studied to determine the constant rate of adsorption using Langmuir adsorption kinetics model. It shows that the rate of
adsorption is very high as compared to the rate of desorption which indicates
MG dye is irreversibly adsorbed on the adsorbent.

34.1   INTRODUCTION

Among the different known types of contamination, water contamination is of incredible worry since water is the prime need of life and very fundamental for the endurance of every single living. It is commendable taking note of that solitary 0.2% of the all-out accessible water on the earth is promptly accessible for use as waterways, lakes, and streams [1].

The demand for protection of our natural water resources from pollution has mounted steadily over the years, and the growth in the demand for clean water has led to the concept of water treatment and reuse. A major problem to water reuse is the occurrence of undesirable organic and inorganic pollutants in amounts greater than permissible level prescribed by environment protection agencies worldwide. The presence of these contaminants causes several problems like objectionable taste and odor and toxicity to various aquatic organisms. Thus, the reduction of these unwanted contaminants to an acceptable concentration level is a major concern and objective of all water and wastewater treatment technology.

Dyes are utilized in coloring, paper, and mash, materials, plastics, calfskin, beauty care products and nourishment businesses. Shading stuff released from these enterprises represents certain perils and ecological issues. Because of regularly developing requests in the material industry, engineered natural colors are broadly utilized for kicking the bucket material filaments, for example, cotton, and polyester [2]. Among the various toxins of an oceanic environment, colors are a huge significant gathering of mechanical synthetics with more than 7,00,000 tons of waste delivered every year [3, 4].

Malachite green (MG) is a one of the mainstream cationic dye and is utilized widely. This dye when released into water, checks the biological activity in aquatic lives and poses a threat to human beings. In spite of the fact that the using MG as an antiseptic and antibacterial agent is notable but it is poisonous and cancer-causing for human being because of the presence of nitrogen [5]. Histopathology examines have uncovered that MG dye causes adverse impacts on the liver, kidney, and digestive tract. It causes sinusoidal blockage and leads to the nuclear alterations [6, 7]. There are different regular methods for removal of dye contamination which includes filtration, oxidation, membrane filtration and adsorption [8]. Among every one of these techniques, adsorption is considered as the most feasible and viable treatment technique for large scale applications.

As of late a ton of work has been distributed on the utilization of farming waste items [9], for example, onion skin, flour squander, paddy husk [10], paddy straw, coffee lees [11], date pits [12], waste slurry [13, 14], fly debris

[15, 16], lignite [17, 18], pine bark, peat, and lignin [19–21] and so on for adsorption considers. A few analysts have assessed the use of ease adsorbents for overwhelming metal expulsion from debased waters [22–24]. Scientist's advantage is becoming being used of other minimal effort and inexhaustibly accessible language material containing cellulose as a forerunner for the planning of activated carbons (AC) [25–27].

Paper enterprises generate lignin as a repetitive waste material in tremendous amounts, and its removal is an extraordinary issue. As lignin is a three-dimensional stretched polymeric structure having a repeating entity of phenols, which degrades gradually then encourages the preparation of char as a significant debasement item on pyrolysis. Lignin is utilized as a potential forerunner for preparing the AC. Industrially prepared AC is costly and probably will not be profitable for wastewater treatment. In this manner for an affordable wastewater treatment reason, the improvement of AC from minimal effort or waste materials procured locally is an intriguing alternative. Consequently, it was thought to use the AC produced from industrial waste lignin (IWL) for the expulsion of different colors and chemical species.

The main goal of the present endeavor is to explain the study of the adsorption equilibrium and kinetics of MG dye against the AC prepared by pyrocatalytic degradation of IWL in the presence of $ZnCl_2$, with an aim to shed some light on the mechanism of adsorption of this pollutant.

34.2 MATERIAL AND METHODS

34.2.1 REAGENT

All reagents utilized in the piece of work were of AR grade including H_2SO_4, $ZnCl_2$, NaOH, HCl, MG. Double distilled water was used for the preparation of all experimental solutions.

34.2.2 PREPARATION OF ADSORBENT

The adsorbent was prepared from IWL as per the following procedure:

- Concentrated dark liquor collected from Ballarpur Paper Industry, MS, India.
- The lignin was precipitated out using dil. H2SO4.
- The lignin was pass through a filter and rinsed with a lot of distilled water.

- The lignin was allowed to dry, and its weight was recorded, it was dried and weighed.
- Oven-dried purified IWL (20 g) was mixed with ZnCl2 activating agent with the impregnation ratio 1:2.
- It was kneaded by adding water and kept at room temperature for 24 hours in an atmospheric air.
- After the completion of 24 Hrs, it was placed in an electric muffle furnace for carbonization process.
- The furnace was warmed at 5°C min–1, up to the last carbonization temperature (800°C), which was held for 1 hr.
- After activation, the activated carbon prepared (ACZN) was taken out from the electric furnace and cooled down up to the room temperature.
- The ACZN was rinsed with 1M HCl until no Zn2+ was distinguished in the filtrate and afterward with distilled water until negative Cl– analysis.
- The ACZN was dried overnight in an electric oven at 110°C followed by pulverization and passed through 72 mesh (A.S.T.M. 70) for characterization investigation.

34.2.3   SURFACE STUDIES OF ADSORBENT

Smart sorb 93 surface area (SA) analyzer was used to analyzing BET SA and complete pore volume of material. The AC_{ZN} was firstly treated in the progression of pure nitrogen gas at 50 ml/min, with the temperature 250°C for 1 hr. prior to the estimations of SA. Pore volume of the sample was estimated at higher partial pressure that is more than 0.35 for multilayer development and filling of pores.

34.2.4   SEM ANALYSIS

The precursor Lignin and AC_{ZN} were analyzed by Scanning Electron Microscope outfitted with an electron test analyzer system (accelerating voltage 30 KV).

34.2.5   PREPARATION OF ADSORBATE

A stock arrangement of 100 mg/L of adsorbate was set up by dissolving the MG dye in 1000 mL of double-distilled water. The solution was shaken gradually to ensure the complete dissolution of the dye powder water. The stock solutions of this adsorbate were then stored in amber colored bottles. This was done to minimize the effect of dissolved oxygen and light on the adsorbate concentration. In both adsorption equilibrium and kinetic investigations, the stock solution was diluted to proper concentrations as and when required with the help of distilled water.

34.2.6   ADSORPTION EXPERIMENTS

34.2.6.1   OPTIMIZATION OF ADSORBENT DOSE

Different doses of AC_{ZN} adsorbent extending from 0.01 g, 0.02 g, 0.025 g, 0.05 g to 0.1 g were mixed with 100 ml aqueous solution of MG dye, and the mixture was agitated in a Remi orbital shaker at $30 \pm 1°C$ for 48 hours. Elmer Lambda 35 UV/VIS spectrophotometer was used to determine the final concentration of MG dye.

The percentage of dye removal was determined by keeping all other factors constant using following equation:

$$\text{Percentage removal} = \frac{(C_o - C_e)}{C_e} \times 100 \tag{1}$$

Here initial and equilibrium concentrations in mg/L of MG dye denoted by C_o and C_e, respectively.

34.2.6.2   OPTIMIZATION OF PH

To optimize the pH, the experiment was conducted from pH range (2 to 11). HCl and NaOH solutions were used to alter the required pH. About 0.05 g of AC was mixed with 100 ml. solution of MG dye. Digital pen type pH meter was used for the pH determination of dye solutions.

34.2.6.3   OPTIMIZATION OF CONTACT TIME

The contact time is optimized by carrying out the adsorption experiments in the time range from 6 hrs. to 72 hrs. All other parameters like initial concentrations, pH, dosage, particle size, and temperature were kept constant.

34.2.6.4   ADSORPTION STUDIES

Adsorption investigations of MG were carried out using AC_{ZN} adsorbent. For adsorption isotherm, the solution of MG dye solution at different concentrations (1 mg/L–180 mg/L) was agitated with the known amount (0.05 g) of AC_{ZN} till the equilibrium was reached. Then the residual concentration of adsorbate was estimated by using UV/Vis spectrophotometer.

The quantity of adsorbed MG dye per gram of AC_{ZN} at the equilibrium, Qe (mg/g) was determined by mass balance as follows:

$$Q_e = \frac{(C_o - C_e)V}{m} \qquad (2)$$

where; C_o is the initial concentration (mg/l) of MG dye solution; C_e is the final concentration (mg/l) of MG dye solution; V is the volume (l) of MG dye solution; m is the mass of adsorbent (g).

34.2.7   KINETIC STUDIES

The kinetics studies of adsorption depict the solute uptake rate, which oversees the residence time of adsorption reaction. It is one of the significant qualities in characterizing the effectiveness of adsorption. In this work, the kinetics of the dye removal was completed to study the nature of prepared low-cost carbon adsorbents.

For kinetics of the adsorption process, the experimental unit was a cylindrical batch reactor having a limit of 5 liters. It was made of Borosil glass material and 8 baffles fitted on its inner wall. The adsorbent-adsorbate system was blended by a two bladed impeller having a length of 7 cm and broadness of 1.5 cm. The impeller used to blend the experimental solutions was manufactured out of a 6 mm glass rod having a basic Teflon or nylon paddle fitted to its lower end so that it used to move freely about the pole of the rod yet did not sneak out. It was additionally kept from sliding upwards during mixing by a mechanical stop which was joined to the glass rod. As from the data available in literature [28], it very well may be inferred that the proportion of paddle length to the diameter of the reactor filled with the solution to be mixed, should lie down in between 0.20 to 0.50 cm. Experimental solution for kinetic runs was three liters in volume and was set up by diluting a suitable quantity of stock solution with boiled and cooled distilled water. The contents were continually mixed at 400 ± 50 rpm and permitted to reach the temperature of the water bath

$(30 \pm 1°C)$, which for the most part took around 5 to 10 minutes. About 0.05 ± 0.0001 g/L of adsorbent for MG was then brought into the solution at a given moment of time. At each ideal time span, 7 ml of experimental solution was pulled back with the help of a syringe, centrifuged, and the concentration of the adsorbate in the aqueous phase was assessed by UV/Vis analysis.

A basic interpretation of the kinetic data dependent on Langmuir theory was utilized, and the rate was expressed as a function of a directly measurable system variable, the fluid phase adsorbate concentration.

34.3   RESULTS AND DISCUSSION

34.3.1   SURFACE STUDIES OF ACTIVATED CARBON

High SA AC_{ZN} with significant pore volume was obtained in the presence of $ZnCl_2$ as an activating agent with $ZnCl_2$-to-lignin impregnation proportion = 2:1 at temperature = 800°C for 1 hr. The AC_{ZN} with a N_2-BET SA of 1183.06 m^2/g and a total pore volume of 0.6481 cm^3/g was obtained.

34.3.2   SEM ANALYSIS

The SEM pictures of AC_{ZN} (see Figure 34.2) shows a large number of pores having different sizes and geometry on the asymmetrical, heterogeneous, rough, and grainy surface of activated carbon prepared from lignin, which may add to the moderately high SA (Figure 34.1).

The pores depict active and dynamic sites for the adsorption process. In addition, these pores may act as binding sites for chemical species.

34.3.3   EFFECT OF ADSORPTION PARAMETERS

34.3.3.1   EFFECT OF ADSORBENT DOSE

The adsorbent dose was ranging from 0.01 g to 0.1 g per 100 ml. It is clear from Figure 34.3 that MG removal increases with increase in adsorbent dose from 0.01 g to 0.05 g. This might be because of the availability of more adsorbent locales. However, no significant changes in percentage removal of MG dye beyond 0.05 g, which might be because of the saturation of the adsorbent surface.

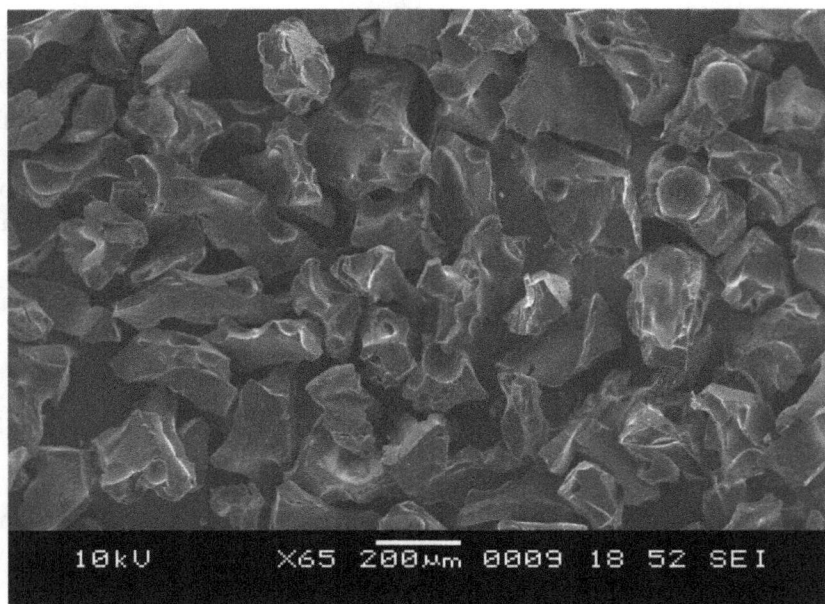

FIGURE 34.1 SEM image of lignin.

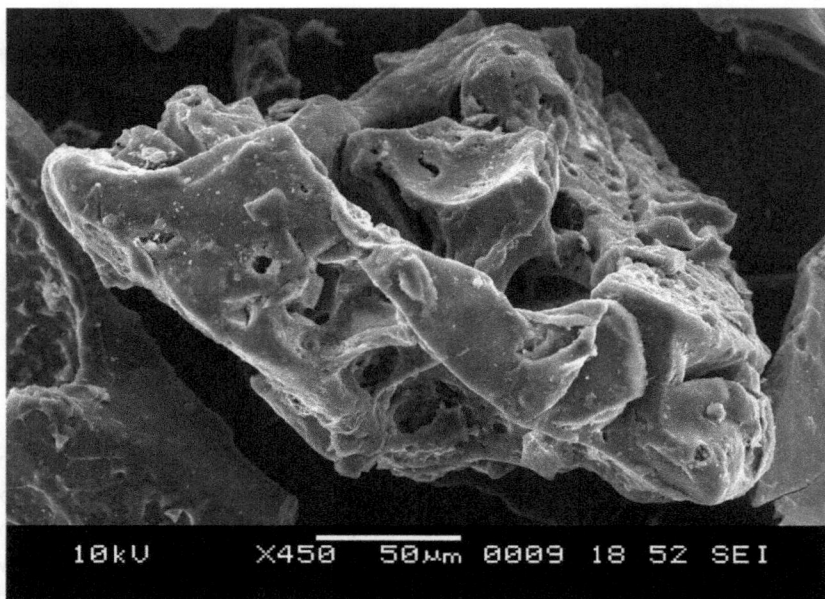

FIGURE 34.2 SEM image of ACZN using ZnCl2 catalyst.

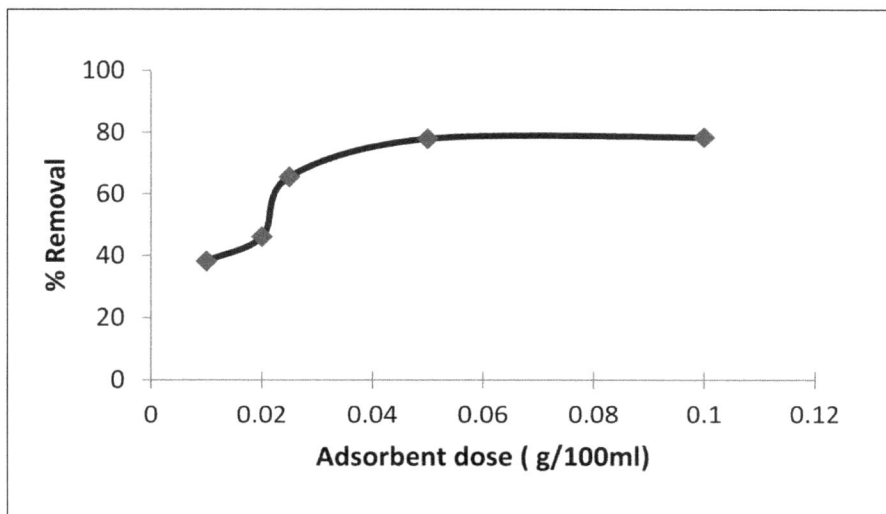

FIGURE 34.3   Effect of adsorbent dose on MG dye removal.

34.3.3.2   EFFECT OF PH

In the case of MG, the adsorption enhances with an increase in pH from 2 to 5. Thereafter, it is clear from Figure 34.4 that there was no remarkable increase in adsorbed quantity of MG due to increase in pH. At lower pH, the surface becomes positively charged, the H^+ ions concentration is very high, and they contend with the dye cations for empty adsorption sites, which causes a decrease in dye removal. At the pH 5 of the solution, the concentration of H^+ ions decrease, which favors the removal of cationic dyes because of increment in electrostatic force of attraction. Hence it has been observed that percentage removal of MG is maximum at pH = 5.

34.3.3.3   EFFECT OF CONTACT TIME

The effect of contact time on MG dye adsorption is shown in Figure 34.5. It was seen that the initial removal percentage of dye is very fast, 36.83% of MG was removed within the initial 6 hrs. and reached to 77.07% at 48 hrs. After this point, there was no significant adsorption. This shows equilibrium was reached following 48 hours. This might be a result of the statement that first all adsorbent locales were empty and the concentration of solute was high, afterwards the dye removal percentage was decreased due to saturation of adsorbent sites followed by adsorption and desorption measure [29].

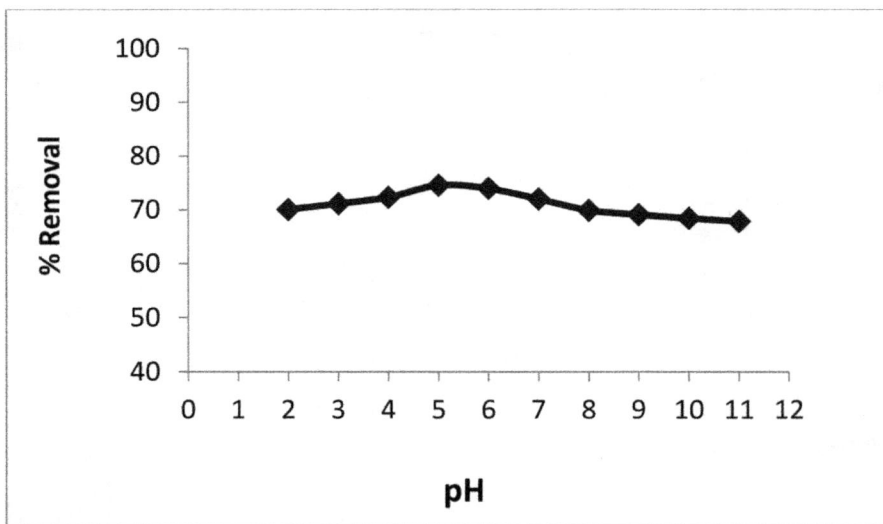

FIGURE 34.4   Effect of pH on MG dye removal.

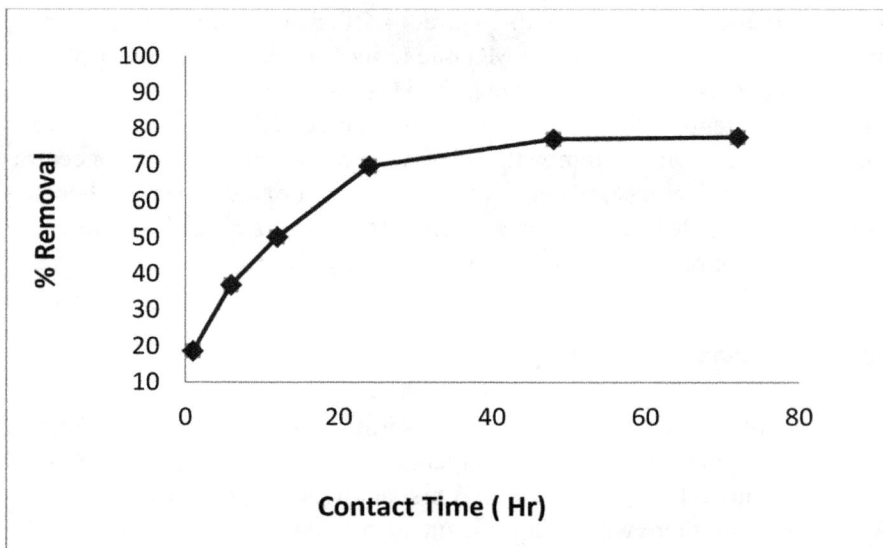

FIGURE 34.5   Effect of contact time on MG dye removal.

34.3.4   ADSORPTION ISOTHERMS

Experimental isotherms are valuable for explaining the limit of an AC for an adsorbate to help for the assessment of practicality of a suitable activated carbon, and for fundamental determination of carbon necessity. Langmuir (Eqn. (3)), Freundlich (Eqn. (4)) and B.E.T. (Eqn. (5)) were applied to this experimental study. The model equations are listed below:

$$C_e/Q_e = (C_e/Q_0) + (1/Q_0 b) \tag{3}$$

$$\log (Qe) = \log K_f + 1/n \log(C_e) \tag{4}$$

$$C_e/Q_e (C_s - C_e) = 1/Q_0 Z + (Z - 1/Q_0 Z) \times C_e/C_s \tag{5}$$

where; Q_e is the amount adsorbed of solute in mg/g at equilibrium; C_e is the solute concentration in mg/L at equilibrium stage; C_s is the solute concentration in mg/L at saturation stage; Q^0 is the monolayer adsorption capacity of adsorbate in mg/g; b is the Langmuir constant; K_f, and n are the Freundlich constants; z is the B.E.T. constant.

According to Langmuir, adsorption is limited to monolayer capacity with all sites having the same adsorption energy with no surface migration of the adsorbed phase. The Freundlich expression is an empirical equation dependent on adsorption on a heterogeneous surface. According to Freundlich, the adsorption takes place only at medium pressure in very dilute solutions at moderately low temperatures.

The Brunauer, Emmett, and Teller equation [30] is an expansion of the Langmuir relationship that accounts for multilayer coverage. Adsorption occurs successively on 2nd, 3rd, 4th, and nth layers with the surface area (SA) available.

The parameters of every one of the three models were determined and summed up from the plots of 1/Ce against 1/Qe, Log Ce against Log Qe and Ce against Ce/Qe as appeared in Figures 34.6–34.8. The adsorption equilibrium information was investigated for their applicability to Langmuir, Freundlich, and B.E.T. plots. The regression co-relation coefficient values, R^2, for every adsorption systems are nearer to unity. In contrast to this, the B.E.T. isotherm equation was seen as adhering linearity over the complete range of adsorbate equilibrium concentration bearing a point or two at extremely low equilibrium concentration.

It was seen that the Langmuir and B.E.T. adsorption isotherm models explained the adsorption capacity (Q^0) precisely with higher regression coefficient, R^2, showing the indistinguishable liking for the adsorbates and no immigration of the adsorbates on the surface of AC. The adsorption capacity of AC was obtained to be 111.11 mg/g and 22.3357 mg/g for Langmuir and Freundlich model separately for expulsion of MG dye (see Table 34.1). In the micromolar concentration range of solute such as those utilized in the present work and which is more applicable to wastewater treatment, evidently the Langmuir and B.E.T equation give identical outcomes.

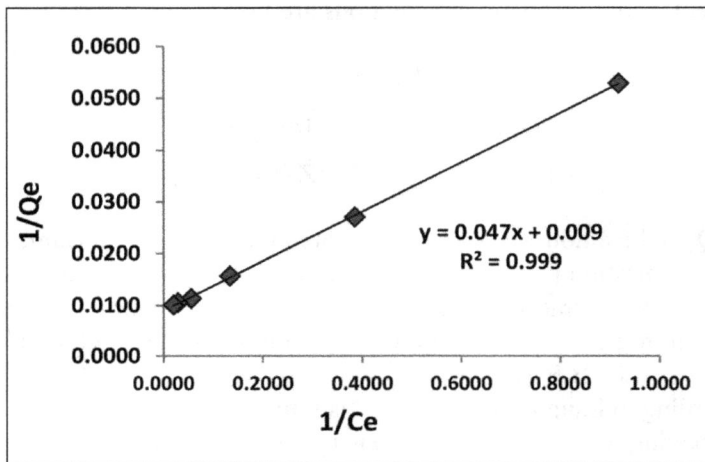

FIGURE 34.6   Langmuir adsorption isotherm of MG dye.

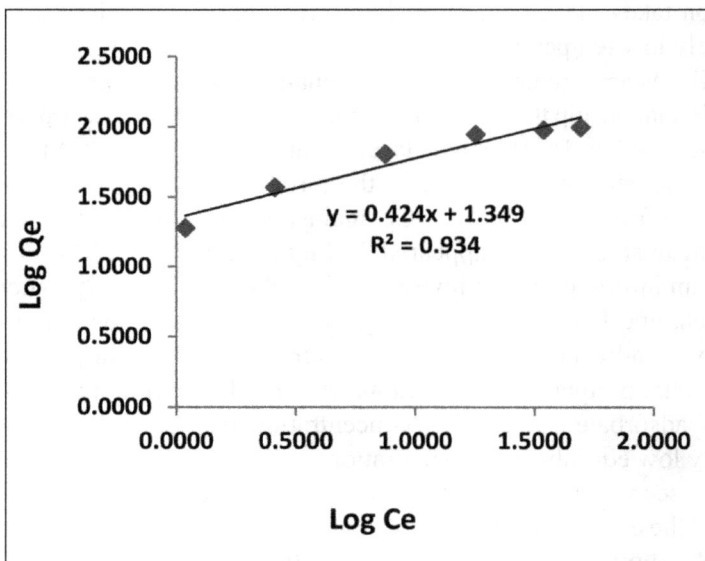

FIGURE 34.7   Freundlich adsorption isotherm of MG dye.

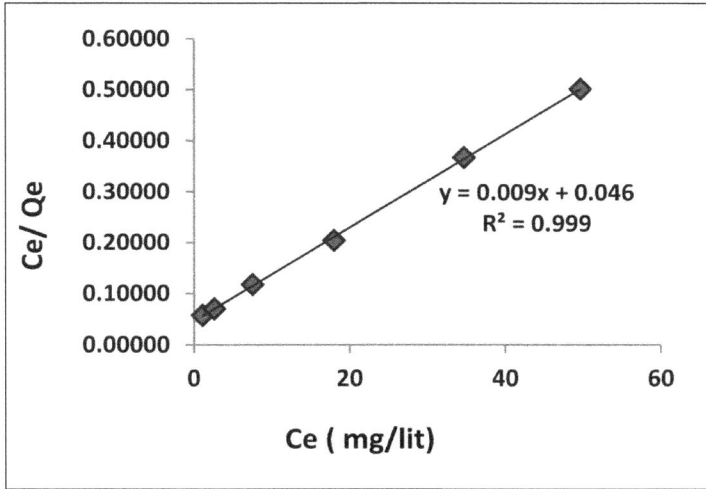

FIGURE 34.8   B.E.T. adsorption isotherm of MG dye.

34.3.5   *ADSORPTION KINETICS STUDY*

In this study, the interpretation of kinetics data dependent on Langmuir theory has been utilized. Langmuir hypothesis expect that the adsorption rate is corresponding to the multiplication of adsorbate concentration in the liquid phase and the fraction of the unfilled adsorbent surface. The portion of the surface secured by the adsorbate; Q will be proportional to the reduction in liquid phase adsorbate concentration; by then:

$$dQ/dt = [Ka.Ct\,(1/Q)] - (Kd.Q) \qquad (6)$$

$$Q = f\,(Co - Ct) \qquad (7)$$

where; Ka is the adsorption constant; and Kd is the desorption constant; C_o, C_t, and C_e are the fluid phase concentrations at 0, t, and equilibrium; f is a constant.

By substituting Eqn. (7) in Eqn. (6) we get:

$$Log\,[(Ct - Ce)/(Ct+a)] = -Ka\,Cet + Log[(C0Ce)/(C0+a)] \qquad (8)$$

where; a is the (C_0/bC_e); b is the Langmuir constant (Ka/Kd).

The graph plotted between C_t/C_0 (adsorbate concentration on the adsorbent, mg/g) versus time. The Ka (adsorption rate constant) and Kd (desorption rate constants) were calculated by designing another graph in between $[(C_t-C_e)/(C_t+a)]$ against time (t) [31].

Adsorption kinetics is utilized to get an idea about the mechanism of adsorption and its rate-controlling steps. The adsorption mechanism includes the chemical reaction related with functional groups situated on the surface of the adsorbent and the adsorbate, temperature, and pH. The graphs of adsorption kinetics studies of MG dye on AC_{ZN} i8s represented in Figures 34.9 and 34.10.

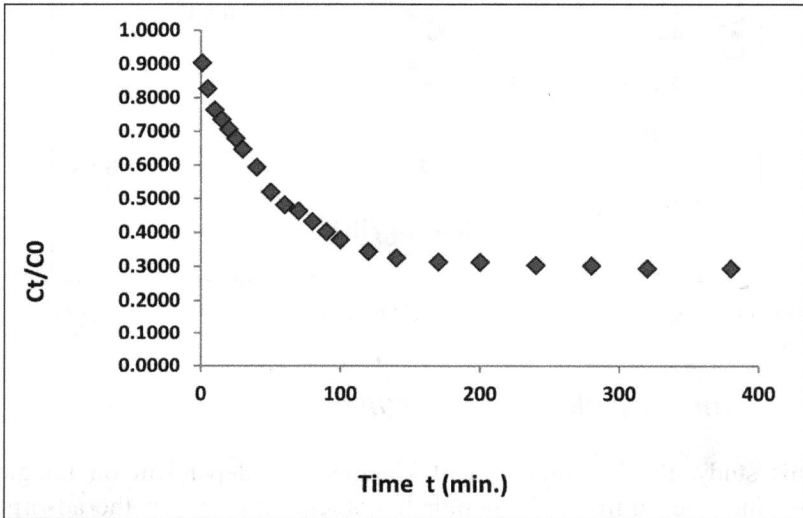

FIGURE 34.9　　Adsorption kinetics plot of MG on AC.

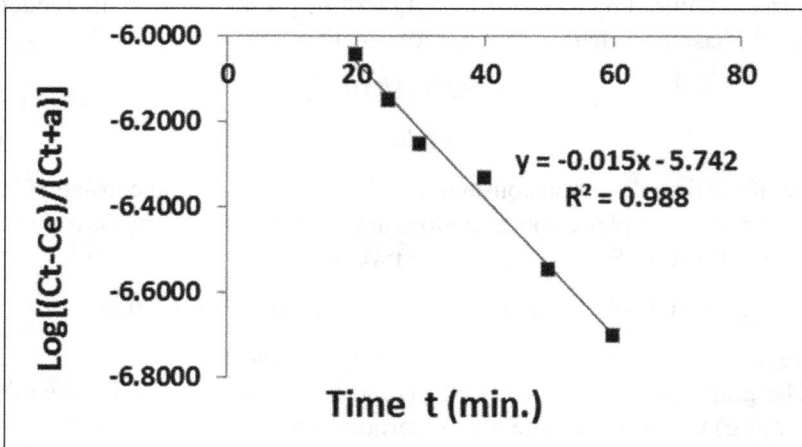

FIGURE 34.10　　Langmuir adsorption kinetics of MG on AC.

An investigation of kinetics of adsorption is attractive because it gives data about the adsorption mechanism, and it is significant for the effectiveness of the procedure. From Figure 34.9, it was seen that the adsorption was very quick in the beginning period and it continuously decreased with time until it came to an equilibrium state. This pattern of adsorption kinetics was a direct outcome of the adsorption of dye on the outer surface of the adsorbent at the starting period of contact time; the MG was adsorbed on the surface of AC_{ZN}. The dye dispersed inside the adsorbent pores and was adsorbed by the interior surface of the AC_{ZN}. The equilibrium time for adsorption of the dyes on various ACs was in the range 320–380 minutes, from that point on, no more adsorption happened with delayed time. In the plot of Langmuir adsorption kinetics (see Figure 34.10) of adsorbate, $\text{Log } [(C_t-C_e)(C_t+a)]$ versus time t, the slope of the plot can be used as a measure of the rate of expulsion of adsorbates from their aqueous solution. The Ka (adsorption rate constant) and K_d (desorption rate constants) were calculated from the same graph [31]. As described above, a simplified understanding of kinetic data dependent on Langmuir theory which has been utilized in the present work. The various adsorption isotherm parameters for adsorption and kinetic study of MG dye onto AC_{ZN} has been summaries in Table 34.1. The rate of adsorption is a lot higher than the desorption rate, from this we can conclude that the adsorbate used in present work is irreversibly adsorbed on the adsorbent.

34.4   CONCLUSION

From the results (see Table 34.1), it has been observed that the prepared ACs using $ZnCl_2$ from IWL has the adsorption capacity of 111.11 mg/g for the expulsion of MG dye.

It has been observed that the value of 1/n is smaller than 1, which be a sign of the good adsorption. In the present work, since the concentration of adsorbate studied is much lower than their saturation concentration, therefore, the BET model resembles the Langmuir adsorption model, hence, the $Q°$ (monolayer adsorption capacity) values of BET are same as that of $Q°$ values of Langmuir adsorption model. The Langmuir adsorption kinetics studies show that the values of kinetics adsorption rate constants (Ka) are very high as compared to the kinetics desorption rate constants (Kd).

This shows that the rate of adsorption is extremely high as compared to the rate of desorption which indicates that the MG dye is irreversibly adsorbed on the adsorbent. From this study, it very well may be concluded

TABLE 34.1 Adsorption Isotherm Parameters for Adsorption and Kinetics Study of MG Adsorption

Langmuir Model			Freundlich Model			B.E.T. Model			Langmuir Kinetic Model		
Qe (mg/g)	B (L/mg)	R^2	K_f (mg/g (L/mg)$^{1/n}$)	1/n	R^2	Qe (mg/g)	Z	R^2	Ka	Kd	R^2
111.11	0.1914	0.99	22.3357	0.424	0.93	111.11	11739.1	0.99	154.24	0.00173	0.98

that the AC prepared from IWL can be potentially utilized as an adsorbent to eliminate MG dye from wastewater.

KEYWORDS

- **activated carbon**
- **adsorption equilibrium**
- **equilibrium isotherms**
- **industrial waste lignin**
- **kinetics**
- **malachite green**

REFERENCES

1. Bhatnagar, A., & Minocha, A. K., (2006). Conventional and Non-conventional adsorbents for removal of pollutants from water: A review. *Ind. J. Chem. Tech., 13*, 203–217.
2. Mane, V. S., Mall, I. D., & Shrivastava, V. C., (2006). *Dyes Pigments*, 2.
3. Neamtu, M., Siminiceanu, I., Yediler, A., & Kettrup, A., (2002). *Dyes Pigments*, 53–93.
4. Jumaasiam, A., Chuab, T. G., Gimbon, J., Choong, T. S., & Azmi, I., (2005). *Desalination, 57*, 186.
5. Bulut, E., Ozacar, M., & Sengil, I. A., (2008). Adsorption of malachite green onto bentonite: Equilibrium and kinetics studies and process design. *Micropor. Matter, 115*, 234–246.
6. Gerundo, N., Alderman, D. J., Clifton-Hadely, R. S., & Feist, S. W., (1991). Pathological effects of repeated doses of malachite green a preliminary study. *J. Fish Dis., 14*, 521–532.
7. Palukurty, M. A., Allu, T. A., Chitturi, & Somalanka, S. R., (2014). *J. Chem. Bio. Phy. Sci., 4*, 1910–1921.
8. Dash, B., (2010). *Project, Competitive Adsorption of Dyes on Activated Carbon*. National Institute of Technology, Rourkela.
9. Kumar, P., & Dara, S. S., (1980). *Prog. Wat. Tech., 13*, 353.
10. Subramani, T., & Revathi, P. K., (2015). Production of Activated carbon from agricultural raw waste. *IOSR J. of Engg., 5*, 54–63.
11. Kawahara, Y., Yamamoto, K., Wakisaka, H., Izutsu, K., Shioya, M., Sakai, T., Takahara, T., & Ishibashi, N., (2009). Carbonaceous adsorbents produced from coffee lees. *J. Mater. Sci., 44*, 1137–1139.
12. Ashour, S. S., (2010). Kinetic and Equilibrium adsorption of methylene blue and Remazol dyes onto steam-activated carbons developed from date pits. *J. Saudi Chem. Soc., 14*, 47–53.
13. Mohan, D., & Pittman, C. U., (2006). Activated carbons and low cost adsorbents for remediation of tri- and hexavalent chromium from water. *J. Hazard. Mater B, 137*, 762–811.
14. Mohan, D., Singh, K. P., & Singh, V. K., (2006). Trivalent chromium removal from wastewater using low cost activated carbon derived from agricultural waste material and activated carbon fabric cloth. *J. Hazard. Mater, 135*, 280–295.

15. Mohan, D., Singh, K. P., & Singh, V. K., (2002). Single- and multi-component adsorption of cadmium and zinc using activated carbon derived from bagasse--an agricultural waste. *Water Res., 36*, 2304–2318.

16. Atkinson, B. W., Bux, F., & Kasan, H. C., (1998). Considerations for application of biosorption technology to remediate metal-contaminated industrial effluents. *Water SA, 24*, 129–135.

17. Lee, S. M., & Davis, A. P., (2001). Removal of Cu(II) and Cd(II) from aqueous solution by seafood processing waste sludge. *Water Res., 35*, 534–540.

18. Liu, C. C., & Wang, M. K., (2005). Removal of nickel from aqueous solution using wine processing waste sludge. *Ind. Eng. Chem Res., 44*, 1438–1445.

19. Chaudhari, A. R., (2004). Thesis of PhD. *Organic Chemistry of Industrial Waste Lignin with a View of Its Utilization*, p. 154.

20. Mohan, D., & Chander, S. J., (2006). Removal and recovery of metal ions from acid mine drainage using ligniteda low cost sorbent. *Hazard. Mater, 137*, 1545–1553.

21. Chaudhari, A. R., Nagpurkar, L. P., & Ekhe, J. D., (2003). Uptake of heavy metal ions by carbonaceous material obtained from industrial waste lignin using microwave irradiation. *Asian Journal of Chemistry, 15*(2), 917–924.

22. Babel, S., & Kurniawan, T. A., (2003). Low-cost adsorbents for heavy metals uptake from contaminated water: A review. *J. Hazard. Mater. B, 97*, 219–243.

23. Bailey, S. E., Trudy, J. O., Bricka, R. M., & Adrian, D. D., (1999). A review of potentially low-cost sorbents. for heavy metals. *Water Res., 33*, 2469–2479.

24. Guo, X., Zhang, & Shan, X., (2008). Removal of Cr(VI) from aqueous solution by Turkish vermiculite: Equilibrium, thermodynamics and kinetic studies. *J. Hazard. Mater., 151*, 138–142.

25. Guo, J., & Lua, A. C., (2002). Characterization of adsorbent prepared from oil palm shell by CO2 activation for removal of gaseous pollutants. *Mater. Lett., 55*, 334–339.

26. Molina-Sabio, M., & Rodriguez-Reinoso, F., (2004). Role of chemical activation in the development of carbon porosity. *Colloid. Surf. A, 241*(1–3), 15–25.

27. Bedmohata, M. A., Chaudhari, A. R., Singh, S. P., & Choudhary, M. D., (2015). Adsorption capacity of activated carbon prepared by chemical activation of lignin for the removal of methylene blue dye. *Int. J. Adv. Res. Chem. Sci., 2*, 1–13.

28. Dean, J. A., (1978). *Lange's Handbook of Chemistry* (12ᵗʰ edn., pp 5.73–5.82). McGraw-Hill: New York.

29. Mishra, P. C., & Patel, R. K., (2009). Removal of lead and zinc ions from water by low cost adsorbents. *J. Haz mater., 168*, 319–325.

30. Weber, T. W., & Chakravorti, R. K., (1974). Pore and solid diffusion models for fixed-bed adsorbers. *J. Am. Inst. Chem. Engg., 2*, 228.

31. Singh, S. P., & Yenkie, M. K. N., (2006). Scavenging of priority organic pollutants from aqueous waste using granular activated carbon. *J Chinese Chem. Soc., 53*, 325–334.

CHAPTER 35

The Potential of Medicinal Plants with Anti-Proteus Activity for the Treatment (and Prevention) of Rheumatoid Arthritis

ALEFIYAH S. BOHRA, S. R. GUPTA, and A. P. KOPULWAR

Department of Biotechnology, Priyadarshini Institute of Engineering and Technology, Nagpur, Maharashtra – 440019, India,
E-mail: ashkopulwar31@gmail.com (A. P. Kopulwar)

ABSTRACT

Rheumatoid arthritis (RA) is an autoimmune disorder related to chronic joint inflammation [1]. The origin of this disorder remains a mystery still; however, the immune system is known to mediate the progression of diseased joints in RA [2]. The main symptoms of this disease are inflammatory joint and pain, which also causes premature mortality, disability, and compromised quality of life. This study was undertaken to explore the photochemical screening and anti-proteus activity for the treatment and prevention of RA by the extract of *Glycyrrhiza glabra* (mulethi), *Piper longum* (pippali), *Terminalia chebula* (hirda), *Commiphara wightii* (guggul), and *Tinospora cordifolia* (guduchi). Plant samples were extracted using the solvent extraction method. In phytochemical screening, all extract showed positive towards steroids and flavonoids, except Piper longum others showed positive results towards phenols. All five extracts were investigated, which have the ability to inhibit the microbial trigger of RA using gram-negative bacteria *Proteus mirabilis*. Antimicrobial activity of the extract was done by cup and plate method on nutrient agar plates [3] among that *P. longum*, *T. chebula*, and *T. cordifolia* showed promising results. Also, the MIC values of these samples were determined. The combination of MIC of *T. chebula* and *T. cordifolia* shows a promising result.

35.1   INTRODUCTION

Rheumatoid arthritis (RA) is an autoimmune disease that is characterized by inflammation at various synovial joints, which is evident by joint swelling, stiffness, deformity, and extreme pain [4]. Morning stiffness which is the symptom of other types of arthritis called osteoarthritis, it wears off within 30 minutes of getting up, but this lasts longer in RA patients. The exact cause of this disease is still unknown; however, the immune system is known to conciliate the progression of synovial diseased joints in RA [2]. RA is an autoimmune disorder, which means our immune system attacks its own healthy cells. In this case, synovial cells are auto immunized resulting in the synovial membrane swelling, hence promotes various causes. The most involved joints in RA are the MCP and PIP joints of the hands, wrists, and MTP joints of the feet [5]. Inflammation and following destruction in synovial joints are kinds of hallmark of RA. It involves interaction between T and B cells, macrophages, mast cells, plasma cells, synovial fibroblasts, and dendritic cells. The cell-to-cell interaction is either directly through cell-cell contact or via cytokines, such as TNFα, IL-1, and IL-6.

The medical approach towards the treatment of RA has improved over the decades. From synthetic steroids to NSAID's (non-steroidal anti-inflammatory drugs) and now the DMARDs (disease-modifying anti-rheumatic drugs). DMARDs became the centerpiece of RA treatment in the 1970s. It has shown an effect on almost 50% of patients [6]. Rather than the synthetic medications, natural components are shown far more effective as it does not show side effect as the synthetic medication. The overall worldwide generality is 0.8% and steadily increases to 5% in women over the age of 60. RA is two to three times more common in women compared to men. It affects 1% of the population in India [7–9]. Surgery is also the one method for the management of RA, the affected joints are replaced by surgery, such as knee replacement [10].

RA infections have been triggered by the bacteria *Proteus mirabilis*. The raised in serum levels of *P. mirabilis* there is a specific cross-reactive antibody that has regularly been announced in individuals suffering from RA. The RA patients with *P. mirabilis* antibodies have cytopathic effects and cross-reactive antibodies on joint tissue. The research has found that *P. mirabilis* bacterium is present in the urine sample of RA patients [11].

The serum of rabbit was collected which is immunized with HLA-DR4. The positive lymphocyte binds specifically to *Proteus* [12]. Amino acid sequence similarities have been identified between the "ESRRAL" amino acid sequence and the EQ/KRRA motif in RA HLA-susceptible antigens present in *P. mirabilis* hemolysin [13].

The plants which were identified in this study were traditionally used in the treatment of RA, such as *piper longum* (pippali), *Glycyrrhizin glabra* (mulethi), *Terminalia chebula* (hirda), *Commiphara wightii* (guggul) and *Tinospora cordifolia* (guduchi). These plants show anti-inflammatory action towards synovial joints, according to the research. Here to screen their potential to block the microbial triggers of RA, the selected plant extracts were tested against *Proteus mirabilis*.

35.2 MATERIAL AND METHOD

35.2.1 PLANT COLLECTION

The plant species tested in the study were *Piper longum* (pippali), *Glycyrrhizin glabra* (mulethi), *Terminalia chebula* (hirda), *Commiphara wightii* (guggul), and *Tinospora cordifolia* (guduchi); which were collected from a local vendor at Nagpur, Maharashtra.

35.2.2 PLANT EXTRACTIONS

35.2.2.1 SAMPLE 1: PIPER LONGUM

About 20-gram dried powder of fruits of pippali was cold macerated for 24 hours in 200 ml of distilled water and then heated on a hot plate for 1 hour, and then the extracted was filtered through filter paper [14]. The water extract was allowed to dry at 60–70°C. Then the dried plant extract was dissolved in a specific quantity of distilled water.

35.2.2.2 SAMPLE 2: GLYCYRRHIZA GLABRA

About 20-gram dried powder of stem of mulethi was extracted by the method of solvent extraction by the Soxhlet apparatus, using 50% methanol as the solvent. Then the extract was allowed to dry at 55–60°C [15]. Then the dried plant extract was dissolved in a specific quantity of distilled water.

35.2.2.3 SAMPLE 3: TERMINALIA CHEBULA

About 20-gram dried powder of the fruit of Hirda plant was extracted using the Soxhlet apparatus where the solvent used was 50% ethanol [16]. The extract was allowed to dry at 70–80°C. Then the dried plant extract was dissolved in a specific quantity of distilled water.

35.2.2.4 SAMPLE 4: COMMIPHARA WIGHTII

About 20-grams of dried powder of guggul gum resin was extracted by the 95% ethanol as the solvent by the Soxhlet apparatus [17]. The solvent which is extracted was allowed to dry at 50–60°C with gradual shaking and residue was weighed. Then the dried plant extract was dissolved in a specific quantity of DMSO.

35.2.2.5 SAMPLE 5: TINOSPORA CORDIFOLIA

About 20-gram of dried powder of stem of guduchi plant was first defatted with petroleum ether, and the extract was prepared with a mixture of ethanol and distilled water (80:20) using a Soxhlet extraction apparatus [18]. The evaporation of the filtrate was carried out at 70–80°C till dryness with gradual shaking and residue was weighed. Then the dried plant extract was dissolved in a specific quantity of DMSO.

35.2.3 QUALITATIVE ANALYSIS

Various photochemical tests were performed to check the presence of certain photochemical in all the extracts, such as carbohydrates, phenols, steroids, etc. (Table 35.1).

35.2.4 QUANTITATIVE ANALYSIS

35.2.4.1 PHENOL TEST

The total phenolic content of extracts of plant samples was estimated using the Folin Ciocalteau reagent using the spectrophotometric method. Test was done in triplicates for confirmation. Absorbance was noted done for 765 nm wavelength. And hence the concentration was calculated using the standard curve using gallic acid [19]. About 300 µl each sample was mixed with 1.5 ml of ferric chloride and 1.2 ml of sodium carbonate; the reaction mixture was allowed to incubate for 30 minutes at room temperature. The absorbance was noted at 765 nm wavelength. The concentration was calculated using the standard phenolic curve.

TABLE 35.1 Preliminary Qualitative Tests

Photochemical	Test
Carbohydrates	Molisch's test [19]
Phenols	$FeCl_3$ test [19]
Alkaloids	Mayer's and Wagner's test [20]
Steroids	Salkowski test [20]
Terpenoids	Salkowski test [20]
Flavonoids	Lead acetate test [20]

35.2.4.2   FLAVONOID CONTENT

The aluminum chloride colorimetric method was used for determining of the Flavonoid content in the plant extract samples [21–24]. About 100 µl of each sample was reacted with 300 µl of methanol, 20 µl of aluminum chloride, 20 µl of potassium acetate, and made up 1 ml volume using double distilled water. The mixture was incubated for 30 minutes at room temperature. The OD was measured at 415 nm, and then concentration was calculated using the standard curve.

35.2.5   ANTI-MICROBIAL SCREENING

35.2.5.1   TEST ORGANISM

The reference strain of *Proteus mirabilis* (MCC 3895) was obtained from National Center for Microbial Resource, Pune. It was received in the form of active culture.

35.2.5.2   ANTIMICROBIAL ACTIVITY

Antimicrobial activity of all the plant extracts was determined using Agar well diffusion method [8]. Approximately a loop full culture of *P. mirabilis* was allowed to grow in 10 ml of fresh nutrient broth until they reached a cell count of 10^8 cells/ml approximately.

About 500 μl of the microbial suspension was poured onto nutrient agar plates and allowed to settle for about 15–20 minutes. Then the holes of diameter 6 mm were punched into the agar plate using sterile punch. About 20 μl plant extract samples (0.5 mg/μl) were pipette into the holes into the inoculated agar plates. Chloramphenicol (25 μg) disc was used as positive control, and 20 μl of distilled water or DMSO were used as negative control respective to their dilution solvent.

35.2.5.3   DETERMINATION OF MINIMUM INHIBITORY CONCENTRATION (MICS)

Minimum inhibitory concentration (MICs) of the plant extract were determined by the cup plate method [8], across a range of doses. The plant extract was diluted in the respective dosage quantity and 20 μl each were tested against *P. mirabilis* in the nutrient agar plates.

35.2.5.4   COMBINATION OF SAMPLES

The combination of MIC concentration of the samples with promising results were tested against *P. mirabilis*. Combination in ratio 1:1, 1:2, 1:3, 2:1, and 3:1 was determined. The first combination was of sample 1 and 3, then sample 1 and 5 and then of sample 3 and 5.

35.3   RESULT

35.3.1   PLANT EXTRACTION YIELDS

35.3.1.1   SAMPLE 1: PIPER LONGUM (PIPPALI)

Extraction of 20 g dried powder of fruit of pippali with distilled water extracts 4.73 g, that is around 23.65% extraction. The resulting dried extract was resuspended in distilled water resulting in extract concentration of 0.5 mg/μl.

35.3.1.2   SAMPLE 2: GLYCYRRHIZA GLABRA (MULETHI)

Extraction of 20 g dried powder of stem of mulethi by solvent extraction using 50% methanol solvent extracts 6.3 g, i.e., 31.5% dried extraction. Then the resulting dried extract was resuspended in distilled water resulting in extract concentration of 0.5 mg/µl.

35.3.1.3   SAMPLE 3: TERMINALIA CHEBULA (HIRDA)

Extraction of 20 g powder of fruit of hirda plant using 50% ethanol as solvent, the extract yield was 10.91 g, i.e., 54.5% dried extraction. Then the resulting dried extract was also resuspended in distilled water, and the extract concentration of 0.5 mg/µl.

35.3.1.4   SAMPLE 4: COMMIPHARA WIGHTII (GUGGUL)

About 20 g dried powder of gum resin of guggul was extracted using 95% ethanol and extract yield was 2.11 g, that means only 10.55% of extract was yield. The resulting dried extract was resuspended in DMSO, resulting in extract concentration of 0.5 mg/µl.

35.3.1.5   SAMPLE 5: TINOSPORA CORDIFOLIA (GUDUCHI)

Dried powder of stem of guduchi weighing 20 g was first defatted using petroleum ether and then extracted using 80% ethanol giving an extract of 0.77 g, which is only 3% yield. The resulting dried extract was resuspended in DMSO, resulting in an extract concentration of 0.5 mg/µl.

35.3.2   PHYTOCHEMICAL ANALYSIS

35.3.2.1   QUALITATIVE ANALYSIS

The medical effects of all plants are because of the phytochemical/bioactive chemical constituents. Samples 3 and 5 contained all tested constituents (Table 35.2). Samples 2 and 4 contained all except carbohydrates. Steroids, flavonoids, and alkaloids are present in all plant samples. Terpenoids and phenols were present in all except plant sample 1.

35.3.2.2 QUANTITATIVE ANALYSIS

The total flavonoid and phenolic content of all the samples were quantified using spectrophotometric method. Tables 35.3 and 35.4 show the phenolic and flavonoid content per 1 gram of extract, respectively.

TABLE 35.2 Preliminary Screening Results

Plant/ Phytochemical	Carbohy-drates	Alkaloids	Flavonoids	Steroids	Terpe-noids	Phenols
Sample 1	−	+	±	+	−	−
Sample 2	−	+	+	+	+	+
Sample 3	+	+	+	+	+	+
Sample 4	−	+	+	+	+	+
Sample 5	+	+	+	+	+	+

Note: +: positive, ±: trace, −: negative.

TABLE 35.3 Total Phenolic Content

Samples	Mean Concentration (mg/g of Extract)
Sample 1	1.70
Sample 2	1.27
Sample 3	2.22
Sample 4	2.37
Sample 5	1.69

TABLE 35.4 Total Flavonoid Content

Samples	Mean Concentration (mg/g of Extract)
Sample 1	24.90
Sample 2	34.77
Sample 3	16.2
Sample 4	73.17
Sample 5	30.95

35.3.3 ANTIMICROBIAL ACTIVITY

The undiluted (decoction) extract of all the samples were used to test zones of inhibition to provide the approximate activity towards the inhibition of *P. mirabilis*. Aliquots (20 µl) of each sample were screened against *P. mirabilis* (Figure 35.1). Promising results were shown by Sample 1, 3 and 5, with the zone of inhibition >10 mm. Each aliquot of 20 µl consists of 10 mg of the sample. The positive standard of chloramphenicol (25 µg).

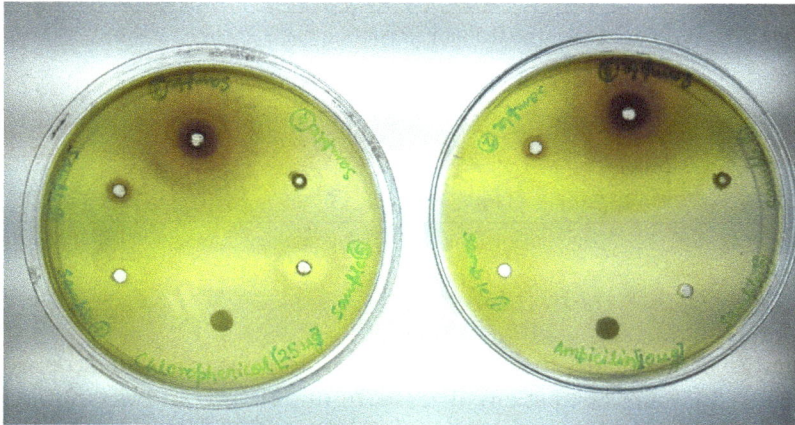

FIGURE 35.1   Zone of inhibition of extract samples.

35.3.3.1   MIC DETERMINATION

Further, the level of anti-proteus activity was determined by the MICs of the sample which showed promising results at the decoction level (Figure 35.2). The concentration was reduced to 0.4 mg/µl, 0.3 mg/µl, 0.2 mg/µl, 0.1 mg/µl, 0.05 mg/µl, and 0.25 mg/µl. Which means that the quantity of 8 mg, 6 mg, 4 mg, 2 mg, 1 mg, and 0.5 mg, respectively was tested against the *P. mirabilis*? The value of MIC's of all extracts is shown in Table 35.5. Moreover, noteworthy was the low MIC value seen for T. cordifolia stem (sample 5) at 0.05 mg/µl concentration.

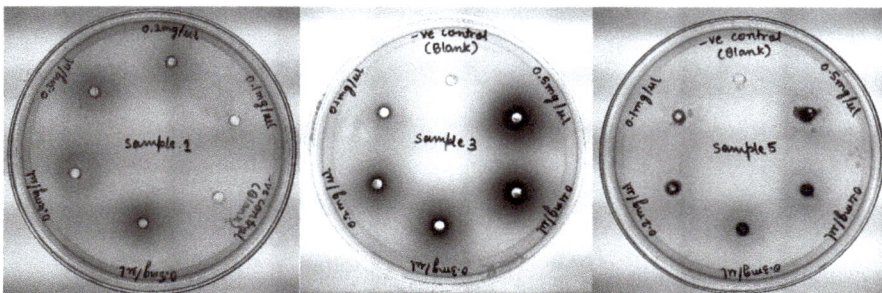

FIGURE 35.2   MIC of samples 1, 3, and 5.

TABLE 35.5   Zone of Inhibition of Samples in mm

Sample\ Concentration	0.5 mg/µl	0.4 mg/µl	0.3 mg/µl	0.2 mg/µl	0.1 mg/µl	0.05 mg/µl	0.25 mg/µl
Sample 1	11 mm	10.5 mm	10 mm	–	–	–	–
Sample 2	–	–	–	–	–	–	–
Sample 3	11.5 mm	11 mm	10.5 mm	–	–	–	–
Sample 4	–	–	–	–	–	–	–
Sample 5	12.5 mm	11 mm	10.5 mm	09 mm	07 mm	06 mm	–

35.3.3.2   COMBINATION OF SAMPLES

Generally, in Ayurveda, the combination of more than one plant decoction is preferred to cure or treat any disease. In this chapter, we determined the anti-proteus activity of a combination of different MIC samples. Table 35.6 shows zone of inhibition value of combination of sample 1, 3 and 5 in the ratio of 1:1, 1:2, 1:3, 2:1 and 3:1 (Figure 35.3). The highest inhibition was determined in the 1:1 combination of samples 5 and 3.

TABLE 35.6   Inhibition of Combination of Samples

Sample\Combination	1:1	1:2	1:3	2:1	3:1
Sample 3: Sample 1	–	9.5 mm	10 mm	–	–
Sample 5: Sample 1	08 mm	8.5 mm	–	12 mm	9.75 mm
Sample 5: Sample 3	12.5 mm	12 mm	08 mm	10 mm	9.5 mm

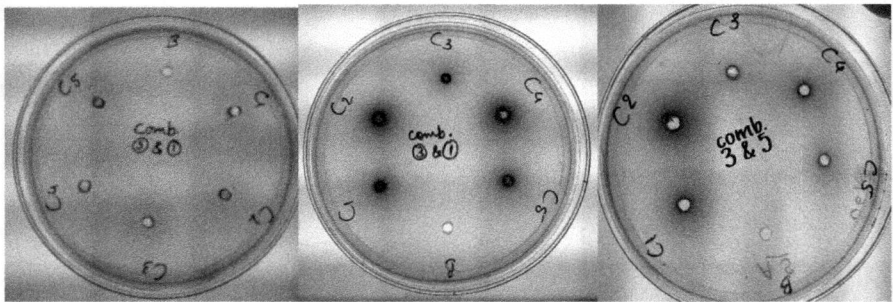

FIGURE 35.3   Combination of samples against P. mirabilis.

35.4 DISCUSSION

Natural remedies are booming day by day over synthetic drugs, and they are way safer than the current drug regimes for the treatment of various chronic diseases. Especially in the disease of RA, this is true. The currently prescribed DMARD's have a great deal of side effects and it mostly focuses on symptoms rather than the cause of the disease. In other words, it only soothes the pain and stiffness but does not stop them from occurring again.

The infection of *P. mirabilis* triggers the RA is now widely accepted due to supporting evidence [26]. The treatment against the Proteus trigger could be the answer, with the inhibition of *P. mirabilis* self-attack to joint tissues would decrease, hence block the disease progression. The result confers here shows the ability of a few medicinal plants to inhibit the infection of *P. mirabilis* in its decoction form. Also, the results were backed by the results of preliminary phytochemical tests and total phenolic and flavonoid test. P. longum, T. cordifolia and T. chebula amongst all 5 plant extracts showed their efficacy towards inhibition of microbial trigger of RA by lower MIC values. And it is excepted as these extracts were substantially stronger than chloramphenicol. The Indian Ayurveda generally treats the diseases using a mixture of decoctions of plants, here we have determined that the MIC of T. chebula and T. cordifolia have a strong inhibition towards *P. mirabilis*.

35.5 CONCLUSION

The results of this study partially confirm the use of plant extracts as a natural treatment for RA and other inflammatory disorders. Also, the results are promising and could be used in further study in the future. It is also interesting to detect the effect of these extracts against the production of cytokines to evaluate as an anti-RA drug.

KEYWORDS

- **anti-proteus**
- **cup plate method**
- **inflammation**
- **medicinal plants**
- **minimum inhibitory concentration**
- **rheumatoid arthritis**

REFERENCES

1. McInnes, I. B., & Schett, G., (2007). Cytokines in the pathogenesis of rheumatoid arthritis. *Nat. Rev. Immunol., 7,* 429–442.
2. Malaviya, A. M., (2006). Cytokine network and its manipulation in rheumatoid arthritis. *J. Assoc. Physicians India., 54,* 15–18.
3. Singh, B. R., (2013). Evaluation of antibacterial activity of *Salvia officinalis* [L.] sage oil on veterinary clinical isolates of bacteria. *Noto-Are: Med.,* pp. 11–22.
4. Feldmann, M., Brennan, F. M., & Maini, R. N., (1996). Rheumatoid arthritis. *Cell, 85,* 307–331.
5. Vosse, D., & De Vlam, K., (2009). Osteoporosis in rheumatoid arthritis and ankylosing spondylitis. *Clinical and Experimental Rheumatology, 27,* 62–69.
6. Breedveld, F. C., Weisman, M. H., & Kavanaugh, A. F., (2006). The PREMIER study: A multicentre, randomized, double-blind clinical trial of combination therapy with adalimumab plus methotrexate versus methotrexate alone or adalimumab alone in patients with early, aggressive rheumatoid arthritis who had not had previous methotrexate treatment. *Arthritis & Rheumatism., 54,* 26–37.
7. Gabriel, S. E., (2001). The epidemiology of rheumatoid arthritis. *Rheum Dis. Clin. North Am., 27,* 269–281.
8. Symmons, D. P., Barrett, E. M., Bankhead, C. R., Scott, D. G., & Silman, A. J., (1994). The incidence of rheumatoid arthritis in the United Kingdom: Results from the Norfolk arthritis register. *Br. J. Rheumatol., 33,* 735–739.
9. Turesson, C., O'Fallon, W. M., Crowson, C. S., Gabriel, S. E., & Matteson, E. L., (2003). Extra-articular disease manifestations in rheumatoid arthritis: Incidence trends and risk factors over 46 years. *Ann, Rheum, Dis., 62,* 722–727.
10. Christie, A., Dagfinrud, H., Engen, M. K., Flaatten, H. I., Ringen, O. H., & Hagen, K. B., (2010). Surgical interventions for the rheumatoid shoulder. *The Cochrane Database of Systematic Reviews, 1,* 188–190.
11. Cock, I. E., Winnett, V., Sirdaarta, J., & Matthews, B., (2015). *The Potential of Selected Australian Medicinal Plants with Anti-Proteus Activity for the Treatment and Prevention of Rheumatoid Arthritis, 11,* 190–208.
12. Ebringer, A., Ptaszynska, T., Corbett, M., Wilson, C., Macafee, Y., & Avakian, H., (1985). Antibodies to proteus in rheumatoid arthritis. *Lancet, 2,* 305–307.
13. Ebringer, A., Cunningham, P., Ahmadi, K., Wrigglesworth, J., Hosseini, R., & Wilson, C., (1992). Sequence similarity between HLA-DR1 and DR4 subtypes associated with rheumatoid arthritis and Proteus/Serratia membrane haemolysins. *Ann. Rheum. Dis., 51,* 1245–1246.
14. Yende, S. R., Sannapuri, Y. D., Vyawahare, N. S., & Harle, U. N., (2010). *Anti-Rheumatoid Activity of Aqueous Extract of Piper Longum on Freunds adjuvant-induced Arthritis in Rats, 1,* 129–133.
15. Sharma, V., Agrawal, R. C., & Pandey, S., (2013). Phytochemical screening and determination of anti-bacterial and anti-oxidant potential of glycyrrhiza glabra. *Root Extracts, 7,* 1552–1558.
16. Seo, J. B., (2012). *Anti-Arthritic and Analgesic Effect of NDI10218, a Standardized Extract of Terminalia Chebula, on Arthritis and Pain Model, 20,* 104–112.

17. Singh, B. J., & Siddiqui, M. Z., (2015). Antimicrobial activity of *Commiphara wightii* gum (Guggul gum) extract against gram-positive and gram-negative bacteria. *Journal of Microbiology and Antimicrobial Agents, 1*, 36–39.

18. Shah, P. A., & Shah, G. B., (2016). Preliminary screening of *Tinospora cordifolia* extract and *Guduchi stava* for anti-rheumatoid activity. *International Journal for Pharmaceutical Research Scholars, 5*, 7–13.

19. Banu, K. S., & Cathrine, L., (2015). General techniques involved in phytochemical analysis. *International Journal of Advanced Research in Chemical Science (IJARCS), 4*, 25–32.

20. Pandey, A., & Tripathi, S., (2014). Concept of standardization, extraction and pre phytochemical screening strategies for herbal drug. *Journal of Pharmacognosy and Phytochemistry, 12*, 115–119.

21. Singleton, V. L., & Rossi, J. A., (1965). Colorimetry of total phenolics with phosphomolybdic phosphotungstic acid reagents. *Am. J. Enol Vitic., 16*, 144–158.

22. Marinova, D., Ribarova, F., & Atanassova, M., (2005). Total phenolic and total flavonoids in Bulgarian fruits and vegetables. *Journal of the University of Chemical Technology and Metallurgy, 40*, 255–260.

23. Chang, C., Yang, M., Wen, H., & Chern, J., (2002). Estimation of total flavonoid content in propolis by two complementary colometries methods. *Journal of Food and Drug Analysis, 10*, 178–182.

24. Pourmorad, F., Hosseinimehr, S. J., & Shahabimajd, N., (2006). Antioxidant activity, phenol and flavonoid contents of some selected Iranian medicinal plants. *African Journal of Biotechnology., 5*, 1142–1145.

25. Miliauskas, G., Venskutonis, P. R., & Van, B. T. A., (2004). Screening of radical scavenging activity of some medicinal and aromatic plant extracts. *Food Chemistry, 85*, 231–237.

26. Ebringer, A., & Rashid, T., (2006). Rheumatoid arthritis is an autoimmune disease triggered by proteus urinary tract infection. *Clin. Dev. Immunol., 13*, 41–48.

CHAPTER 36

Water Footprint of Agriculture

ASHISH P. PRADHANE[1] and RASHMI P. NIMBALKAR[2]

[1]Department of Chemical Engineering, Visvesaraya National Institute of Technology, Nagpur, Maharashtra, India

[2]Department of Chemical Engineering, Priyadarshini Institute of Engineering and Technology, Rashtrasant Tukadoji Maharaj Nagpur University, Nagpur, Maharashtra, India,
E-mail: rashminagmote25@gmail.com

ABSTRACT

Water scarcity is an extremely worrisome issue to deal with, and it is expected to worsen more in the coming years, primarily due to estimated climatic changes and the humongous human population. The availability of water and its accessibility are the most important factors considering agricultural production. Water footprint (WF) is a simple yet effective tool useful for the quantification of virtual water contents of agricultural products. The WF comprises three components, which are green, blue, and gray, each having its own significance in making the complete assessment. The concept of WF is extremely important in taking consciousness about the content of water in agricultural products.

The concept of WF and its application in the field of agriculture were reviewed in the following book chapter. The methodology of WF was briefly described, and the WF assessment at the global level was also briefly studied. Shortcomings in the analyzes were observed in various studies, and improvements to overcome them were also discussed.

36.1 INTRODUCTION

Water is undoubtedly the most important natural resource, and it is one of the reasons for the existence of life on planet Earth. As we all are aware, a minuscule fraction of water available on Earth is of usable form. Water, along with helping life to sustain, majorly contributes to agriculture and industries. The global water usage of agricultural and industrial sectors is 70% and 22%, respectively, which indeed is a huge number [1]. Nature has been successful in keeping the amount of freshwater almost constant over such a humongous duration of time by recycling it back through the atmosphere. The population has exploded in such a manner that almost every resource on Earth is in crisis, and water is no exception. The depletion of freshwater sources is occurring at an alarming rate because of the explosion of population, changing lifestyles, and incessant industrialization [2]. The projection for the global population by the year 2050 is 9.2 billion [3], which will increase freshwater demand for production purposes up to 400% [4]. It is also estimated that more than in 2050, 40% of the global population will be residing in areas will extreme paucity of freshwater [5]. Consumptive water usage and pollution are exerting tremendous pressure on freshwater resources. Freshwater withdrawal worldwide has increased almost sevenfold in the 20th century [6], while more than 80% of the wastewater produced globally is released into water bodies without any pre-treatment [7].

36.1.1   CONCEPT OF WATER FOOTPRINT (WF)

There is a dire need to raise the issue of freshwater availability at a global level as there is concern about fulfilling the future requirements of food and fiber in regions with limited water resources [8]. Efforts have been taken in the past to address this issue. Water Framework Directive (Directive 2000/60/EC), one of the most advanced and recent legislative frameworks for the protection of water globally, was enacted by the European Union (EU) to set targets for preserving freshwater resources to the member states [9]. Projects such as E4WATER (2016) [52] and EcoWater (2014) [53] dealing with sustainable assessment and management of freshwater in agriculture and industries, have also been funded by the Union. EcoWater project targets assessment of economic and environmental efficiency of different water-friendly practices, supporting the decision making in diverse water consumption systems [10]. A recently developed methodology, known as water footprint (WF), is seen as to tool for analysis and quantification of water consumption and to improve the understanding of the connection between the production activities and the increasing stress on water, both in direct and indirect relation with products and services [11]. The concept of WF is briefly reviewed in the following section.

The concept of ecological footprint, a measure of the human requirement of nature (specifically, biologically productive area) was introduced in the early 1900s. A similar concept, known as WF was introduced by Hoekstra [8] to facilitate the direct or indirect evaluation of water consumption or pollution by agriculture and industries. Since the introduction of WF methodology, several studies have been conducted in the field of agriculture, specifically in field production and processing phases. In recent times, the WF was even standardized by ISO standard 14046 [12]. Two main approaches of assessing WF can be found in the literature; the volumetric approach (developed by WF Network) and the Life Cycle Analysis approach (developed by the LCA community) [13]. The key difference in both approaches lies in the emphasis given to either product or water management. Life Cycle Analysis is product-focused, while the volumetric approach is water management focused [11]. Only the volumetric approach will be reviewed in this book chapter as it is well established and emphasizes water management.

The concept of WF enables us to answer a variety of questions related to freshwater availability and consumption, such as water dependence of different operations of an organization, capability of existing regulations to protect freshwater, the possibility of reducing the WF of an individual, a society, or an organization, etc. One can express the WF in different units for answering different questions. For example, the WF of a product (also known as virtual water content) which is the sum of WFs of the processes required for the production of the product, can be expressed in m^3 water/ton product, [14] while the WF of within a specific geographical location; a sum of WFs of all the processes occurring in that location; is expressed in m^3 water/area [14].

The WF can be distinguished into three components: blue, green, and gray, as shown in Figure 36.1. The blue WF describes the consumption of ground or surface water. It can be quantified as the difference between the ground and surface water consumed and the water returned to the same sources. Ground and surface water are considered as renewable resources of water, while the deep aquifers are considered as non-renewable resources of water since the rate at which they replenish is insignificant [15]. The green WF represents the consumption of rainwater or soil water. The green WF can be classified into two components: The productive component, which is nothing but transpiration from plants and the non-productive component as evaporation [16]. Thus, for agriculture and forestry, green WF refers to the total rainwater evaporated from fields and/or transpired from the plants and the water contained in harvested crops [17]. The gray WF indicates freshwater pollution and can be defined as the quantity of water required to assimilate pollutants based on existing water quality and background concentrations [17]. Thus, the WF is indicative of both consumptive and degradative usage of firewater. Though we can quantify all the three components separately; theoretically we may sum the components to yield a single indicator [11].

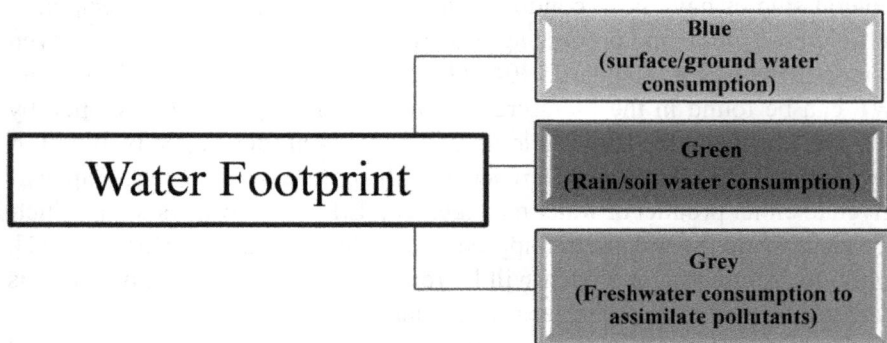

FIGURE 36.1   Components of the water footprint.

$$WF = WF_{green} + WF_{blue} + WF_{grey} \qquad (1)$$

The concept of WF differs from conventional measures of water withdrawal from water bodies (blue WF) as:

i.   The water footprint is a much broader concept expressing the net water consumption, unlike the conventional measures expressing gross water consumption.
ii. The water footprint can be used to measure water consumption over supply chains.

In a nutshell, WF should not be understood as one aggregate number but should be understood as a multidimensional indicator of water consumption, and it is of prime importance in developing strategies for sustainable usage of freshwater [13].

36.1.2   COMPONENTS OF THE WATER FOOTPRINT (WF) IN AGRICULTURE

As discussed earlier, many environmental factors, the explosion of population and international forces are posing extremely difficult challenges to agriculture. To fulfill the humongous requirements of food and fiber, major changes in traditional agricultural practices are required. About 90% of the total freshwater consumption across the globe during the last century was accounted for agriculture [18]. Apart from the explosion of population, our efforts in the noble cause of reducing the percentage of the global population living in hunger are also imparting tremendous stress on water resources. A shift of diet preference of wealthier population towards the meat also adds to the existing stress on water resources, as around 2.5–10% more energy is necessary to yield the equivalent amount of calories and proteins from livestock as compared to grains. One may expect the situation in the future is expected to get even worse if we continue consuming water irresponsibly. We must develop strategies for sustainable consumption of freshwater, and *WF* methodology will play an important role in it.

The green *WF* is a representation of a major share of water consumption in agricultural production phases. We must consider rainfall evaporation from the soil, transpiration from crops, absorption by roots, and water contained in harvested products to evaluate the green *WF* for agricultural products. Its calculation is usually done following the work of Allan et al. [19]. Green *WF* has no opportunity cost, and it can be estimated for agricultural processes based on evapotranspiration (ET) and effective rainfall. During the period of crop growth, green water footprint (WF_{green}) is equivalent to whichever is lesser between ET and effective rainfall (P_{eff}) [11, 20].

$$WF_{green} = P_{eff} \qquad \text{when } P_{eff} \leq ET \qquad (2)$$

and

$$WF_{green} = ET \qquad \text{when } P_{eff} > ET \qquad (3)$$

The actual ET, which is dependent on factors such as climate, crop features, and availability of soil-water, can be determined as [14, 19]:

$$ET = K_c \times K_s \times ET_0 \qquad (4)$$

where; K_c represents the crop coefficient; K_s represents a transpiration reduction factor; and ET_0 represents the reference ET. K_s is a dimensionless factor which is dependent on the soil-water content and has a value between zero and one. The crop coefficient is a function and time and has different values in different season stages. K_c remains constant during the initial and middle stages; it is assumed to increase linearly in the developmental stage and then decreases linearly in the late-season stage [14]. On the other hand, K_s is calculated daily, and as mentioned earlier, as a function of maximum possible and actual soil-water content, specifically in the root region. Assuming S and

S_{max} to represent the actual and maximum soil-water content, respectively and p to represent the extractable fraction of S_{max}, without being subjected to water stress, K_s can be determined as [14]:

$$K_s = \begin{cases} \dfrac{S}{(1-p).S_{max}} & if\ (1-p).S_{max} > S \\ 1 & \text{otherwise} \end{cases} \qquad (5)$$

We must also consider the blue water footprint (WF_{blue}) for agricultural analysis, which is nothing but the surface or groundwater provided to agricultural fields in the form of irrigation and for processing phases such as washing. It has a high opportunity cost, and thus a reduction in it will result in reduced production costs and environmental impacts [11]. The water evaporation during and after irrigation is not considered while evaluating the blue WF. Blue WF evaluation is dependent on factors such as the crop itself, its tolerance to the water-deficient environment, the efficiency of irrigation and green WF. If green water is available in sufficient quantity, there is no requirement of blue water. So, if ET is greater than rainfall (P_{eff}) [20], the difference between them indicates the blue water footprint (WF_{blue}) in case of lack in details about irrigation and crop. If we are aware of the details, the blue WF can be quantified in detail [9, 11].

$$WF_{blue} = ET - P_{eff} \qquad \text{when } P_{eff} \leq ET \qquad (6)$$

and

$$WF_{blue} = 0 \qquad \text{when } P_{eff} > ET \text{ and in case of lack of} \qquad (7)$$
$$\text{irrigation and crop details}$$

As discussed earlier, the gray WF represents the amount of water consumed for the assimilation of pollutants and not the real amount required during a production process. It represents the amount of water required for restoring the quality of polluted water. The greywater footprint can be estimated as [11, 21]:

$$W_{Fgrey} = L_{add}/(C_{max} - C_{nat}) \qquad (8)$$

where; L_{add} is the load of pollutants in Gg/yr; and

$$L_{add} = L - C_{nat} \cdot Q_{act} \qquad (9)$$

where; C_{max} is the maximum admissible pollutant concentration in mg/dm^3; C_{nat} is the natural pollutant concentration in mg/dm^3; Q_{act} is the actual volume of basin discharge in m^3/yr.

Natural pollutant concentration, though it is not, is often considered zero for the sake of simplicity [11], [14, 22]. In the agricultural field, the major contributor to the gray *WF* is fertilizers' nutrient loads, most commonly nitrogen as it is mostly sprayed on the field [14]. The value of the maximum permissible concentration of a particular substance varies with the local conditions and thus to bring uniformity, recommended limits by either US Environmental Protection Agency (US-EPA) as 45 mg/dm^3 of NO$_3$ or EU as 50 mg/dm^3 of the same, are used.

36.1.3 THE METHODOLOGY OF WATER FOOTPRINT (WF) EVALUATION

As discussed earlier, the *WF* of any product refers to the quantity of freshwater consumption for production over the whole supply chain. Evaluation of *WF* requires the knowledge of the location and measurement of the *WF* of a specific product, process, producers, and/or consumers and assessment of its sustainability, which is followed up by a response strategy [9]. Evaluation of the *WF* helps us understand how freshwater is being consumed by human activities and production processes and its impact on issues such as water stress and its pollution. It also helps us in developing sustainable strategies from the perspective of the environment, society, and economy and thus comprising the three pillars of sustainability.

Measurement of the *WF* of a product can be done comprising all facets of a specific production activity or process or its supply chain. Like life cycle assessment (LCA), *WF* evaluation can be done in four phases:

i. **Phase I:** Establishing the goals and scope.
ii. **Phase II:** Accounting of the water footprint.
iii. **Phase III:** Sustainability evaluation of the water footprint.
iv. **Phase IV:** Response formulation of the Water footprint.

As shown in Figure 36.2, establishing goals and objectives is the first phase in *WF* evaluation. Though there exists a single main objective for every study, there can be multiple goals for the same. For example, the main objective, which should be stated clearly, can either be creating awareness in society, identification of the hotspots, or assisting policy formulation, there can exist multiple goals such as the study of the *WF* of a product, production processes, consumers, etc. One must even specify the specific focus, such as green, blue, or gray footprints, or direct/indirect footprint while conducting the study.

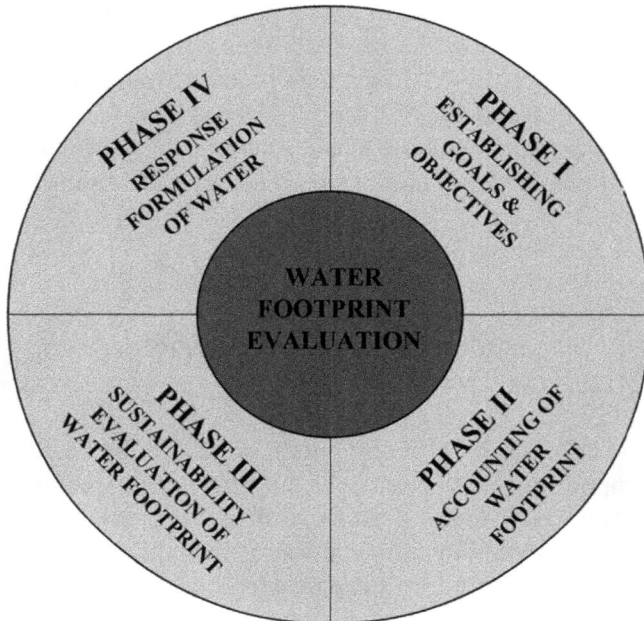

FIGURE 36.2   Phases of water footprint evaluation.

The second phase of the *WF* evaluation is accounting. For this purpose, boundaries of the *WF* should be well defined, and there must be a clear distinction between areas identified for inclusion and exclusion. The following questions must also be answered during the accounting phase:

- What components of water footprint, green or blue/gray, must be given the emphasis?
- What should be the period of consideration for the data to be included in the study?
- Which footprint should be considered, direct or indirect? Can both be considered?
- What should be the point of cut-off when the whole supply chain is considered?
- Should the consumption of a part of the nation or the entire nation be considered during the accounting of national Footprint?

The third and most important phase of *WF* evaluation is sustainability evaluation. The points to consider during sustainability evaluation depends upon the perspective of the study. For example, from a geographic perspective, the sustainability of the aggregated *WF*, involving every human activity and production process in that particular geographic location must be considered. For a product/process perspective, only the *WF* of that specific product or process should be considered for sustainability evaluation [9]. There are two elements when we talk about the contribution, first is the contribution of the specific product/process to the aggregated *WF* and the second is its contribution to the global *WF*. The sustainability evaluation depends very closely on the first phase of establishing goals and objectives. Specification of the scope depends on the established goals. One must answer the following questions while considering the geographic perspective [9]:

- Which of the water footprint, green, blue, or gray should be considered while evaluating sustainability?
- Which dimension of sustainability, environment, society, or economy, should be considered?
- Should the hotspots be only identified or analyzed in detail for primary or secondary impacts on it?

The following questions arise when one considers the product/process perspective [9]:

- Does the water footprint of the product/process contribute to the global water footprint unnecessarily?
- Does the water footprint of the product/process contribute to some specific hotspots?

The final phase in *WF* evaluation is the response formulation, which is again related to the first phase. In this phase, the policies, or strategies to lessen the *WF* will be constituted. Based on the scope of the study; responsible individuals or organizations should be informed about the necessary steps to reduce the *WF*. One must refer to 'The Water Footprint Assessment Manual' [9] for further understanding and knowing the detailed steps for calculating the *WF* of a variety of entities.

36.1.4   *WATER FOOTPRINT (WF) ASSESSMENT ON A GLOBAL STAGE*

The first estimate at a global level for water consumption for a variety of crops in different countries was made by Hoekstra and Hung [23] though it was not in the form of its components (green, blue, and gray). Since then, several studies have been presented on a global scale [14, 24, 25]. During the period 1997–2001, the water consumption globally for the production of crops, as per Hoekstra and Chapagain [26] was 6390 Gm^3/yr. The virtual water content of animals was found to be greater than the plants on a general basis [11, 23, 27, 28]. The average WF for maize, wheat, and rice, measured by Chapagain and Hoekstra [26], was 900, 1300 and 3000 m^3/t, respectively. On the other hand, the average WF for meat, pork, and beef was 3900, 4900, and 15,500 m^3/t, which is significantly higher. These global averages do not reflect the actual scenario as there is a significant effect of geographical locations, climatic conditions, technologies, and yield on these values, resulting in different values of WF at different locations. These evaluations are also supported by recent research measuring the WF of animal products. For example, Palhares and Pezzopane [28] presented a comparison of WF of dairy farms in conventional (1422 m^3/kg ECM milk) and organic farming (1510 m^3/kg ECM milk); Pahlow et al. [29] reported the WF of farmed fishes and crustaceans as 1974 m^3/t with components as 83% green, 9% blue and 8% gray. de Miguel et al. [30] reported the WF for production and processing of pork as 19.5 billion m^3/yr with components as 82% green, 10% gray and 8% blue.

A variety of agricultural products and were studied by Mekonnen and Hoekstra [14] and Hoekstra [9] for evaluating the WF. Wheat was found to have the highest value, 1827 m^3/t, rice was found to have the value close to the average of WF for crops, 1644 m^3/t and maize was found to have the lowest value of WF. A similar result for rice, 1675 m^3/t, was obtained by Chapagain and Hoekstra [25]. Vegetables and sugar crops were observed to have less WF of around 300 and 200 m^3/t, respectively. Fruits and crops were observed to have the value of WF values of 1000 and 2400 m^3/t, respectively, while that for pulses, nuts, and spices were observed to be in the range of 4000 to 9000 m^3/t. Though these numbers indicate the WF at a global level, there is inaccuracy in the results because of uncertainties and lesser geospatial specificity [11]. Thus, there is a necessity of modification in the water trade analyzes at the global level.

The first step in this direction was taken by arguing that the total WF should not be considered as a sum of its components as every component differs in the opportunity costs and impacts [31]. The question of evaluating all the three components separately even started to rise as there was a significant difference in the value of global WF during 1996–2005 (30% to be specific), reported by Mekonnen [32] and Chapagain and Hoekstra [26].

The three components were estimated as 74% green, 15% gray and 11% blue [32]. As per the same study, Agriculture was observed to contribute 91% to the total value, while industrial goods and domestic requirements were 5% and 4%, respectively. The average global WF of a consumer was 1385 m³/yr, but the WF for consumers in the US (2842 m³/yr per Capita) was twice the average global value and the same was just 36–44% (500–600 m³/yr per Capita) of the average global value in many African countries [11, 32].

India is the largest consumer of the groundwater in the world as around 90% of ground and surface water is utilized for irrigated agriculture [33]. In India, the mean value of the total WF of agricultural consumption was quantified as 777 billion m³/yr in the last century, [34]. However, the same was found to be 967.167 billion m³/yr during the year 2007–2008 [35]. There is a depletion of groundwater resources in many areas [36], specifically in the Indus-Ganga plain where double cropping of rice-wheat is practiced widely [37]. Environmental changes in the future, such as temperature increase, meltwater changes, uncertain patterns of rainfall are expected to have severe implications on Indian agriculture. Quantification of WF may help in improving its sustainability by forecasting the potential implications.

The total WF of Indian agriculture can be divided into two components, external, and internal. Internal component, which represents agricultural products consumed within the country, was observed to cover around 85% of the total WF, while the rest is covered by the external component (representing agricultural products exported) [34]. The significant factor influencing the WF was the staple food of each state of India, which was majorly wheat and rice with variable consumption of oil and sugar. Based on it, the WF of different states of India has been determined [34, 37]. Cereals were found to leave the maximum WF among all crops with a value of 647.032 billion m³ followed by pulses (152.857 billion m³), oilseeds (103.053 billion m³), fruits, and vegetables (43.204 billion m³) lastly cash crops (16.002 billion m³).

As per Jayaram [35], rice, and wheat were the two among the cereals to have maximum internal water use. Rice consumed around 56% while wheat consumed around 23% during the year 2007–2008. In the same year, Soyabean (58%) and tur (20%) were the most internal water-consuming pulses. Among oilseeds, groundnut consumed around 38% and rapeseed and mustard collectively consumed around 35%. As mentioned earlier, most of the earlier reported values of the global WF did not represent the variability at the global scale and thus does not help in drawing any conclusion. Statistical analysis shows that there was too much approximation of the literature data and *WF* had a lesser impact than expected. Mekonnen et al. [38] obtained a close result in a study fixated on Latin America and the Caribbean. Readers are advised to refer to the review presented by Lovarelli et al. [11]. It is an excellent review of 96 scientific articles on the indicator of *WF* for Agricultural productions, covering the important results, and the analysis of strengths and weaknesses.

36.1.5 *CHALLENGES IN WATER FOOTPRINT (WF) MANAGEMENT*

Being the most important reason behind the existence of life on earth, it has been the center of attraction in the international agenda. There are numerous areas, facing the challenge of water scarcity [39]. There will be a strong impact on the water consumption of agriculture in the future, considering the estimated rise in the global population growth rate [40] and corresponding higher food and fiber demands. In addition, due to the problem of scarcity and drought [41], there is an expectation of excessive use of water for irrigation owing to the increase in competition among the different sectors of the economy. The development of measures for optimizing the water consumption efficiency is of prime importance considering the humongous volumes of water consumption in agricultural productions [42]. The losses due to ET are recovered by irrigation in order to achieve maximum production in its growing environment [42, 43].

Irrigation is also utilized in pesticide application as a solvent for diluting the chemicals, for protecting sensitive crops from frost, for providing nutrients by dissolving them in water, for removal of excess of salinity from the soil, modifying its pH and for improving physical properties [42]. Recently, there has been an emergence of biofuel technology, and it is competing with food production. This trend, if continues, will cause a significant reduction in water and land availability for food production and ultimately the food availability [44].

The increasing CO_2 concentration in the atmosphere has led to global warming and has resulted in changes in precipitation patterns. These changes, along with the ET fluxes, runoff, precipitation intensity, frequency of excessive precipitation, soil moisture, etc., and many other changes and anticipated in the future and they will have a strong impact on crop production [45]. A significant increase in events of excessive precipitation has been estimated by Sillmann and Roeckner [46] in regions which are relatively wet than the other regions in the present climate and time. In addition, there will be an increase in the evaporative demands of the crops due to the combined effect of water unavailability and high temperatures in the regions where the annual/seasonal precipitation is falling, resulting in decreased crop yield and productivity for crops whose growth is temperature constrained [42]. Contrastingly, several studies show a positive impact of increased levels of CO_2 levels in the atmosphere [47–49] (known as the CO_2 fertilization effect), as the stomatal conductance and rate of transpiration of crops [50] is significantly reduced in higher CO_2 concentrations, resulting in improved root uptake capacity [51] and water use efficiency.

As the water demand and availability balance have reached its critical limit in numerous regions worldwide and the humongous demands of water and food in the future is expected, a sustainable approach for *WF* management is of extreme importance. There is a need for the development of methodologies that will enhance the crop yield and reduce the *WF* of non-essential activities.

36.2 SUMMARY

Water is the most important resource for all living organisms to survive and for developing any human activity. The uneven distribution of water resources in space and time on earth is worrisome. It is estimated that up to 2/3rd of the global population would suffer from water scarcity in the coming years. The availability of water is of utmost importance in achieving the desired agricultural productions, both in terms of quality and quantity. With the expected increase in the global population, the magnitude of these problems will be amplified, if corrective measures are not taken. To formulate and practice the corrective measures, one must have a detailed knowledge of how water consumption in agriculture takes place, in other words, one must know the *WF* of the process of agriculture and its resulting products. The *WF* is a simple yet effective tool for sustainable water resource management. Once the *WF* for the process/product is correctly estimated, the process can be optimized by minimizing the *WF* of non-essential activities. Studies have indicated the importance of the correct estimation of *WF*s in its management. In recent times, a lot of improvement has been done in improving the estimations of the *WF*, which will mean the accurate implementation of corrective measures.

KEYWORDS

- **agricultural and industrial sectors**
- **European Union**
- **evapotranspiration**
- **global water**
- **life cycle assessment**
- **methodology**

REFERENCES

1. *United Nations World Water Assessment Program*, (2009). Water in a Changing World-The United Nations World Water Development Report 3. UNESCO, 48223.
2. Manzardo, A., Ren, J., Piantella, A., Mazzi, A., Fedele, A., & Scipioni, A., (2014). Integration of water footprint accounting and costs for optimal chemical pulp supply mix in paper industry. *J. Clean. Prod., 72*, 167–173.
3. WWAP, (2007). *World Population Prospects the 2006 Revision, Highlights, Working Paper No. ESA/P/WP. 202*. Department of Economic and Social Affairs, Population Division, United Nations, New York.
4. UN, WWAP (United Nations World Water Assessment Program), (2015). *The United Nations World Water Development Report 2015: Water for a Sustainable World*. Paris, UNESCO.
5. WWAP, (2014). *The United Nations World Water Development Report 2014-Water and Energy* (Vols. 1, 27).
6. Gleick, P. H., (2000). A look at twenty-first century water resources development. *Water Int., 25*(1), 127–138.
7. UNESCO; UN-Water. WWAP (United Nations World Water Assessment Program), (2017). *The United Nations World Water Development Report 2017 Wastewater: The Untapped Resource*. Paris, UNESCO.
8. Hoekstra, A. Y., (2008). *Water neutral: reducing and offsetting the impacts of water footprints, Value of Water Research Report Series No.28*, UNESCO-IHE, Delft, the Netherlands. www.waterfootprint.org 34/Business water footprint accounting Hoekstra.
9. Hoekstra, Arjen Y., Chapagain, Ashok K., Aldaya, Maite M., & Mekonnen, Mesfin, (2011). *The Water Footprint Assessment Manual: Setting the Global Standard*. Daugherty Water for Food Global Institute: Faculty Publications. 77. https://digitalcommons.unl.edu/wffdocs/77 (accessed on 31 July 2021).
10. Levidow, L., Lindgaard-Jørgensen, P., Nilsson, Å., Skenhall, S. A., & Assimacopoulos, D., (2016). Process eco-innovation: Assessing meso-level eco-efficiency in industrial water-service systems. *J. Clean. Prod., 110*, 54–65.
11. Lovarelli, D., Bacenetti, J., & Fiala, M., (2016). Water footprint of crop productions: A review. *Sci. Total Environ., 548, 549*, 236–251.

12. ISO, (2014). *Environmental Management. ISO 14046*. Water footprint-principles, require-ments, and guidelines.
13. Vanham, D., & Bidoglio, G., (2015). *A Review on the Indicator Water Footprint for the EU28, 26*(2013), 61–75.
14. Mekonnen, M. M., & Hoekstra, A. Y., (2011). The green, blue and grey water footprint of crops and derived crop products. *Hydrol. Earth Syst. Sci., 15*(5), 1577–1600.
15. FAO, (2003). *Review of World Water Resources by Country: 2*. Concepts and Definitions.
16. Falkenmark, M., & Rockström, J., (2006). The new blue and green water paradigm: Breaking new ground for water resources planning and management. *J. Water Resour. Plan. Manag., 132*(3), 129–132.
17. Hoekstra, A. Y., (2015). The water footprint of industry. In: *Assessing and Measuring Environmental Impact and Sustainability* (pp. 221–254). Elsevier Inc.
18. Shiklomanov, I. A., (2000). Appraisal and assessment of world water resources. *Water Int., 25*(1), 11–32.
19. Allen, R. G., Pereira, L. S., Raes, D., & Smith, M., (1998). *Crop Evapotranspiration-Guidelines for Computing Crop Water Requirements*. FAO Irrigation and Drainage Paper: 56.
20. Bocchiola, D., Nana, E., & Soncini, A., (2013). Impact of climate change scenarios on crop yield and water footprint of maize in the po valley of Italy. *Agric. Water Manag., 116*, 50–61.
21. Hoekstra, A., (2010). The Water Footprint: Water in the Supply Chain., No. 93, 12–13.
22. Aldaya, M. M., & Hoekstra, A. Y., (2010). The water needed for Italians to eat pasta and pizza. *Agric. Syst., 103*(6), 351–360.
23. Hoekstra, A. Y., (2003). Virtual water trade: A quantification of virtual water flows between nations in relation to international crop trade, *Proceedings of the International Expert Meeting on Virtual Water Trade 12*, Delft, 25–47, https://ci.nii.ac.jp/naid/10029232667/en/ (accessed on 31 July 2021).
24. Chapagain, A. K., Hoekstra, A. Y., Savenije, H. H. G., & Gautam, R., (2006). The water footprint of cotton consumption: An assessment of the impact of worldwide consumption of cotton products on the water resources in the cotton-producing countries. *Ecol. Econ., 60*(1), 186–203.
25. Chapagain, A. K., & Hoekstra, A. Y., (2010). *The Green, Blue and Grey Water Footprint of Rice from Both a Production and Consumption Perspective* (pp. 1–62). Research report no. 40. water, no. 40.
26. Chapagain, A. K., & Hoekstra, A. Y., (2004). *Water Footprints of Nations: Main Report Value of Water Research Report Series No. 16* (Vol. 1).
27. Gerbens-Leenes, W., & Hoekstra, A. Y., (2012). The water footprint of sweeteners and bio-ethanol. *Environ. Int., 40*(1), 202–211.
28. Palhares, J. C. P., & Pezzopane, J. R. M., (2015). Water footprint accounting and scarcity indicators of conventional and organic dairy production systems. *J. Clean. Prod., 93*, 299–307.
29. Pahlow, M., Van, O. P. R., Mekonnen, M. M., & Hoekstra, A. Y., (2015). Increasing pressure on freshwater resources due to terrestrial feed ingredients for aquaculture production. *Sci. Total Environ., 536*, 847–857.
30. De Miguel, Á., Hoekstra, A. Y., & García-Calvo, E., (2015). Sustainability of the water footprint of the Spanish pork industry. *Ecol. Indic., 57*, 465–474.

31. Hoekstra, A. Y., & Chapagain, A. K., (2008). Globalization of Water: Sharing the planet's freshwater resources. *Glob. Water Shar. Planet's Freshw. Resour.*, 1–208.
32. Mekonnen, M. (2011). *Spatially and Temporally Explicit Water Footprint Accounting*, University of Twente, Enschede, The Netherlands. https://doi.org/10.3990/1.9789036532211.
33. *AQUASTAT – FAO's Global Information System on Water and Agriculture*, Food and agriculture Organization of the United Nations, (27th March 2019). http://www.fao.org/aquastat/en/ (accessed on 31 July 2021).
34. Kampman, D., (2007). The water footprint of India. Value Water Res. Rep. Ser., no. 32, 152.
35. Jayaram, K., (2016). *A Water Footprint Analysis for Agriculture in India*. Indian Agricultural Research Institute New Delhi.
36. Rodell, M., Velicogna, I., & Famiglietti, J. S., (2009). Satellite-based estimates of groundwater depletion in India. *Nature, 460*(7258), 999–1002.
37. Harris, F., Green, R. F., Joy, E. J. M., Kayatz, B., Haines, A., & Dangour, A. D., (2017). The water use of Indian diets and socio-demographic factors related to dietary blue water footprint. *Sci. Total Environ., 587, 588*, 128–136.
38. Mekonnen, M. M., Pahlow, M., Aldaya, M. M., Zarate, E., & Hoekstra, A. Y., (2015). Sustainability, efficiency, and equitability of water consumption and pollution in Latin America and the Caribbean. *Sustain., 7*(2), 2086–2112.
39. Alcamo, J., Döll, P., Kaspar, F., & Siebert, S., (1997). *Global Change and Global Scenarios of Water Use and Availability: An Application of Water GAP 1.0.* (pp. 1–47) Univ. Kassel, Ger.
40. United Nations Department of Economic and Social Affairs/Population Division, (2013). *World Population Prospects: The 2012 Revision: Demographic Profiles* (Vol. II).
41. Jiménez Cisneros, B. E., Oki, T., Arnell, N. W., Benito, G., Cogley, J. G., Döll, P., Jiang, T., & Mwakalila, S. S. (2014). Freshwater resources. In: Climate Change 2014: *Impacts, Adaptation, and Vulnerability. Part A: Global and Sectoral Aspects. Contribution of Working Group II to the Fifth Assessment Report of the Intergovernmental Panel on Climate Change* [Field, C.B., Barros, V. R., Dokken, D. J., Mach, K. J., Mastrandrea, M. D., Bilir, T. E., Chatterjee, M., Ebi, K. L., Estrada, Y. O., Genova, R. C., Girma, B., Kissel, E. S., Levy, A. N., MacCracken, S., Mastrandrea, P. R., & White, L. L. (eds.). Cambridge University Press, Cambridge, United Kingdom and New York, NY, USA, pp. 229–269.
42. Mancosu, N., Snyder, R. L., Kyriakakis, G., & Spano, D., (2015). Water scarcity and future challenges for food production. *Water (Switzerland), 7*(3), 975–992.
43. Doorenbos, J., & Pruitt, W. O., (1977). Guidelines for predicting crop water requirements. *FAO Irrig. Drain. Pap., 24*, 144.
44. Sulle, E., & Nelson, F., (2009). *Biofuels Land Access and Rural Livelihood in Tanzania*, IIED, London. ISBN: 978-1-84369-749-7.
45. Bates, B., Kundzewicz, Z. W., Wu, S., Burkett, V., Doell, P., Gwary, D., Hanson, C., et al., (2008). Climate change and water. *Technical Paper of the Intergovernmental Panel on Climate Change*.
46. Sillmann, J., & Roeckner, E., (2008). Indices for extreme events in projections of anthropogenic climate change. *Clim. Change, 86*(1, 2), 83–104.
47. Kimball, B. A., Kobayashi, K., & Bindi, M., (2002). Responses of agricultural crops to free-air CO_2 enrichment. *Advances Agron., 77*, 293–368.

48. Jablonski, L. M., Wang, X., & Curtis, P. S., (2002). Plant reproduction under elevated CO_2 conditions: A meta-analysis of reports on 79 crops and wild species. *New Phytol., 156*(1), 9–26.

49. Ainsworth, E. A., & Long, S. P., (2005). What have we learned from 15 years of free-air CO_2 enrichment (FACE)? A meta-analytic review of the responses of photosynthesis, canopy properties and plant production to rising CO_2. *New Phytol., 165*(2), 351–372.

50. Olesen, J., & Bindi, M., (2002). Consequences of climate change for European agricultural productivity, land use and policy. *Eur. J. Agron., 16*, 239–262.

51. Tubiello, F. N., & Ewert, F., (2002). Simulating the effects of elevated CO_2 on crops: Approaches and applications for climate change. *Eur. J. Agron., 18*(1, 2), 57–74.

52. Lagerås, P. (2003). Approaches and Methods for Commissioned Archaeology in Wetlands: Experience from the E4 Project in Skåne, Southern Sweden. *European Journal of Archaeology, 6*(3), 231–249. doi:10.1179/eja.2003.6.3.231.

53. Simon A. Parsons, (2000). The effect of domestic ion-exchange water softeners on the microbiological quality of drinking water, *Water Research, 34*(8), 2369–2375, ISSN 0043-1354, https://doi.org/10.1016/S0043-1354(99)00407-8.

Index